Lecture Notes in Mathematics 1671

Springer
Berlin
Heidelberg
New York
Barcelona
Budapest
Hong Kong
London
Milan
Paris
Santa Clara
Singapore
Tokyo

Serge Bouc

Green Functors and *G*-sets

Springer

Author

Serge Bouc
Equipe des groupes finis
CNRS UMR 9994
UFR de Mathématiques
Université Paris 7 – Denis Diderot
2, Place Jussieu
F-75251 Paris, France
e-mail: bouc@math.jussieu. fr

Cataloging-in-Publication Data applied for

Die Deutsche Bibliothek - CIP-Einheitsaufnahme

Bouc, Serge:
Green functors and G-sets / Serge Bouc. - Berlin ; Heidelberg ; New York ; Barcelona ; Budapest ; Hong Kong ; London ; Milan ; Paris ; Santa Clara ; Singapore ; Tokyo : Springer, 1997
(Lecture notes in mathematics ; 1671)
ISBN 3-540-63550-5

Mathematics Subject Classification (1991): 19A22, 20C05, 20J06, 18D35

ISSN 0075-8434
ISBN 3-540-63550-5 Springer-Verlag Berlin Heidelberg New York

Typesetting: Camera-ready TEX output by the author
SPIN: 10553356 46/3142-543210 - Printed on acid-free paper

Contents

Introduction

The theory of Mackey functors has been developed during the last 25 years in a series of papers by various authors (J.A. Green [8], A. Dress [5], T. Yoshida [17], J. Thévenaz and P. Webb [13],[15],[14], G. Lewis [6]). It is an attempt to give a single framework for the different theories of representations of a finite group and its subgroups.

The notion of Mackey functor for a group G can be essentially approached from three points of view: the first one ([8]), which I call "naive", relies on the poset of subgroups of G. The second one ([5],[17]) is more "categoric", and relies on the category of G-sets. The third one ([15]) is "algebraic", and defines Mackey functors as modules over the Mackey algebra.

Each of these points of view induces its own natural definitions, and the reason why this subject is so rich is probably the possibility of translation between them. For instance, the notion of minimal subgroup for a Mackey functor comes from the first definition, the notion of induction of Mackey functors is quite natural with the second, and the notion of projective Mackey functor is closely related to the third one.

The various rings of representations of a group (linear, permutation, p-permutation...), and cohomology rings, are important examples of Mackey functors, having moreover a product (tensor product or cup-product). This situation has been axiomatized, and those functors have been generally called G-functors in the literature, or Green functors.

This definition of a Green functor for a group G is a complement to the "naive" definition of a Mackey functor: to each subgroup of G corresponds a ring, and the various rings are connected by operations of transfer and restriction, which are compatible with the product through Frobenius relations.

The object of this work is to give a definition of Green functors in terms of G-sets, and to study various questions raised by this new definition. From that point of view, a Green functor is a generalized ring, in the sense that the theory of Green functors for the trivial group is the theory of ordinary rings. Now ring theory gives a series of directions for possible generalizations, and I will treat some cases here (tensor product, bimodules, Morita theory, commutants, simple modules, centres).

The first chapter deals only with Mackey functors: my purpose was not to give a full exposition of the theory, and I just recall the possible equivalent definitions, as one can find for instance in the article of Thevenaz and Webb ([15]). I show next how to build Mackey functors "with values in the Mackey functors", leading to the functors $\mathcal{H}(M,N)$ and $M\hat{\otimes}N$, which will be an essential tool: they are analogous to the homomorphisms modules and tensor products for ordinary modules. Those constructions already appear in Sasaki ([12]) and Lewis ([6]). The notion of n-linear map can be generalized in the form of n-linear morphism of Mackey functors. The reader may find that this part is a bit long: this is because I have tried here to give complete proofs, and as the subject is rather technical, this requires many details.

Chapter 2 is devoted to the definition of Green functors in terms of G-sets, and to the proof of the equivalence between this definition and the classical one. It is then possible to define a module over a Green functor in terms of G-sets. I treat next the fundamental case of the Burnside functor, which plays for Green functors the role of the ring **Z** of integers.

In chapter 3, I build a category $\mathcal{C}_A$ associated to a Green functor A, and show that the category of A-modules is equivalent to the category of representations of $\mathcal{C}_A$. This category is a generalization of a construction of Lindner ([9]) for Mackey functors, and of the category of permutation modules studied by Yoshida ([17]) for cohomological Mackey functors.

Chapter 4 describes the algebra associated to a Green functor: this algebra enters the scene if one looks for G-sets Ω such that the evaluation functor at Ω is an equivalence of categories between the category of representations of $\mathcal{C}_A$ and the category of $\mathrm{End}_{\mathcal{C}_A}(\Omega)$-modules. This algebra generalizes the Mackey algebra defined by Thevenaz and Webb ([15]) and the Hecke algebra of Yoshida ([17]). It is possible to give a definition of this algebra by generators and relations.

This algebra depends on the set Ω, but only up to Morita equivalence. Chapter 5 is devoted to the relation between those Morita equivalences and the classical notion of relative projectivity of a Green functor with respect to a G-set (see for instance the article of Webb [16]). More generally, I will deduce some progenerators for the category of A-modules.

Chapter 6 introduces some tools giving new Green functors from known ones: after a neat description of the Green functors $\mathcal{H}(M, M)$, I define the opposite functor of a Green functor, which leads to the notion of right module over a Green functor. A natural example is the dual of a left module. The notion of tensor product of Green functors leads naturally to the definition of bimodule, and the notion of commutant to a definition of the Mackey functors $\mathcal{H}_A(M, N)$ and $M\hat{\otimes}_A N$.

Those constructions are the natural framework for Morita contexts, in chapter 7. The usual Morita theory can be generalized without difficulty to the case of Green functors for a given group G.

The chapters 8,9, and 10 examine the relations between Green functors and bisets: this notion provides a single framework for induction, restriction, inflation, and coinflation of Mackey functors (see [2]).

In chapter 8, I show how the composition with U, if U is a G-set-H, gives a Green functor $A \circ U$ for the group H starting with a Green functor A for the group G. This construction passes down to the associated categories, so there is a corresponding functor from $\mathcal{C}_{A\circ U}$ to $\mathcal{C}_A$. This gives a functor between the categories of representations, which can also be obtained by composition with U. I study next the functoriality of these constructions with respect to U, and give the example of induction and restriction.

Chapter 9 is devoted to the construction of the associated adjoint functors: I build a left and a right adjoint to the functors of composition with a biset $M \mapsto M \circ U$ for Mackey functors, and I give the classical examples of induction, restriction and inflation, and also the less well-known example of coinflation.

Chapter 10 is the most technical of this work: I show how the previous left adjoint functors give rise to Green functors, and I study the associated functors and their adjoints between the corresponding categories of modules. An important consequence of this is the compatibility of left adjoints of composition with tensor products, which proves that if there is a surjective Morita context for two Green functors A and B for the group G, then there is one for all the residual rings $\overline{A}(H)$ and $\overline{B}(H)$, for any subgroup H of G.

In chapter 11, I classify the simple modules over a Green functor, and describe their structure. Applying those results to the Green functor $A\hat{\otimes}A^{op}$, I obtain a new proof of the theorem of Thévenaz classifying the simple Green functors. Finally, I study how the simple modules (or similarly defined modules) behave with respect to the constructions $\mathcal{H}(-,-)$ and $-\hat{\otimes}-$.

Chapter 12 gives two possible generalizations of the notion of centre of a ring, one in terms of commutants, the other in terms of natural transformations of functors. The first one gives a decomposition of any Green functor using the idempotents of the Burnside ring, and shows that up to (usual) Morita equivalence, it is possible to consider only the case of Green functors which are projective relative to certain sets of solvable π-subgroups. The second one keeps track of the blocks of the associated algebras. Then I give the example of the fixed points functors, and recover the isomorphism between the center of Yoshida algebra and the center of the group algebra. Next, the example of the Burnside ring leads to the natural bijection between the p-blocks of the group algebra and the blocks of the p-part of the Mackey algebra.

Chapter 1

Mackey functors

All the groups and sets with group action considered in this book will be finite.

1.1 Equivalent definitions

Throughout this section, I denote by G a (finite) group and R a ring, that may be non-commutative. First I will recall briefly the three possible definitions of Mackey functors: the first one is due to Green ([8]), the second to Dress ([5]), and the third to Thévenaz and Webb ([15]).

1.1.1 Definition in terms of subgroups

One of the possible definitions of Mackey functors is the following:

A Mackey functor for the group G, with values in the category R-**Mod** of R-modules, consists of a collection of R-modules $M(H)$, indexed by the subgroups H of G, together with maps $t_K^H : M(K) \to M(H)$ and $r_K^H : M(H) \to M(K)$ whenever K is a subgroup of H, and maps $c_{c,H} : M(H) \to M(^xH)$ for $x \in G$, such that:

- If $L \subseteq K \subseteq H$, then $t_K^H t_L^K = t_L^H$ and $r_L^K r_K^H = r_L^H$.
- If $x, y \in G$ and $H \subseteq G$, then $c_{y,^xH} c_{x,H} = c_{yx,H}$.
- If $x \in G$ and $H \subseteq G$, then $c_{x,H} t_K^H = t_{^xK}^{^xH} c_{x,K}$ and $c_{x,K} r_K^H = r_{^xK}^{^xH} c_{x,H}$. Moreover $c_{x,H} = Id$ if $x \in H$.
- (Mackey axiom) If $L \subseteq H \supseteq K$, then

$$r_L^H t_K^H = \sum_{x \in L\backslash H/K} t_{L\cap {^xK}}^L c_{x,L^x\cap K} r_{L^x\cap K}^K$$

The maps t_K^H are called *transfers* or *traces*, and the maps r_K^H are called *restrictions*.

A morphism θ from a Mackey functor M to a Mackey functor N consists of a collection of morphisms of R-modules $\theta_H : M(H) \to N(H)$, for $H \subseteq G$, such that if

$K \subseteq H$ and $x \in G$, the squares

$$\begin{CD} M(K) @>\theta_K>> N(K) \\ @V{t_K^H}VV @VV{t_K^H}V \\ M(H) @>>\theta_H> N(H) \end{CD} \qquad \begin{CD} M(K) @>\theta_K>> N(K) \\ @A{r_K^H}AA @AA{r_K^H}A \\ M(H) @>>\theta_H> N(H) \end{CD} \qquad \begin{CD} M(H) @>\theta_H>> N(H) \\ @V{c_{x,H}}VV @VV{c_{x,H}}V \\ M({}^xH) @>>\theta_{{}^xH}> N({}^xH) \end{CD}$$

are commutative.

1.1.2 Definition in terms of G-sets

If K and H are subgroups of G, then the morphisms of G-sets from G/K to G/H are in one to one correspondence with the classes xH, where $x \in G$ is such that $K^x \subseteq H$. This observation provides a way to extend a Mackey functor M to any G-set X, by choosing a system of representatives of orbits $G\backslash X$, and defining

$$M(X) = \bigoplus_{x \in G\backslash X} M(G_x)$$

There is a way to make this equality functorial in X, and this leads to the following definition:

Definition: *Let R be a ring. If G is a (finite) group, let G-**set** be the category of finite sets with a left G action. A Mackey functor for the group G, with values in R-**Mod**, is a bifunctor from G-**set** to R-**Mod**, i.e. a couple of functors (M^*, M_*), with M^* contravariant and M_* covariant, which coincide on objects (i.e. $M^*(X) = M_*(X) = M(X)$ for any G-set X). This bifunctor is supposed to have the two following properties:*

- *(M1) If X and Y are G-sets, let i_X and i_Y be the respective injections from X and Y into $X \coprod Y$, then the maps $M^*(i_X) \oplus M^*(i_Y)$ and $M_*(i_X) \oplus M_*(i_Y)$ are mutual inverse R-module isomorphisms between $M(X \coprod Y)$ and $M(X) \oplus M(Y)$.*
- *(M2) If*

$$\begin{CD} T @>\gamma>> Y \\ @V{\delta}VV @VV{\alpha}V \\ Z @>>\beta> X \end{CD}$$

is a cartesian (or pull-back) square of G-sets, then $M^*(\beta).M_*(\alpha) = M_*(\delta).M^*(\gamma)$.

A morphism θ from the Mackey functor M to the Mackey functor N is a natural transformation of bifunctors, consisting of a morphism $\theta_X : M(X) \to N(X)$ for any G-set X, such that for any morphism of G-sets $f : X \to Y$, the squares

$$\begin{CD} M(X) @>\theta_X>> N(X) \\ @V{M_*(f)}VV @VV{N_*(f)}V \\ M(Y) @>>\theta_Y> N(Y) \end{CD} \qquad\qquad \begin{CD} M(X) @>\theta_X>> N(X) \\ @A{M^*(f)}AA @AA{N^*(f)}A \\ M(Y) @>>\theta_Y> N(Y) \end{CD}$$

are commutative.

I will denote by $Mack_R(G)$ or $Mack(G)$ the category of Mackey functors for G over R.

Conversely, if M is a Mackey functor in the sense of this second definition, then one can build a Mackey functor M_1 in the first sense by setting

$$M_1(H) = M(G/H) \qquad t_K^H = M_*(p_K^H) \qquad r_K^H = M^*(p_K^H) \qquad c_{x,H} = M_*(\gamma_{x,H})$$

where $p_K^H : G/K \to G/H$ is the natural projection, and $\gamma_{x,H} : G/H \to G/{}^xH$ is the map $gH \mapsto gx^{-1}.{}^xH = gHx^{-1}$

1.1.3 Definition as modules over the Mackey algebra

There is a third definition of a Mackey functor, using the Mackey algebra $\mu_R(G)$: consider first the algebra $\mu(G)$ over $\mathbf{Z}$: it is the algebra generated by the elements t_K^H, r_K^H, and $c_{x,H}$, where H and K are subgroups of G such that $K \subseteq H$, and $x \in G$, with the following relations:

$$t_K^H t_L^K = t_L^H \ \forall\ L \subseteq K \subseteq H$$

$$r_L^K r_K^H = r_L^H \ \forall\ L \subseteq K \subseteq H$$

$$c_{y,{}^xH} c_{x,H} = c_{yx,H} \ \forall\ x, y, H$$

$$t_H^H = r_H^H = c_{h,H} \ \forall\ h, H\ h \in H$$

$$c_{x,H} t_K^H = t_{{}^xK}^{{}^xH} c_{x,K} \ \forall\ x, K, H$$

$$c_{x,K} r_K^H = r_{{}^xK}^{{}^xH} c_{x,H} \ \forall\ x, K, H$$

$$\sum_H t_H^H = \sum_H r_H^H = 1$$

$$r_K^H t_L^H = \sum_{x \in K\backslash H/L} t_{K\cap {}^xL}^K c_{x,K^x\cap L} r_{K^x \cap L}^L \ \forall\ K \subseteq H \supseteq L$$

any other product of r_H^K, t_H^K and $c_{g,H}$ being zero.

A Mackey functor M for the first definition gives a module $\widetilde{M}$ for the "algebra" $\mu_R(G) = R \otimes_Z \mu(G)$ (which is not really an algebra if R is not commutative) defined by

$$\widetilde{M} = \bigoplus_{H \subseteq G} M(H)$$

and a morphism $f : M \to N$ of Mackey functors gives a morphism of $\mu_R(G)$-modules $\tilde{f} : \widetilde{M} \to \widetilde{N}$.

It it then possible to define a Mackey functor as a $\mu_R(G)$-module, and a morphism of Mackey functors as a morphism of $\mu_R(G)$-modules: if M is a $\mu_R(G)$-module, then M corresponds to a Mackey functor M_1 in the first sense, defined by $M_1(H) = t_H^H M$, the maps t_K^H, r_K^H and $c_{x,H}$ being defined as the multiplications by the corresponding elements of the Mackey algebra.

1.2 The Mackey functors $M \mapsto M_Y$

If Y is a G-set, and M is a Mackey functor for G, then let M_Y be the Mackey functor defined by

$$M_Y(X) = M(X \times Y)$$

and for a map of G-sets $f : X \to X'$, by

$$(M_Y)^*(f) = M^*(f \times Id)$$

$$(M_Y)_*(f) = M_*(f \times Id)$$

This construction is functorial in Y: if Y' is another G-set, and if g is a morphism of G-sets from Y to Y', then there is a morphism of Mackey functors $M_g : M_Y \to M_{Y'}$, defined over the G-set X by

$$M_{g,X} = M_*(Id \times g) : M_Y(X) \to M_{Y'}(X)$$

To see this, let $f : X \to X'$ be a map of G-sets. Then the square

$$\begin{array}{ccc} M_Y(X) & \xrightarrow{M_{g,X}} & M_{Y'}(X) \\ {\scriptstyle (M_Y)_*(f)} \downarrow & & \downarrow {\scriptstyle (M_{Y'})_*(f)} \\ M_Y(X') & \xrightarrow[M_{g,X'}]{} & M_{Y'}(X') \end{array}$$

is commutative, since $M_*(f \times Id) \circ M_*(Id \times g) = M_*(f \times g) = M_*(Id \times g) \circ M_*(f \times Id)$. The square

$$\begin{array}{ccc} M_Y(X) & \xrightarrow{M_{g,X}} & M_{Y'}(X) \\ {\scriptstyle (M_Y)^*(f)} \uparrow & & \uparrow {\scriptstyle (M_{Y'})^*(f)} \\ M_Y(X') & \xrightarrow[M_{g,X'}]{} & M_{Y'}(X') \end{array}$$

is also commutative, because the square

$$\begin{array}{ccc} X \times Y & \xrightarrow{X \times g} & X \times Y' \\ {\scriptstyle f \times Id} \downarrow & & \downarrow {\scriptstyle f \times Id} \\ X' \times Y & \xrightarrow[Id \times g]{} & X' \times Y' \end{array}$$

is cartesian.

There is also a morphism M^g from $M_{Y'}$ to M_Y defined over X by

$$M^g_X = M^*(Id \times g) : M_{Y'}(X) \to M_Y(X)$$

In other words, I have defined a bifunctor from G-**set** to the category $Mack_R(G)$ of Mackey functors for G over R, which is equivalent to $\mu_R(G)$-**Mod.** I will check the conditions (M1) and (M2) for this bifunctor, proving that $Y \mapsto M_Y$ is a Mackey functor with values in the category of Mackey functors.

For the condition (M1), let Y and Y' be G-sets, and i and i' be the respective injections from Y and Y' into $Y\coprod Y'$. If X is a G-set, then

$$(M_{i,X}\oplus M_{i',X})\circ(M^i_X\oplus M^{i'}_X) = \big(M_*(Id\times i)\oplus M_*(Id\times i')\big)\circ\big(M^*(Id\times i)\oplus M^*(Id\times i')\big) = Id$$

and it it also clear that

$$(M^i_X \oplus M^{i'}_X)\circ(M_{i,X}\oplus M_{i',X}) = Id$$

For the condition (M2), let

$$\begin{array}{ccc} T & \xrightarrow{\gamma} & Y \\ \delta\big\downarrow & & \big\downarrow\alpha \\ Z & \xrightarrow[\beta]{} & U \end{array}$$

be a cartesian square. Then for any G-set X

$$M^\beta_X \circ M_{\alpha,X} = M^*(Id\times\beta)\circ M_*(Id\times\alpha)$$

As the square

$$\begin{array}{ccc} X\times T & \xrightarrow{Id\times\gamma} & X\times Y \\ Id\times\delta\big\downarrow & & \big\downarrow Id\times\alpha \\ X\times Z & \xrightarrow[Id\times\beta]{} & X\times U \end{array}$$

is cartesian, I have

$$M^*(Id\times\beta)\circ M_*(Id\times\alpha) = M_*(Id\times\delta)\circ M^*(Id\times\gamma)$$

and so

$$M^\beta_X \circ M_{\alpha,X} = M_{\delta,X}\circ M^\gamma_X$$

So I have defined a Mackey functor with values in the Mackey functors.

If now A is a ring, and F is a functor from $Mack_R(G)$ to A-**Mod**, then I will get a Mackey functor $\overline{F}$ with values in A-**Mod**, defined for G-sets X and Y, and for a map f of G-sets from X to Y, by

$$\overline{F}(X) = F(M_X) \qquad \overline{F}^*(f) = F(M^f_X) \qquad \overline{F}_*(f) = F(M_{f,X})$$

1.3 Construction of $\mathcal{H}(M,N)$ and $M\hat{\otimes}N$

From now on, the ring R will be commutative. There are two important examples of the previous construction: in both cases, let M and N be Mackey functors for G with values in R-**Mod**. For the first construction, I consider the functor

$$F = \mathrm{Hom}_{Mack(G)}(N,-)$$

and I can define the Mackey functor $\mathcal{H}(N,M)$, with values in R-**Mod**, by the following formulae

$$\mathcal{H}(N,M)(X) = \mathrm{Hom}_{Mack(G)}(N,M_X)$$

$$\mathcal{H}(N,M)^*(f) = \mathrm{Hom}_{Mack(G)}(N, M_X^f) \qquad \mathcal{H}(N,M)_*(f) = \mathrm{Hom}_{Mack(G)}(N, M_{f,X})$$

It is a kind of "internal homomorphisms" in the category of Mackey functors.

For the second construction, I need to observe that the Mackey algebra admits a natural anti-automorphism $m \mapsto \widehat{m}$, defined by

$$\widehat{t_K^H} = r_K^H \quad \widehat{r_K^H} = t_K^H \quad \widehat{c_{x,H}} = c_{x^{-1},H}$$

This allows to view any left $\mu_R(G)$-module as a right $\mu_R(G)$-module. Now being given a Mackey functor N, I can consider the functor

$$F : M \mapsto \widetilde{M} \otimes_{\mu_R(G)} \widetilde{N}$$

The functor $\overline{F}$ is then denoted by $M\hat{\otimes}N$: it is a Mackey functor with values in R-**Mod**, such that for any G-set X

$$(M\hat{\otimes}N)(X) = \widetilde{M} \otimes_{\mu_R(G)} \widetilde{N_X}$$

I will call it the tensor product of the Mackey functors M and N.

1.4 Identification of $\mathcal{H}(M,N)$

If H is a subgroup of G, then there is a restriction functor from $Mack(G)$ to $Mack(H)$, and an induction functor from $Mack(H)$ to $Mack(G)$, defined over a G-set X by

$$(\mathrm{Res}_H^G M)(X) = M(\mathrm{Ind}_H^G X) \qquad (\mathrm{Ind}_H^G N)(Y) = N(\mathrm{Res}_H^G Y)$$

As functors between categories of Mackey functors, they are mutual left and right adjoints (see [14]). Moreover, the isomorphism of G-sets

$$X \times (G/H) \simeq \mathrm{Ind}_H^G \mathrm{Res}_H^G X$$

gives the following isomorphism of Mackey functors

$$M_{G/H} \simeq \mathrm{Ind}_H^G \mathrm{Res}_H^G M$$

Those remarks prove that

$$\mathcal{H}(M,N)(H) = \mathrm{Hom}_{Mack(G)}(M, N_{G/H}) \simeq \mathrm{Hom}_{Mack(H)}(\mathrm{Res}_H^G M, \mathrm{Res}_H^G N)$$

I will translate the operations of transfer and restriction for $\mathcal{H}(M,N)$ using the previous identification.

I must first state precisely the adjunction

$$\mathrm{Hom}_{Mack(G)}(M, N_{G/H}) \simeq \mathrm{Hom}_{Mack(H)}(\mathrm{Res}_H^G M, \mathrm{Res}_H^G N)$$

A morphism $\phi \in \mathrm{Hom}_{Mack(H)}(\mathrm{Res}_H^G M, \mathrm{Res}_H^G N)$ gives in particular for any subgroup K of G, a morphism ϕ_K from $M(K)$ to $N(K)$. The adjunction now gives a morphism ϕ^* from M to $\mathrm{Ind}_H^G \mathrm{Res}_H^G N$ defined on the subgroup L of G by the morphism ϕ_L^* from $M(L)$ to $\mathrm{Ind}_H^G \mathrm{Res}_H^G N(L) \simeq \oplus_{x \in H\backslash G/L} N(H \cap {}^xL)$ given by

$$\phi_L^*(m) = \oplus_{x \in H\backslash G/L} \phi_{H \cap^x L}(x r_{H^x \cap L}^L(m))$$

Conversely, if $\psi \in \mathrm{Hom}_{Mack(G)}(M, \mathrm{Ind}_H^G \mathrm{Res}_H^G N)$, then ψ gives morphisms

$$\psi_L : M(L) \to \oplus_{x\in H\backslash G/L} N(H \cap {}^xL)$$

so I get a morphism from $M(L)$ to $N(H \cap {}^xL)$.
The adjunction gives a morphism ψ_* from $\mathrm{Res}_H^G M$ to $\mathrm{Res}_H^G N$: if $K \subseteq H$, the morphism $\psi_{*,K}$ from $M(K)$ to $N(K)$ is the component on the double class $H.1.K = H$ in the above map.

Now if $H \supseteq H' \supseteq K$, and if $m \in M(K)$, I have

$$\phi_K^*(m) = \oplus_{x\in H\backslash G/K} \phi_{H\cap^x K}(x r^K_{H^x\cap K}(m)) \in N_{G/H}(K)$$

Restricting to $N_{G/H'}(K)$, I get

$$r^H_{H'} \phi_K^*(m) \in \oplus_{y\in H'\backslash G/K} N(H' \cap {}^yK)$$

The component associated to y is obtained as follows: if x is the representative of $H\backslash G/K$, then there exists $h \in H$ and $k \in K$ such that $y = hxk$. The y component is then equal to $r^{H\cap^y K}_{H'\cap^y K}\big(h\phi_{H\cap^x K}(x r^K_{H^x\cap K}(m))\big)$.
In particular, the component $y = 1$ comes from the component $x = 1$, and the morphism from $\mathrm{Res}^G_{H'} M$ to $\mathrm{Res}^G_{H'} N$ associated to $r^H_{H'}\phi$ is defined by

$$(r^H_{H'}\phi)_K(m) = r^K_K \phi_K(r^K_K(m)) = \phi_K(m)$$

If now $H \subseteq H' \supseteq K$, and if $m \in M(K)$, then $(t^{H'}_H \phi^*)_K(m)$ is an element of $N_{G/H'}(K) = \oplus_{y\in H'\backslash G/K} N(H \cap {}^xK)$, and the component $y = 1$ comes from the doubles classes $x \in H\backslash H'/K$. The homomorphism from $\mathrm{Res}^G_{H'} M$ to $\mathrm{Res}^G_{H'} N$ associated to $t^{H'}_H \phi$ is then defined by

$$(t^{H'}_H \phi)_K(m) = \sum_{x\in H\backslash H'/K} x^{-1} t^{^xK}_{H\cap^x K} \phi_{H\cap^x K} x r^K_{H^x\cap K}(m)$$

that is

$$(t^{H'}_H \phi)_K(m) = \sum_{x\in H\backslash H'/K} t^K_{H^x\cap K} x^{-1} \phi_{H\cap^x K} x r^K_{H^x\cap K}(m)$$

Finally, it is clear that if $x \in G$, then the morphism ${}^x\phi$ from $\mathrm{Res}^G_{^xH} M$ to $\mathrm{Res}^G_{^xH} N$ is defined for $K \subseteq {}^xH$ and $m \in M(K)$ by

$$({}^x\phi)_K(m) = x\phi_{K^x}(x^{-1}m)$$

The next proposition is a summary of the previous remarks:

Proposition 1.4.1: Let M and N be Mackey functors for the group G. Then for any subgroup H of G,.

$$\mathcal{H}(M,N)(H) \simeq \mathrm{Hom}_{Mack(H)}(\mathrm{Res}^G_H M, \mathrm{Res}^G_H N)$$

An element ϕ in $\mathcal{H}(M,N)(H)$ is defined by a sequence of morphisms $\phi_L \in \mathrm{Hom}_R\big(M(L), N(L)\big)$, for $L \subseteq H$. Then:

- **If $H \subseteq H' \supseteq K$, then**

$$(t_H^{H'}\phi)_K(m) = \sum_{x \in H\backslash H'/K} t^K_{H^x\cap K} x^{-1}\phi_{H\cap {}^xK} x r^K_{H^x\cap K}(m)$$

- **If $K \subseteq H' \subseteq H$, then**

$$(r^H_{H'}\phi)_K(m) = \phi_K(m)$$

- **If $x \in G$ and $K \subseteq {}^xH$, then**

$$({}^x\phi)_K(m) = x\phi_{K^x}(x^{-1}m)$$

Remark: This proposition shows that $\mathcal{H}(M,N)$ coincides with the construction called the exponent by Sasaki (see [12]).

1.5 Identification of $M\hat{\otimes}N$

Let M and N be Mackey functors for G.
a) The definition of $M\hat{\otimes}N$ shows that

$$M\hat{\otimes}N(H) = \widetilde{M} \otimes_{\mu_R(G)} \widetilde{N_{G/H}}$$

But $N_{G/H} \simeq \mathrm{Ind}_H^G \mathrm{Res}_H^G N$, and the restriction functor from $Mack_R(G)$ to $Mack_R(H)$ comes from the natural morphism i_H^G from the algebra $\mu_R(H)$ to $\mu_R(G)$. In other words, if M is a Mackey functor for G, then

$$\widetilde{\mathrm{Res}_H^G M} = i_H^G(1_{\mu_R(H)}).\widetilde{M}$$

The induction functor from $Mack_R(H)$ to $Mack_R(G)$ is left adjoint to it. Then for any Mackey functor L for H, I have

$$\widetilde{\mathrm{Ind}_H^G L} = \mu_R(G) \otimes_{\mu_R(H)} \widetilde{L}$$

This proves that

$$M\hat{\otimes}N(H) = \widetilde{M} \otimes_{\mu_R(G)} \mu_R(G) \otimes_{\mu_R(H)} \widetilde{\mathrm{Res}_H^G N} = \widetilde{M} \otimes_{\mu_R(H)} i_H^G(1_{\mu_R(H)}).\widetilde{N}$$

or

$$M\hat{\otimes}N(H) = \widetilde{M} i_H^G(1_{\mu_R(H)}) \otimes_{\mu_R(H)} i_H^G(1_{\mu_R(H)}).\widetilde{N} = \widetilde{\mathrm{Res}_H^G M} \otimes_{\mu_R(H)} \widetilde{\mathrm{Res}_H^G N}$$

since $i_H^G(1_{\mu_R(H)})$ is invariant under the antiautomorphism of $\mu_R(G)$.
b) I also claim that

$$\widetilde{M} \otimes_{\mu_R(G)} \widetilde{N} \simeq \left(\oplus_{H\subseteq G} M(H) \otimes_R N(H)\right)/\mathcal{I}$$

where $\mathcal{I}$ is the R-submodule generated by the elements of the form

$$t_K^L m \otimes_R n - m \otimes_R r_K^L n \quad \text{for} \quad K \subseteq L \subseteq G,\ m \in M(K),\ n \in N(L)$$

$$m\otimes_R t_L^K n - r_L^K m\otimes_R n \quad \text{for} \quad L\subseteq K\subseteq G,\ m\in M(K),\ n\in N(L)$$

$$gm\otimes_R gn - m\otimes_R n \quad \text{for} \quad K\subseteq G,\ m\in M(K),\ n\in N(K),\ g\in G$$

To see this, let P be the right hand side, and π be the projection from $\oplus_{H\subseteq G}M(H)\otimes_R N(H)$ onto P. There is a map θ from $\oplus_{H\subseteq G}M(H)\otimes_R N(H)$ to $\widetilde{M}\otimes_{\mu_R(G)}\widetilde{N}$, sending $m\otimes_R n\in M(H)\otimes N(H)$ to $m\otimes_{\mu_R(G)} n\in \widetilde{M}\otimes_{\mu_R(G)}\widetilde{N}$, since $M(H)\subseteq \widetilde{M}$ and $N(H)\subseteq\widetilde{N}$. It is easy to see that this factors through P, since for instance

$$\theta(t_K^L m\otimes_R n - m\otimes_R r_K^L n) = t_K^L m\otimes_{\mu_R(G)} n - m\otimes_{\mu_R(G)} r_K^L n = \ldots$$

$$\ldots = m.r_K^L\otimes_{\mu_R(G)} n - m\otimes_{\mu_R(G)} r_K^L.n = 0$$

Conversely, let θ' be the map from $\widetilde{M}\otimes_{\mu_R(G)}\widetilde{N}$ to P which sends $m\otimes_{\mu_R(G)} n$ to the image in P of $\oplus_{H\subseteq G}t_H^H m\otimes_R t_H^H n$. This map is well defined, because if $a = t_B^A x r_{B^x}^C\in\mu_R(G)$, then

$$\theta'(m.a\otimes_{\mu_R(G)} n) = \theta'(t_{B^x}^C x^{-1} r_A^B m\otimes_{\mu_R(G)} n) = \ldots$$
$$\ldots = \pi(\oplus_{H\subseteq G}t_H^H t_{B^x}^C x^{-1} r_A^B m\otimes_R t_H^H n) = \pi(t_{B^x}^C x^{-1} r_A^B m\otimes_R t_C^C n)$$

I have also

$$\theta'(m\otimes_{\mu_R(G)} a.n) = \theta'(m\otimes_{\mu_R(G)} t_B^A x r_{B^x}^C n) = \pi(t_A^A m\otimes_R t_B^A x r_{B^x}^C n)$$

But $t_{B^x}^C x^{-1} r_A^B m\otimes_R t_C^C n - t_A^A m\otimes_R t_B^A x r_{B^x}^C n\in\mathcal{I}$, and

$$\theta'(m.a\otimes_{\mu_R(G)} n) = \theta'(m\otimes_{\mu_R(G)} a.n)$$

and this proves that θ' is well defined.

Now if $m\in M(H)$ and $n\in N(H)$, then

$$\theta'\theta(\pi(m\otimes_R n)) = \pi(\oplus_{K\subseteq G}t_K^K m\otimes_R t_K^K n) = \pi(t_H^H m\otimes_R t_H^H n) = \pi(m\otimes n)$$

So $\theta'\theta$ is the identity.
And if $m\in\widetilde{M}$ and $n\in\widetilde{N}$, then $m = \sum_{H\subseteq G} t_H^H m$, and $n = \sum_{K\subseteq G} t_K^K n$, which gives

$$m\otimes_{\mu_R(G)} n = \sum_{\substack{H\subseteq G\\ K\subseteq G}} t_H^H m\otimes_{\mu_R(G)} t_K^K n = \ldots$$

$$\ldots = \sum_{\substack{H\subseteq G\\ K\subseteq G}} m.r_H^H\otimes_{\mu_R(G)} t_K^K.n = \sum_{H\subseteq G} m.r_H^H\otimes_{\mu_R(G)} t_H^H.n$$

On the other hand

$$\theta\theta'(m\otimes_{\mu_R(G)} n) = \sum_{H\subseteq G} t_H^H m\otimes_{\mu_R(G)} t_H^H n$$

So $\theta\theta'$ is also equal to identity, and this proves the claim.

c) Points a) and b) show that

$$(M\hat{\otimes}N)(H) = \widetilde{\mathrm{Res}_H^G M}\otimes_{\mu_R(H)}\widetilde{\mathrm{Res}_H^G N} = (\oplus_{K\subseteq H}M(K)\otimes_R N(K))/\mathcal{J}$$

where $\mathcal{J}$ is the submodule generated by

$$t_K^L m \otimes_R n - m \otimes_R r_K^L n \quad \text{for} \quad K \subseteq L \subseteq H,\ m \in M(K),\ n \in N(L)$$

$$m \otimes_R t_L^K n - r_L^K m \otimes_R n \quad \text{for} \quad L \subseteq K \subseteq H,\ m \in M(K),\ n \in N(L)$$

$$hm \otimes_R hn - m \otimes_R n \quad \text{for} \quad K \subseteq H,\ m \in M(K),\ n \in N(K),\ h \in H$$

d) In order to obtain formulae for transfer and restriction in $M\hat{\otimes}N$, I must first describe precisely the isomorphism θ of a) giving $M\hat{\otimes}N(H)$ as

$$\widetilde{M} \otimes \widetilde{\mathrm{Ind}_H^G \mathrm{Res}_H^G N} \simeq \widetilde{\mathrm{Res}_H^G M} \otimes_{\mu_R(H)} \widetilde{\mathrm{Res}_H^G N}$$

If K is a subgroup of H, then $M(K)$ (resp. $N(K)$) maps into $\widetilde{\mathrm{Res}_H^G M}$ (resp. into $\widetilde{\mathrm{Res}_H^G N}$), so there is a morphism π from $\oplus_{K \subseteq H} M(K) \otimes_R N(K)$ to the right hand side. This morphism is surjective, because if $m \in M(K)$ and $m' \in M(K')$ for $K \neq K'$, then $m \otimes m' = 0$ in $\widetilde{\mathrm{Res}_H^G M} \otimes_{\mu_R(H)} \widetilde{\mathrm{Res}_H^G N}$.
Now $M(K)$ maps into $\widetilde{M}$, and $N(K)$ maps into $\mathrm{Ind}_H^G \mathrm{Res}_H^G N(K)$, so into $\widetilde{\mathrm{Ind}_H^G \mathrm{Res}_H^G N}$. So there is a morphism π' from $\oplus_{K \subseteq H} M(K) \otimes_R N(K)$ to the left hand side. The above isomorphism is the given (from right to left) by

$$\theta^{-1}(\pi(m \otimes n)) = \pi'(m \otimes n)$$

If now K is any subgroup of G, then

$$N_{G/H}(K) = \mathrm{Ind}_H^G \mathrm{Res}_H^G N(K) \simeq \oplus_{x \in H\backslash G/K} N(H \cap {}^xK)$$

If $n \in N(H \cap {}^xK) \subseteq N_{G/H}(K)$, let n' be the element $x^{-1}n \in N(H^x \cap K)$, viewed inside $N_{G/H}(H^x \cap K)$. In $N_{G/H}(K)$, I have then

$$n = t_{H^x \cap K}^K n'$$

Then if $m \in M(K)$, I have the following equalities in $\widetilde{M} \otimes_{\mu_R(G)} \widetilde{\mathrm{Ind}_H^G \mathrm{Res}_H^G N}$

$$m \otimes_{\mu_R(G)} n = m \otimes_{\mu_R(G)} t_{H^x \cap K}^K n' = m.t_{H^x \cap K}^K \otimes_{\mu_R(G)} n' = \ldots$$

$$\ldots = r_{H^x \cap K}^K m \otimes_{\mu_R(G)} x^{-1} n = x r_{H^x \cap K}^K m \otimes_{\mu_R(G)} n$$

and this element is in the component $M(H \cap {}^xK) \otimes_R N(H \cap {}^xK)$ of $\widetilde{\mathrm{Res}_H^G M} \otimes_{\mu_R(H)} \widetilde{\mathrm{Res}_H^G N}$. It is the image under θ of $\pi'(m \otimes n)$.
Those precisions give explicit formulae of transfer and restriction for $M\hat{\otimes}N$: let $H \subseteq H'$ be subgroups of G, and K be a subgroup of H. If $[m \otimes n]_K$ denotes the image in $M\hat{\otimes}N(H)$ of $m \otimes n \in M(K) \otimes_R N(K)$, then $\theta^{-1}([m \otimes n]_K)$ is the image of $m \otimes n$ in $\widetilde{M} \otimes \widetilde{\mathrm{Ind}_H^G \mathrm{Res}_H^G N}(H)$, which is a quotient of $\oplus_{L \subseteq G} \oplus_{x \in H\backslash G/L} M(L) \otimes_R N(H \cap {}^xL)$:it is the image of the element $m \otimes n$ of the component corresponding to $L = K$ and $x = H.1.K$.
By the transfer from H to H', this element maps to the image of the element

$$m \otimes t_{H \cap {}^xL}^{H' \cap {}^xL} n = m \otimes n$$

of the component corresponding to $L = K$ and $x = H'.1.L$, which in turn is the image under θ^{-1} of the element $[m \otimes n]_K$ of the component $K \subseteq H'$ of $M\hat{\otimes}N(H')$. Finally

$$t_H^{H'}[m \otimes n]_K = [m \otimes n]_K$$

If now K is a subgroup of H', then $\theta^{-1}([m \otimes n]_K)$ is the image of the element $m \otimes n$ of the component $L = K$ and $x = H'.1.K$ of $\widetilde{M} \otimes_{\mu_R(G)} N_{G/H'}$. Its restriction to H is the image of

$$\oplus_{x \in H\backslash H'/K} m \otimes x r^K_{H^x \cap K} n$$

corresponding to the double classes $H\backslash H'.1.K/K$. The image under θ of this element is the element

$$\oplus_{x \in H\backslash H'/K} x r^K_{H^x \cap K} m \otimes x r^K_{H^x \cap K} n$$

In other words

$$r_H^{H'}([m \otimes n])_K = \sum_{x \in H\backslash H'/K} [x r^K_{H^x \cap K} m \otimes x r^K_{H^x \cap K} n]_{H \cap {}^x K}$$

Finally, if $x \in G$, conjugation by x from $M\hat{\otimes}N(H)$ to $M\hat{\otimes}N({}^xH)$ is given by

$$x[m \otimes n]_K = [xm \otimes xn]_{{}^xK}$$

I have proved the following:

Proposition 1.5.1: Let M and N be Mackey functors for the group G. Then for any subgroup H of G

$$(M\hat{\otimes}N)(H) \simeq \Big(\oplus_{K \subseteq H} M(K) \otimes_R N(K)\Big)/\mathcal{J}$$

where $\mathcal{J}$ is the R-submodule generated by

$$t_K^L m \otimes_R n - m \otimes_R r_K^L n \text{ for } K \subseteq L \subseteq H,\ m \in M(K),\ n \in N(L)$$

$$m \otimes_R t_L^K n - r_L^K m \otimes_R n \text{ for } L \subseteq K \subseteq H,\ m \in M(K),\ n \in N(L)$$

$$hm \otimes_R hn - m \otimes_R n \text{ for } K \subseteq H,\ m \in M(K),\ n \in N(K),\ h \in H$$

Let $[m \otimes n]_K$ denote the image in $M\hat{\otimes}N(H)$ of $m \otimes n \in M(K) \otimes_R N(K)$. Then

- **If $H \subseteq H' \supseteq K$, then**

$$r_H^{H'}\Big([m \otimes n]_K\Big) = \sum_{x \in H\backslash H'/K} [x r^K_{H^x \cap K} m \otimes x r^K_{H^x \cap K} n]_{H \cap {}^x K}$$

- **If $K \subseteq H \subseteq H'$, then**

$$t_H^{H'}\Big([m \otimes n]_K\Big) = [m \otimes n]_K$$

- **If $x \in G$, then**

$${}^x\Big([m \otimes n]_K\Big) = [{}^x m \otimes {}^x n]_{{}^x K}$$

1.6 Another identification of $M\hat{\otimes}N$

The first identification of $M\hat{\otimes}N$ gives $M\hat{\otimes}N(H)$ as a quotient of $\oplus_{K\subseteq H}M(K)\otimes_R N(K)$. This expression is not very useful to evaluate $M\hat{\otimes}N(X)$ for an arbitrary G-set X.

If K is a subgroup of H, there is a corresponding "G-set over G/H", defined as the natural projection $p_K^H : G/K \to G/H$. More generally, a G-set Y over the G-set X is a couple (Y,ϕ), where Y is a G-set and ϕ a map of G-sets from Y to X. A morphism from (Y,ϕ) to (Y',ϕ') is a morphism of G-sets f from Y to Y' such that $\phi' f = \phi$.

The next idea is then to consider for any G-set X

$$T(X) = \Big(\bigoplus_{Y\xrightarrow{\phi}X} M(Y)\otimes_R N(Y)\Big)/\mathcal{J}$$

where $\mathcal{J}$ is the R-submodule generated by the elements

$$M^*(f)(m')\otimes n - m'\otimes N_*(f)(n),\ M_*(f)(m)\otimes n' - m\otimes N^*(f)(n')$$

$$\forall f:(Y,\phi)\to(Y',\phi'),\ \forall m\in M(Y),\ m'\in M(Y'), n\in N(Y),\ n'\in N(Y')$$

(Note that there is no serious set theoretical problem in this definition, since the category of finite G-sets over X admits a small equivalent subcategory).

If $m\otimes n\in M(Y)\otimes_R N(Y)$, I will denote by $[m\otimes n]_{(Y,\phi)}$ its image in $T(X)$.

Lemma 1.6.1: With those definitions

$$T(G/H)\simeq M\hat{\otimes}N(H) \qquad \forall H\subseteq G$$

Proof: There is indeed a natural map θ from $M\hat{\otimes}N(H)$ to $T(G/H)$, which sends $[m\otimes n]_K$ to $[m\otimes n]_{(G/K,p_K^H)}$: this map is well defined, because $M_*(p_K^H) = t_K^H$ and $M^*(p_K^H) = r_K^H$, whereas $M_*(x)$ is conjugation by x, and $M^*(x)$ is conjugaison by x^{-1}.
Conversely, let (Y,ϕ) be a G-set over G/H. As G is transitive on G/H, there is a system of representatives S of $G\backslash Y$ contained in $\phi^{-1}(H)$. In particular $G_s\subseteq H$ for all $s\in S$. I denote by i_s the injection $g.G_s\mapsto g.s$ from G/G_s to Y, and for $m\in M(Y)$ (resp. $n\in N(Y)$), I write $m_s = M^*(i_s)(m)$ (resp. $n_s = N^*(i_s)(n)$). The map $m\mapsto \oplus_{s\in S}m_s$ is an isomorphism from $M(Y)$ to $\oplus_{s\in S}M(G_s)$. I define a map θ' from $T(X)$ to $M\hat{\otimes}N(H)$ by

$$\theta'([m\otimes n]_{(Y,\phi)}) = \sum_{s\in S}[m_s\otimes n_s]_{G_s}$$

The map θ' does not depend on the choice of S inside $\phi^{-1}(H)$: if S' is another system, then for any $s'\in S'$, there exists $s\in S$ and $h_s\in H$ such that $s' = h_s.s$. In those conditions, I have $m_{s'} = h_s m_s$, and $[m_{s'}\otimes n_{s'}]_{G_{s'}} = [h_s m_s\otimes h_s n_s]_{G_{s'}} = [m_s\otimes n_s]_{G_s}$ in $M\hat{\otimes}N(H)$.

The map θ' is also well defined: if $f:(Y,\phi)\to(Y',\phi')$ is a morphism of sets over G/H, I have to check that

$$\theta'\Big(M_*(f)(m)\otimes n' - m\otimes N^*(f)(n')\Big) = 0$$

and that

$$\theta'\Big(M^*(f)(m')\otimes n - m'\otimes N_*(f)(n)\Big) = 0$$

Let $S' \subseteq \phi'^{-1}(H)$ be the system of representatives of $G\backslash Y'$ chosen for Y'. Then for any $s \in S$, there exists a unique $t_s \in S'$ and an element $h_s \in H$ such that $f(s) = h_s.t_s$. In those conditions

$$\theta'([M_*(f)(m)\otimes n']_{(Y',\phi')}) = \oplus_{s'\in S'}[M^*(i_{s'})M_*(f)(m)\otimes N^*(i_{s'})(n)]_{G_{s'}}$$

Let P be the pull-back of Y and $G/G_{s'}$, such that the square

$$\begin{array}{ccc} P & \longrightarrow & Y \\ \downarrow & & \downarrow f \\ G/G_{s'} & \xrightarrow[i_{s'}]{} & Y' \end{array}$$

is cartesian.
Then $P \simeq \{(g.G_{s'}, y) \in G/G_s \times Y \mid f(y) = g.s'\}$. In particular, if $s' \notin f(Y)$, then P is empty and $M^*(i_{s'})M_*(f) = 0$ in this case.
And if $f(y) = g.s'$, let $s \in S \cap (G.s)$. Then there exists g_1 such that $y = g_1.s$, so $f(s) = g_1^{-1}g.s'$, and this proves in particular that $s' = t_s$. Conversely, if $t_s = s'$ and $y = g.s$, then $f(y) = gh_s.t_s$ and $(gh_sG_{s'}, y) \in P$. So P may be identified as the union of the orbits $G.s$ such that $t_s = s'$. I have the following cartesian square

$$\begin{array}{ccc} \coprod_{t_s=s'} G/G_s & \xrightarrow{\coprod i_s} & Y \\ {\scriptstyle\coprod \gamma_s}\downarrow & & \downarrow f \\ G/G_{s'} & \xrightarrow[i_{s'}]{} & Y' \end{array}$$

where $\gamma_s(g.G_s) = gh_s.G_{s'}$. This shows that

$$\theta'\Big([M_*(f)(m)\otimes n']_{(Y',\phi')}\Big) = \sum_{s\in S}[M_*(\gamma_s)M^*(i_s)(m)\otimes N^*(i_{t_s})(n')]_{G_{t_s}}$$

But $M_*(\gamma_s) = h_s t_{G_s}^{G_{t_s}}$, and so in $M\hat{\otimes}N(G_{t_s})$

$$[M_*(\gamma_s)M^*(i_s)(m)\otimes N^*(i_{t_s})(n')]_{G_{t_s}} = [M^*(i_s)(m)\otimes r_{G_s}^{G_{s'}}h_s^{-1}N^*(i_{t_s})(n')]_{G_s}$$

and moreover

$$r_{G_s}^{G_{s'}}h_s^{-1}N^*(i_{t_s}) = N^*(\gamma_s)N^*(i_{t_s}) = N^*(i_{t_s}\gamma_s) = N^*(fi_s) = N^*(i_s)N^*(f)$$

whence finally

$$\theta'([M_*(f)(m)\otimes n']_{(Y',\phi')}) = \sum_{s\in S}[M^*(i_s)(m)\otimes N^*(i_s)N^*(f)(n')]_{G_s} = \dots$$

$$\dots = \theta'([m\otimes N^*(f)(n')]_{(Y,\phi)})$$

A similar argument shows that

$$\theta'\Big(M^*(f)(m')\otimes n - m'\otimes N_*(f)(n)\Big) = 0$$

and so θ' is well defined.
It is clear that $\theta'\theta$ is equal to identity of $M\hat{\otimes}N(H)$. Conversely,

$$\theta\theta'([m\otimes n]_{(Y,\phi)}) = \sum_{s\in S}[M^*(i_s)(m)\otimes N^*(i_s)(n)]_{(G/G_s,\phi i_s)}$$

But Y is isomorphic to the disjoint union of the G/G_s, so

$$m = \sum_{s\in S} M_*(i_s)M^*(i_s)(m)$$

so in $T(X)$, I have

$$[m\otimes n]_{(Y,\phi)} = \sum_{s,t\in S}[M^*(i_s)(m)\otimes M^*(i_s)M_*(i_t)M^*(i_t)(n)]_{(G/G_s,\phi i_s)}$$

and as M is a Mackey functor, the product $M^*(i_s)M_*(i_t)$ is zero if $s\neq t$, and equal to the identity if $s=t$. So $\theta\theta'$ is equal to the identity, and this proves the lemma. ■

I can now give T a structure of bifunctor: let X and X' be G-sets, and f be a morphism from X to X'. Then any set (Y,ϕ) over X defines a set $(Y,f\phi)$ over X': I can define

$$T_*(f)\Big([m\otimes n]_{(Y,\phi)}\Big) = [m\otimes n]_{(Y,f\phi)}$$

The map $T_*(f)$ is well defined, because if $g:(Y,\phi)\to(Y',\phi')$, if $m\in M(Y)$ and $n'\in N(Y')$, then

$$T_*(f)\Big([M_*(g)(m)\otimes n']_{(Y',\phi')}\Big) = [M_*(g)(m)\otimes n']_{(Y',f\phi')} = \ldots$$

$$\ldots = [m\otimes N^*(g)(n')]_{(Y,f\phi)} = T_*(f)\Big([m\otimes N^*(g)(n')]_{(Y,\phi)}\Big)$$

A similar argument shows that

$$T_*(f)\Big([M^*(g)(m')\otimes n]_{(Y,\phi)}\Big) = T_*(f)\Big([m'\otimes N_*(g)(n)]_{(Y',\phi')}\Big)$$

so $T_*(f)$ is well defined. It is clear moreover that this turns T into a covariant functor. Conversely, (Y',ϕ') is a set over X', then let (Y,ϕ) be the pull-back of X and Y', giving a cartesian square

$$\begin{array}{ccc} Y & \xrightarrow{a} & Y' \\ {\scriptstyle\phi}\downarrow & & \downarrow{\scriptstyle\phi'} \\ X & \xrightarrow[f]{} & X' \end{array}$$

The set (Y,ϕ) is by definition the inverse image of (Y',ϕ') under f.
Now let

$$T^*(f)\Big([m\otimes n]_{(Y',\phi')}\Big) = [M^*(a)(m)\otimes N^*(a)(n)]_{(Y,\phi)}$$

The map T^* is well defined: if g is a morphism from (Y', ϕ') to (Y'_1, ϕ'_1), then I have the following commutative diagram

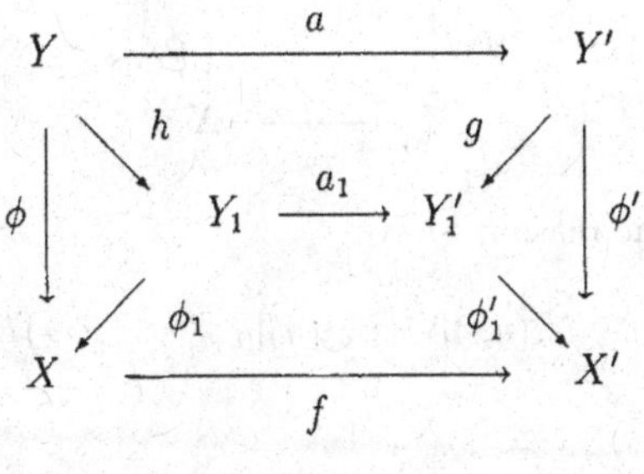

There is a map h because the squares (Y, Y', X, X') and (Y_1, Y'_1, X, X') are cartesian by construction. Moreover

$$\phi'_1 ga = \phi' a = f\phi$$

Then as (Y_1, Y'_1, X, X') is cartesian, there exists a unique morphism h from Y to Y_1 such that $\phi = \phi_1 h$ and $ga = a_1 h$. The square (Y, Y', Y_1, Y'_1) is then cartesian, for its composition with the cartesian square (Y_1, Y'_1, X, X') gives the cartesian square (Y, Y', X, X').
In those conditions, if $m' \in M(Y')$ and $n'_1 \in N(Y'_1)$, I have

$$T^*\Big([M_*(g)(m') \otimes n'_1]_{(Y'_1,\phi'_1)}\Big) = [M^*(a_1)M_*(g)(m') \otimes N^*(a_1)(n'_1)]_{(Y_1,\phi_1)} = \dots$$

$$\dots = [M_*(h)M^*(a)(m') \otimes N^*(a_1)(n'_1)]_{(Y_1,\phi_1)} = [M^*(a)(m') \otimes N^*(h)N^*(a_1)(n'_1)]_{(Y,\phi)} = \dots$$

$$\dots = [M^*(a)(m') \otimes N^*(a_1 h)(n'_1)]_{(Y,\phi)} = [M^*(a)(m') \otimes N^*(ga)(n'_1)]_{(Y,\phi)} = \dots$$

$$\dots = [M^*(a)(m') \otimes N^*(a)N^*(g)(n'_1)]_{(Y,\phi)} = T^*\Big([m' \otimes N^*(g)(n'_1)]_{(Y',\phi')}\Big)$$

A similar argument proves that

$$T^*([M^*(g)(m'_1) \otimes n']_{(Y',\phi')}) = T^*([m'_1 \otimes N_*(g)(n')]_{(Y'_1,\phi'_1)})$$

so the map $T^*(f)$ is well defined. It is also clear that T becomes a contravariant functor.

In order to prove that the bifunctor T is a Mackey functor, I must check the conditions (M1) and (M2). Let $X = Y \coprod Z$ be the disjoint union of Y and Z. I denote by i_Y and i_Z the respective injections from Y and Z to X. If (U, ϕ) is a G-set over X, let $U_Y = \phi^{-1}(Y)$ (resp. $U_Z = \phi^{-1}(Z)$), let ϕ_Y (resp. ϕ_Z) be the restriction of ϕ to U_Y (resp. to U_Z), and let u_Y (resp. u_Z) be the injection from U_Y into U. Then U is the disjoint union of U_Y and U_Z.
Let $[m \otimes n]_{(U,\phi)} \in T(X)$. Since M and N are Mackey functors, I know that $m = \Big(M_*(u_Y)M^*(u_Y) + M_*(u_Z)M^*(u_Z)\Big)(m)$ and $n = \Big(N_*(u_Y)N^*(u_Y) + N_*(u_Z)N^*(u_Z)\Big)(n)$. Moreover

$$[M_*(u_Y)M^*(u_Y)(m) \otimes N_*(u_Y)N^*(u_Y)(n)]_{(U,\phi)} = \dots$$

$$\dots = [M^*(u_Y)(m) \otimes N^*(u_Y)N_*(u_Y)N^*(u_Y)(n)]_{(U_Y,\phi u_Y)} = \dots$$

$$\dots = [M^*(u_Y)(m) \otimes N^*(u_Y)(n)]_{(U_Y, i_Y\phi_Y)} = T_*(i_Y)T^*(i_Y)\Big([m \otimes n]_{(U,\phi)}\Big)$$

as the square

$$\begin{array}{ccc} U_Y & \xrightarrow{u_Y} & U \\ {\scriptstyle \phi_Y}\downarrow & & \downarrow{\scriptstyle \phi} \\ Y & \xrightarrow[i_Y]{} & X \end{array}$$

is cartesian. For the same reason, I have

$$[M_*(u_Z)M^*(u_Z)(m) \otimes N_*(u_Z)N^*(u_Z)(n)]_{(U,\phi)} = T_*(i_Z)T^*(i_Z)\Big([m \otimes n]_{(U,\phi)}\Big)$$

Since finally

$$[M_*(u_Y)M^*(u_Y)(m) \otimes N_*(u_Z)N^*(u_Z)(n)]_{(U,\phi)} = \dots$$

$$\dots = [M^*(u_Y)(m) \otimes N^*(u_Z)N_*(u_Y)N^*(u_Y)(n)]_{(U_Y,\phi u_Y)} = 0$$

I have also

$$[M_*(u_Z)M^*(u_Z)(m) \otimes N_*(u_Y)N^*(u_Y)(n)]_{(U,\phi)} = 0$$

and

$$[m \otimes n]_{(U,\phi)} = T_*(i_Y)T^*(i_Y)\Big([m \otimes n]_{(U,\phi)}\Big) + T_*(i_Z)T^*(i_Z)\Big([m \otimes n]_{(U,\phi)}\Big)$$

This proves one half of (M1).
I don't need to check that $T^*(i_Y)T_*(i_Y)$ is the identity of $T(Y)$, if I know that (M2) is true for T: in that case indeed, this follows from the fact that since i_Y is injective, then the square

$$\begin{array}{ccc} Y & \xrightarrow{Id} & Y \\ {\scriptstyle Id}\downarrow & & \downarrow{\scriptstyle i_Y} \\ Y & \xrightarrow[i_Y]{} & X \end{array}$$

is cartesian. In the same way, the cartesian square

$$\begin{array}{ccc} \emptyset & \longrightarrow & Z \\ \downarrow & & \downarrow{\scriptstyle i_Z} \\ Y & \xrightarrow[i_Y]{} & X \end{array}$$

and the fact that $T(\emptyset) = 0$ show that $T^*(i_Y)T_*(i_Z) = 0$. This proves (M1) for T, assuming (M2).
To prove (M2), let

$$\begin{array}{ccc} V & \xrightarrow{a} & Y \\ {\scriptstyle b}\downarrow & & \downarrow{\scriptstyle c} \\ Z & \xrightarrow[d]{} & X \end{array}$$

be a cartesian square, let (U,ϕ) be a G-set over Y, and let $[m\otimes n]_{(U,\phi)}\in T(Y)$. Let W, $\phi': W\to V$ and $a': W\to U$ such that the square

$$\begin{array}{ccc} W & \xrightarrow{a'} & U \\ {\scriptstyle \phi'}\downarrow & & \downarrow{\scriptstyle \phi} \\ V & \xrightarrow[a]{} & Y \end{array}$$

is cartesian. Then

$$T_*(b)T^*(a)\Big([m\otimes n]_{(U,\phi)}\Big) = [M^*(a')(m)\otimes N^*(a')(n)]_{(W,b\phi')}$$

But the square

$$\begin{array}{ccc} W & \xrightarrow{a'} & U \\ {\scriptstyle b\phi'}\downarrow & & \downarrow{\scriptstyle c\phi} \\ Z & \xrightarrow[d]{} & X \end{array}$$

is composed of the two previous ones. It is then cartesian, and

$$T^*(d)T_*(c)\Big([m\otimes n]_{(U,\phi)}\Big) = T^*(d)\Big([m\otimes n]_{(U,c\phi)}\Big) = \ldots$$

$$\ldots = [M^*(a')(m)\otimes N^*(a')(n)]_{(W,b\phi')}$$

and this proves (M2) for the functor T. So T is a Mackey functor.

To prove that $T = M\hat{\otimes}N$, all I have to do is to check that when f is the natural projection p_K^H from G/K to G/H, for $K\subseteq H$, then $T_*(f)$ is equal up to the isomorphisms of lemma 1.6.1, to the transfer t_K^H of $M\hat{\otimes}N$, and that $T^*(f)$ is equal to the restriction r_K^H. Let (Y,ϕ) be a G-set over G/K, and let $S\subseteq\phi^{-1}(K)$ be a system of representatives of the orbits of Y. Then if $[m\otimes n]_{(Y,\phi)}\in T(G/K)$, I have

$$\theta'\Big([m\otimes n]_{(Y,\phi)}\Big) = \sum_{s\in S}[m_s\otimes n_s]_{G_s}\in M\hat{\otimes}N(K)$$

So

$$t_K^H\theta'\Big([m\otimes n](Y,\phi)\Big) = \sum_{s\in S}[m_s\otimes n_s]_{G_s}\in M\hat{\otimes}N(H)$$

But I have also

$$T_*(p_K^H)\Big([m\otimes n]_{(Y,\phi)}\Big) = [m\otimes n]_{(Y,p_K^H\phi)}$$

and $S\subseteq\phi^{-1}(K)\subseteq(p_K^H\phi)^{-1}(H)$. So

$$\theta' T_*(p_K^H)\Big([m\otimes n]_{(Y,\phi)}\Big) = \sum_{s\in S}[m_s\otimes n_s]_{G_s} = \theta'\Big([m\otimes n]_{(Y,\phi)}\Big)$$

Now if (Y,ϕ) is a G-set over G/H, let Z be the pull-back of Y and G/K. I have the following cartesian square

$$\begin{array}{ccc} Z & \xrightarrow{a} & Y \\ {\scriptstyle b}\downarrow & & \downarrow{\scriptstyle \phi} \\ G/K & \xrightarrow[p_K^H]{} & G/H \end{array}$$

As p_K^H is surjective, so is a. Then if $S \subseteq \phi^{-1}(H)$ is a system of representatives of the orbits of Y, and if $[m \otimes n]_{(Y,\phi)} \in T(G/H)$, I have

$$r_K^H \theta'\Big([m \otimes n]_{(Y,\phi)}\Big) = r_K^H\Big(\sum_{s \in S}[m_s \otimes n_s]_{G_s}\Big) = \sum_{\substack{s \in S \\ x \in K\backslash H/G_s}} [x r_{K^x \cap G_s}^{G_s} m_s \otimes x r_{K^x \cap G_s}^{G_s} n_s]_{K \cap {}^x G_s}$$

But to $s \in S$ and $x \in K\backslash H/G_s$, I can associate the couple $(K, xs) \in Z$, and as a is surjective, it is easy to see that I obtain that way a system of representatives $S' \subseteq (\phi a)^{-1}(H)$ of orbits of Z. Then

$$T^*\Big([m \otimes n]_{(Y,\phi)}\Big) = [M^*(a)(m) \otimes N^*(a)(n)]_{(Z,b)}$$

whence

$$\theta' T^*\Big([m \otimes n]_{(Y,\phi)}\Big) = \sum_{s' \in S'} [M^*(a)(m)_{s'} \otimes N^*(a)(n)_{s'}]_{G_{s'}}$$

Here $M^*(a)(m)_{s'} = M^*(i_{s'})M^*(a)(m) = M^*(a i_{s'})(m)$, denoting by i'_s the injection from $G/G_{s'}$ into Z associated to s'. If $s' = (K, xs)$, then $G_{s'} = K \cap {}^x G_s$ and

$$a i_{s'}(gG_{s'}) = a(gs') = a\Big((gK, gxs)\Big) = gxs = i_s(gxG_s) = i_s x p_{G_{K^x \cap G_s}}^{G_s}(gG_{s'})$$

so $a i_{s'} = i_s x p_{G_{K^x \cap G_s}}^{G_s}$, and this gives

$$\theta' T^*\Big([m \otimes n]_{(Y,\phi)}\Big) = \sum_{(K,xs) \in S'} [x r_{K^x \cap G_s}^{G_s} M^*(i_s)(m) \otimes x r_{K^x \cap G_s}^{G_s} N^*(i_s)(n)]_{K \cap {}^x G_s}$$

and finally, I have

$$\theta' T^*\Big([m \otimes n]_{(Y,\phi)}\Big) = r_K^H \theta'\Big([m \otimes n]_{(Y,\phi)}\Big)$$

I still have to check the compatibility of the isomorphisms θ and θ' with G-conjugations. Let $H \subseteq G$ and $x \in G$. If (Y, ϕ) is a G-set over G/H, let $S \subseteq \phi^{-1}(H)$ be a system of representatives of $G\backslash Y$. Let m_x be the conjugation $gH \mapsto (gx^{-1})^x H$ from G/H to $G/{}^x H$. Then if $s \in S$, I have

$$m_x \phi(xs) = m_x(xH) = {}^x H$$

so $xS = \{xs \mid s \in S\}$ is such that $m_x \phi(xS) \subseteq {}^x H$, and then

$$\theta' T_*(m_x)\Big([m \otimes n]_{(Y,\phi)}\Big) = \sum_{s' \in xS} [m_{s'} \otimes n_{s'}]_{G'_s}$$

But if $s' = xs$, then $G_{s'} = {}^x G_s$, and $m_{s'} = M^*(i_{s'})(m)$, where $i_{s'}$ is the injection $gG_{s'} \mapsto gs'$ from $G/G_{s'}$ to Y. But

$$i_{s'} m_x(gG_s) = i_{s'}(gx^{-1}G_{s'}) = gx^{-1}s' = gs = i_s(gG_s)$$

So $i'_s m_x = i_s$. But then $i'_s = i_s(m_x)^{-1}$ and

$$m_{s'} = M^*\Big((m_x)^{-1}\Big)M^*(i_s)(m) = M_*(m_x)(m_s)$$

It follows that

$$\theta' T_*(m_x)\Big([m \otimes n]_{(Y,\phi)}\Big) = \sum_{s \in S}[M_*(m_x)(m_s) \otimes N_*(m_x)(n_s)]_{{}^x G_s}$$

On the other hand

$$\theta'\Big([m\otimes n]_{(Y,\phi)}\Big) = \sum_{s\in S}[m_s\otimes n_s]_{G_s}$$

so

$$(M\hat{\otimes}N)_*(m_x)\,\Big(\theta'\Big([m\otimes n]_{(Y,\phi)}\Big)\Big) = \sum_{s\in S}[M_*(m_x)(m_s)\otimes N_*(m_x)(n_s)]_{{}^xG_s}$$

and this shows that

$$\theta' T_*(m_x)\Big([m\otimes n]_{(Y,\phi)}\Big) = (M\hat{\otimes}N)_*(m_x)\,\Big(\theta'\Big([m\otimes n]_{(Y,\phi)}\Big)\Big)$$

I finally proved the following:

Proposition 1.6.2: Let M and N be Mackey functors for the group G. If X is a G-set, then

$$(M\hat{\otimes}N)(X) \simeq \left(\bigoplus_{Y\xrightarrow{\phi}X} M(Y)\otimes_R N(Y)\right)/\mathcal{J}$$

where $\mathcal{J}$ is the R-submodule generated by

$M_*(f)(m)\otimes_R n' - m\otimes_R N^*(f)(n')$ **for** $f:(Y,\phi)\to(Y',\phi')$, $m\in M(Y)$, $n'\in N(Y')$

$M^*(f)(m')\otimes_R n' - m\otimes_R N_*(f)(n)$ **for** $f:(Y,\phi)\to(Y',\phi')$, $m'\in M(Y')$, $n\in N(Y)$

If $[m\otimes n]_{(Y,\phi)}$ denotes the image in $M\hat{\otimes}N(X)$ of $m\otimes n\in M(Y)\otimes_R N(Y)$, for $Y\xrightarrow{\phi}X$, then

- **If $g: X\to X'$, then**

$$(M\hat{\otimes}N)_*(g)\Big([m\otimes n]_{(Y,\phi)}\Big) = [m\otimes n]_{(Y,g\phi)}$$

- **If $g: X'\to X$, let Y', ϕ', and a such that the square**

$$\begin{array}{ccc} Y' & \xrightarrow{\;a\;} & Y \\ {\scriptstyle\phi'}\downarrow & & \downarrow{\scriptstyle\phi} \\ X' & \xrightarrow[\;g\;]{} & X \end{array}$$

is cartesian. Then

$$(M\hat{\otimes}N)^*(g)\Big([m\otimes n]_{(Y,\phi)}\Big) = [M^*(a)(m)\otimes N^*(a)(n)]_{(Y',\phi')}$$

Remark: This proposition shows that the tensor product of Mackey functors is the same as the box product defined by Lewis (see [6]).

1.7 Functoriality

The previous constructions have obvious functorial properties: the correspondence

$$(M, N) \mapsto \mathcal{H}(M, N)$$

is a functor in two variables, contravariant in M and covariant in N, and the correspondence

$$(M, N) \mapsto M\hat{\otimes}N$$

is a covariant functor in M and N: if f (resp. g) is a morphism of Mackey functors from M' to M (resp. from N to N'), and if Y is a G-set, I can define a morphism $\mathcal{H}(f, g)_Y$ from $\mathcal{H}(M, N)(Y)$ to $\mathcal{H}(M', N')(Y)$ by

$$\mathcal{H}(f, g)_Y(h) = g_{(Y)} \circ h \circ f \quad \text{for} \quad h \in \mathcal{H}(M, N)(Y) = \mathrm{Hom}_{Mack(G)}(M, N_Y)$$

where $g_{(Y)}$ is the morphism from N_Y to N'_Y defined by

$$(g_{(Y)})_X = g_{XY} : N_Y(X) = N(XY) \to N'_Y(X) = N'(XY)$$

It is easy to check that if $u : Y \to Y'$ is a map of G-sets, then the square

$$\begin{array}{ccc}
\mathcal{H}(M,N)(Y) & \xrightarrow{\mathcal{H}(M,N)_*(u)} & \mathcal{H}(M,N)(Y') \\
{\scriptstyle \mathcal{H}(f,g)_Y}\downarrow & & \downarrow{\scriptstyle \mathcal{H}(f,g)_{Y'}} \\
\mathcal{H}(M',N')(Y) & \xrightarrow[\mathcal{H}(M',N')_*(u)]{} & \mathcal{H}(M',N')(Y')
\end{array}$$

is commutative, as well as the square

$$\begin{array}{ccc}
\mathcal{H}(M,N)(Y') & \xrightarrow{\mathcal{H}(M,N)^*(u)} & \mathcal{H}(M,N)(Y) \\
{\scriptstyle \mathcal{H}(f,g)_{Y'}}\downarrow & & \downarrow{\scriptstyle \mathcal{H}(f,g)_Y} \\
\mathcal{H}(M',N')(Y') & \xrightarrow[\mathcal{H}(M',N')^*(u)]{} & \mathcal{H}(M',N')(Y)
\end{array}$$

So I have a morphism of Mackey functors $\mathcal{H}(f, g)$ from $\mathcal{H}(M, N)$ to $\mathcal{H}(M', N')$.

Now if f (resp. g) is a morphism from M to M' (resp. from N to N'), then I can define a morphism $(f\hat{\otimes}g)_Y$ from $M\hat{\otimes}N(Y)$ to $M'\hat{\otimes}N'(Y)$: if (Z, ϕ) is a G-set over Y, let

$$(f\hat{\otimes}g)_Y\big([m \otimes n]_{(Z,\phi)}\big) = [f_Z(m) \otimes g_Z(n)]_{(Y,\phi)}$$

If f and g are morphisms of Mackey functors, then these maps are well defined, and if $u : Y \to Y'$ is a morphism of G-sets, then the squares

$$\begin{array}{ccc}
M\hat{\otimes}N(Y) & \xrightarrow{(M\hat{\otimes}N)_*(u)} & M\hat{\otimes}N(Y') \\
{\scriptstyle (f\hat{\otimes}g)_Y}\downarrow & & \downarrow{\scriptstyle (f\hat{\otimes}g)_{Y'}} \\
M'\hat{\otimes}N'(Y) & \xrightarrow[(M'\hat{\otimes}N')_*(u)]{} & M'\hat{\otimes}N'(Y')
\end{array}$$

and

$$\begin{array}{ccc} M\hat{\otimes}N(Y') & \xrightarrow{(M\hat{\otimes}N)^*(u)} & M\hat{\otimes}N(Y) \\ {\scriptstyle (f\hat{\otimes}g)_{Y'}}\downarrow & & \downarrow{\scriptstyle (f\hat{\otimes}g)_Y} \\ M'\hat{\otimes}N'(Y') & \xrightarrow[(M'\hat{\otimes}N')^*(u)]{} & M'\hat{\otimes}N'(Y) \end{array}$$

are commutative, turning $f\hat{\otimes}g$ into a morphism of Mackey functors from $M\hat{\otimes}N$ to $M'\hat{\otimes}N'$.

1.8 n-fold tensor product

1.8.1 Definition

The second identification of the tensor products can be extended to the case of the n-fold tensor product, defined as follows: let n be an integer, greater or equal to 2, and let M_i for $i \in \{1,\ldots,n\}$ be Mackey functors for the group G. If X is a G-set, let

$$(M_1\hat{\otimes}\ldots\hat{\otimes}M_n)(X) = \bigoplus_{Y \xrightarrow{\phi} X} \big(M_1(Y)\otimes\ldots\otimes M_n(Y)\big)/ < \mathcal{J}_i \mid 1 \leq i \leq n >$$

where $\mathcal{J}_i$ is the R-submodule generated by the elements of the form

$$m'_1\otimes\ldots m'_{i-1}\otimes M_{i,*}(f)(m_i)\otimes m'_{i+1}\otimes\ldots\otimes m'_n - \ldots$$
$$\ldots M_1^*(f)(m'_1)\otimes\ldots\otimes M_{i-1}^*(f)(m'_{i-1})\otimes m_i\otimes M_{i+1}^*(f)(m'_{i+1})\otimes\ldots\otimes M_n^*(f)(m'_n)$$

for $f : (Y\phi) \to (Y',\phi')$, $m'_j \in M_j(Y')$ if $j \neq i$ and $m_i \in M(Y)$. If (Y,ϕ) is a G-set over X, and if $m_i \in M_i(Y)$, for $i \in \{1,\ldots,n\}$, I denote by $[m_1\otimes\ldots\otimes m_n]_{(Y,\phi)}$ the image of $m_1\otimes\ldots\otimes m_n$ in $(M_1\hat{\otimes}\ldots\hat{\otimes}M_n)(X)$. Then:

Proposition 1.8.1: Let M_i, $i \in \{1,\ldots,n\}$ be Mackey functors for G. Let $L_1 = M_1$, and for $i \in \{2,\ldots,n\}$, let $L_i = L_{i-1}\hat{\otimes}M_i$. Then for any G-set X, there is an isomorphism

$$L_n(X) \simeq (M_1\hat{\otimes}\ldots\hat{\otimes}M_n)(X)$$

which turns $M_1\hat{\otimes}\ldots\hat{\otimes}M_n$ into a Mackey functor, in the following way:

- **If $f : X \to X'$ is a morphism of G-sets, if (Y,ϕ) is a G-set over X, and if $m_i \in M_i(Y)$, for $i \in \{1,\ldots,n\}$, then**

$$(M_1\hat{\otimes}\ldots\hat{\otimes}M_n)_*(f)\big([m_1\otimes\ldots\otimes m_n]_{(Y,\phi)}\big) = [m_1\otimes\ldots\otimes m_n]_{(Y,f\phi)}$$

- **If $g : X' \to X$, let Y', a and ϕ' such that the square**

$$\begin{array}{ccc} Y' & \xrightarrow{a} & Y \\ {\scriptstyle \phi'}\downarrow & & \downarrow{\scriptstyle \phi} \\ X' & \xrightarrow[g]{} & X \end{array}$$

is cartesian. Then

$$(M_1\hat{\otimes}\ldots\hat{\otimes}M_n)^*(g)\big([m_1\otimes\ldots\otimes m_n]_{(Y,\phi)}\big) = [M_1^*(a)(m_1)\otimes\ldots\otimes M_n^*(a)(m_n)]_{(Y',\phi')}$$

Proof: By induction on n, the case $n = 2$ being a consequence of the identification of $M\hat{\otimes}N$. By definition of $L_n = L_{n-1}\hat{\otimes}M_n$, I have

$$L_n(X) = \Big(\bigoplus_{Y \xrightarrow{\phi} X} L_{n-1}(Y) \otimes M_n(Y)\Big)/\mathcal{J}$$

where $\mathcal{J}$ is the submodule generated by

$$L_{n-1,*}(f)(l) \otimes m' - l \otimes M_n^*(f)(m')$$

$$L_{n-1}^*(f)(l') \otimes m - l' \otimes M_{n,*}(f)(m)$$

for $f : (Y, \phi) \to (Y', \phi')$, $m \in M(Y)$, $m' \in M(Y')$, $l \in L_{n-1}(Y)$, $l' \in L_{n-1}(Y')$. By induction hypothesis, I know that $L_{n-1}(Y)$ is generated by the elements

$$[m_1 \otimes \ldots \otimes m_{n-1}]_{(Z,\psi)}$$

where (Z, ψ) is a G-set over Y. I define a map $\theta : L_n(X) \to (M_1\hat{\otimes}\ldots\hat{\otimes}M_n)(X)$ in the following way: if (Y, ϕ) is a G-set over X, if $m \in M(Y)$, if (Z, ψ) is a G-set over Y, and if $l = [m_1 \otimes \ldots \otimes m_{n-1}]_{(Z,\psi)}$, let

$$\theta\Big([l \otimes m]_{(Y,\phi)}\Big) = [m_1 \otimes \ldots \otimes m_{n-1} \otimes M_n^*(\psi)(m)]_{(Z,\phi\psi)}$$

I must check that this map is well defined: let $f : (Y, \phi) \to (Y', \phi')$ be a morphism of G-sets over X. Then, by induction hypothesis,

$$L_{n-1,*}(f)(l) = [m_1 \otimes \ldots \otimes m_{n-1}]_{(Z,f\psi)}$$

So for $m' \in M_n(Y')$

$$\theta\Big(L_{n-1,*}(f)(l) \otimes m'\Big) = [m_1 \otimes \ldots \otimes m_{n-1} \otimes M_n^*(f\psi)(m')]_{(Z,\phi' f\psi)}$$

whereas

$$\theta\Big(l \otimes M_n^*(f)(m')\Big) = [m_1 \otimes \ldots \otimes m_{n-1} \otimes M_n^*(\psi)M^*(f)(m')]_{(Z,\phi\psi)}$$

It follows that

$$\theta\Big((L_{n-1})_*(f)(l) \otimes m'\Big) = \theta\Big(l \otimes M^*(f)(m')\Big)$$

since $\phi' f = \phi$ and $M_n^*(f\psi) = M_n^*(\psi)M_n^*(f)$.
Similarly, if $g : (Y', \phi') \to (Y, \phi)$, let Z', a and ψ' such that the square

$$\begin{array}{ccc} Z' & \xrightarrow{a} & Z \\ {\scriptstyle \psi'}\downarrow & & \downarrow{\scriptstyle \psi} \\ Y' & \xrightarrow[g]{} & Y \end{array}$$

is cartesian. Then by induction hypothesis, I have

$$L_{n-1}^*(g)(l) = [M_1^*(a)(m_1) \otimes \ldots \otimes M_{n-1}^*(a)(m_{n-1})]_{(Z',\psi')}$$

and so

$$\theta\Big(L_{n-1}^*(g)(l)\otimes m'\Big) = [M_1^*(a)(m_1)\otimes\ldots\otimes M_{n-1}^*(a)(m_{n-1})\otimes M_n^*(\psi')(m')]_{(Z',\phi'\psi')}$$

But in $(M_1\hat{\otimes}\ldots\hat{\otimes}M_n)(X)$, the right hand side is equal to

$$[m_1\otimes\ldots\otimes m_{n-1}\otimes M_{n,*}(a)M_n^*(\psi')(m')]_{(Z,\phi\psi)}$$

On the other hand

$$\theta\Big(l\otimes M_{n,*}(g)(m')\Big) = [m_1\otimes\ldots\otimes m_{n-1}\otimes M_n^*(\psi)(M_n)_*(g)(m')]_{(Z,\phi\psi)}$$

and I have

$$\theta\Big(L_{n-1}^*(g)(l)\otimes m'\Big) = \theta\Big(l\otimes M_{n,*}(g)(m')\Big)$$

since $M_{n,*}(a)M_n^*(\psi') = M_n^*(\psi)M_{n,*}(g)$ in the above cartesian square. Thus θ is well defined.

I can now define a map θ' in the other direction: if

$$[m_1\otimes\ldots\otimes m_n]_{(Y,\phi)}\in M_1\hat{\otimes}\ldots\hat{\otimes}M_n(X)$$

let $l = m_1\otimes\ldots\otimes m_{n-1}$, so I can consider the element $[l]_{(Y,\phi)}\in L_{n-1}(Y)$, and let

$$\theta'\Big([m_1\otimes\ldots\otimes m_n]_{(Y,\phi)}\Big) = \Big[[l]_{(Y,Id)}\otimes m\Big]_{(Y,\phi)}\in L_n(X)$$

This map is well defined: if i is an integer lower or equal to $n-1$, and if $u\in\mathcal{J}_i$, then u is a linear combination of elements of the form $v-w$, with

$$v = m_1'\otimes\ldots m_{i-1}'\otimes M_{i,*}(f)(m_i)\otimes m_{i+1}'\otimes\ldots\otimes m_n'$$

$$w = M_1^*(f)(m_1')\otimes\ldots\otimes M_{i-1}^*(f)(m_{i-1}')\otimes m_i\otimes M_{i+1}^*(f)(m_{i+1}')\otimes\ldots\otimes M_n^*(f)(m_n')$$

for $f:(Y,\phi)\to(Y',\phi')$, and elements $m_j'\in M_j(Y')$ if $j\neq i$, and $m_i\in M_i(Y)$. But then if $l = m_1'\otimes\ldots m_{i-1}'\otimes M_{i,*}(f)(m_i)\otimes m_{i+1}'\otimes\ldots\otimes m_{n-1}'$, and if

$$l'' = M_1^*(f)(m_1')\otimes\ldots\otimes M_{i-1}^*(f)(m_{i-1}')\otimes m_i\otimes M_{i+1}^*(f)(m_{i+1}')\otimes\ldots\otimes M_n^*(f)(m_{n-1}')$$

I have

$$\theta'\Big([w]_{(Y,\phi)}\Big) = \Big[[l'']_{(Y,Id)}\otimes M_n^*(f)(m_n')\Big]_{(Y,\phi)}$$

and in $L_n(X)$, this element is equal to $\Big[L_{n-1,*}(f)\Big([l'']_{(Y,Id)}\Big)\otimes m_n\Big]_{(Y',\phi')}$. But

$$L_{n-1,*}(f)\Big([l'']_{(Y,Id)}\Big) = [l'']_{(Y,f)}$$

and in $L_{n-1}(Y')$, this element is equal to $[l']_{(Y',Id)}$. Finally

$$\theta'\Big([w]_{(Y,\phi)}\Big) = \Big[[l']_{(Y',Id)}\otimes m_n'\Big]_{(Y',\phi')} = \theta'\Big([v]_{(Y',\phi')}\Big)$$

Now if $i = n$, and if

$$v = m_1'\otimes\ldots\otimes m_{n-1}'\otimes M_{n,*}(f)(m_n)$$

$$w = M_1^*(f)(m_1') \otimes \ldots \otimes M_{n-1}^*(f)(m_{n-1}') \otimes m_n$$

let

$$l' = m_1' \otimes \ldots \otimes m_{n-1}'$$

$$l'' = M_1^*(f)(m_1') \otimes \ldots \otimes M_{n-1}^*(f)(m_{n-1}')$$

I have

$$\theta'\Big([w]_{(Y,Id)}\Big) = \Big[[l'']_{(Y,Id)} \otimes m_n\Big]_{(Y,\phi)}$$

But as the square

$$\begin{array}{ccc} Y & \xrightarrow{f} & Y' \\ {\scriptstyle Id}\downarrow & & \downarrow{\scriptstyle Id} \\ Y & \xrightarrow[f]{} & Y' \end{array}$$

is cartesian, I have

$$L_{n-1}^*(f)\Big([l']_{(Y',Id)}\Big) = [l'']_{(Y,Id)}$$

and so

$$\theta'\Big([w]_{(Y,Id)}\Big) = \Big[L_{n-1}^*(f)\Big([l']_{(Y',Id)}\Big) \otimes m_n\Big]_{(Y,\phi)}$$

In $L_n(X)$, this element is equal to

$$\Big[[l']_{(Y',Id)} \otimes M_{n,*}(f)(m_n)\Big]_{(Y',\phi')} = \theta'\Big([v]_{(Y',\phi')}\Big)$$

So the map θ' is well defined.
Now let (Y,ϕ) be a G-set over X, and (Z,ϕ) be a G-set over Y. If $m_i \in M_i(Z)$ for $i \leq n-1$ and $m_n \in M_n(Y)$, let $l = m_1 \otimes \ldots \otimes m_{n-1}$. I have

$$\theta'\theta\Big(\Big[[l]_{(Z,\psi)} \otimes m_n\Big]_{(Y,\phi)}\Big) = \theta'\Big([l \otimes M_n^*(\psi)(m_n)]_{(Z,\phi\psi)}\Big) = \Big[[l]_{(Z,Id)} \otimes M_n^*(\psi)(m_n)\Big]_{(Z,\phi\psi)}$$

and in $L_n(X)$, this is equal to

$$\Big[L_{n-1,*}(\psi)\Big([l]_{(Z,Id)}\Big) \otimes m_n\Big]_{(Y,\phi)}$$

Finally, as $L_{n-1,*}(\psi)\Big([l]_{(Z,Id)}\Big) = [l]_{(Z,\psi)}$, I have $\theta'\theta = Id$.
Conversely, if $m_i \in M_i(Y)$ for $i \leq n$, and if $l = m_1 \otimes \ldots \otimes m_{n-1}$, then

$$\theta\theta'\Big([l \otimes m_n]_{(Y,\phi)}\Big) = \theta\Big(\Big[[l]_{(Y,Id)} \otimes m_n\Big]_{(Y,\phi)}\Big) = [l \otimes m_n]_{(Y,\phi)}$$

and this proves that θ and θ' are mutual inverse isomorphisms, which completes the proof of the proposition. ■

1.8.2 Universal property

The analogy between the tensor product of Mackey functors and the usual tensor product will become clear with the notion of n-linear morphism, defined as follows:

Definition: *Let $M_1, \ldots, M_n$ and P be Mackey functors for the group G. An n-linear morphism L from $M_1, \ldots, M_n$ to P is by definition a correspondence which to any n-tuple of G sets $X_1, \ldots, X_n$ associates an n-linear map*

$$L_{X_1,\ldots,X_n} : M_1(X_1) \times \ldots \times M_n(X_n) \to P(X_1 \times \ldots \times X_n)$$

in a functorial way: if $f_i : X_i \to X_i'$ for $1 \leq i \leq n$ are morphisms of G-sets, then the squares

$$\begin{array}{ccc} M_1(X_1) \times \ldots \times M_n(X_n) & \xrightarrow{L_{X_1,\ldots,X_n}} & P(X_1 \times \ldots \times X_n) \\ {\scriptstyle M_{1,*}(f_1) \times \ldots \times M_{n,*}(f_n)} \downarrow & & \downarrow {\scriptstyle P_*(f_1 \times \ldots \times f_n)} \\ M_1(X_1') \times \ldots \times M_n(X_n') & \xrightarrow[L_{X_1',\ldots,X_n'}]{} & P(X_1' \times \ldots \times X_n') \end{array}$$

and

$$\begin{array}{ccc} M_1(X_1) \times \ldots \times M_n(X_n) & \xrightarrow{L_{X_1,\ldots,X_n}} & P(X_1 \times \ldots \times X_n) \\ {\scriptstyle M_1^*(f_1) \times \ldots \times M_n^*(f_n)} \uparrow & & \uparrow {\scriptstyle P^*(f_1 \times \ldots \times f_n)} \\ M_1(X_1') \times \ldots \times M_n(X_n') & \xrightarrow[L_{X_1',\ldots,X_n'}]{} & P(X_1' \times \ldots \times X_n') \end{array}$$

are commutative.

I denote by $\mathcal{L}(M_1, \ldots, M_n; P)$ the set of n-linear morphisms from $M_1, \ldots, M_n$ to P, with its natural structure of R-module. This construction is clearly functorial, contravariant in $M_1, \ldots M_n$, and covariant in P.
If $X_1, \ldots, X_n$ are G-sets, I define a map

$$\pi_{X_1,\ldots,X_n} : M_1(X_1) \times \ldots \times M_n(X_n) \to (M_1 \hat{\otimes} \ldots \hat{\otimes} M_n)(X_1 \times \ldots \times X_n)$$

by the formulae

$$\pi_{X_1,\ldots,X_n}(m_1, \ldots, m_n) = \left[M_1^*(p_1)(m_1) \otimes \ldots \otimes M^*(p_n)(m_n)\right]_{(X_1 \times \ldots \times X_n, Id)}$$

where for all i, the map p_i is the i-th projection from $X_1 \times \ldots \times X_n$ onto X_i. Then

Proposition 1.8.2: The previous equalities define an element

$$\pi \in \mathcal{L}(M_1, \ldots, M_n; M_1 \hat{\otimes} \ldots \hat{\otimes} M_n)$$

which has moreover the following universal property: if P is a Mackey functor for the group G, and if $f \in \mathcal{L}(M_1, \ldots, M_n; P)$, then there exists a unique morphism of Mackey functors $\tilde{f} : M_1 \hat{\otimes} \ldots \hat{\otimes} M_n \to P$ such that for any G-sets $X_1, \ldots, X_n$

$$f_{X_1,\ldots,X_n} = \tilde{f}_{X_1 \times \ldots \times X_n} \pi_{X_1,\ldots,X_n}$$

Conversely, if $\tilde{f}$ is a morphism of Mackey functors from $M_1\hat{\otimes}\ldots\hat{\otimes}M_n$ to P, then this formula defines an n-linear morphism from $M_1,\ldots,M_n$ to P, and this correspondence induces an isomorphism of R-modules

$$\mathrm{Hom}_{Mack_R(G)}(M_1\hat{\otimes}\ldots\hat{\otimes}M_n, P)\simeq\mathcal{L}(M_1,\ldots,M_n;P)$$

which is natural with respect to $M_1,\ldots,M_n,P$.

Proof: 1) First I have to check that π is an n-linear morphism from $M_1,\ldots,M_n$ to $M_1\hat{\otimes}\ldots\hat{\otimes}M_n$. I suppose given for an integer i a morphism $f_i: X_i\rightarrow X_i'$. Let then $f=(Id_{X_1\times\ldots\times X_{i-1}})\times f_i\times(Id_{X_{i+1}\times\ldots\times X_n})$, and $\Lambda=(M_1\hat{\otimes}\ldots\hat{\otimes}M_n)_*(f)$. I have

$$\Lambda\pi_{X_1,\ldots,X_n}(m_1,\ldots,m_n)=\Lambda\Big(\big[M_1^*(p_1)(m_1)\otimes\ldots\otimes M_n^*(p_n)(m_n)\big]_{(X_1\times\ldots\times X_n,Id)}\Big)=\ldots$$

$$\ldots=\big[M_1^*(p_1)(m_1)\otimes\ldots\otimes M_n^*(p_n)(m_n)\big]_{(X_1\times\ldots\times X_n,f)}$$

On the other hand, if p_i' is the projection from $X_1\times\ldots\times X_i'\times\ldots\times X_n'$ onto its i-th component, I have

$$\pi_{X_1,\ldots,X_i',\ldots,X_n}\Big(m_1,\ldots\otimes M_{i,*}(f_i)(m_i),\ldots,m_n\Big)=\ldots$$

$$\ldots=\big[M_1^*(p_1')(m_1)\otimes\ldots\otimes M_i^*(p_i')M_{i,*}(f_i)(m_i)\otimes\ldots\otimes M^*(p_n')(m_n)\big]_{(X_1\times\ldots\times X_i'\times\ldots\times X_n,Id)}$$

But the square

$$\begin{array}{ccc}
X_1\times\ldots\times X_i\times\ldots\times X_n & \stackrel{f}{\longrightarrow} & X_1\times\ldots\times X_i'\times\ldots\times X_n\\
{\scriptstyle p_i}\downarrow & & \downarrow{\scriptstyle p_i'}\\
X_i & \underset{f_i}{\longrightarrow} & X_i'
\end{array}$$

is cartesian, so $M_i^*(p_i')M_{i,*}(f_i)=M_{i,*}(f)M_i^*(p_i)$, and then

$$\pi_{X_1,\ldots,X_i',\ldots,X_n}\Big(m_1,\ldots,M_{i,*}(f_i)(m_i),\ldots,m_n\Big)=\ldots$$

$$\ldots=\big[M_1^*(f)M_1^*(p_1')(m_1)\otimes\ldots\otimes M_i^*(p_i)(m_i)\otimes\ldots\otimes M_n^*(f)M^*(p_n')(m_n)\big]_{(X_1\times\ldots\times X_n,f)}$$

in $(M_1\hat{\otimes}\ldots\hat{\otimes}M_n)(X_1\times\ldots\times X_i'\times\ldots\times X_n)$. Finally as for all $j\neq i$, I have $p_j'f=p_j$, it follows that

$$\pi_{X_1,\ldots,X_i',\ldots,X_n}\Big(m_1,\ldots,M_{i,*}(f_i)(m_i),\ldots,m_n\Big)=\ldots$$

$$\ldots=\big[M_1^*(p_1)(m_1)\otimes\ldots\otimes M_i^*(p_i)(m_i)\otimes\ldots\otimes M^*(p_n)(m_n)\big]_{(X_1\times\ldots\times X_n,f)}$$

which proves that π is covariant with respect to the i-th factor.

Conversely, if $m_i'\in M_i(X_i')$, I have

$$\pi_{X_1,\ldots,X_n}\Big(m_1,\ldots,M_i^*(f_i)(m_i'),\ldots,m_n\Big)=\ldots$$

$$\ldots = \Big[M^*(p_1)(m_1)\otimes\ldots\otimes M_i^*(p_i)M_i^*(f_i)(m_i')\otimes\ldots\otimes M_n^*(p_n)(m_n)\Big]_{(X_1\times\ldots\times X_n,Id)}$$

On the other hand, let $\Lambda' = (M_1\hat{\otimes}\ldots\hat{\otimes}M_n)^*(f)$. I have

$$\Lambda'\pi_{X_1,\ldots,X_i',\ldots,X_n}(m_1,\ldots,m_i',\ldots,m_n) = \ldots$$

$$\ldots = \Lambda'\Big(\Big[M_1^*(p_1)(m_1)\otimes\ldots\otimes M_i^*(p_i)(m_i')\otimes\ldots\otimes M_n^*(p_n)(m_n)\Big]_{(X_1\times\ldots\times X_i'\times\ldots\times X_n,Id)}\Big)$$

As the square

$$\begin{array}{ccc} X_1\times\ldots\times X_n & \xrightarrow{f} & X_1\times\ldots\times X_i'\times\ldots\times X_n \\ {\scriptstyle Id}\downarrow & & \downarrow{\scriptstyle Id} \\ X_1\times\ldots\times X_n & \xrightarrow[f]{} & X_1\times\ldots\times X_i'\times\ldots\times X_n \end{array}$$

is cartesian, this is also equal to

$$\Big[M_1^*(f)M_1^*(p_1)(m_1)\otimes\ldots\otimes M_i^*(f)M_i^*(p_i)(m_i')\otimes\ldots\otimes M_n^*(f)M_n^*(p_n)(m_n)\Big]_{(X_1\times\ldots\times X_n,Id)}$$

Finally as $p_jf = p_j$ for $j\neq i$, and as $p_if = f_ip_i$, it follows that π is contravariant with respect to the i-th factor, and hence it is an n-linear morphism.

2) I must now prove the universal property: let P be a Mackey functor, and $f\in\mathcal{L}(M_1,\ldots,M_n;P)$. If X is a G-set, I define a morphism $\tilde{f}_X$ from $(M_1\hat{\otimes}\ldots\hat{\otimes}M_n)(X)$ to $P(X)$ in the following way: if (Y,ϕ) is a G-set over X, and if $m_i\in M_i(Y)$ for all i, let

$$\tilde{f}_X\big([m_1\otimes\ldots\otimes m_n]_{(Y,\phi)}\big) = P_*(\phi)P^*(\delta_{n,Y})f_{Y,\ldots,Y}(m_1,\ldots,m_n)$$

where $\delta_{n,Y}$ denotes the diagonal injection $y\mapsto(y,\ldots,y)$ from Y into Y^n.
First I have to check that $\tilde{f}$ is well defined. I suppose that $g:(Y,\phi)\to(Y',\phi')$ is a morphism of G-sets over X, that $m_j'\in M_j(Y')$ for $j\neq i$, and that $m_i\in M_i(Y)$. Then

$$\tilde{f}_X\Big(\big[m_1'\otimes\ldots\otimes M_{i,*}(g)(m_i)\otimes\ldots\otimes m_n'\big]_{(Y',\phi')}\Big) = \ldots$$

$$\ldots = P_*(\phi')P^*(\delta_{n,Y'})f_{Y',\ldots,Y'}\big(m_1',\ldots,M_{i,*}(g)(m_i),\ldots,m_n'\big)$$

Since f is an n-linear morphism, setting $h = Id^{i-1}\times g\times Id^{n-i-1}$, I have

$$f_{Y',\ldots,Y'}\big(m_1',\ldots,M_{i,*}(g)(m_i),\ldots,m_n'\big) = \ldots$$

$$\ldots = P_*(h)f_{Y',\ldots,Y',Y,Y',\ldots,Y'}(m_1',\ldots,m_{i-1}',m_i,m_{i+1}',\ldots,m_n')$$

Let k be the map from Y to $Y'^{i-1}YY'^{n-i}$ defined by

$$k(y) = \big(g(y),\ldots,g(y),y,g(y),\ldots,g(y)\big)$$

The square

$$\begin{array}{ccc} Y'^{i-1}YY'^{n-i} & \xrightarrow{h} & Y'^n \\ {\scriptstyle k}\uparrow & & \uparrow{\scriptstyle \delta_{n,Y'}} \\ Y & \xrightarrow[g]{} & Y' \end{array}$$

is cartesian, hence $P^*(\delta_{n,Y'})P_*(h) = P_*(g)P^*(k)$. Setting $l = g^{i-1} \times Id \times g^{n-i-1}$, I have $k = l\delta_{n,Y}$, so $P^*(k) = P^*(\delta_{n,Y})P^*(l)$. Using again the n-linearity of f, I have

$$P^*(l)f_{Y',\ldots,Y',Y,Y',\ldots,Y'}(m'_1,\ldots,m'_{i-1},m_i,m'_{i+1},\ldots,m'_n) = \ldots$$

$$\ldots = f_{Y,\ldots,Y}\Big(M_1^*(g)(m'_1),\ldots,M_i^*(g)(m'_i),m_i,M_{i+1}^*(g)(m'_{i+1}),\ldots,M_n^*(g)(m'_n)\Big)$$

and denoting by F this expression, I have finally

$$\tilde{f}_X\Big(\big[m'_1\otimes\ldots\otimes M_{i,*}(g)(m_i)\otimes\ldots\otimes m'_n\big]_{(Y',\phi')}\Big) = P_*(\phi')P_*(g)P^*(\delta_{n,Y})F = \ldots$$

$$\ldots = \tilde{f}_X\Big(\big[M_1^*(g)(m'_1)\otimes\ldots\otimes M_{i-1}^*(g)(m'_{i-1})\otimes m_i\otimes M_{i+1}^*(g)(m'_{i+1})\otimes\ldots\otimes M_n^*(g)(m'_n)\big]_{(Y,\phi)}\Big)$$

which proves that $\tilde{f}$ is well defined.

3) I must now check that $\tilde{f}$ is a morphism of Mackey functors. Let g be a morphism of G-sets from X to X'. If (Y,ϕ) is a G-set over X, and if $m_i \in M_i(Y)$ for $1 \le i \le n$, then

$$(M_1\hat{\otimes}\ldots\hat{\otimes}M_n)_*(g)\Big([m_1\otimes\ldots\otimes m_n]_{(Y,\phi)}\Big) = [m_1\otimes\ldots\otimes m_n]_{(Y,g\phi)}$$

so

$$\tilde{f}_{X'}\Big((M_1\hat{\otimes}\ldots\hat{\otimes}M_n)_*(g)\Big([m_1\otimes\ldots\otimes m_n]_{(Y,\phi)}\Big)\Big) = P_*(g)P_*(\phi)P^*(\delta_{n,Y})f_{Y,\ldots,Y}(m_1,\ldots,m_n)$$

that is

$$\tilde{f}_{X'}\Big((M_1\hat{\otimes}\ldots\hat{\otimes}M_n)_*(g)\Big([m_1\otimes\ldots\otimes m_n]_{(Y,\phi)}\Big)\Big) = P_*(g)\tilde{f}_X\Big([m_1\otimes\ldots\otimes m_n]_{(Y,\phi)}\Big)$$

which proves that $\tilde{f}$ is covariant.

If now g is a morphism from X' to X, let Y', ϕ' and a such as the square

$$\begin{array}{ccc} Y' & \xrightarrow{a} & Y \\ {\scriptstyle \phi'}\downarrow & & \downarrow{\scriptstyle \phi} \\ X' & \xrightarrow[g]{} & X \end{array}$$

is cartesian. Then

$$(M_1\hat{\otimes}\ldots\hat{\otimes}M_n)^*(g)\Big([m_1\otimes\ldots\otimes m_n]_{(Y,\phi)}\Big) = \Big[M_1^*(a)(m_1)\otimes\ldots\otimes M_n^*(a)(m_n)\Big]_{(Y',\phi')}$$

The image under $\tilde{f}_{X'}$ of this element is then

$$P_*(\phi')P^*(\delta_{n,Y'})f_{Y',\ldots,Y'}\Big(M_1^*(a)(m_1),\ldots,M_n^*(a)(m_n)\Big)$$

and as f is n-linear, I have

$$f_{Y',\ldots,Y'}\Big(M_1^*(a)(m_1),\ldots,M_n^*(a)(m_n)\Big) = P^*(a^n)f_{Y,\ldots,Y}(m_1,\ldots,m_n)$$

Moreover $a^n\delta_{n,Y'} = \delta_{n,Y}a$, and $P_*(\phi')P^*(a) = P^*(g)P_*(\phi)$ in the previous cartesian square. Thus

$$\tilde{f}_{X'}\Big((M_1\hat{\otimes}\ldots\hat{\otimes}M_n)^*(g)\Big([m_1\otimes\ldots\otimes m_n]_{(Y,\phi)}\Big)\Big) = \ldots$$

$$\ldots = P^*(g)P_*(\phi)P^*(\delta_{n,Y})f_{Y,\ldots,Y}(m_1,\ldots,m_n) = P^*(g)\tilde{f}_X\Big([m_1\otimes\ldots\otimes m_n]_{(Y,\phi)}\Big)$$

showing that $\tilde{f}$ is contravariant, hence a morphism of Mackey functors.

4) Finally, let $X_1,\ldots,X_n$ be G sets, and $m_i \in M_i(X_i)$ for $1 \le i \le n$. Then, setting $X = X_1 \times \ldots \times X_n$, I have

$$\tilde{f}_X\pi_{X_1,\ldots,X_n}(m_1,\ldots,m_n) = \tilde{f}_X\Big(\big[M_1^*(p_1)(m_1)\otimes\ldots\otimes M_n^*(p_n)(m_n)\big]_{(X,Id)}\Big) = \ldots$$

$$\ldots = P^*(\delta_{n,X})f_{X,\ldots,X}\Big(M_1^*(p_1)(m_1),\ldots,M_n^*(p_n)(m_n)\Big)$$

The n-linearity of f gives then

$$f_{X,\ldots,X}\Big(M_1^*(p_1)(m_1),\ldots,M_n^*(p_n)(m_n)\Big) = P^*(p_1\times\ldots\times p_n)f_{X_1,\ldots,X_n}(m_1,\ldots,m_n)$$

and as $(p_1\times\ldots\times p_n)\delta_{n,X} = Id_X$, I have

$$\tilde{f}_{X_1\times\ldots\times X_n}\pi_{X_1,\ldots,X_n} = f_{X_1,\ldots,X_n}$$

5) Conversely, if $\tilde{f}$ is an n-linear morphism from $M_1\hat{\otimes}\ldots\hat{\otimes}M_n$ to P, and if I define $f_{X_1,\ldots,X_n}$ by the above formulae, I get an element f of $\mathcal{L}(M_1,\ldots,M_n;P)$: to see this, let $i\in\{1,\ldots,n\}$, let $g_i : X_i \to X_i'$ be a morphism of G-sets, and let $m_j \in M_j(X_j)$ for $1\le j\le n$. Setting then $X_j' = X_j$ and $g_j = Id$ for $j \ne i$, and denoting by X (resp. X') the product $X_1\times\ldots\times X_n$ (resp. $X_1'\times\ldots\times X_n'$), and p_j (resp. p_i') the projection from X (resp. X') onto X_i (resp. onto X_i'), I have

$$f_{X_1',\ldots,X_n'}\Big(M_{1,*}(g_1)(m_1),\ldots,M_{n,*}(g_n)(m_n)\Big) = \ldots$$

$$\ldots = \tilde{f}_{X'}\Big(\big[M_1^*(p_1')M_{1,*}(g_1)(m_1)\otimes\ldots\otimes M_n^*(p_n')M_{n,*}(g_n)(m_n)\big]_{(X',Id)}\Big)$$

Since $p_j' = p_j$ and $g_j = Id$ for $j\ne i$, this is also equal to

$$\tilde{f}_{X'}\Big(\big[M_1^*(p_1)(m_1)\otimes\ldots\otimes M_i^*(p_i')M_{i,*}(g_i)(m_i)\otimes\ldots\otimes M_n^*(p_n)(m_n)\big]_{(X',Id)}\Big)$$

Denoting by g the product $g_1\times\ldots\times g_n$, the square

$$\begin{array}{ccc} X & \xrightarrow{g} & X' \\ {\scriptstyle p_i}\downarrow & & \downarrow{\scriptstyle p_i'} \\ X_i & \xrightarrow[g_i]{} & X_i' \end{array}$$

is cartesian (because $g_j = Id$ for $j \ne i$), so

$$M_i^*(p_i')M_{i,*}(g_i) = M_{i,*}(g)M_i^*(p_i)$$

On the other hand, setting $m_j' = M_j^*(p_j)(m_j)$, for $1\le j\le n$, I have the following equality in $(M_1\hat{\otimes}\ldots\hat{\otimes}M_n)(X')$

$$\big[m_1'\otimes\ldots\otimes M_{i,*}(g)(m_i')\otimes\ldots\otimes m_n'\big]_{(X',Id)} = \ldots$$

$$\ldots = \left[M_1^*(g)(m_1')\otimes\ldots\otimes M_{i-1}^*(g)(m_{i-1}')\otimes m_i'\otimes M_{i+1}^*(g)(m_{i+1}')\otimes\ldots\otimes M_n^*(g)(m_n')\right]_{(X,g)}$$

Moreover

$$M_j^*(g)(m_j') = M_j^*(g)M_j^*(p_j)(m_j) = M_j^*(p_jg)(m_j) = M_j^*(p_j)(m_j)$$

if $j \neq i$, because then $p_jg = p_j$. Finally

$$f_{X_1',\ldots,X_n'}\Big(M_{1,*}(g_1)(m_1),\ldots,M_{n,*}(g_n)(m_n)\Big) = \tilde{f}_{X'}\Big(\left[M_1^*(p_1)(m_1)\otimes\ldots\otimes M_n^*(p_n)(m_n)\right]_{(X,g)}\Big)$$

But

$$\left[M_1^*(p_1)(m_1)\otimes\ldots\otimes M_n^*(p_n)(m_n)\right]_{(X,g)} = \ldots$$

$$\ldots = (M_1\hat{\otimes}\ldots\hat{\otimes}M_n)(g)\Big(\left[M_1^*(p_1)(m_1)\otimes\ldots\otimes M_n^*(p_n)(m_n)\right]_{(X,Id)}\Big)$$

and $\tilde{f}$ is a morphism of Mackey functors. Thus

$$f_{X_1',\ldots,X_n'}\Big(M_{1,*}(g_1)(m_1),\ldots,M_{n,*}(g_n)(m_n)\Big) = \ldots$$

$$\ldots = P_*(g)\tilde{f}_X\Big(\left[M_1^*(p_1)(m_1)\otimes\ldots\otimes M_n^*(p_n)(m_n)\right]_{(X,Id)}\Big) = P_*(g)f_{X_1,\ldots,X_n}(m_1,\ldots,m_n)$$

which proves that f is covariant with respect to the i-th factor.
Now if g_i is a morphism from X_i' to X_i, and keeping the other notations above, I have

$$P^*(g)f_{X_1,\ldots,X_n}(m_1,\ldots,m_n) = P^*(g)\tilde{f}_X\Big(\left[M_1^*(p_1)(m_1)\otimes\ldots\otimes M_n^*(p_n)(m_n)\right]_{(X,Id)}\Big)$$

and as $\tilde{f}$ is a morphism of Mackey functors, this is equal to

$$\tilde{f}_{X'}\bigg((M_1\hat{\otimes}\ldots\hat{\otimes}M_n)^*(g)\Big(\left[M_1^*(p_1)(m_1)\otimes\ldots\otimes M_n^*(p_n)(m_n)\right]_{(X,Id)}\Big)\bigg)$$

The square

$$\begin{array}{ccc} X' & \xrightarrow{g} & X \\ {\scriptstyle Id}\downarrow & & \downarrow{\scriptstyle Id} \\ X' & \xrightarrow[g]{} & X \end{array}$$

being cartesian, I have

$$(M_1\hat{\otimes}\ldots\hat{\otimes}M_n)^*(g)\Big(\left[M_1^*(p_1)(m_1)\otimes\ldots\otimes M_n^*(p_n)(m_n)\right]_{(X,Id)}\Big) = \ldots$$

$$\ldots\left[M_1^*(g)M_1^*(p_1)(m_1)\otimes\ldots\otimes M_n^*(g)M_n^*(p_n)(m_n)\right]_{(X',Id)}$$

But for $1 \leq j \leq n$, I have $p_jg = g_jp_j$. Hence

$$P^*(g)f_{X_1,\ldots,X_n}(m_1\otimes\ldots\otimes m_n) = \ldots$$

$$\ldots\tilde{f}_{X'}\Big(\left[M_1^*(p_1)M_1^*(g_1)(m_1)\otimes\ldots\otimes M_n^*(p_n)M_n^*(g_n)(m_n)\right]_{(X',Id)}\Big) = \ldots$$

$$\ldots = f_{X_1',\ldots,X_n'}\Big(M_1^*(g_1)(m_1),\ldots,M_n^*(g_n)(m_n)\Big)$$

which proves that f is contravariant with respect to the i-th factor, and that it is an n-linear morphism.

6) To complete the proof, it remains to see that the above correspondences between $\mathcal{L}(M_1, \ldots, M_n; P)$ and $\mathrm{Hom}_{Mack(G)}(M_1\hat{\otimes}\ldots\hat{\otimes}M_n, P)$, which are clearly natural with respect to $M_1, \ldots, M_n, P$, are bijective.
Let $f \in \mathcal{L}(M_1, \ldots, M_n; P)$. It corresponds to $\tilde{f} \in \mathrm{Hom}_{Mack(G)}(M_1\hat{\otimes}\ldots\hat{\otimes}M_n, P)$, which in turn correspond to $f' \in \mathcal{L}(M_1, \ldots, M_n; P)$, and I must prove that $f = f'$. Let $X_1, \ldots, X_n$ be G-sets, and $m_i \in M_i(X_i)$, for $1 \leq i \leq n$. Then, setting $X = X_1 \times \ldots \times X_n$

$$f'_{X_1,\ldots,X_n}(m_1, \ldots, m_n) = \tilde{f}_X \pi_{X_1,\ldots,X_n}(m_1, \ldots, m_n) = \ldots$$
$$\ldots = \tilde{f}_X\Big(\big[M_1^*(p_1)(m_1) \otimes \ldots \otimes M_n^*(p_n)(m_n)\big]_{(X,Id)}\Big) = \ldots$$
$$\ldots = P^*(\delta_{n,X}) f_{X,\ldots,X}\Big(M_1^*(p_1)(m_1), \ldots, M_n^*(p_n)(m_n)\Big)$$

Since f is n-linear by hypothesis, this is also

$$P^*(\delta_{n,X})P^*(p_1 \times \ldots \times p_n) f_{X_1,\ldots,X_n}(m_1, \ldots, m_n)$$

and as $(p_1 \times \ldots \times p_n)\delta_{n,X}$ is the identity of X, I have

$$f'_{X_1,\ldots,X_n}(m_1, \ldots, m_n) = f_{X_1,\ldots,X_n}(m_1, \ldots, m_n)$$

and $f = f'$.
Conversely, if $\tilde{f}$ is an element in $\mathrm{Hom}_{Mack(G)}(M_1\hat{\otimes}\ldots\hat{\otimes}M_n, P)$, it corresponds to $f \in \mathcal{L}(M_1, \ldots, M_n; P)$, which in turn corresponds to $\tilde{f}' \in \mathrm{Hom}_{Mack(G)}(M_1\hat{\otimes}\ldots\hat{\otimes}M_n, P)$, and I must show that $\tilde{f} = \tilde{f}'$. Let X be a G-set, let (Y, ϕ) be a G-set over X, and for any $i \in \{1, \ldots, n\}$, let $m_i \in M_i(Y)$. Then

$$\tilde{f}'_X\big([m_1 \otimes \ldots \otimes m_n]_{(Y,\phi)}\big) = P_*(\phi)P^*(\delta_{n,Y}) f_{Y,\ldots,Y}(m_1, \ldots, m_n) = \ldots$$
$$\ldots = P_*(\phi)P^*(\delta_{n,Y})\tilde{f}_{Y^n}\pi_{Y,\ldots,Y}(m_1, \ldots, m_n) = \ldots$$
$$\ldots = P_*(\phi)P^*(\delta_{n,Y})\tilde{f}_{Y^n}\Big(\big[M_1^*(p_1)(m_1) \otimes \ldots \otimes M_n^*(p_n)(m_n)\big]_{(Y^n,Id)}\Big)$$

Since $\tilde{f}$ is by hypothesis a morphism of Mackey functors, this is equal to

$$P_*(\phi)\tilde{f}_Y\Big((M_1\hat{\otimes}\ldots\hat{\otimes}M_n)^*(\delta_{n,Y})\Big(\big[M_1^*(p_1)(m_1) \otimes \ldots \otimes M_n^*(p_n)(m_n)\big]_{(Y^n,Id)}\Big)\Big)$$

But the square

$$\begin{array}{ccc} Y & \xrightarrow{\delta_{n,Y}} & Y^n \\ {\scriptstyle Id}\downarrow & & \downarrow{\scriptstyle Id} \\ Y & \xrightarrow[\delta_{n,Y}]{} & Y^n \end{array}$$

is cartesian. So

$$(M_1\hat{\otimes}\ldots\hat{\otimes}M_n)^*(\delta_{n,Y})\Big(\big[M_1^*(p_1)(m_1) \otimes \ldots \otimes M_n^*(p_n)(m_n)\big]_{(Y^n,Id)}\Big) = \ldots$$

$$\dots = \Big[M_1^*(\delta_{n,Y})M_1^*(p_1)(m_1)\otimes\dots\otimes M_n^*(\delta_{n,Y})M_n^*(p_n)(m_n)\Big]_{(Y,Id)}$$

and for all $i \in \{1,\dots,n\}$, I have $p_i\delta_{n,Y} = Id$. Finally

$$\tilde{f}'_X\Big([m_1\otimes\dots\otimes m_n]_{(Y,\phi)}\Big) = P_*(\phi)\tilde{f}_Y\Big([m_1\otimes\dots\otimes m_n]_{(Y,Id)}\Big)$$

Using again the fact that $\tilde{f}$ is a morphism of Mackey functors, I get

$$\tilde{f}'_X\Big([m_1\otimes\dots\otimes m_n]_{(Y,\phi)}\Big) = \tilde{f}_X\Big((M_1\hat{\otimes}\dots\hat{\otimes}M_n)_*(\phi)\Big([m_1\otimes\dots\otimes m_n]_{(Y,Id)}\Big)\Big) = \dots$$

$$\dots = \tilde{f}_X\Big([m_1\otimes\dots\otimes m_n]_{(Y,\phi)}\Big)$$

which proves that $\tilde{f}' = \tilde{f}$, and completes the proof of the proposition. ■

Let $f \in \mathcal{L}(M_1,\dots,M_n;P)$ be an n-linear morphism, associated by the previous proposition to a unique morphism of Mackey functors $\tilde{f}$ from $M_1\hat{\otimes}\dots\hat{\otimes}M_n$ to P. The formula

$$\tilde{f}_X\Big([m_1\otimes\dots\otimes m_n]_{(Y,\phi)}\Big) = P_*(\phi)P^*(\delta_{n,Y})f_{Y,\dots,Y}(m_1,\dots,m_n)$$

shows that $\tilde{f}$ is entirely determined by the knowledge, for any G-set Y, of the n-linear map $\hat{f}_Y : M_1(Y)\times\dots\times M_n(Y)\to P(Y)$ defined by

$$\hat{f}_Y(m_1,\dots,m_n) = P^*(\delta_{n,Y})f_{Y,\dots,Y}(m_1,\dots,m_n)$$

Indeed, knowing $\hat{f}$, I can recover $\tilde{f}$ by

$$\tilde{f}_X\Big([m_1\otimes\dots\otimes m_n]_{(Y,\phi)}\Big) = P_*(\phi)\hat{f}_Y(m_1,\dots,m_n)$$

This formula will give the conditions on the maps $\hat{f}_Y$ so that the map $\tilde{f}$ is well defined and is a morphism of Mackey functors.
In order $\tilde{f}$ to be well defined, if $g : (Y',\phi')\to(Y,\phi)$ is a morphism of G-sets over X, if $m_j \in M_j(Y)$ for $j\neq i$ and $m'_i \in M_i(Y')$, the elements

$$[m_1\otimes\dots\otimes m_{i-1}\otimes M_{i,*}(g)(m'_i)\otimes m_{i+1}\otimes\dots\otimes m_n]_{(Y,\phi)} = \dots$$

$$\dots = \Big[M_1^*(g)(m_1)\otimes\dots\otimes M_{i-1}^*(g)(m_{i-1})\otimes m'_i\otimes M_{i+1}^*(g)(m_{i+1})\otimes\dots\otimes M_n^*(g)(m_n)\Big]_{(Y',\phi')}$$

must have the same image under $\tilde{f}$, and since $P_*(\phi') = P_*(\phi)P_*(g)$, this gives

$$P_*(\phi)\hat{f}_Y\Big(m_1,\dots,m_{i-1},M_{i,*}(g)(m'_i),m_{i+1},\dots,m_n\Big) = \dots$$

$$= P_*(\phi)P_*(g)\hat{f}_{Y'}\Big(M_1^*(g)(m_1),\dots,M_{i-1}^*(g)(m_{i-1}),m'_i,M_{i+1}^*(g)(m_{i+1}),\dots,M_n^*(g)(m_n)\Big)$$

This equality will result from the case $X = Y$ and $\phi = Id$, which says that for any $g : Y'\to Y$

$$(A_i)\qquad \hat{f}_Y\Big(m_1,\dots,m_{i-1},M_{i,*}(g)(m'_i),m_{i+1},\dots,m_n\Big) = \dots$$

$$\dots\quad = P_*(g)\hat{f}_{Y'}\Big(M_1^*(g)(m_1),\dots,M_{i-1}^*(g)(m_{i-1}),m'_i,M_{i+1}^*(g)(m_{i+1}),\dots,M_n^*(g)(m_n)\Big)$$

If equality (A_i) is true for all $i \in \{1,\dots,n\}$, then $\tilde{f}$ is well defined.
In those conditions, it is clear moreover that $\tilde{f}$ is covariant: if $g : X \to X'$ is a morphism of G-sets, then for any G-set (Y,ϕ) over X and for any $m_j \in M_j(Y)$, I have

$$(M_1\hat{\otimes}\dots\hat{\otimes}M_n)_*(g)\Big([m_1\otimes\dots\otimes m_n]_{(Y,\phi)}\Big) = [m_1\otimes\dots\otimes m_n]_{(Y,g\phi)}$$

so that

$$\tilde{f}_{X'}\Big((M_1\hat{\otimes}\dots\hat{\otimes}M_n)_*(g)\Big([m_1\otimes\dots\otimes m_n]_{(Y,\phi)}\Big)\Big) = P_*(g)P_*(\phi)\hat{f}_Y(m_1,\dots,m_n) = \dots$$

$$\dots = P_*(g)\tilde{f}_X\Big([m_1\otimes\dots\otimes m_n]_{(Y,\phi)}\Big)$$

I still have to express the fact that $\tilde{f}$ must be contravariant. Let $g : X' \to X$ be a morphism of G-sets, and suppose the square

$$\begin{array}{ccc} Y' & \xrightarrow{a} & Y \\ {\scriptstyle \phi'}\downarrow & & \downarrow{\scriptstyle \phi} \\ X' & \xrightarrow[g]{} & X \end{array}$$

is cartesian. If $m_i \in M_i(Y)$ for $1 \le i \le n$, then

$$(M_1\hat{\otimes}\dots\hat{\otimes}M_n)^*(g)\Big([m_1\otimes\dots\otimes m_n]_{(Y,\phi)}\Big) = \Big[M_1^*(a)(m_1)\otimes\dots\otimes M_n^*(a)(m_n)\Big]_{(Y',\phi')}$$

Taking the image under $\tilde{f}_{X'}$ gives

$$P_*(\phi')\hat{f}_{Y'}\Big(M_1^*(a)(m_1),\dots,M_n^*(a)(m_n)\Big)$$

that must be equal to

$$P^*(g)P_*(\phi)\hat{f}_Y(m_1,\dots,m_n)$$

Since moreover $P^*(g)P_*(\phi) = P_*(\phi')P^*(a)$, this equality will result from the case $X = Y$, $X' = Y'$, $\phi = \phi' = Id$ and $a = g$. This gives

$$(B) \qquad \hat{f}_{Y'}\Big(M_1^*(g)(m_1),\dots,M_n^*(g)(m_n)\Big) = P^*(g)\hat{f}_Y(m_1,\dots,m_n)$$

for any $g : Y' \to Y$. The next proposition is a summary of conditions (A_i) and (B) in the case $n = 2$

Proposition 1.8.3: Let M, N and P be Mackey functors for the group G. The following data are equivalent:

- **A bilinear morphism f from M, N to P.**
- **For any G-set Y, a bilinear morphism $\hat{f}_Y : M(Y)\times N(Y) \to P(Y)$, such that for all $g : Y' \to Y$, $m \in M(Y)$, $n \in N(Y')$, $m' \in M(Y')$, and $n' \in N(Y')$**

i) $$\hat{f}_Y\Big(M_*(g)(m')\times n\Big) = P_*(g)\hat{f}_{Y'}\Big(m', N^*(g)(n)\Big)$$

ii) $$\hat{f}_Y\Big(m, N_*(g)(n')\Big) = P_*(g)\hat{f}_{Y'}\Big(M^*(g)(m), n'\Big)$$

iii) $$\hat{f}_{Y'}\Big(M^*(g)(m), N^*(g)(n)\Big) = P^*(g)\hat{f}_Y(m,n)$$

1.9 Commutativity and associativity

It should be clear from the previous sections that the tensor product is commutative and associative:

Proposition 1.9.1: Let M, N, and P be Mackey functors for the group G. Then there are isomorphisms of Mackey functors

$$M\hat{\otimes}N \simeq N\hat{\otimes}M$$

$$(M\hat{\otimes}N)\hat{\otimes}P \simeq M\hat{\otimes}(N\hat{\otimes}P)$$

which are natural in M, N and P.

Proof: Let X be a G-set, and (Y,ϕ) be a G-set over X. It is clear that the map

$$[m\otimes n]_{(Y,\phi)} \in M\hat{\otimes}N(X) \mapsto [n\otimes m]_{(Y,\phi)} \in N\hat{\otimes}M(X)$$

is well defined, and induces the desired isomorphism between $M\hat{\otimes}N$ and $N\hat{\otimes}M$.
The second assertion can be proved by observing that, for any Mackey functor Q, morphisms from each of its sides to Q are in one to one correspondence with trilinear morphisms from M, N, P to Q. The desired isomorphism is then a consequence of this universal correspondence.
Another (less canonical) way to see this isomorphism, is to use commutativity:

$$M\hat{\otimes}(N\hat{\otimes}P) \simeq (N\hat{\otimes}P)\hat{\otimes}M = N\hat{\otimes}P\hat{\otimes}M$$

and to observe that the n-fold tensor product is clearly independent, up to isomorphism, of the order of its factors. ■

1.10 Adjunction

I have moreover the following adjunction properties:

Proposition 1.10.1: Let M, N, and P be Mackey functors for the group G. Then there exists an isomorphism

$$\mathcal{H}(M\hat{\otimes}N,P) \simeq \mathcal{H}\big(N,\mathcal{H}(M,P)\big)$$

natural in M, N, and P.

Proof: Let X be a G-set. Then

$$\mathcal{H}\big(N,\mathcal{H}(M,P)\big)(X) = \mathrm{Hom}_{Mack(G)}\big(N,\mathcal{H}(M,P)_X\big)$$

On the other hand

$$\mathcal{H}(M,P)_X(Y) = \mathcal{H}(M,P)(YX) = \mathrm{Hom}_{Mack(G)}(M,P_{YX})$$

whereas

$$\mathcal{H}(M,P_X)(Y) = \mathrm{Hom}_{Mack(G)}(M,(P_X)_Y)$$

But $(P_X)_Y(Z) = P_X(ZY) = P(ZYX)$, and those isomorphisms are easily seen to induce a natural isomorphism of Mackey functors $(P_X)_Y \simeq P_{YX}$. Those isomorphisms of Mackey functors induce in turn an isomorphism

$$\mathcal{H}(M,P)_X \simeq \mathcal{H}(M,P_X)$$

It follows that

$$\mathcal{H}\big(N,\mathcal{H}(M,P)\big)(X) \simeq \mathrm{Hom}_{Mack(G)}\big(N,\mathcal{H}(M,P_X)\big)$$

But since

$$\mathcal{H}(M\hat{\otimes}N,P)(X) = \mathrm{Hom}_{Mack(G)}(M\hat{\otimes}N,P_X)$$

it is enough to have a natural isomorphism

$$\mathrm{Hom}_{Mack(G)}(M\hat{\otimes}N,P) \simeq \mathrm{Hom}_{Mack(G)}\big(N,\mathcal{H}(M,P)\big)$$

This is equivalent to say that the functor $N \mapsto M\hat{\otimes}N$ is left adjoint to the functor $P \mapsto \mathcal{H}(M,P)$. But a morphism from $M\hat{\otimes}N$ to P corresponds naturally to a bilinear morphism from M, N to P. To prove the proposition, it is enough to build a natural bijection between $\mathrm{Hom}_{Mack(G)}\big(N,\mathcal{H}(M,P)\big)$ and $\mathcal{L}(M,N;P)$.

But a morphism α from N to $\mathcal{H}(M,P)$ is determined by a collection α_Y, indexed by G-sets, of morphisms from $N(Y)$ to

$$\mathcal{H}(M,P)(Y) = \mathrm{Hom}_{Mack(G)}(M,P_Y)$$

The morphism α_Y is determined by morphisms $(\alpha_Y(n))_X$ from $M(X)$ to $P_Y(X) = P(XY)$, for any G-sets X, and any $n \in N(Y)$. I have then a morphism $\tilde{\alpha}_{X,Y}$ from $M(X)\otimes_R N(Y)$ to $P(XY)$, defined by

$$\tilde{\alpha}_{X,Y}(m\otimes n) = (\alpha_Y(n))_X(m)$$

and it is easy to see that α is a morphism of Mackey functors from N to $\mathcal{H}(M,P)$ if and only if the element $\tilde{\alpha}$ defined that way is a bilinear morphism from M, N to P.

Conversely, if I have a bilinear morphism $\tilde{\alpha}$ from M, N to P, reading from right to left the above formula, I can define a morphism

$$\big(\alpha_Y(n)\big)_X : N(Y) \to \mathrm{Hom}_R\big(M(X),P(XY)\big)$$

which induces a morphism of Mackey functors from N to $\mathcal{H}(M,P)$.

Those constructions are clearly inverse to each other, and natural in M, N, and P. This completes the proof of the proposition. ∎

Chapter 2

Green functors

2.1 Definitions

The classical definition of a Green functor is "a Mackey functor with a compatible ring structure": a Green functor A for the group G over the ground ring R is a Mackey functor, such that for any subgroup H of G, the R-module $A(H)$ has a structure of R-algebra (associative, with unit), which is compatible with the Mackey structure, in the following sense:

- If $x \in G$, and K is a subgroup of G, then the conjugation by x is a morphism of rings (with unit) from $A(K)$ to $A({}^xK)$.
- If $H \subseteq K$ are subgroups of G, then r_H^K is a morphism of rings (with unit) from $A(K)$ to $A(H)$.
- In the same conditions, if $a \in A(K)$ and $b \in A(H)$, then

$$a.t_H^K(b) = t_H^K(r_H^K(a).b)$$
$$t_H^K(b).a = t_H^K(b.r_H^K(a))$$

There is an evident notion of morphism of Green functors: a morphism ϕ from the Green functor A to the Green functor B is a morphism of Mackey functors such that, for any subgroup H of G, the morphism ϕ_H is a morphism of rings. The morphism ϕ is said to be unitary if the morphism ϕ_H preserves unit for all H. It is actually enough that the morphism ϕ_G preserves unit, since

$$\phi_H(1_{A(H)}) = \phi_H(r_H^G 1_{A(G)}) = r_H^G \phi_G(1_{A(G)}) = r_H^G 1_{B(G)} = 1_{B(H)}$$

A module over the Green functor A, or A-module, is defined similarly as a Mackey functor M for the group G, such that for any subgroup H of G, the module $M(H)$ has a structure of $A(H)$-module (with unit). This structure must be compatible with the Mackey structure, in the following sense:

- If $x \in G$ and $K \subseteq G$, let $m \mapsto {}^xm$ be the conjugation by x from $M(K)$ to $M({}^xK)$. If $a \in A(K)$ and $m \in M(K)$, then ${}^x(a.m) = {}^x(a).{}^x(m)$
- If $H \subseteq K$ are subgroups of G, if $a \in A(K)$ and $m \in M(K)$, then $r_H^K(a.m) = r_H^K(a).r_H^K(m)$.

- In the same conditions, if $a \in A(K)$ and $m \in M(H)$, then
$$a.t_H^K(m) = t_H^K(r_H^K(a).m)$$
and if $a \in A(H)$ and $m \in M(K)$, then
$$t_H^K(a).m = t_H^K(a.r_H^K(m))$$

A morphism ϕ from the A-module M to the A-module N is a morphism of Mackey functors from M to N such that for any subgroup H of G, the morphism ϕ_H is a morphism of $A(H)$-modules.

An important example of Green functor is obtained for any Mackey functor from the isomorphism
$$\mathcal{H}(M,M)(H) \simeq \mathrm{Hom}_{Mack(H)}(\mathrm{Res}_H^G M, \mathrm{Res}_H^G M)$$
In particular, the composition of morphisms turns $\mathcal{H}(M,M)(H)$ into a ring with unit, and $\mathcal{H}(M,M)$ is actually a Green functor.

Proposition 2.1.1: If M is a Mackey functor for the group G, then $\mathcal{H}(M,M)$ has a natural structure of Green functor.

Proof: Let ϕ be an endomorphism of the Mackey functor $\mathrm{Res}_H^G M$. Then ϕ is determined by morphisms $\phi_L : M(L) \to M(L)$, for $L \subseteq H$. The product of ϕ and ψ is defined by
$$(\phi\psi)_L = \phi_L \psi_L$$
If K is a subgroup of H, and if L is a subgroup of K, then
$$(r_K^H \phi)_L = \phi_L$$

It is then clear that r_K^H is a morphism of rings (with unit) from $\mathrm{End}_{Mack(H)}(\mathrm{Res}_H^G M)$ to $\mathrm{End}_{Mack(K)}(\mathrm{Res}_K^G M)$.
Moreover, if $x \in G$, and if $L \subseteq {}^xH$, then denoting by c_x the conjugation by x, I have
$$({}^x\phi)_L = c_x \phi_{L^x} c_x^{-1}$$

and it is also clear that the conjugation by x is also a morphism of rings with unit from $\mathrm{End}_{Mack(H)}(\mathrm{Res}_H^G M)$ to $\mathrm{End}_{Mack({}^xH)}(\mathrm{Res}_{{}^xH}^G M)$.
If now $H \subseteq K$, if $a \in \mathrm{End}_{Mack(K)}(\mathrm{Res}_K^G M)$ and if $b \in \mathrm{End}_{Mack(H)}(\mathrm{Res}_H^G M)$, then for any subgroup L of G
$$\left(a.t_H^K(b)\right)_L = a_L \circ \left(t_H^K b\right)_L = a_L \circ \sum_{x \in L\backslash K/H} t_{L\cap {}^xH}^L c_x b_{L^x \cap H} c_x^{-1} r_{L\cap {}^xH}^L$$

As a is an endomorphism of $\mathrm{Res}_K^G M$, as $L \subseteq K$, and as $x \in K$, I have
$$a_L \circ t_{L\cap {}^xH}^L c_x = t_{L\cap {}^xH}^L c_x a_{L^x \cap H}$$

so
$$\left(a.t_H^K(b)\right)_L = \sum_{x \in L\backslash K/H} t_{L\cap {}^xH}^L c_x a_{L^x\cap H} b_{L^x \cap H} c_x^{-1} r_{L\cap {}^xH}^L = \dots$$

$$\dots = \sum_{x\in L\backslash K/H} t^L_{L\cap^x H} c_x(ab)_{L^x\cap H} c_x^{-1} r^L_{L\cap^x H} = \Big(t^K_H(r^K_H(a).b)\Big)_L$$

hence $a.t^K_H(b) = t^K_H(r^K_H(a).b)$. The equality $t^K_H(b).a = t^K_H(b.r^K_H(a))$ follows similarly from the equality

$$c_x^{-1} r^L_{L\cap^x H} \circ a_L = a_{L^x\cap H} c_x^{-1} r^L_{L\cap^x H}$$

and this proves the proposition. ■

When G is a group and k a ring, a kG-module is a k-module with a morphism of rings $kG \to \mathrm{End}_k(M)$. Reformulating this assertion in the case of Green functors, replacing "ring" by "Green functor" and $\mathrm{End}_k(M)$ by $\mathcal{H}(M,M)$, I get the following proposition

Proposition 2.1.2: Let M be a Mackey functor, and A a Green functor for the group G. Then it is equivalent to give M a structure of A-module, or to give a unitary morphism of Green functors from A to $\mathcal{H}(M,M)$.

Proof:1) Let M be an A-module. If $L \subseteq H$ are subgroups of G, and if $a \in A(H)$, I define an endomorphism $\phi_H(a)_L$ of $M(L)$ by setting

$$\phi_H(a)_L(m) = (r^H_L a).m$$

and I claim that I obtain that way an endomorphism $\phi_H(a)$ of $\mathrm{Res}^G_H M$. Indeed, if $L \subseteq K \subseteq H$, and if $m \in M(K)$, I have

$$\phi_H(a)_L(r^K_L m) = r^H_L a.r^K_L m = r^K_L(r^H_K(a).m) = r^K_L\Big(\phi_H(a)_K(m)\Big)$$

and this shows that $\phi_H(a)$ commutes with restrictions. Similarly, if $x \in H$, if $L \subseteq H$, and if $m \in M(L)$, then

$$\phi_H(a)_{^xL}(^xm) = r^H_{^xL}(a).^xm = {^x r^H_L}(a).^xm = {^x\Big(r^H_L(a).m\Big)} = {^x\Big(\phi_H(a)_L(m)\Big)}$$

since if $x \in H$ and $a \in A(H)$, then $^xa = a$. Hence $\phi_H(a)$ also commutes with conjugations by elements of H.
Finally, if $L \subseteq K \subseteq H$, and if $m \in M(L)$, then

$$\phi_H(a)_K(t^K_L m) = r^H_K a.t^K_L m = t^K_L\Big(r^K_L r^H_K(a).m\Big) = t^K_L(r^H_L(a).m) = t^K_L\Big(\phi_H(a)_L(m)\Big)$$

and this shows that $\phi_H(a)$ is an endomorphism of $\mathrm{Res}^G_H M$.
It is clear that $\phi_H(1_{A(H)})$ is the identity endomorphism of $\mathrm{Res}^G_H M$, since $r^H_L(1_{A(H)}) = 1_{A(L)}$, and that ϕ_H is a morphism of rings with unit from $A(H)$ to $\mathrm{End}_{Mack(H)}(\mathrm{Res}^G_H M)$, since moreover for $a, a' \in A(H)$

$$\Big(\phi_H(a)\phi_H(a')\Big)_L(m) = \phi_H(a)_L\Big(r^H_L(a').m\Big) = \dots$$

$$\dots = r^H_L(a)r^H_L(a').m = r^H_L(aa').m = \phi_H(aa')_L(m)$$

I also claim that the maps ϕ_H define a morphism of Mackey functors from A to $\mathcal{H}(M,M)$: indeed, let $L \subseteq H' \subseteq H$ be subgroups of G, let $a \in A(H)$, and $m \in M(L)$. Then

$$\phi_{H'}(r^H_{H'}a)_L(m) = r^{H'}_L r^H_{H'}(a).m = r^H_L(a).m = \phi_H(a)_L(m)$$

and this shows that $\phi_{H'}(r^H_{H'}a)_L = \phi_H(a)_L$ for any subgroup L of H', or that $\phi_{H'}(r^H_{H'}a) = r^H_{H'}\phi_H(a)$.
Now if $x \in G$, if $L \subseteq H \subseteq G$, if $a \in A(H)$ and $m \in M(L)$, then

$$\phi_{{}^xH}({}^xa)_{{}^xL}({}^xm) = r^{{}^xH}_{{}^xL}({}^xa).{}^xm = {}^x\big(r^H_L(a).m\big) = {}^x\big(\phi_H(a)_L(m)\big)$$

and ${}^x\big(\phi_H(a)\big) = \phi_{{}^xH}({}^xa)$.
Finally if $L \subseteq H' \supseteq H$, if $a \in A(H)$, and $m \in M(L)$, then

$$\phi_{H'}(t^{H'}_H a)_L(m) = r^{H'}_L t^{H'}_H(a).m = \Big(\sum_{x\in L\backslash H'/H} t^L_{L\cap {}^xH}\, {}^x r^H_{L^x\cap H}(a)\Big).m$$

But

$$\big(t^L_{L\cap {}^xH}\, {}^x r^H_{L^x\cap H}(a)\big).m = t^L_{L\cap {}^xH}\Big({}^x r^H_{L^x\cap H}(a).r^L_{L\cap {}^xH}(m)\Big) = t^L_{L\cap {}^xH}\, {}^x\Big(r^H_{L^x\cap K}(a).(r^L_{L\cap {}^xH}(m))^x\Big)$$

where I denote by $v \mapsto v^x$ the conjugation by x^{-1}. This equality can also be written as

$$\big(t^L_{L\cap {}^xH}\, {}^x r^H_{L^x\cap H}(a)\big).m = t^L_{L\cap {}^xH} c_x \phi_H(a)_{L^x\cap H}\big(c_x^{-1} r^L_{L\cap {}^xH}(m)\big)$$

and finally

$$\phi_{H'}(t^{H'}_H a)_L = \sum_{x\in L\backslash H'/H} t^L_{L\cap {}^xH} c_x \phi_H(a)_{L^x\cap H} c_x^{-1} r^L_{L\cap {}^xH}$$

which proves that

$$\phi_{H'}(t^{H'}_H a) = t^{H'}_H \phi_H(a)$$

so ϕ is a morphism of Mackey functors. So to any A-module M, I know how to associate a morphism of Green functors from A to $\mathcal{H}(M,M)$.

2) Conversely, let M be a Mackey functor, and ψ be a morphism of Green functors from A to $\mathcal{H}(M,M)$. Then for any subgroup H of G, I have a morphism of rings with unit ψ_H from $A(H)$ to $\mathrm{End}_{Mack(H)}(\mathrm{Res}^G_H M)$. Then, for any $a \in A(H)$ and any subgroup K of H, I have an endomorphism $\psi_H(a)_K$ of the R-module $M(K)$. Then for $m \in M(H)$ and $a \in A(H)$, I set

$$a.m = \psi_H(a)_H(m)$$

I obtain a structure of $A(H)$-module (with unit) on $M(H)$: indeed

$$1_{A(H)}.m = \psi_H(1_{A(H)})_H(m) = m$$

because ψ preserves the unit. Moreover, the product $(a,m) \mapsto a.m$ is distributive with respect to addition in $M(H)$, because $\psi_H(a)$ is an endomorphism of $M(H)$. The product is also distributive with respect to addition in $A(H)$, because ψ_H is a morphism of R-modules from $A(H)$ to $\mathrm{End}_{Mack(H)}(\mathrm{Res}^G_H M)$. Finally, as ψ_H is multiplicative, it follows that for $a, a' \in A(H)$ and $m \in M(H)$, I have $(aa').m = a.(a'm)$.
Those structures of $A(H)$-module on $M(H)$ are compatible with the Mackey functor structures of A and M: indeed, if $x \in G$, if $K \subseteq G$, if $a \in A(K)$ and if $m \in M(K)$, then

$${}^xa.{}^xm = \psi_{{}^xK}({}^xa)_{{}^xK}({}^xm)$$

But $\psi_{^xK}(^xa) = {}^x\big(\psi_K(a)\big)$ because ψ commutes with conjugations. But for any subgroup L of ^xK, I have

$$\Big({}^x\big(\psi_K(a)\big)\Big)_L = c_x\psi_K(a)_{L^x}c_x^{-1}$$

and for $L = {}^xK$ this gives

$${}^xa.{}^xm = c_x\psi_K(a)_K c_x^{-1}({}^xm) = c_x\psi_K(a)_K(m) = {}^x(a.m)$$

Similarly, if $H \subseteq K$, if $a \in A(K)$ and $m \in M(H)$, then

$$a.t_H^K(m) = \psi_K(a)_K(t_H^K m) = t_H^K\big(\psi_K(a)_H(m)\big)$$

because $\psi_K(a)$ is an endomorphism of the Mackey functor $\mathrm{Res}_K^G M$. But

$$\psi_K(a)_H = \big(r_H^K\psi_K(a)\big)_H = \big(\psi_H(r_H^K(a))\big)_H$$

because ψ is a morphism of Mackey functors from A to $\mathcal{H}(M,M)$. Finally,

$$a.t_H^K(m) = t_H^K\Big(\big(\psi_H(r_H^K(a))\big)_H(m)\Big) = t_H^K(r_H^K(a).m)$$

Now if $a \in A(H)$ and $m \in M(K)$, then

$$t_H^K(a).m = \psi_K(t_H^K(a))_K(m) = \big(t_H^K(\psi_H(a)\big)_K(m)$$

and moreover

$$\big(t_H^K(\psi_H(a)\big)_K = \sum_{x\in K\backslash K/H} t_{K\cap {}^xH}^K c_x\psi_H(a)_{K^x\cap H}c_x^{-1}r_{K\cap {}^xH}^K = t_H^K\psi_H(a)_H r_H^K$$

Finally

$$t_H^K(a).m = \big(t_H^K\psi_H(a)_H r_H^K\big)(m) = t_H^K\big(\psi_H(a)_H r_H^K(m)\big) = t_H^K\big(a.r_H^K(m)\big)$$

and this proves that M is an A-module.

3) The last observation is that the above correspondences are mutually inverse: if M is an A-module, I associate to M the morphism ϕ defined by

$$\phi_H(a)_L(m) = (r_L^H a).m$$

and starting from ϕ, I define an A-module structure by

$$a.m = \phi_H(a)_H(m) = (r_H^H a).m = a.m$$

and I recover the initial A-module structure.
Conversely, if ψ is a morphism of Green functors from A to $\mathcal{H}(M,M)$, I associate to ψ the A-module structure defined by

$$a.m = \psi_H(a)_H(m)$$

and this in turn corresponds to the morphism ϕ of Green functors defined by

$$\phi_H(a)_L(m) = (r_L^H a).m = \big(\psi_L(r_L^H(a))\big)_L(m) = \big(r_L^H\psi_H(a)\big)_L(m) = \psi_H(a)_L(m)$$

and this completes the proof. ■

2.2 Definition in terms of G-sets

The previous section leads to a definition of Green functors and their modules in terms of G-sets. Indeed, if A is a Green functor and M an A-module, then by the previous proposition there is a morphism of Green functors

$$A \to \mathcal{H}(M, M)$$

Using adjunction, and commutativity of the tensor product, I get a morphism

$$A\hat{\otimes}M \to M$$

which in turn corresponds to a bilinear morphism $\lambda_M \in \mathcal{L}(A, M; M)$, determined by maps

$$\lambda_{X,Y} : A(X) \times M(Y) \to M(X \times Y)$$

for any G-sets X and Y. I will use intensively the following multiplicative notation, for $a \in A(X)$ and $m \in M(Y)$

$$a \times m = \lambda_{X,Y}(a, m)$$

Those remarks and notations lead to the following definition, akin to the classical definition of an algebra:

Definition: *Let R be a commutative ring. A Green functor A (over R) for the group G is a Mackey functor (over R) endowed for any G-sets X and Y with bilinear maps*

$$A(X) \times A(Y) \to A(X \times Y)$$

denoted by $(a, b) \mapsto a \times b$ which are bifunctorial, associative, and unitary, in the following sense:

- *(Bifunctoriality) If $f : X \to X'$ and $g : Y \to Y'$ are morphisms of G-sets, then the squares*

$$\begin{array}{ccc} A(X) \times A(Y) & \xrightarrow{\times} & A(X \times Y) \\ {\scriptstyle A_*(f) \times A_*(g)} \downarrow & & \downarrow {\scriptstyle A_*(f \times g)} \\ A(X') \times A(Y') & \xrightarrow[\times]{} & A(X' \times Y') \end{array}$$

$$\begin{array}{ccc} A(X) \times A(Y) & \xrightarrow{\times} & A(X \times Y) \\ {\scriptstyle A^*(f) \times A^*(g)} \uparrow & & \uparrow {\scriptstyle A^*(f \times g)} \\ A(X') \times A(Y') & \xrightarrow[\times]{} & A(X' \times Y') \end{array}$$

  are commutative.

- *(Associativity) If X, Y and Z are G-sets, then the square*

$$\begin{array}{ccc} A(X) \times A(Y) \times A(Z) & \xrightarrow{Id_{A(X)} \times (\times)} & A(X) \times A(Y \times Z) \\ {\scriptstyle (\times) \times Id_{A(Z)}} \downarrow & & \downarrow {\scriptstyle \times} \\ A(X \times Y) \times A(Z) & \xrightarrow[\times]{} & A(X \times Y \times Z) \end{array}$$

  is commutative, up to identifications $(X \times Y) \times Z \simeq X \times Y \times Z \simeq X \times (Y \times Z)$.

- *(Unitarity) If $\bullet$ denotes the G-set with one element, there exists an element $\varepsilon \in A(\bullet)$ such that for any G-set X and for any $a \in A(X)$*

$$A_*(p_X)(a \times \varepsilon) = a = A_*(q_X)(\varepsilon \times a)$$

denoting by p_X (resp. q_X) the (bijective) projection from $X \times \bullet$ (resp. from $\bullet \times X$) to X.

If A and B are Green functors for the group G, a morphism f (of Green functors) from A to B is a morphism of Mackey functors such that for any G-sets X and Y, the square

$$\begin{array}{ccc} A(X) \times A(Y) & \xrightarrow{\times} & A(X \times Y) \\ {\scriptstyle f_X \times f_Y}\downarrow & & \downarrow{\scriptstyle f_{X\times Y}} \\ B(X) \times B(Y) & \xrightarrow[\times]{} & B(X \times Y) \end{array}$$

is commutative.
If moreover $f_\bullet : A(\bullet) \to B(\bullet)$ maps the unit of A to the unit of B, then I will say that f is unitary.
I will denote by $Green_R(G)$ or $Green(G)$ the category of Green functors for G over R, and unitary morphisms between them.

Remarks: 1) The expression "up to identifications $(X \times Y) \times Z \simeq X \times Y \times Z \simeq X \times (Y \times Z)$" means more precisely that the diagrams

$$\begin{array}{ccccc} \big(A(X)\times A(Y)\big)\times A(Z) & \xrightarrow{\simeq} & A(X)\times A(Y)\times A(Z) & \xrightarrow{\simeq} & A(X)\times\big(A(Y)\times A(Z)\big) \\ {\scriptstyle (\times)\times Id}\downarrow & & & & \downarrow{\scriptstyle Id\times(\times)} \\ A(X\times Y)\times A(Z) & & & & A(X)\times A(Y\times Z) \\ {\scriptstyle \times}\downarrow & & & & \downarrow{\scriptstyle \times} \\ A\big((X\times Y)\times Z\big) & \underset{\simeq}{\xrightarrow{A_*(\alpha_{X,Y,Z})}} & A(X\times Y\times Z) & \underset{\simeq}{\xrightarrow{A^*(\beta_{X,Y,Z})}} & A\big(X\times(Y\times Z)\big) \end{array}$$

are commutative.
Here $\alpha_{X,Y,Z}$ is the canonical bijection $\big((x,y),z\big) \mapsto (x,y,z)$ from $(X \times Y) \times Z$ to $X \times Y \times Z$, and $\beta_{X,Y,Z}$ is the bijection $\big(x,(y,z)\big) \mapsto (x,y,z)$ from $X \times (Y \times Z)$ to $X \times Y \times Z$. From now on, I will always forget those isomorphisms, and I will write directly the commutative square of the definition.
2) Similarly, I will always write $a \times \varepsilon = a = \varepsilon \times a$, identifying $(X \times \bullet)$ and $(\bullet \times X)$ with X.

There is an analogous definition for a module over the Green functor A:

Definition: *Let A be a Green functor over R for the group G. A module M over A is a Mackey functor M (over R), endowed for any G-sets X and Y with R-bilinear maps*

$$A(X) \times M(Y) \to M(X \times Y)$$

denoted by $a \times m \mapsto a \times m$, which are bifunctorial, associative and unitary in the following sense:

- *(Bifunctoriality) If $f : X \to X'$ and $g : Y \to Y'$ are morphisms of G-sets, then the squares*

$$\begin{array}{ccc} A(X) \times M(Y) & \xrightarrow{\times} & M(X \times Y) \\ {\scriptstyle A_*(f) \times M_*(g)} \downarrow & & \downarrow {\scriptstyle M_*(f \times g)} \\ A(X') \times M(Y') & \xrightarrow[\times]{} & M(X' \times Y') \end{array}$$

$$\begin{array}{ccc} A(X) \times M(Y) & \xrightarrow{\times} & M(X \times Y) \\ {\scriptstyle A^*(f) \times M^*(g)} \uparrow & & \uparrow {\scriptstyle A^*(f \times g)} \\ A(X') \times M(Y') & \xrightarrow[\times]{} & M(X' \times Y') \end{array}$$

are commutative.

- *(Associativity) If X, Y and Z are G-sets, then the square*

$$\begin{array}{ccc} A(X) \times A(Y) \times M(Z) & \xrightarrow{Id_{A(X)} \times (\times)} & A(X) \times M(Y \times Z) \\ {\scriptstyle (\times) \times Id_{M(Z)}} \downarrow & & \downarrow {\scriptstyle \times} \\ A(X \times Y) \times M(Z) & \xrightarrow[\times]{} & M(X \times Y \times Z) \end{array}$$

is commutative, up to identifications $(X \times Y) \times Z \simeq X \times Y \times Z \simeq X \times (Y \times Z)$.

- *(Unitarity) For any G-set X and any $m \in M(X)$*

$$M_*(q_X)(\varepsilon \times m) = m$$

If M and N are modules over A, a morphism (of A-modules) f from M to N is a morphism of Mackey functors such that for any G-sets X and Y, the square

$$\begin{array}{ccc} A(X) \times M(Y) & \xrightarrow{\times} & M(X \times Y) \\ {\scriptstyle Id \times f_Y} \downarrow & & \downarrow {\scriptstyle f_{X \times Y}} \\ A(X) \times N(Y) & \xrightarrow[\times]{} & N(X \times Y) \end{array}$$

is commutative.
*I will denote by A-**Mod** the category of A-modules.*

Remark: Just as before, I will identify $(\bullet \times X)$ and X, and write $\varepsilon \times m = m$.

2.3 Equivalence of the two definitions

Let A be a Green functor in the classical sense. It is clear that A is also an A-module in the classical sense, since the module $A(H)$ has a natural structure of $A(H)$-module. Then proposition 2.1.2 proves that there exists a morphism from A to $\mathcal{H}(A, A)$, hence by adjunction a morphism from $A\hat{\otimes}A$ to A, and also then a bilinear morphism from A, A to A: to be concrete, if H and K are subgroups of G, the maps

$$A(G/H)\times A(G/K)\to A(G/H\times G/K)\simeq \bigoplus_{x\in H\backslash G/K} A\big(G/(H\cap {}^xK)\big)$$

are defined for $a\in A(G/H)=A(H)$ and $b\in A(G/K)=A(K)$ by

$$a\times b=\bigoplus_{x\in H\backslash G/K} r^H_{H\cap^xK}(a).r^{^xK}_{H\cap^xK}(^xb)$$

This follows easily from the proofs of propositions 1.10.1 and 2.1.2: proposition 2.1.2 shows that A is a $\mathcal{H}(A,A)$ module. If H is a subgroup of G, and if $a\in A(H)$, then a defines an endomorphism of the Mackey functor $\mathrm{Res}^G_H A$ by

$$\forall L\subseteq H,\ b\in A(L)\mapsto r^H_L(a).b\in A(L)$$

By adjunction, I can associate to a a morphism from A to $\mathrm{Ind}^G_H\mathrm{Res}^G_H A=A_{G/H}$ defined for $K\subseteq G$ by

$$b\in A(K)\mapsto \bigoplus_{x\in H\backslash G/K} r^H_{H\cap^xK}(a).r^{^xK}_{H\cap^xK}(^xb)\in \mathrm{Ind}^G_H\mathrm{Res}^G_H A(K)=\bigoplus_{x\in H\backslash G/K} A(H\cap {}^xK)$$

and the formula for $a\times b$ follows.

Similarly, if M is a module over A in the classical sense, and if $m\in M(K)$, I have

$$a\times m=\bigoplus_{x\in H\backslash G/K} r^H_{H\cap^xK}(a).r^{^xK}_{H\cap^xK}(^xm) \tag{2.1}$$

By proposition 1.8.3, I know that such a bilinear morphism is determined by R-bilinear maps $A(Y)\times M(Y)\to M(Y)$, that I will denote by $(a,m)\mapsto a.m$, defined by

$$a.m=M^*(\delta_{2,Y})(a\times m)$$

This notation is coherent, for in the case $Y=G/H$, the map

$$A^*(\delta_{2,G/H}): A(G/H\times G/H)\simeq \bigoplus_{x\in H\backslash G/H} A(H\cap {}^xH)\to A(H)$$

is precisely the projection on the component associated to the double class $H.1.H=H$. Formula (2.1) now gives

$$M^*(\delta_{2,Y})(a\times m)=r^H_{H\cap H}(a).r^H_{H\cap H}(m)=a.m$$

The ring structures of the $A(H)$, and the $A(H)$-module structures of the $M(H)$, admit a unique extension to maps $A(Y)\times M(Y)\to M(Y)$ defined for any G-sets Y, and having properties $i)$, $ii)$, and $iii)$ of proposition 1.8.3: indeed, if Y is a G-set, let $[G\backslash Y]$ be a system of representatives of orbits of G on Y. For $x\in[G\backslash Y]$, let i_x be the injection from G/G_x into Y defined by

$$i_x(gG_x)=gx$$

Then as Y is the disjoint union of its orbits, for any $a\in A(Y)$ and any $m\in M(Y)$, I have

$$a=\sum_{x\in[G\backslash Y]} A_*(i_x)A^*(i_x)(a),\quad m=\sum_{x\in[G\backslash Y]} M_*(i_x)M^*(i_x)(m)$$

Now condition *i*) of proposition 1.8.3 shows that for $x, y \in [G\backslash Y]$, $a' \in A(G_x)$ and $m' \in M(G_y)$

$$A^*(i_x)(a').M^*(i_y)(m') = M_*(i_x)\Big(a'.M_*(i_x)M^*(i_y)(m')\Big)$$

But the product $M_*(i_x)M^*(i_y)$ is zero if $x \neq y$, and equal to the identity of $M(G_x)$ otherwise. It follows that

$$a.m = \sum_{x\in[G\backslash Y]} M_*(i_x)\Big(A^*(i_x)(a).M^*(i_x)(m)\Big) \tag{2.2}$$

and the product $(a, m) \to a.m$ is entirely determined (in particular, the above formula does not depend on the choice of the system of representatives of the orbits of Y).

Notation: *If f is a map from a set X to a set Y, I will denote it by $f = \binom{x}{f(x)}$. For instance, the first projection from $X \times Y$ onto X will be denoted by $\binom{(x,y)}{x}$ or $\binom{xy}{x}$, and the second projection by $\binom{xy}{y}$.*

With this notation, propositions 1.8.2 and 1.8.3 show that the product $(a, m) \mapsto a \times m$ can be recovered from the product $(a, m) \mapsto a.m$ using the following formula

$$a \times m = A^*\begin{pmatrix} xy \\ x \end{pmatrix}(a).M^*\begin{pmatrix} xy \\ y \end{pmatrix}(m) \tag{2.3}$$

Associativity of the product $\times$ now follows from associativity of the product ".", and from condition *iii*) of proposition 1.8.3, since for $a \in A(X)$, $b \in A(Y)$, and $m \in M(Z)$, I have

$$a \times (b \times m) = A^*\begin{pmatrix} xyz \\ x \end{pmatrix}.M^*\begin{pmatrix} xyz \\ yz \end{pmatrix}(b \times m)$$

But condition iii) shows that

$$M^*\begin{pmatrix} xyz \\ yz \end{pmatrix}(b\times m) = M^*\begin{pmatrix} xyz \\ yz \end{pmatrix}\left(A^*\begin{pmatrix} yz \\ y \end{pmatrix}(b).M^*\begin{pmatrix} yz \\ z \end{pmatrix}(m)\right) = \ldots$$

$$\ldots = A^*\begin{pmatrix} xyz \\ yz \end{pmatrix}A^*\begin{pmatrix} yz \\ y \end{pmatrix}(a).M^*\begin{pmatrix} xyz \\ yz \end{pmatrix}M^*\begin{pmatrix} yz \\ z \end{pmatrix}(m) = \ldots$$

$$\ldots = A^*\begin{pmatrix} xyz \\ y \end{pmatrix}(a).M^*\begin{pmatrix} xyz \\ z \end{pmatrix}(m)$$

hence finally

$$a \times (b \times m) = A^*\begin{pmatrix} xyz \\ x \end{pmatrix}(a).\left(A^*\begin{pmatrix} xyz \\ y \end{pmatrix}(b).M^*\begin{pmatrix} xyz \\ z \end{pmatrix}(m)\right)$$

For the same reason, I have

$$(a \times b) \times m = \left(A^*\begin{pmatrix} xyz \\ x \end{pmatrix}(a).A^*\begin{pmatrix} xyz \\ y \end{pmatrix}(b)\right).M^*\begin{pmatrix} xyz \\ z \end{pmatrix}(m)$$

The existence of unit for the product $\times$ also follows from formula (2.3), since denoting by ε the unit of $A(G) = A(G/G) = A(\bullet)$, I have

$$\varepsilon \times m = A^*\begin{pmatrix} y \\ \bullet \end{pmatrix}(\varepsilon).M^*\begin{pmatrix} y \\ y \end{pmatrix}(m) = A^*\begin{pmatrix} y \\ \bullet \end{pmatrix}(\varepsilon).m$$

But formula (2.2) shows that $\varepsilon_Y = A^*\begin{pmatrix} y \\ \bullet \end{pmatrix}$ is a unit for the product ".", since

$$\varepsilon_Y.m = \sum_{x\in[G\backslash Y]} M_*(i_x)\Big(A^*(i_x)A^*\begin{pmatrix} y \\ \bullet \end{pmatrix}.M^*(i_x)(m)\Big)$$

But now $A^*(i_x)A^*\begin{pmatrix} y \\ \bullet \end{pmatrix} = A^*\Big(\begin{pmatrix} y \\ \bullet \end{pmatrix} i_x\Big)$, and

$$\begin{pmatrix} y \\ \bullet \end{pmatrix} i_x(gG_x) = \begin{pmatrix} y \\ \bullet \end{pmatrix}(gx) = \bullet$$

so $\begin{pmatrix} y \\ \bullet \end{pmatrix} i_x$ is the only map from G/G_x to $\bullet = G/G$, and then $A^*\Big(\begin{pmatrix} y \\ \bullet \end{pmatrix} i_x\Big) = r^G_{G_x}$. Since $r^G_{G_x}(\varepsilon)$ is the unit of $A(G_x)$, I have

$$\varepsilon_Y.m = \sum_{x\in[G\backslash Y]} M_*(i_x)\Big(r^G_{G_x}(\varepsilon).M^*(i_x)(m)\Big) = \sum_{x\in[G\backslash Y]} M_*(i_x)M^*(i_x)(m) = m$$

A similar argument shows that ε_Y is also a right unit for the product "." of $A(Y)$. Conversely, I have seen that

$$a.m = M^*(\delta_{2,Y})(a \times m) = M^*\begin{pmatrix} y \\ yy \end{pmatrix}(a \times m) \tag{2.4}$$

and it is easy to deduce from this equation that bifunctoriality and associativity of the product $\times$ imply associativity of the product ".", since for instance

$$a.(b.m) = M^*\begin{pmatrix} y \\ yy \end{pmatrix}\big(a \times (b.m)\big) = \ldots$$

$$\ldots = M^*\begin{pmatrix} y \\ yy \end{pmatrix}\Big(a\times M^*\begin{pmatrix} y \\ yy \end{pmatrix}(b\times m)\Big) = M^*\begin{pmatrix} y \\ yy \end{pmatrix} M^*\begin{pmatrix} y_1y_2 \\ y_1y_2y_2 \end{pmatrix}\big(a\times(b\times m)\big) = \ldots$$

$$\ldots = M^*\begin{pmatrix} y \\ yyy \end{pmatrix}\big(a \times (b \times m)\big)$$

whereas by a similar computation

$$(a.b).m = M^*\begin{pmatrix} y \\ yyy \end{pmatrix}\big((a \times b) \times m\big)$$

And now if ε is the unit for the product $\times$, then ε_Y is a unit for the product ".", since

$$\varepsilon_Y.m = M^*\begin{pmatrix} y \\ yy \end{pmatrix}\Big(A^*\begin{pmatrix} y \\ \bullet \end{pmatrix}(\varepsilon)\times m\Big) = M^*\begin{pmatrix} y \\ yy \end{pmatrix} M^*\begin{pmatrix} y_1y_2 \\ y_2 \end{pmatrix}(\varepsilon\times m) = \ldots$$

$$\ldots = M^*\begin{pmatrix} y \\ yy \end{pmatrix} M^*\begin{pmatrix} y_1y_2 \\ y_2 \end{pmatrix}(m) = M^*\begin{pmatrix} y \\ y \end{pmatrix}(m) = m$$

A similar argument proves that ε_Y is also a right unit for the product "." of $A(Y)$. ∎

So the definitions of a Green functor and of a module over a Green functor in terms of G-sets are equivalent to the classical ones. It also follows from equalities (2.3) and (2.4) that the definitions of morphisms of Green functors and of morphisms of modules over a Green functor coincide: for example, if f is a morphism in the classical sense from the Green functor A to the Green functor B, then f_H is a ring morphism from $A(H)$ to $B(H)$, for any subgroup H of G. It follows easily that for any G-set X, the morphism $f_X : A(X) \to B(X)$ is compatible with the product ".". Then formula (2.3) gives for $M = A$, $a \in A(X)$ and $a' \in A(Y)$

$$a \times a' = A^*\begin{pmatrix} xy \\ x \end{pmatrix}(a).A^*\begin{pmatrix} xy \\ y \end{pmatrix}(a')$$

Taking image under $f_{X\times Y}$ of this equality gives

$$f_{X\times Y}(a \times a') = f_{X\times Y}\left(A^*\begin{pmatrix} xy \\ x \end{pmatrix}(a).A^*\begin{pmatrix} xy \\ y \end{pmatrix}(a')\right) = \ldots$$

$$\ldots = f_{X\times Y}\left(A^*\begin{pmatrix} xy \\ x \end{pmatrix}(a)\right).f_{X\times Y}\left(A^*\begin{pmatrix} xy \\ y \end{pmatrix}(a')\right)$$

But as f is a morphism of Mackey functors, I have

$$f_{X\times Y}A^*\begin{pmatrix} xy \\ x \end{pmatrix} = B^*\begin{pmatrix} xy \\ x \end{pmatrix} f_X$$

and

$$f_{X\times Y}A^*\begin{pmatrix} xy \\ y \end{pmatrix} = B^*\begin{pmatrix} xy \\ y \end{pmatrix} f_Y$$

Thus

$$f_{X\times Y}(a \times a') = B^*\begin{pmatrix} xy \\ x \end{pmatrix} f_X(a).B^*\begin{pmatrix} xy \\ y \end{pmatrix} f_Y(a') = f_X(a) \times f_Y(a')$$

and f is a morphism in the sense of the new definition. Equality (2.4) proves similarly that a morphism in the sense of the new definition is also a morphism in the classical sense.

2.4 The Burnside functor

In this section, I suppose for simplicity that $R = \mathbf{Z}$. All the results can be extended to the case of an arbitrary commutative ring by "tensoring everything with R", and replacing tensor products over $\mathbf{Z}$ by tensor products over R.

2.4.1 The Burnside functor as Mackey functor

The classical definition of the Burnside functor b (over $\mathbf{Z}$) is the following: for a subgroup H of G, the module $b(H)$ is the Grothendieck group of the category of H-sets, for relations given by decomposition into disjoint union. Restriction and induction of H-sets give a Mackey functor structure on b. The direct product of sets turns $b(H)$ into a (commutative) ring, and it is easy to check that b is actually a Green functor.

But this definition does not give the structure of $b(X)$, when X is an arbitrary G-set. The next proposition answers this question. If X is a G-set, I denote by G-**set**$\downarrow_X$ the category with G-sets over X as objects, and maps of G-sets over X as morphisms.

Lemma 2.4.1: Let H be a subgroup of G. The functor mapping a G-set (Y,ϕ) over G/H to the H-set $\phi^{-1}(H)$ is an equivalence of categories from G-set$\downarrow_{G/H}$ to H-set.The inverse equivalence is the induction functor from H-set to G-set$\downarrow_{G/H}$.

Proof: Let (Y,ϕ) be a G-set over G/H. Then it is clear that the inverse image $\phi^{-1}(H)$ of the element $H\in G/H$ is an H-set. If $f:(Y,\phi)\to(Y',\phi')$ is a morphism of G-sets over G/H, then $\phi=\phi' f$, so

$$f\big(\phi^{-1}(H)\big)\subseteq\phi'^{-1}(H)$$

and so I do have a functor F from G-**set**$\downarrow_{G/H}$ to H-**set**.
Conversely, if X is an H-set, then there exists a unique morphism of H-sets $\binom{x}{\bullet}$: $X\to\bullet$. The image of this morphism under the induction functor from H to G gives a G-set

$$\phi: F'(X)=\mathrm{Ind}_H^G X\to\mathrm{Ind}_H^G\bullet\simeq G/H$$

over G/H, and it is clear that F' is a functor from H-**set** to G-**set**$\downarrow_{G/H}$. I will now check that F and F' are mutual inverse equivalences of categories.
Let me first recall that $\mathrm{Ind}_H^G X$ is the quotient set of $G\times X$ by the equivalence relation identifying (gh,x) and (g,hx) for any $h\in H$. The group G acts on the left by $g'.\overline{(g,x)}=\overline{(g'g,x)}$, where I denote by $\overline{(g,x)}$ the equivalence class of (g,x). The above map ϕ is then given by $\phi\big(\overline{(g,x)}\big)=gH$. If f is a morphism of H-sets from X to X', then $F'(f)$ is given by

$$F'(f)\big(\overline{(g,x)}\big)=\overline{(g,f(x))}$$

But then $\phi\big(\overline{(g,x)}\big)=H$ if and only if $g\in H$. It it then clear that the map from $FF'(X)=\{\overline{(h,x)}\mid h\in H,\ x\in X\}$ to X which sends $\overline{(h,x)}$ to hx is an isomorphism from the functor FF' onto the identity functor of H-**set**.
Conversely, if (Y,ϕ) is a G-set over G/H, then the map from $\mathrm{Ind}_H^G\phi^{-1}(H)$ to Y sending $\overline{(g,y)}$ to gy is surjective, because any orbit of G on Y meets $\phi^{-1}(H)$. It is also injective, for if $gy=g'y'$ with $\phi(y)=\phi(y')=H$, then $gH=g'H$, and there exists $h\in H$ such that $g'=gh$, and $gy=ghy'$, i.e. $y=hy'$. Then $\overline{(g',y')}=\overline{(g,hy')}=\overline{(g,y)}$. It is easy to see that this is an isomorphism of functors from $F'F$ to the identity functor of G-**set**$\downarrow_{G/H}$, and this completes the proof of the lemma. ■

As a consequence, if H is a subgroup of G, then $b(H)$ is isomorphic to the Grothendieck group of G-**set**$\downarrow_{G/H}$, for relations given by decomposition into disjoint union of G-sets over G/H. This is actually true for any G-set:

Proposition 2.4.2: Let X be a G-set. Then $b(X)$ is isomorphic to the Grothendieck group of G-set$\downarrow_X$, for relations given by decomposition into disjoint union. Moreover, if (Y,ϕ) is a G-set over X,

- **If $f : X \to X'$ is a morphism of G-sets, then**

$$b_*\Big((Y,\phi)\Big) = (Y, f\phi)$$

- **If $f : X' \to X$ is a morphism of G-sets, then $b_*\Big((Y,\phi)\Big)$ is the pull-back (Y',ϕ') of (Y,ϕ) along f, obtained by filling the cartesian square**

$$\begin{array}{ccc} Y' & \xrightarrow{a} & Y \\ {\scriptstyle \phi'}\downarrow & & \downarrow{\scriptstyle \phi} \\ X' & \xrightarrow[f]{} & X \end{array}$$

Remark: In this proposition, I identify (Y,ϕ) with its image in the Grothendieck group.

Proof: I can consider for a while this proposition as a definition of b. It is clear that the definitions of b_* and b^* can be extended by linearity to group homomorphisms between $b(X)$ and $b(X')$, turning b into a bifunctor defined on G-**set**, with values in abelian groups (i.e. **Z**-modules). If I can prove that b is a Mackey functor, and that it coincides as a Mackey functor with the Burnside functor defined on transitive G-sets, the proposition will follow.

First it is clear that the Grothendieck group of G-**set**$\downarrow_X$ is additive with respect to X: if (U,ϕ) is a G-set over $X \coprod Y$, then U is the disjoint union of $\phi^{-1}(X)$ and $\phi^{-1}(Y)$, which are G-sets respectively over X and Y. Conversely, the disjoint union of a G-set over X and a G-set over Y is a G-set over $X \coprod Y$. Additivity of b follows.

Now let

$$\begin{array}{ccc} T & \xrightarrow{\gamma} & X \\ {\scriptstyle \delta}\downarrow & & \downarrow{\scriptstyle \alpha} \\ Y & \xrightarrow[\beta]{} & Z \end{array}$$

be a cartesian square, and (U,ϕ) be a G-set over X. Let V, a and b such that the square

$$\begin{array}{ccc} V & \xrightarrow{a} & U \\ {\scriptstyle b}\downarrow & & \downarrow{\scriptstyle \phi} \\ T & \xrightarrow[\gamma]{} & X \end{array}$$

is cartesian. Then $b^*(\gamma)\Big((U,\phi)\Big) = (V,b)$, so

$$b_*(\delta)b^*(\gamma)\Big((U,\phi)\Big) = (V,\delta b)$$

On the other hand, I have $b_*(\alpha)\Big((U,\phi)\Big) = (U,\alpha\phi)$, and as the square

$$\begin{array}{ccc} V & \xrightarrow{a} & U \\ {\scriptstyle \delta b}\downarrow & & \downarrow{\scriptstyle \alpha\phi} \\ Y & \xrightarrow[\beta]{} & Z \end{array}$$

is cartesian, because it is composed of the two above squares, I have

$$b^*(\beta)b_*(\alpha)\big((U,\phi)\big) = (V,\delta b)$$

which proves that $b^*(\beta)b_*(\alpha) = b_*(\delta)b^*(\gamma)$, and so b is a Mackey functor.

Now let $H \subseteq K$ be subgroups of G, and L be a subgroup of H. I denote by F_H and F'_H the equivalences between the categories associated to H in lemma 2.4.1. Then the image of the H-set H/L under the functor F'_H is the G-set

$$G/L \xrightarrow{\pi_L^H} G/H$$

over G/H, denoting by π_L^H the natural projection from G/L onto G/H. The image of that set under $b_*(\pi_H^K)$ is the set

$$G/L \xrightarrow{\pi_H^K\pi_L^H} G/K$$

that is the set $(G/L, \pi_L^K)$ over G/K. The image of that set under the functor F_K is the set K/L, which is the induced set of H/L from H to K.
If now L is a subgroup of K, then the image of K/L under F'_K is the G-set $(G/L, \pi_L^K)$, which is mapped under $b^*(\pi_H^K)$ to the set (P,ϕ) filling the cartesian square

$$\begin{array}{ccc} P & \longrightarrow & G/L \\ {\scriptstyle\phi}\downarrow & & \downarrow{\scriptstyle\pi_L^K} \\ G/H & \underset{\pi_H^K}{\longrightarrow} & G/K \end{array}$$

Then

$$P \simeq \coprod_{x\in H\backslash K/L} G/(H\cap {}^xL)$$

and the image of this under F_H is

$$\coprod_{x\in H\backslash K/L} H/(H\cap {}^xL) \simeq \mathrm{Res}_H^K K/L$$

Finally, if $x \in G$ and if L is a subgroup of K, then the image of K/L under F'_K is the set $(G/L, \pi_L^K)$. Let c_x be the conjugation by x, from G/K to $G/{}^xK$. The image of the previous set under $b_*(c_x)$ is $(G/L, c_x\pi_L^K) = (G/L, \pi_{{}^xL}^{{}^xK} c_x)$, which is mapped by $F_{{}^xK}$ to ${}^xK/{}^xL$, and this completes the proof of the proposition. ∎

2.4.2 The Burnside functor as Green functor

The Green functor structure of the Burnside functor is also very natural:

Proposition 2.4.3: Let X and Y be G-sets. If $E = (U,\phi)$ (resp. $F = (V,\psi)$) is a G-set over X (resp. over Y), I denote by $E \times F$ the G-set $(U\times V, \phi\times\psi)$ over $X \times Y$. Then the product $\times$ can be extended by linearity to a product from $b(X) \otimes b(Y)$ to $b(X\times Y)$, and this turns b into a Green functor, with the set $(\bullet, Id)$ over $\bullet$ as unit $\varepsilon \in b(\bullet)$.

Proof: It is clear that the product $\big((U,\phi),(V,\psi)\big)$ is left distributive with respect to disjoint union of G-sets over X, and right distributive with respect to disjoint union of G-sets over Y. It follows that the product $\times$ can be extended by bilinearity to a product $b(X)\otimes b(Y)\to b(X\times Y)$.
This extended product is clearly associative, and admits as a unit the trivial G-set over itself.
The product $\times$ is also clearly bifunctorial: if $f : X \to X'$ and $g : Y \to Y'$ are morphisms of G-sets, then

$$b_*(f)\big((U,\phi)\big)\times b_*(g)\big((V,\psi)\big) = (U,f\phi)\times(V,g\psi) = (U\times V, f\phi\times g\psi) = \dots$$

$$\dots = \big(U\times V,(f\times g)(\phi\times\psi)\big) = b_*(f\times g)\big((U,\phi)\times(V,\psi)\big)$$

Now if $f : X'\to X$ and $g : Y'\to Y$ are morphisms of G-sets, and if the squares

$$\begin{array}{ccc} U' & \xrightarrow{f'} & U \\ {\scriptstyle\phi'}\downarrow & & \downarrow{\scriptstyle\phi} \\ X' & \xrightarrow[f]{} & X \end{array} \qquad\qquad \begin{array}{ccc} V' & \xrightarrow{g'} & V \\ {\scriptstyle\psi'}\downarrow & & \downarrow{\scriptstyle\psi} \\ Y' & \xrightarrow[g']{} & Y \end{array}$$

are cartesian, then the product square

$$\begin{array}{ccc} U'\times V' & \xrightarrow{f'\times g'} & U\times V \\ {\scriptstyle\phi'\times\psi'}\downarrow & & \downarrow{\scriptstyle\phi\times\psi} \\ X'\times Y' & \xrightarrow[f\times g]{} & X\times Y \end{array}$$

is also cartesian, and then

$$b^*(f\times g)\big((U,\phi)\times(V,\psi)\big) = (U'\times V',\phi'\times\psi') = b^*(f)\big((U,\phi)\big)\times b^*(g)\big((V,\psi)\big)$$

This shows that the product $\times$ is bifunctorial, and proves the proposition. ∎

This product $\times$ gives in turn a product "." from $b(X)\otimes b(X)$ to $b(X)$, defined for two G-sets (U,ϕ) and (V,ψ) over X by

$$(U,\phi).(V,\psi) = b^*\begin{pmatrix} x \\ xx \end{pmatrix}\big((U\times V,\phi\times\psi)\big)$$

Let P be the pull-back of (U,ϕ) and (V,ψ), such that the square

$$\begin{array}{ccc} P & \xrightarrow{\phi'} & V \\ {\scriptstyle\psi'}\downarrow & & \downarrow{\scriptstyle\psi} \\ U & \xrightarrow[\phi]{} & X \end{array}$$

is cartesian. Then the square

$$\begin{array}{ccc} P & \xrightarrow{\begin{pmatrix} p \\ \phi'(p)\psi'(p) \end{pmatrix}} & U \times V \\ {\scriptstyle \phi\psi' = \psi\phi'} \downarrow & & \downarrow {\scriptstyle \phi \times \psi} \\ X & \xrightarrow[\begin{pmatrix} x \\ xx \end{pmatrix}]{} & X \times X \end{array}$$

is also cartesian, and it follows that

$$(U, \phi).(V, \psi) = \left(P, \begin{pmatrix} p \\ \phi'(p)\psi'(p) \end{pmatrix}\right)$$

This equality proves that the product "." coincides for $X = G/H$ with the product of H-sets, up to identifications of lemma 2.4.1: indeed, if K and L are subgroups of H, then $F'_H(H/K) = (G/L, \pi^H_K)$ and $F'_H(H/L) = (G/L, \pi^H_L)$, and the pull-back of those two sets is

$$\coprod_{x \in K \backslash H/L} G/(K \cap {}^xL)$$

which is mapped under F_H to the set

$$\coprod_{x \in K \backslash H/L} H/(K \cap {}^xL) \simeq (H/K) \times (H/L)$$

2.4.3 The Burnside functor as initial object

The Burnside functor plays the same role for Green functors as the ring $\mathbf{Z}$ does for rings with unit: it is an initial object of the category of Green functors:

Proposition 2.4.4: Let A be a Green functor for the group G. Then there exists a unique (unitary) morphism of Green functors from b to A.

Proof: Let f be a unitary morphism from b to A, and X be a G-set. If (U, ϕ) is a G-set over X, then denoting by ε_b the unit of b, I have

$$(U, \phi) = b_*(\phi) b^* \begin{pmatrix} u \\ \bullet \end{pmatrix} (\varepsilon_b)$$

Indeed, the square

$$\begin{array}{ccc} U & \xrightarrow{\begin{pmatrix} u \\ \bullet \end{pmatrix}} & \bullet \\ {\scriptstyle Id} \downarrow & & \downarrow {\scriptstyle Id} \\ U & \xrightarrow[\begin{pmatrix} u \\ \bullet \end{pmatrix}]{} & \bullet \end{array}$$

is cartesian, and its right column is equal to ε. Hence $b^*\binom{u}{\bullet}(\varepsilon) = (U, Id)$, which proves the above formula.
Now if f is a unitary morphism of Green functors, I must have

$$f_X\Big((U,\phi)\Big) = A_*(\phi)A^*\begin{pmatrix} u \\ \bullet \end{pmatrix}(\varepsilon_A)$$

and this proves that f is unique.
Conversely, this formula defines a unitary morphism of Green functors: first it is a morphism of Mackey functors, for if $g : X \to X'$ is a morphism of G-sets, then

$$b_*\Big((U,\phi)\Big) = (U, g\phi)$$

Thus

$$f_{X'}b_*\Big((U,\phi)\Big) = A_*(g)A_*(\phi)A^*\begin{pmatrix} u \\ \bullet \end{pmatrix}(\varepsilon_A) = A_*f_X\Big((U,\phi)\Big)$$

Similarly, if $g : X' \to X$ is a morphism of G-sets, then $b^*\Big((U,\phi)\Big)$ is obtain by filling the cartesian square

$$\begin{array}{ccc} Y' & \xrightarrow{a} & Y \\ {\scriptstyle\phi'}\downarrow & & \downarrow{\scriptstyle\phi} \\ X' & \xrightarrow[g]{} & X \end{array}$$

Then

$$f_{X'}b^*\Big((U,\phi)\Big) = f_{X'}\Big((U',\phi')\Big) = A_*(\phi')A^*\begin{pmatrix} y' \\ \bullet \end{pmatrix}(\varepsilon_A) = A_*(\phi')A^*(a)A^*\begin{pmatrix} y \\ \bullet \end{pmatrix}(\varepsilon_A) = \dots$$

$$\dots = A^*(g)A_*(\phi)A^*\begin{pmatrix} y \\ \bullet \end{pmatrix}(\varepsilon_A) = A^*(g)f_X\Big((U,\phi)\Big)$$

It is also clear that $f_\bullet\Big((\bullet, Id)\Big) = \varepsilon_A$. Finally, if Y is a G-set, and (V,ψ) a G-set over Y, then

$$f_{X\times Y}\Big((U\times V, \phi\times\psi)\Big) = A_*(\phi\times\psi)A^*\begin{pmatrix} uv \\ \bullet \end{pmatrix}(\varepsilon_A)$$

But up to identification of $\bullet\times\bullet$ with $\bullet$, I have

$$\begin{pmatrix} uv \\ \bullet \end{pmatrix} = \begin{pmatrix} u \\ \bullet \end{pmatrix} \times \begin{pmatrix} v \\ \bullet \end{pmatrix} \qquad \text{and} \qquad \varepsilon_A = \varepsilon_A \times \varepsilon_A$$

and as the product $\times$ is bifunctorial

$$A^*\begin{pmatrix} uv \\ \bullet \end{pmatrix}(\varepsilon_A) = A^*\begin{pmatrix} u \\ \bullet \end{pmatrix}(\varepsilon_A) \times A^*\begin{pmatrix} v \\ \bullet \end{pmatrix}(\varepsilon_A)$$

For the same reason, I have then

$$f_{X\times Y}\Big((U\times V, \phi\times\psi)\Big) = A_*(\phi)A^*\begin{pmatrix} u \\ \bullet \end{pmatrix}(\varepsilon_A) \times A_*(\psi)A^*\begin{pmatrix} v \\ \bullet \end{pmatrix}(\varepsilon_A) = \dots$$

$$\dots = f_X\Big((U,\phi)\Big) \times f_Y\Big((V,\psi)\Big)$$

which proves that f is a morphism of Green functors, and completes the proof of the proposition. ■

In particular, any Mackey functor M is a b-module, since there is a unitary morphism from b to $\mathcal{H}(M, M)$. So Mackey functors can be identified to modules over the Burnside functor. This structure of b-module for a Mackey functor M is quite easy to describe: if X and Y are G-sets, and if (U, ϕ) is a G-sets over X, I have seen that

$$(U,\phi) = b_*(\phi)b^*\begin{pmatrix} u \\ \bullet \end{pmatrix}(\varepsilon_b)$$

So if $m \in M(Y)$, as the product $\times$ must be bifunctorial, I must have

$$(U,\phi)\times m = M_*\begin{pmatrix} uy \\ \phi(u)y \end{pmatrix} M^*\begin{pmatrix} uy \\ \bullet y \end{pmatrix}(\varepsilon_b \times m) = M_*\begin{pmatrix} uy \\ \phi(u)y \end{pmatrix} M^*\begin{pmatrix} uy \\ y \end{pmatrix}(m)$$

2.4.4 The Burnside functor as unit

The next proposition confirms the analogy between the Burnside functor and the ring **Z**:

Proposition 2.4.5: Let M be a Mackey functor for the group G. Then there exists isomorphisms of Mackey functors

$$b\hat{\otimes}M \simeq M\hat{\otimes}b \simeq M$$

$$\mathcal{H}(b, M) \simeq M$$

which are natural in M.

Proof: The first isomorphism is a consequence of the second one and of proposition 1.10.1: indeed, if $\mathcal{H}(b, M) \simeq M$, then for any Mackey functor N, I have

$$\mathcal{H}(N, M) \simeq \mathcal{H}\big(N, \mathcal{H}(b, M)\big) \simeq \mathcal{H}(b\hat{\otimes}N, M)$$

and those isomorphisms are natural in M and N. Evaluating this isomorphism at the trivial G-set gives the functorial isomorphism

$$\mathrm{Hom}_{Mack(G)}(N, -) \simeq \mathrm{Hom}_{Mack(G)}(b\hat{\otimes}N, -)$$

and Yoneda's lemma now proves the first isomorphism.

To prove the second one, I observe first that

$$\mathrm{Hom}_{Mack(G)}(b, M) \simeq M(\bullet)$$

Indeed, I saw in the previous proof that if (U, ϕ) is a G-set over X, then in $b(X)$, I have

$$(U,\phi) = b_*(\phi)b^*\begin{pmatrix} u \\ \bullet \end{pmatrix}(\varepsilon_b)$$

So if θ is a morphism of Mackey functors from b to M, I have

$$\theta_X\big((U,\phi)\big) = M_*(\phi)M^*\begin{pmatrix} u \\ \bullet \end{pmatrix}\theta_\bullet(\varepsilon_b)$$

and θ is then determined by the element $\theta_\bullet(\varepsilon_b)$ of $M(\bullet)$. Conversely, if $m \in M(\bullet)$, then it is easy to check that the equality

$$\theta_X\big((U,\phi)\big) = M_*(\phi)M^*\binom{u}{\bullet}(m)$$

defines a morphism θ from b to M.
This remark shows that for any G-set X

$$\mathcal{H}(b,M)(X) = \mathrm{Hom}_{Mack(G)}(b, M_X) \simeq M_X(\bullet) \simeq M(X)$$

and it is easy to see that this isomorphism is actually bifunctorial in X, and so it is an isomorphism of Mackey functors from $\mathcal{H}(b,M)$ to M. ∎

Remark: The previous argument shows also that the isomorphism $\theta : b\hat{\otimes}M \to M$ is obtained in the following way: if X is a G-set, if (Y,ϕ) is a G-set over X, if (Z,ψ) is a G-set over Y, viewed as an element of $b(Y)$, and if $m \in M(Y)$, then

$$\theta_X\Big([(Z,\psi)\otimes m]_{(Y,\phi)}\Big) = M_*(\phi\psi)M^*(\psi)(m)$$

Chapter 3

The category associated to a Green functor

3.1 Examples of modules over a Green functor

Let A be a Green functor for the group G, and X be a G-set. If M is an A-module, then M is in particular a Mackey functor. I know then how to build the Mackey functor M_X, and I can try to turn it into a module over A: if Y and Z are G-sets, if $a \in A(Y)$ and $m \in M_X(Z) = M(ZX)$, then the product $a \times m$ is in $M(YZX) \simeq M_X(YZ)$. I can view that product as a definition of an A-module structure for M_X:

Lemma 3.1.1: Let A be a Green functor, let M be an A-module, and X be a G-set. If Y and Z are G-sets, the product

$$A(Y) \times M_X(Z) \to M_X(Y \times Z): \ (a, m) \mapsto a \times m$$

turns M_X into an A-module. The construction $M \mapsto M_X$ is an endofunctor from the category A-Mod of A-modules, which is its own left and right adjoint.

Proof: The first assertion comes from the fact that M is a module over the Green functor A. Moreover, the construction $M \mapsto M_X$ is clearly functorial in M. For the adjunction property, let N be an A-module, and θ be a morphism of A-modules from M_X to N. Then for any G-set Y, I have a morphism θ_Y from $M_X(Y) = M(YX)$ to $N(Y)$. So I have a morphism θ_{YX} from $M(YX^2)$ to $N(YX)$. I deduce a morphism θ'_Y from $M(Y)$ to $N(YX)$ by setting for $m \in M(Y)$

$$\theta'_Y(m) = \theta_{YX} M_* \begin{pmatrix} yx \\ yxx \end{pmatrix} M^* \begin{pmatrix} yx \\ y \end{pmatrix} (m)$$

Those definitions turn θ' into a morphism of Mackey functors from M to N_X: if $f : Y \to Z$ is a morphism of G-sets, then

$$N_{X,*}(f)\theta'_Y(m) = N_* \begin{pmatrix} yx \\ f(y)x \end{pmatrix} \theta_{YX} M_* \begin{pmatrix} yx \\ yxx \end{pmatrix} M^* \begin{pmatrix} yx \\ y \end{pmatrix} (m)$$

As θ is a morphism of Mackey functors, I have

$$N_* \begin{pmatrix} yx \\ f(y)x \end{pmatrix} \theta_{YX} = \theta_{ZX} M_* \begin{pmatrix} yx_1x_2 \\ f(y)x_1x_2 \end{pmatrix}$$

and so

$$N_{X,*}(f)\theta'_Y(m) = \theta_{ZX} M_* \begin{pmatrix} yx \\ f(y)xx \end{pmatrix} M^* \begin{pmatrix} yx \\ y \end{pmatrix} (m)$$

On the other hand

$$\theta'_Z M_*(f)(m) = \theta_{ZX} M_* \begin{pmatrix} zx \\ zxx \end{pmatrix} M^* \begin{pmatrix} zx \\ z \end{pmatrix} M_*(f)(m)$$

But the square

$$\begin{array}{ccc} YX & \xrightarrow{\begin{pmatrix} yx \\ f(y)x \end{pmatrix}} & ZX \\ {\scriptstyle \begin{pmatrix} yx \\ y \end{pmatrix}} \downarrow & & \downarrow {\scriptstyle \begin{pmatrix} zx \\ z \end{pmatrix}} \\ Y & \xrightarrow[f]{} & Z \end{array}$$

is cartesian, so $M^* \left(\begin{smallmatrix} zx \\ z \end{smallmatrix}\right) M_*(f) = M_* \left(\begin{smallmatrix} yx \\ f(y)x \end{smallmatrix}\right) M^* \left(\begin{smallmatrix} yx \\ y \end{smallmatrix}\right)$, and

$$\theta'_Z M_*(f)(m) = \theta_{ZX} M_* \begin{pmatrix} zx \\ zxx \end{pmatrix} M_* \begin{pmatrix} yx \\ f(y)x \end{pmatrix} M^* \begin{pmatrix} yx \\ y \end{pmatrix} (m) = \ldots$$

$$\ldots = \theta_{ZX} M_* \begin{pmatrix} yx \\ f(y)xx \end{pmatrix} M^* \begin{pmatrix} yx \\ y \end{pmatrix} (m) = N_{X,*}(f)\theta'_Y(m)$$

hence θ'_Y is covariant in Y.

Similarly

$$N_X^*(f)\theta'_Z = N^* \begin{pmatrix} yx \\ f(y)x \end{pmatrix} \theta_{ZX} M_* \begin{pmatrix} zx \\ zxx \end{pmatrix} M^* \begin{pmatrix} zx \\ z \end{pmatrix}$$

As θ is a morphism of Mackey functors

$$N^* \begin{pmatrix} yx \\ f(y)x \end{pmatrix} \theta_{ZX} = \theta_{YX} M^* \begin{pmatrix} yx_1x_2 \\ f(y)x_1x_2 \end{pmatrix}$$

But the square

$$\begin{array}{ccc} YX & \xrightarrow{\begin{pmatrix} yx \\ yxx \end{pmatrix}} & YX^2 \\ {\scriptstyle \begin{pmatrix} yx \\ f(y)x \end{pmatrix}} \downarrow & & \downarrow {\scriptstyle \begin{pmatrix} yx_1x_2 \\ f(y)x_1x_2 \end{pmatrix}} \\ ZX & \xrightarrow[\begin{pmatrix} zx \\ zxx \end{pmatrix}]{} & ZX^2 \end{array}$$

is cartesian, so $M^* \left(\begin{smallmatrix} yx_1x_2 \\ f(y)x_1x_2 \end{smallmatrix}\right) M_* \left(\begin{smallmatrix} zx \\ zxx \end{smallmatrix}\right) = M_* \left(\begin{smallmatrix} yx \\ yxx \end{smallmatrix}\right) M^* \left(\begin{smallmatrix} yx \\ f(y)x \end{smallmatrix}\right)$, and

$$N^* \begin{pmatrix} yx \\ f(y)x \end{pmatrix} \theta_{ZX} = \theta_{YX} M_* \begin{pmatrix} yx \\ yxx \end{pmatrix} M^* \begin{pmatrix} yx \\ f(y)x \end{pmatrix} M^* \begin{pmatrix} zx \\ z \end{pmatrix} = \ldots$$

$$\ldots = \theta_{YX} M_* \begin{pmatrix} yx \\ yxx \end{pmatrix} M^* \begin{pmatrix} yx \\ f(y) \end{pmatrix}$$

On the other hand

$$\theta'_Y M^*(f) = \theta_{YX} M_* \begin{pmatrix} yx \\ yxx \end{pmatrix} M^* \begin{pmatrix} yx \\ y \end{pmatrix} M^*(f) = \theta_{YX} M_* \begin{pmatrix} yx \\ yxx \end{pmatrix} M^* \begin{pmatrix} yx \\ f(y) \end{pmatrix}$$

and this proves that θ' is a morphism of Mackey functors.

Moreover, if U is a G-set, and if $a \in A(U)$, then

$$\theta'_{U\times Y}(a \times m) = \theta_{UYX} M_* \begin{pmatrix} uyx \\ uyxx \end{pmatrix} M^* \begin{pmatrix} uyx \\ uy \end{pmatrix} (a \times m)$$

As M is an A-module

$$M^* \begin{pmatrix} uyx \\ uy \end{pmatrix} (a \times m) = a \times M^* \begin{pmatrix} yx \\ y \end{pmatrix} (m)$$

and

$$M_* \begin{pmatrix} uyx \\ uyxx \end{pmatrix} \Big(a \times M^* \begin{pmatrix} yx \\ y \end{pmatrix} (m)\Big) = a \times M_* \begin{pmatrix} yx \\ yxx \end{pmatrix} M^* \begin{pmatrix} yx \\ y \end{pmatrix} (m)$$

Hence

$$\theta'_{U\times Y}(a \times m) = \theta_{UYX}\Big(a \times M_* \begin{pmatrix} yx \\ yxx \end{pmatrix} M^* \begin{pmatrix} yx \\ y \end{pmatrix} (m)\Big)$$

Finally as θ is a morphism of A-modules, I have

$$\theta_{UYX}\Big(a \times M_* \begin{pmatrix} yx \\ yxx \end{pmatrix} M^* \begin{pmatrix} yx \\ y \end{pmatrix} (m)\Big) = a \times \theta_{YX} M_* \begin{pmatrix} yx \\ yxx \end{pmatrix} M^* \begin{pmatrix} yx \\ y \end{pmatrix} (m) = a \times \theta'_Y(m)$$

and θ' is a morphism of A-modules.

Conversely, if θ' is a morphism of A-modules from M to N_X, I define for any G-set Y a map $\theta"_Y$ from $M(YX)$ to $N(Y)$, by setting for $m \in M(YX)$

$$\theta"_Y(m) = N_* \begin{pmatrix} yx \\ y \end{pmatrix} N^* \begin{pmatrix} yx \\ yxx \end{pmatrix} \theta'_{YX}(m)$$

A similar argument shows that $\theta"$ is a morphism of A-modules from M_X to N. Moreover, the constructions $\theta \mapsto \theta'$ and $\theta' \mapsto \theta$ are mutual inverse isomorphisms between $\mathrm{Hom}_{A\text{-}Mod}(M_X, N)$ and $\mathrm{Hom}_{A\text{-}Mod}(M, N_X)$: indeed, if θ is a morphism of A-modules from M_X to N, associated to a morphism θ' from M to N_X, associated in turn to $\theta"$, then for any G-set Y

$$\theta"_Y = N_* \begin{pmatrix} yx \\ y \end{pmatrix} N^* \begin{pmatrix} yx \\ yxx \end{pmatrix} \theta_{YX^2} M_* \begin{pmatrix} yx_1x_2 \\ yx_1x_2x_2 \end{pmatrix} M^* \begin{pmatrix} yx_1x_2 \\ yx_1 \end{pmatrix}$$

As θ is a morphism of Mackey functors

$$N^* \begin{pmatrix} yx \\ yxx \end{pmatrix} \theta_{YX^2} = \theta_{YX} M^* \begin{pmatrix} yx_1x_2 \\ yx_1x_1x_2 \end{pmatrix}$$

and for the same reason

$$N_* \begin{pmatrix} yx \\ y \end{pmatrix} \theta_{YX} = \theta_Y M_* \begin{pmatrix} yx_1x_2 \\ yx_2 \end{pmatrix}$$

and finally

$$\theta"_Y = \theta_Y M_*\begin{pmatrix} yx_1x_2 \\ yx_2 \end{pmatrix} M^*\begin{pmatrix} yx_1x_2 \\ yx_1x_1x_2 \end{pmatrix} M_*\begin{pmatrix} yx_1x_2 \\ yx_1x_2x_2 \end{pmatrix} M^*\begin{pmatrix} yx_1x_2 \\ yx_1 \end{pmatrix}$$

But the square

$$\begin{array}{ccc} YX & \xrightarrow{\begin{pmatrix} yx \\ yxx \end{pmatrix}} & YX^2 \\ {\scriptstyle\begin{pmatrix} yx \\ yxx \end{pmatrix}}\Big\downarrow & & \Big\downarrow{\scriptstyle\begin{pmatrix} yx_1x_2 \\ yx_1x_1x_2 \end{pmatrix}} \\ YX^2 & \xrightarrow[\begin{pmatrix} yx_1x_2 \\ yx_1x_2x_2 \end{pmatrix}]{} & YX^3 \end{array}$$

is cartesian, so

$$M^*\begin{pmatrix} yx_1x_2 \\ yx_1x_1x_2 \end{pmatrix} M_*\begin{pmatrix} yx_1x_2 \\ yx_1x_2x_2 \end{pmatrix} = M_*\begin{pmatrix} yx \\ yxx \end{pmatrix} M^*\begin{pmatrix} yx \\ yxx \end{pmatrix}$$

and it follows that

$$\theta"_Y = \theta_Y M_*\begin{pmatrix} yx_1x_2 \\ yx_2 \end{pmatrix} M_*\begin{pmatrix} yx \\ yxx \end{pmatrix} M^*\begin{pmatrix} yx \\ yxx \end{pmatrix} M^*\begin{pmatrix} yx_1x_2 \\ yx_1 \end{pmatrix} = \theta_Y$$

since $\begin{pmatrix} yx_1x_2 \\ yx_2 \end{pmatrix}\begin{pmatrix} yx \\ yxx \end{pmatrix}$ and $\begin{pmatrix} yx_1x_2 \\ yx_1 \end{pmatrix}\begin{pmatrix} yx \\ yxx \end{pmatrix}$ are both equal to identity.

A similar argument shows that if I start with a morphism θ' from M to N_X, associated to the morphism $\theta"$ from M_X to N, then θ' is the morphism associated to $\theta"$ by the first construction.

Now the lemma follows, because the previous constructions are clearly functorial in M and N. ■

Lemma 3.1.2: Let A be a Green functor, and M be an A-module. Then

$$\mathrm{Hom}_{A-\mathcal{M}od}(A, M) \simeq M(\bullet)$$

Proof: Let f be a morphism of A-modules from A to M. Then, for any G-set X, the square

$$\begin{array}{ccc} A(X)\times A(\bullet) & \xrightarrow{\times} & A(X) \\ {\scriptstyle Id\times f_\bullet}\Big\downarrow & & \Big\downarrow{\scriptstyle f_X} \\ A(X)\times M(\bullet) & \xrightarrow[\times]{} & M(X) \end{array}$$

has to be commutative. In particular, for all $a \in A(X)$, I must have

$$f_X(a) = f_X(a \times \varepsilon) = a \times f_\bullet(\varepsilon)$$

so f is determined by the element $f_\bullet(\varepsilon) \in M(\bullet)$. Conversely, if $m \in M(\bullet)$, it is clear that the equation

$$f_X(a) = a \times m$$

defines a morphism f a morphism from A to M. ■

Remark: I have already proved this lemma in the case $A = b$, by a slightly different method.

Proposition 3.1.3: Let A be a Green functor.

- **If M is an A-module, and X is a G-set, then**

$$\mathrm{Hom}_{A\text{-}Mod}(A_X, M) \simeq M(X)$$

- **In particular, if X and Y are G-sets, then**

$$\mathrm{Hom}_{A\text{-}Mod}(A_X, A_Y) \simeq A(X \times Y)$$

Proof: The second assertion is a consequence of the first one, and of the fact that $A_Y(X) = A(X \times Y)$. And the first assertion follows from the previous lemmas, since

$$\mathrm{Hom}_{A\text{-}Mod}(A_X, M) \simeq \mathrm{Hom}_{A\text{-}Mod}(A, M_X) \simeq M_X(\bullet) \simeq M(X)$$

3.2 The category C_A

Let $f \to \alpha(f)$ be the isomorphism of the previous proposition. Translating through f the composition product

$$\mathrm{Hom}_{A\text{-}Mod}(A_Y, A_Z) \times \mathrm{Hom}_{A\text{-}Mod}(A_X, A_Y) \to \mathrm{Hom}_{A\text{-}Mod}(A_X, A_Z)$$

leads to the product $\circ_Y$

$$\circ_Y : A(X \times Y) \times A(Y \times Z) \to A(X \times Z)$$

defined by setting, for $a \in A(X \times Y)$ and $a' \in A(Y \times Z)$

$$a \circ_Y a' = \alpha\Big(\alpha^{-1}(a') \circ \alpha^{-1}(a)\Big)$$

With this definition, I have

Lemma 3.2.1: Let $a \in A(X \times Y)$ and $a' \in A(Y \times Z)$. Then

$$a \circ_Y a' = A_* \begin{pmatrix} xyz \\ xz \end{pmatrix} A^* \begin{pmatrix} xyz \\ xyyz \end{pmatrix} (a \times a')$$

Proof: Let $a \in A(X \times Y)$. Then a determines a morphism h of A-modules from A to A_{XY}, such that if U is a G-set and $b \in A(U)$, then

$$h_U(b) = b \times a \in A(UXY) = A_{XY}(U)$$

I need to state precisely the homomorphism $f = \alpha^{-1}(h)$ from A_X to A_Y. It is determined by its evaluations on G-sets, and the previous lemmas show that for a G-set U and $b \in A_X(U) = A(U \times X)$, I have

$$f_U(b) = A_{Y,*} \begin{pmatrix} ux \\ u \end{pmatrix} A_Y^* \begin{pmatrix} ux \\ uxx \end{pmatrix} h_{UX}(b)$$

Now it follows that

$$f_U(b) = A_*\begin{pmatrix} uxy \\ uy \end{pmatrix} A^*\begin{pmatrix} uxy \\ uxxy \end{pmatrix}(b \times a)$$

Similarly, any $a' \in A(Y \times Z)$ is associated to a morphism f' from A_Y to A_Z defined for $b' \in A(U \times Y) = A_Y(U)$ by

$$f'_U(b') = A_*\begin{pmatrix} uyz \\ uz \end{pmatrix} A^*\begin{pmatrix} uyz \\ uyyz \end{pmatrix}(b' \times a')$$

The composite $f' \circ f$ is such that

$$(f' \circ f)_U(b) = A_*\begin{pmatrix} uyz \\ uz \end{pmatrix} A^*\begin{pmatrix} uyz \\ uyyz \end{pmatrix}\left(A_*\begin{pmatrix} uxy \\ uy \end{pmatrix} A^*\begin{pmatrix} uxy \\ uxxy \end{pmatrix}(b \times a) \times a'\right)$$

which may also be written, since A is a Green functor

$$(f' \circ f)_U(b) = A_*\begin{pmatrix} uyz \\ uz \end{pmatrix} A^*\begin{pmatrix} uyz \\ uyyz \end{pmatrix} A_*\begin{pmatrix} uxy_1y_2z \\ uy_1y_2z \end{pmatrix} A^*\begin{pmatrix} uxy_1y_2z \\ uxxy_1y_2z \end{pmatrix}(b \times a \times a')$$

But the square

$$\begin{array}{ccc} UXYZ & \xrightarrow{\begin{pmatrix} uxyz \\ uxyyz \end{pmatrix}} & UXY^2Z \\ {\scriptstyle\begin{pmatrix} uxyz \\ uyz \end{pmatrix}}\downarrow & & \downarrow{\scriptstyle\begin{pmatrix} uxy_1y_2z \\ uy_1y_2z \end{pmatrix}} \\ UYZ & \xrightarrow[\begin{pmatrix} uyz \\ uyyz \end{pmatrix}]{} & UY^2Z \end{array}$$

is cartesian, and

$$A^*\begin{pmatrix} uyz \\ uyyz \end{pmatrix} A_*\begin{pmatrix} uxy_1y_2z \\ uy_1y_2z \end{pmatrix} = A_*\begin{pmatrix} uxyz \\ uyz \end{pmatrix} A^*\begin{pmatrix} uxyz \\ uxyyz \end{pmatrix}$$

Finally

$$(f' \circ f)_U(b) = A_*\begin{pmatrix} uyz \\ uz \end{pmatrix} A_*\begin{pmatrix} uxyz \\ uyz \end{pmatrix} A^*\begin{pmatrix} uxyz \\ uxyyz \end{pmatrix} A^*\begin{pmatrix} uxy_1y_2z \\ uxxy_1y_2z \end{pmatrix}(b \times a \times a') = \ldots$$

$$\ldots = A_*\begin{pmatrix} uxyz \\ uz \end{pmatrix} A^*\begin{pmatrix} uxyz \\ uxxyyz \end{pmatrix}(b \times a \times a')$$

The morphism $f" = f' \circ f$ from A_X to A_Z is then associated to the morphism $h" = \alpha(f' \circ f)$ from A to A_{XZ} defined for $c \in A(U)$ by

$$h"_U(c) = f"_{UX} A_*\begin{pmatrix} ux \\ uxx \end{pmatrix} A^*\begin{pmatrix} ux \\ u \end{pmatrix}(c)$$

and $h"$ is associated to the element $a" = h"_\bullet(\varepsilon)$ of $A(XZ)$. But

$$h"_\bullet(\varepsilon) = f"_X A_*\begin{pmatrix} x \\ xx \end{pmatrix} A^*\begin{pmatrix} x \\ \bullet \end{pmatrix}(\varepsilon)$$

Finally

$$a" = A_*\begin{pmatrix} x_1x_2yz \\ x_1z \end{pmatrix} A^*\begin{pmatrix} x_1x_2yz \\ x_1x_2x_2yyz \end{pmatrix}\left(A_*\begin{pmatrix} x \\ xx \end{pmatrix} A^*\begin{pmatrix} x \\ \bullet \end{pmatrix}(\varepsilon)\times a\times a'\right)$$

As A is a Green functor, I have

$$A_*\begin{pmatrix} x \\ xx \end{pmatrix} A^*\begin{pmatrix} x \\ \bullet \end{pmatrix}(\varepsilon)\times a\times a' = A_*\begin{pmatrix} x_1x_2y_1y_2z \\ x_1x_1x_2y_1y_2z \end{pmatrix} A^*\begin{pmatrix} x_1x_2y_1y_2z \\ x_2y_1y_2z \end{pmatrix}(a\times a')$$

But the square

$$\begin{array}{ccc} XYZ & \xrightarrow{\begin{pmatrix} xyz \\ xxyyz \end{pmatrix}} & X^2Y^2Z \\ {\scriptstyle\begin{pmatrix} xyz \\ xxyz \end{pmatrix}}\Big\downarrow & & \Big\downarrow{\scriptstyle\begin{pmatrix} x_1x_2y_1y_2z \\ x_1x_1x_2y_1y_2z \end{pmatrix}} \\ X^2YZ & \xrightarrow[\begin{pmatrix} x_1x_2yz \\ x_1x_2x_2yyz \end{pmatrix}]{} & X^3Y^2Z \end{array}$$

is commutative. So

$$A^*\begin{pmatrix} x_1x_2yz \\ x_1x_2x_2yyz \end{pmatrix} A_*\begin{pmatrix} x_1x_2y_1y_2z \\ x_1x_1x_2y_1y_2z \end{pmatrix} = A_*\begin{pmatrix} xyz \\ xxyz \end{pmatrix} A^*\begin{pmatrix} xyz \\ xxyyz \end{pmatrix}$$

It follows that

$$a" = A_*\begin{pmatrix} x_1x_2yz \\ x_1z \end{pmatrix} A_*\begin{pmatrix} xyz \\ xxyz \end{pmatrix} A^*\begin{pmatrix} xyz \\ xxyyz \end{pmatrix} A^*\begin{pmatrix} x_1x_2y_1y_2z \\ x_2y_1y_2z \end{pmatrix}(a\times a') = \ldots$$

$$\ldots = A_*\begin{pmatrix} xyz \\ xz \end{pmatrix} A^*\begin{pmatrix} xyz \\ xyyz \end{pmatrix}(a\times a')$$

which proves the lemma. ∎

The previous lemma leads to the following definition:

Definition: *Let A be a Green functor for the group G. I denote by C_A the category defined as follows:*

- *The objects of C_A are the finite G-sets.*
- *If X and Y are G-sets, then*

$$\mathrm{Hom}_{C_A}(X,Y) = A(Y\times X)$$

- *If X, Y, and Z are G-sets, if $a\in A(Y\times X) = \mathrm{Hom}_{C_A}(X,Y)$ and if $a'\in A(Z\times Y) = \mathrm{Hom}_{C_A}(Y,Z)$, then the composite morphism $a'\circ a\in \mathrm{Hom}_{C_A}(X,Z)$ is defined by*

$$a'\circ a = a'\circ_Y a$$

This definition is made on purpose so that the following proposition holds:

Proposition 3.2.2: The category $\mathcal{C}_A$ is an R-additive category, and the correspondence which maps a G-set X to the A-module A_X is a fully faithful contravariant functor from $\mathcal{C}_A$ to A-Mod.

Proof: First the category $\mathcal{C}_A$ is a pre-additive category (or "*Ab*-category" in the sense of Mac Lane [10]), in that for any objects X and Y of $\mathcal{C}_A$, the set $\mathrm{Hom}_{\mathcal{C}_A}(X,Y) = A(Y \times X)$ has a natural structure of abelian group, for which the composition product is biadditive, and even R-bilinear. Moreover, the category $\mathcal{C}_A$ has a zero object, namely the empty set: for any G-set X, the product $X \times \emptyset$ is empty, and as $A(\emptyset) = \{0\}$, I have

$$\mathrm{Hom}_{\mathcal{C}_A}(\emptyset, X) = \mathrm{Hom}_{\mathcal{C}_A}(X, \emptyset) = \{0\}$$

The existence of biproducts in $\mathcal{C}_A$ (in the sense of Mac Lane [10] VIII.2) will then follow from the axioms of Mackey functors, and from the following lemma:

Lemma 3.2.3: Let $f : X \to Y$ be a morphism of G-sets. I set

$$f_* = A_*\begin{pmatrix} x \\ f(x)x \end{pmatrix} A^*\begin{pmatrix} x \\ \bullet \end{pmatrix}(\varepsilon) \in A(Y \times X)$$

$$f^* = A_*\begin{pmatrix} x \\ xf(x) \end{pmatrix} A^*\begin{pmatrix} x \\ \bullet \end{pmatrix}(\varepsilon) \in A(X \times Y)$$

I denote by I_* (resp. I^*) the correspondence which maps a G-set X to the object X of $\mathcal{C}_A$, and the morphism $f : X \to Y$ in G-set to $f_* \in \mathrm{Hom}_{\mathcal{C}_A}(X,Y)$ (resp. to $f^* \in \mathrm{Hom}_{\mathcal{C}_A}(Y,X)$). Then I_* (resp. I^*) is a covariant (resp. contravariant) functor from G-set to $\mathcal{C}_A$.

Admitting this lemma for a while, I see indeed that if i_X and i_Y are the respective injections from X and Y to $Z = X \coprod Y$, then in $\mathcal{C}_A$, I have

$$i_X^* \circ i_{X,*} = i_X^* \circ_Z i_{X,*} = A_*\begin{pmatrix} x_1zx_2 \\ x_1x_2 \end{pmatrix} A^*\begin{pmatrix} x_1zx_2 \\ x_1zzx_2 \end{pmatrix}(i_X^* \times i_{X,*})$$

Moreover

$$i_X^* \times i_{X,*} = A_*\begin{pmatrix} x \\ xi_X(x) \end{pmatrix} A^*\begin{pmatrix} x \\ \bullet \end{pmatrix}(\varepsilon) \times A_*\begin{pmatrix} x \\ i_X(x)x \end{pmatrix} A^*\begin{pmatrix} x \\ \bullet \end{pmatrix}(\varepsilon) = \ldots$$

$$\ldots = A_*\begin{pmatrix} x_1x_2 \\ x_1i_X(x_1)i_X(x_2)x_2 \end{pmatrix} A^*\begin{pmatrix} x_1x_2 \\ \bullet \end{pmatrix}(\varepsilon)$$

But as i_X is injective, the square

$$\begin{array}{ccc} X & \xrightarrow{\begin{pmatrix} x \\ xx \end{pmatrix}} & X^2 \\ {\scriptstyle\begin{pmatrix} x \\ xi_X(x)x \end{pmatrix}}\downarrow & & \downarrow{\scriptstyle\begin{pmatrix} x_1x_2 \\ x_1i_X(x_1)i_X(x_2)x_2 \end{pmatrix}} \\ XZX & \xrightarrow[\begin{pmatrix} x_1zx_2 \\ x_1zzx_2 \end{pmatrix}]{} & XZ^2X \end{array}$$

is cartesian, and then

$$A^*\begin{pmatrix} x_1zx_2 \\ x_1zzx_2 \end{pmatrix} A_*\begin{pmatrix} x_1x_2 \\ x_1i_X(x_1)i_X(x_2)x_2 \end{pmatrix} = A_*\begin{pmatrix} x \\ xi_X(x)x \end{pmatrix} A^*\begin{pmatrix} x \\ xx \end{pmatrix}$$

so that

$$i_X^*\circ i_{X,*} = A_*\begin{pmatrix} x_1zx_2 \\ x_1x_2 \end{pmatrix} A_*\begin{pmatrix} x \\ xi_X(x)x \end{pmatrix} A^*\begin{pmatrix} x \\ xx \end{pmatrix} A^*\begin{pmatrix} x_1x_2 \\ \bullet \end{pmatrix}(\varepsilon) = A_*\begin{pmatrix} x \\ xx \end{pmatrix} A^*\begin{pmatrix} x \\ \bullet \end{pmatrix}(\varepsilon)$$

and this is the unit $1_{A(X^2)} = Id_* = Id^*$ of $\mathrm{End}_{C_A}(X)$.

Similarly

$$i_{X,*}\circ i_X^* = i_{X,*}\circ_X i_X^* = A_*\begin{pmatrix} z_1xz_2 \\ z_1z_2 \end{pmatrix} A^*\begin{pmatrix} z_1xz_2 \\ z_1xxz_2 \end{pmatrix}(i_{X,*}\times i_X^*)$$

Moreover

$$i_{X,*}\times i_X^* = A_*\begin{pmatrix} x \\ i_X(x)x \end{pmatrix} A^*\begin{pmatrix} x \\ \bullet \end{pmatrix}(\varepsilon)\times A_*\begin{pmatrix} x \\ xi_X(x) \end{pmatrix} A^*\begin{pmatrix} x \\ \bullet \end{pmatrix}(\varepsilon) = \ldots$$

$$\ldots = A_*\begin{pmatrix} x_1x_2 \\ i_X(x_1)x_1x_2i_X(x_2) \end{pmatrix} A^*\begin{pmatrix} x_1x_2 \\ \bullet \end{pmatrix}(\varepsilon)$$

As the square

$$\begin{array}{ccc} X & \xrightarrow{\begin{pmatrix} x \\ xx \end{pmatrix}} & X^2 \\ {\scriptstyle\begin{pmatrix} x \\ i_X(x)xi_X(x) \end{pmatrix}}\downarrow & & \downarrow{\scriptstyle\begin{pmatrix} x_1x_2 \\ i_X(x_1)x_1x_2i_X(x_2) \end{pmatrix}} \\ ZXZ & \xrightarrow[\begin{pmatrix} z_1xz_2 \\ z_1xxz_2 \end{pmatrix}]{} & ZX^2Z \end{array}$$

is cartesian, I have also

$$A^*\begin{pmatrix} z_1xz_2 \\ z_1xxz_2 \end{pmatrix} A_*\begin{pmatrix} x_1x_2 \\ i_X(x_1)x_1x_2i_X(x_2) \end{pmatrix} = A_*\begin{pmatrix} x \\ i_X(x)xi_X(x) \end{pmatrix} A^*\begin{pmatrix} x \\ xx \end{pmatrix}$$

and it follows that

$$i_{X,*}\circ i_X^* = A_*\begin{pmatrix} z_1xz_2 \\ z_1z_2 \end{pmatrix} A_*\begin{pmatrix} x \\ i_X(x)xi_X(x) \end{pmatrix} A^*\begin{pmatrix} x \\ xx \end{pmatrix} A^*\begin{pmatrix} x_1x_2 \\ \bullet \end{pmatrix}(\varepsilon) = \ldots$$

$$\ldots = A_*\begin{pmatrix} x \\ i_X(x)i_X(x) \end{pmatrix} A^*\begin{pmatrix} x \\ \bullet \end{pmatrix}(\varepsilon) = A_*\begin{pmatrix} z \\ zz \end{pmatrix} A_*\begin{pmatrix} x \\ i_X(x) \end{pmatrix} A^*\begin{pmatrix} x \\ \bullet \end{pmatrix}(\varepsilon)$$

The same computation shows that

$$i_{Y,*}\circ i_Y^* = A_*\begin{pmatrix} z \\ zz \end{pmatrix} A_*\begin{pmatrix} y \\ i_Y(y) \end{pmatrix} A^*\begin{pmatrix} y \\ \bullet \end{pmatrix}(\varepsilon)$$

But as A is a Mackey functor, I have

$$A^*\begin{pmatrix} z \\ \bullet \end{pmatrix} = A_*(i_X)A^*(i_X)A^*\begin{pmatrix} z \\ \bullet \end{pmatrix} + A_*(i_Y)A^*(i_Y)A^*\begin{pmatrix} z \\ \bullet \end{pmatrix}$$

Since $A^*(i_X)A^*\binom{z}{\bullet} = A^*\binom{x}{\bullet}$ and $A^*(i_Y)A^*\binom{z}{\bullet} = A^*\binom{y}{\bullet}$, and since $i_X = \binom{x}{i_X(x)}$, I finally have

$$i_{X,*}\circ i_X^* + i_{Y,*}\circ i_Y^* = A_*\binom{z}{zz}A^*\binom{z}{\bullet}(\varepsilon) = 1_{A(Z^2)}$$

which proves that the disjoint union of G-sets is a biproduct (or direct sum) in $\mathcal{C}_A$. ∎

Proof of lemma 3.2.3: Let X, Y and Z be G-sets, let f be a morphism of G-sets from X to Y, and let g be a morphism from Y to Z. By definition of f_*, g_* and their product in $\mathcal{C}_A$, I have

$$g_*\circ f_* = A_*\binom{zyx}{zx}A^*\binom{zyx}{zyyx}\left[A_*\binom{y}{g(y)y}A^*\binom{y}{\bullet}(\varepsilon)\times A_*\binom{x}{f(x)x}A^*\binom{x}{\bullet}(\varepsilon)\right]$$

As A is a Green functor, the expression inside hooks can also be written

$$A_*\binom{yx}{g(y)yf(x)x}A^*\binom{yx}{\bullet}(\varepsilon)$$

Moreover, since the square

$$\begin{array}{ccc}
X & \xrightarrow{\binom{x}{f(x)x}} & YX \\
{\scriptstyle\binom{x}{gf(x)f(x)x}}\big\downarrow & & \big\downarrow{\scriptstyle\binom{yx}{g(y)yf(x)x}} \\
ZYX & \xrightarrow[\binom{zyx}{zyyx}]{} & ZY^2X
\end{array}$$

is cartesian, it follows that

$$g_*\circ f_* = A_*\binom{zyx}{zx}A_*\binom{x}{gf(x)f(x)x}A^*\binom{x}{f(x)x}A^*\binom{yx}{\bullet}(\varepsilon) = \ldots$$

$$\ldots = A_*\binom{x}{gf(x)x}A^*\binom{x}{\bullet}(\varepsilon) = (g\circ f)_*$$

To prove that I_* is a covariant functor, is suffices then to observe that the identity morphism of A_X is mapped to the morphism $h = \alpha(Id)$ from A to A_{X^2} defined by

$$h_U = Id_{UX}A_*\binom{ux}{uxx}A^*\binom{ux}{u} = A_*\binom{ux}{uxx}A^*\binom{ux}{u}$$

And the morphism h corresponds in turn to the element $h_\bullet(\varepsilon)$ of $A(X^2)$. But

$$h_\bullet(\varepsilon) = A_*\binom{x}{xx}A^*\binom{x}{\bullet}(\varepsilon) = 1_{A(X^2)}$$

proving that $1_{A(X^2)}$ is the identity morphism of X in $\mathcal{C}_A$.

Similarly

$$f^*\circ g^* = A_*\binom{xyz}{xz}A^*\binom{xyz}{xyyz}\left[A_*\binom{x}{xf(x)}A^*\binom{x}{\bullet}(\varepsilon)\times A_*\binom{y}{yg(y)}A^*\binom{y}{\bullet}(\varepsilon)\right] = \ldots$$

$$\ldots = A_*\begin{pmatrix} xyz \\ xz \end{pmatrix} A^*\begin{pmatrix} xyz \\ xyyz \end{pmatrix} A_*\begin{pmatrix} xy \\ xf(x)yg(y) \end{pmatrix} A^*\begin{pmatrix} xy \\ \bullet \end{pmatrix}(\varepsilon)$$

As the square

$$\begin{array}{ccc} X & \xrightarrow{\begin{pmatrix} x \\ xf(x) \end{pmatrix}} & XY \\ \begin{pmatrix} x \\ xf(x)gf(x) \end{pmatrix}\Big\downarrow & & \Big\downarrow\begin{pmatrix} xy \\ xf(x)yg(y) \end{pmatrix} \\ XYZ & \xrightarrow[\begin{pmatrix} xyz \\ xyyz \end{pmatrix}]{} & XY^2Z \end{array}$$

is cartesian, I have

$$f^* \circ g^* = A_*\begin{pmatrix} xyz \\ xz \end{pmatrix} A_*\begin{pmatrix} x \\ xf(x)gf(x) \end{pmatrix} A^*\begin{pmatrix} x \\ xf(x) \end{pmatrix} A^*\begin{pmatrix} xy \\ \bullet \end{pmatrix}(\varepsilon) = \ldots$$

$$\ldots = A_*\begin{pmatrix} x \\ xgf(x) \end{pmatrix} A^*\begin{pmatrix} x \\ \bullet \end{pmatrix}(\varepsilon) = (g \circ f)^*$$

and this proves that I^* is a contravariant functor, and completes the proof of the lemma. ∎

3.3 A-modules and representations of C_A

Definition: *If R is a commutative ring, and $\mathcal{C}$ is an R-additive category, a representation of $\mathcal{C}$ over R is an R-additive functor from $\mathcal{C}$ to R-**Mod**. If F and F' are two representations of $\mathcal{C}$, a morphism from F to F' is a morphism of functors from F to F'. I denote by $Funct_R(\mathcal{C})$ the category of representations of $\mathcal{C}$ over R.*

Let M be an A-module. By composition of the contravariant functor $\mathrm{Hom}_{A-Mod}(-,M)$ with the additive contravariant functor $X \mapsto A_X$ from C_A to A-**Mod**, I obtain a representation F_M of C_A over R, defined for a G-set X by

$$F_M(X) = \mathrm{Hom}_{C_A}(A_X, M)$$

But I have seen (see prop 3.1.3) that if M is an A-module, then $\mathrm{Hom}_A(A_X, M)$ identifies with $M(X)$: a morphism f from A_X to M is determined by the element

$$m = f_X A_*\begin{pmatrix} x \\ xx \end{pmatrix} A^*\begin{pmatrix} x \\ \bullet \end{pmatrix}(\varepsilon)$$

of $M(X)$: if U is a G-set, and if $a \in A_X(U) = A(UX)$, then

$$f_U(a) = M_*\begin{pmatrix} ux \\ u \end{pmatrix} M^*\begin{pmatrix} ux \\ uxx \end{pmatrix}(a \times m) \tag{3.1}$$

Then if X and Y are G-sets, and if $\phi \in A(Y \times X)$ is a morphism in C_A from X to Y, it is associated by proposition 3.1.3 to a morphism ψ from A_Y to A_X defined for $a \in A(U \times Y) = A_Y(U)$ by

$$\psi_U(a) = A_{X,*}\begin{pmatrix} uy \\ u \end{pmatrix} A_X^*\begin{pmatrix} uy \\ uyy \end{pmatrix}(a \times \phi) = A_*\begin{pmatrix} uyx \\ ux \end{pmatrix} A^*\begin{pmatrix} uyx \\ uyyx \end{pmatrix}(a \times \phi)$$

Let $m \in M(X)$. It is associated to a morphism from A_X to M defined by equation (3.1). By composition with ψ, I get a morphism $f\psi$ from A_Y to M, which is determined by the element

$$m' = (f\psi)_Y A_* \begin{pmatrix} y \\ yy \end{pmatrix} A^* \begin{pmatrix} y \\ \bullet \end{pmatrix} (\varepsilon)$$

and m' is also by definition the image of m under $F_M(\phi)$. As $\psi_Y A_* \binom{y}{yy} A^* \binom{y}{\bullet} (\varepsilon)$ is the element determining the morphism ϕ, equation (3.1) gives

$$m' = F_M(\phi)(m) = f_Y(\phi) = M_* \begin{pmatrix} yx \\ y \end{pmatrix} M^* \begin{pmatrix} yx \\ yxx \end{pmatrix} (\phi \times m)$$

This is the expression of the morphism $F_M(\phi) : M(X) \to M(Y)$, image of ϕ under F_M.

The analogy of those formulae with the definition of the composition product $\circ_X$ in $\mathcal{C}_A$ explains the following notation:

Notation: *Let A be a Green functor, and M be an A-module. If X and Y are G-sets, if $a \in A(Y \times X)$ and $m \in M(X)$, I set*

$$a \circ_X m = M_* \begin{pmatrix} yx \\ y \end{pmatrix} M^* \begin{pmatrix} yx \\ yxx \end{pmatrix} (a \times m)$$

With this notation, I see that

$$F_M(\phi)(m) = \phi \circ_X m$$

and now say that F_M is an R-additive functor is equivalent to say that this product $\circ_X$ is bilinear, associative and unitary (in an obvious sense).

Remark: The previous notation is coherent with the initial one: if M is the functor A_Z, for a G-set Z, then for $a \in A(YX)$ and $b \in M(X) = A(XZ)$ I find

$$a \circ_X b = A_{Z,*} \begin{pmatrix} yx \\ y \end{pmatrix} A_Z^* \begin{pmatrix} yx \\ yxx \end{pmatrix} (a \times b) = A_* \begin{pmatrix} yxz \\ yz \end{pmatrix} A^* \begin{pmatrix} yxz \\ yxxz \end{pmatrix} (a \times b)$$

and this coincides with the initial definition of the product $a \circ_X b$

Now to any A-module M, I have associated a representation F_M of $\mathcal{C}_A$. This clearly defines a functor from A-**Mod** to $Funct_R(\mathcal{C}_A)$: if $f : M \to N$ is a morphism of A-modules, composition with f induces a natural transformation from $\mathrm{Hom}_{A\text{-}Mod}(-, M)$ to $\mathrm{Hom}_{A\text{-}Mod}(-, N)$, and so a natural transformation from F_M to F_N. Conversely, let F be a representation of $\mathcal{C}_A$ over R. If X is a G-set, I set

$$M_F(X) = F(X)$$

It is an R-module. If $f : X \to Y$ is a morphism of G-sets, I set

$$M_{F,*}(f) = F(f_*) \qquad M_F^*(f) = F(f_*)$$

Those definitions turn M_F into a bifunctor over G-**set**, with values in R-**Mod**. Moreover as F is additive, it follows that M_F has property (M1) of Mackey functors, i.e. that M_F transforms disjoint unions into direct sums. The following lemma shows that M_F has also property (M2), hence that it is a Mackey functor:

Lemma 3.3.1: Let

$$\begin{array}{ccc} T & \xrightarrow{\gamma} & Y \\ {\scriptstyle\delta}\downarrow & & \downarrow{\scriptstyle\alpha} \\ Z & \xrightarrow[\beta]{} & X \end{array}$$

be a cartesian square of G-sets. Then in $\mathcal{C}_A$, I have

$$\beta^* \circ \alpha_* = \delta_* \circ \gamma^*$$

Proof: By definition

$$\beta^* \circ \alpha_* = \beta^* \circ_X \alpha_* = A_* \begin{pmatrix} zxy \\ zy \end{pmatrix} A^* \begin{pmatrix} zxy \\ zxxy \end{pmatrix} (\beta^* \times \alpha_*)$$

and moreover

$$\beta^* \times \alpha_* = A_* \begin{pmatrix} z \\ z\beta(z) \end{pmatrix} A^* \begin{pmatrix} z \\ \bullet \end{pmatrix} (\varepsilon) \times A_* \begin{pmatrix} y \\ \alpha(y)y \end{pmatrix} A^* \begin{pmatrix} y \\ \bullet \end{pmatrix} (\varepsilon) = \ldots$$

$$\ldots = A_* \begin{pmatrix} zy \\ z\beta(z)\alpha(y)y \end{pmatrix} A^* \begin{pmatrix} zy \\ \bullet \end{pmatrix} (\varepsilon)$$

By hypothesis, the square

$$\begin{array}{ccc} T & \xrightarrow{\gamma} & Y \\ {\scriptstyle\delta}\downarrow & & \downarrow{\scriptstyle\alpha} \\ Z & \xrightarrow[\beta]{} & X \end{array}$$

is cartesian, and so is the square

$$\begin{array}{ccc} T & \xrightarrow{\begin{pmatrix} t \\ \delta(t)\gamma(t) \end{pmatrix}} & ZY \\ {\scriptstyle\begin{pmatrix} t \\ \delta(t)\,\beta\delta(t)\,\gamma(t) \end{pmatrix}}\downarrow & & \downarrow{\scriptstyle\begin{pmatrix} zy \\ z\beta(z)\alpha(y)y \end{pmatrix}} \\ ZXY & \xrightarrow[\begin{pmatrix} zxy \\ zxxy \end{pmatrix}]{} & ZX^2Y \end{array}$$

which gives

$$A^* \begin{pmatrix} zxy \\ zxxy \end{pmatrix} A_* \begin{pmatrix} zy \\ z\beta(z)\alpha(y)y \end{pmatrix} = A_* \begin{pmatrix} t \\ \delta(t)\,\beta\delta(t)\,\gamma(t) \end{pmatrix} A^* \begin{pmatrix} t \\ \delta(t)\gamma(t) \end{pmatrix}$$

It follows that

$$\beta^* \circ \alpha_* = A_* \begin{pmatrix} zxy \\ zy \end{pmatrix} A_* \begin{pmatrix} t \\ \delta(t)\,\beta\delta(t)\,\gamma(t) \end{pmatrix} A^* \begin{pmatrix} t \\ \delta(t)\gamma(t) \end{pmatrix} A^* \begin{pmatrix} zy \\ \bullet \end{pmatrix} (\varepsilon) = \ldots$$

$$\ldots = A_*\begin{pmatrix} t \\ \delta(t)\gamma(t)\end{pmatrix} A^*\begin{pmatrix} t \\ \bullet\end{pmatrix}(\varepsilon)$$

On the other hand

$$\delta_* \circ \beta^* = \delta_* \circ_T \beta^* = A_*\begin{pmatrix} zty \\ zy\end{pmatrix} A^*\begin{pmatrix} zty \\ ztty\end{pmatrix}(\delta_* \times \gamma^*)$$

Moreover

$$\delta_* \times \gamma^* = A_*\begin{pmatrix} t \\ \delta(t)t\end{pmatrix} A^*\begin{pmatrix} t \\ \bullet\end{pmatrix}(\varepsilon) \times A_*\begin{pmatrix} t \\ t\gamma(t)\end{pmatrix} A^*\begin{pmatrix} t \\ \bullet\end{pmatrix}(\varepsilon) = \ldots$$

$$\ldots = A_*\begin{pmatrix} t_1t_2 \\ \delta(t_1)t_1t_2\gamma(t_2)\end{pmatrix} A^*\begin{pmatrix} t_1t_2 \\ \bullet\end{pmatrix}(\varepsilon)$$

But the square

$$\begin{array}{ccc} T & \xrightarrow{\begin{pmatrix} t \\ tt\end{pmatrix}} & T^2 \\ {\scriptstyle\begin{pmatrix} t \\ \delta(t)t\gamma(t)\end{pmatrix}}\Big\downarrow & & \Big\downarrow{\scriptstyle\begin{pmatrix} t_1t_2 \\ \delta(t_1)t_1t_2\delta(t_2)\end{pmatrix}} \\ ZTY & \xrightarrow[\begin{pmatrix} zty \\ ztty\end{pmatrix}]{} & ZT^2Y \end{array}$$

is cartesian. Hence

$$A^*\begin{pmatrix} zty \\ ztty\end{pmatrix} A_*\begin{pmatrix} t_1t_2 \\ \delta(t_1)t_1t_2\gamma(t_2)\end{pmatrix} = A_*\begin{pmatrix} t \\ \delta(t)t\gamma(t)\end{pmatrix} A^*\begin{pmatrix} t \\ tt\end{pmatrix}$$

and it follows that

$$\delta_* \circ \beta^* = A_*\begin{pmatrix} zty \\ zy\end{pmatrix} A_*\begin{pmatrix} t \\ \delta(t)t\gamma(t)\end{pmatrix} A^*\begin{pmatrix} t \\ tt\end{pmatrix} A^*\begin{pmatrix} t_1t_2 \\ \bullet\end{pmatrix}(\varepsilon) = A_*\begin{pmatrix} t \\ \delta(t)\gamma(t)\end{pmatrix} A^*\begin{pmatrix} t \\ \bullet\end{pmatrix}(\varepsilon)$$

which proves the lemma. ■

Finally, I can turn M_F into an A-module: is X and Y are G-sets, if $a \in A(X)$ and $m \in M_F(Y) = F(Y)$, then the element $A_*\begin{pmatrix} xy \\ xyy\end{pmatrix} A^*\begin{pmatrix} xy \\ x\end{pmatrix}(a)$ is in $A(XY^2) = \mathrm{Hom}_{\mathcal{C}_A}(Y, XY)$, and I can consider its image under F, which is a morphism from $F(Y)$ to $F(XY)$. Then I set

$$a \times m = F\left(A_*\begin{pmatrix} xy \\ xyy\end{pmatrix} A^*\begin{pmatrix} xy \\ x\end{pmatrix}(a)\right)(m) \in F(XY) = M_F(XY)$$

Lemma 3.3.2: Those definitions turn M_F into an A-module.

Proof: I must check that the product is bifunctorial, associative, and unitary. To simplify the proofs, I will first check that in the case $F = F_N$ for an A-module N, then M_F is equal to the A-module N.

It is clear indeed that for any G-set X, I have

$$M_{F_N}(X) = F_N(X) = N(X)$$

On the other hand, if $f : X \to Y$ is a morphism of G-sets, and if $n \in N(X)$, then

$$M_{F_N,*}(f)(n) = F_N(f_*)(n) = f_* \circ_X n = N_*\begin{pmatrix} yx \\ y \end{pmatrix} N^*\begin{pmatrix} yx \\ yxx \end{pmatrix}(f_* \times n)$$

But

$$f_* \times n = A_*\begin{pmatrix} x \\ f(x)x \end{pmatrix} A^*\begin{pmatrix} x \\ \bullet \end{pmatrix}(\varepsilon) \times n = N_*\begin{pmatrix} x_1x_2 \\ f(x_1)x_1x_2 \end{pmatrix} N^*\begin{pmatrix} x_1x_2 \\ x_2 \end{pmatrix}(n)$$

As the square

$$\begin{array}{ccc} X & \xrightarrow{\begin{pmatrix} x \\ xx \end{pmatrix}} & X^2 \\ {\scriptstyle\begin{pmatrix} x \\ f(x)x \end{pmatrix}}\Big\downarrow & & \Big\downarrow{\scriptstyle\begin{pmatrix} x_1x_2 \\ f(x_1)x_1x_2 \end{pmatrix}} \\ YX & \xrightarrow[\begin{pmatrix} yx \\ yxx \end{pmatrix}]{} & YX^2 \end{array}$$

is cartesian, I have

$$N^*\begin{pmatrix} yx \\ yxx \end{pmatrix} N_*\begin{pmatrix} x_1x_2 \\ f(x_1)x_1x_2 \end{pmatrix} = N_*\begin{pmatrix} x \\ f(x)x \end{pmatrix} N^*\begin{pmatrix} x \\ xx \end{pmatrix}$$

and so

$$M_{F_N,*}(f)(n) = N_*\begin{pmatrix} yx \\ y \end{pmatrix} N_*\begin{pmatrix} x \\ f(x)x \end{pmatrix} N^*\begin{pmatrix} x \\ xx \end{pmatrix} N^*\begin{pmatrix} x_1x_2 \\ x_2 \end{pmatrix}(n) = \ldots$$

$$\ldots = N_*\begin{pmatrix} x \\ f(x) \end{pmatrix} N^*\begin{pmatrix} x \\ x \end{pmatrix}(n) = N_*(f)(n)$$

If now $n \in N(Y)$, I have

$$M^*_{F_N}(f)(n) = F_N(f^*)(n) = f^* \circ_Y n = N_*\begin{pmatrix} xy \\ x \end{pmatrix} N^*\begin{pmatrix} xy \\ xyy \end{pmatrix}(f^* \times n)$$

Moreover

$$f^* \times n = A_*\begin{pmatrix} x \\ xf(x) \end{pmatrix} A^*\begin{pmatrix} x \\ \bullet \end{pmatrix}(\varepsilon) \times n = N_*\begin{pmatrix} xy \\ xf(x)y \end{pmatrix} N^*\begin{pmatrix} xy \\ y \end{pmatrix}(n)$$

But the square

$$\begin{array}{ccc} X & \xrightarrow{\begin{pmatrix} x \\ xf(x) \end{pmatrix}} & XY \\ {\scriptstyle\begin{pmatrix} x \\ f(x) \end{pmatrix}}\Big\downarrow & & \Big\downarrow{\scriptstyle\begin{pmatrix} xy \\ xf(x)y \end{pmatrix}} \\ XY & \xrightarrow[\begin{pmatrix} xy \\ xyy \end{pmatrix}]{} & XY^2 \end{array}$$

is cartesian. It follows that

$$N^*\begin{pmatrix} xy \\ xyy \end{pmatrix} N_*\begin{pmatrix} xy \\ xf(x)y \end{pmatrix} = N_*\begin{pmatrix} x \\ xf(x) \end{pmatrix} N^*\begin{pmatrix} x \\ xf(x) \end{pmatrix}$$

and then

$$M^*_{F_N}(f)(n) = N_*\begin{pmatrix} xy \\ x \end{pmatrix} N_*\begin{pmatrix} x \\ xf(x) \end{pmatrix} N^*\begin{pmatrix} x \\ xf(x) \end{pmatrix} N^*\begin{pmatrix} xy \\ y \end{pmatrix}(n) = \ldots$$

$$\ldots = N_*\begin{pmatrix} x \\ x \end{pmatrix} N^*\begin{pmatrix} x \\ f(x) \end{pmatrix}(n) = N^*(f)(n)$$

Now for the product, if $a \in A(X)$ and $n \in M_{F_N}(Y) = N(Y)$, then

$$a \times n = F_N\left(A_*\begin{pmatrix} xy \\ xyy \end{pmatrix} A^*\begin{pmatrix} xy \\ x \end{pmatrix}(a)\right)(n) = A_*\begin{pmatrix} xy \\ xyy \end{pmatrix} A^*\begin{pmatrix} xy \\ x \end{pmatrix}(a) \circ_Y n = \ldots$$

$$\ldots = N_*\begin{pmatrix} xy_1y_2 \\ xy_1 \end{pmatrix} N^*\begin{pmatrix} xy_1y_2 \\ xy_1y_2y_2 \end{pmatrix}\left[A_*\begin{pmatrix} xy \\ xyy \end{pmatrix} A^*\begin{pmatrix} xy \\ x \end{pmatrix}(a) \times n\right]$$

The expression inside hooks is also equal to

$$N_*\begin{pmatrix} xy_1y_2 \\ xy_1y_1y_2 \end{pmatrix} N^*\begin{pmatrix} xy_1y_2 \\ xy_2 \end{pmatrix}(a \times n)$$

As the square

$$\begin{array}{ccc} XY & \xrightarrow{\begin{pmatrix} xy \\ xyy \end{pmatrix}} & XY^2 \\ {\scriptstyle\begin{pmatrix} xy \\ xyy \end{pmatrix}}\downarrow & & \downarrow{\scriptstyle\begin{pmatrix} xy_1y_2 \\ xy_1y_2y_2 \end{pmatrix}} \\ XY^2 & \xrightarrow[\begin{pmatrix} xy_1y_2 \\ xy_1y_1y_2 \end{pmatrix}]{} & XY^3 \end{array}$$

is cartesian, I have

$$N_*\begin{pmatrix} xy_1y_2 \\ xy_1y_1y_2 \end{pmatrix} N^*\begin{pmatrix} xy_1y_2 \\ xy_1y_2y_2 \end{pmatrix} = N_*\begin{pmatrix} xy \\ xyy \end{pmatrix} N^*\begin{pmatrix} xy \\ xyy \end{pmatrix}$$

Hence

$$a \times n = N_*\begin{pmatrix} xy_1y_2 \\ xy_1 \end{pmatrix} N_*\begin{pmatrix} xy \\ xyy \end{pmatrix} N^*\begin{pmatrix} xy \\ xyy \end{pmatrix} N^*\begin{pmatrix} xy_1y_2 \\ xy_2 \end{pmatrix}(a \times n) = \ldots$$

$$\ldots = N_*\begin{pmatrix} xy \\ xy \end{pmatrix} N^*\begin{pmatrix} xy \\ xy \end{pmatrix}(a \times n) = a \times n$$

So I recover the initial product for the A-module N. Let me record this result in the following lemma:

Lemma 3.3.3: Let A be a Green functor, and N be an A-module. If X and Y are G-sets, if $a \in A(X)$ and $n \in N(Y)$, then

$$a \times n = A_*\begin{pmatrix} xy \\ xyy \end{pmatrix} A^*\begin{pmatrix} xy \\ x \end{pmatrix}(a) \circ_Y n$$

I must now prove that the product I have defined on M_F is bifunctorial. Let $f : X \to X'$ and $g : Y \to Y'$ be morphisms of G-sets. If $a \in A(X)$ and $m \in M(Y)$, then by definition of M_F

$$A_*(f)(a) \times M_{F,*}(g)(m) = F\left(A_*\begin{pmatrix} x'y' \\ x'y'y' \end{pmatrix} A^*\begin{pmatrix} x'y' \\ x' \end{pmatrix} A_*(f)(a)\right)\left(F(g_*)(m)\right) = \dots$$

$$\dots = F\left(A_*\begin{pmatrix} x'y' \\ x'y'y' \end{pmatrix} A^*\begin{pmatrix} x'y' \\ x' \end{pmatrix} A_*(f)(a) \circ_{Y'} g_*\right)(m)$$

On the other hand

$$M_{F,*}(f \times g)(a \times m) = F\left((f \times g)_*\right) F\left(A_*\begin{pmatrix} xy \\ xyy \end{pmatrix} A^*\begin{pmatrix} xy \\ x \end{pmatrix}(a)\right)(m) = \dots$$

$$\dots = F\left((f \times g)_* \circ_{XY} A_*\begin{pmatrix} xy \\ xyy \end{pmatrix} A^*\begin{pmatrix} xy \\ x \end{pmatrix}(a)\right)(m)$$

So I have to show that if I set

$$\phi = A_*\begin{pmatrix} x'y' \\ x'y'y' \end{pmatrix} A^*\begin{pmatrix} x'y' \\ x' \end{pmatrix} A_*(f)(a) \circ_{Y'} g_*$$

$$\psi = (f \times g)_* \circ_{XY} A_*\begin{pmatrix} xy \\ xyy \end{pmatrix} A^*\begin{pmatrix} xy \\ x \end{pmatrix}(a)$$

then $F(\phi)(m) = F(\psi)(m)$ for any R-additive functor F on $\mathcal{C}_A$ and any $m \in F(Y)$. This is equivalent to say that for any F, I have $F(\phi) = F(\psi)$, and the following lemma proves that this is equivalent to say that $\phi = \psi$:

Lemma 3.3.4: Let X and Y be G-sets. Let $\phi \neq \psi$ be distinct elements of $A(YX) = \mathrm{Hom}_{\mathcal{C}_A}(X, Y)$. Let F be the functor F_{A_X}. Then $F(\phi) \neq F(\psi)$.

Proof: Let F be the functor F_{A_X}. Let $1_{A(X^2)} \in F(X)$. Then

$$F(\phi)(1_{A(X^2)}) = \phi \circ_X 1_{A(X^2)} = \phi \neq \psi = F(\psi)(1_{A(X^2)})$$

and so $F(\phi) \neq F(\psi)$. ■

Now to prove that M_F is covariant, I must prove that

$$\phi = A_*\begin{pmatrix} x'y' \\ x'y'y' \end{pmatrix} A^*\begin{pmatrix} x'y' \\ x' \end{pmatrix} A_*(f)(a) \circ_{Y'} g_* = \psi = (f \times g)_* \circ_{XY} A_*\begin{pmatrix} xy \\ xyy \end{pmatrix} A^*\begin{pmatrix} xy \\ x \end{pmatrix}(a)$$

but this equation is the case $F = F_{A_{Y'}}$. And in that case, I have seen that $M_F = A_{Y'}$, and the product $\times$ on this functor is covariant. It follows that the product $\times$ on M_F is covariant for any F.

Similarly, to prove that the product on M_F is contravariant, I consider in the same conditions elements $a' \in A(Y')$ and $m' \in M_F(Y') = F(Y')$. I have to show that

$$A^*(f)(a') \times M_F^*(g)(m') = M_F^*(f \times g)(a' \times m') \tag{3.2}$$

The left hand side is equal to

$$F\left(A_*\begin{pmatrix} xy \\ xyy \end{pmatrix} A^*\begin{pmatrix} xy \\ x \end{pmatrix} A^*(f)(a')\right) F(g^*)(m') = \ldots$$

$$\ldots = F\left(A_*\begin{pmatrix} xy \\ xyy \end{pmatrix} A^*\begin{pmatrix} xy \\ f(x) \end{pmatrix} (a') \circ_{Y'} g^*\right)(m')$$

whereas the right hand side is

$$F\big((f\times g)^*\big) F\left(A_*\begin{pmatrix} x'y' \\ x'y'y' \end{pmatrix} A^*\begin{pmatrix} x'y' \\ x' \end{pmatrix} (a')\right)(m') = \ldots$$

$$\ldots = F\left((f\times g)^* \circ_{X'Y'} A_*\begin{pmatrix} x'y' \\ x'y'y' \end{pmatrix} A^*\begin{pmatrix} x'y' \\ x' \end{pmatrix} (a')\right)(m')$$

The previous lemma now proves that equality (3.2) is a consequence of the following one

$$A_*\begin{pmatrix} xy \\ xyy \end{pmatrix} A^*\begin{pmatrix} xy \\ f(x) \end{pmatrix} (a') \circ_{Y'} g^* = (f\times g)^* \circ_{X'Y'} A_*\begin{pmatrix} x'y' \\ x'y'y' \end{pmatrix} A^*\begin{pmatrix} x'y' \\ x' \end{pmatrix} (a')$$

which is the case $F = F_{A_Y}$. As $M_{F_{A_Y}} = A_Y$, and as the product on A_Y is bifunctorial, this equality holds, and the product on M_F is contravariant.

Similarly, to check the associativity of the product on M_F, I consider three G-sets X, Y, Z, and elements $a \in A(X)$, $b \in A(Y)$, and $m \in M_F(Z) = F(Z)$. Then

$$a \times (b \times m) = F\left(A_*\begin{pmatrix} xyz \\ xyzyz \end{pmatrix} \begin{pmatrix} xyz \\ x \end{pmatrix} (a)\right)(b \times m) = \ldots$$

$$\ldots = F\left(A_*\begin{pmatrix} xyz \\ xyzyz \end{pmatrix} \begin{pmatrix} xyz \\ x \end{pmatrix} (a)\right) F\left(A_*\begin{pmatrix} yz \\ yzz \end{pmatrix} A^*\begin{pmatrix} yz \\ y \end{pmatrix} (b)\right)(m) = \ldots$$

$$\ldots = F\left(A_*\begin{pmatrix} xyz \\ xyzyz \end{pmatrix} \begin{pmatrix} xyz \\ x \end{pmatrix} (a) \circ_{YZ} A_*\begin{pmatrix} yz \\ yzz \end{pmatrix} A^*\begin{pmatrix} yz \\ y \end{pmatrix} (b)\right)(m)$$

On the other hand

$$(a \times b) \times m = F\left(A_*\begin{pmatrix} xyz \\ xyzz \end{pmatrix} A^*\begin{pmatrix} xyz \\ xy \end{pmatrix} (a \times b)\right)(m)$$

Now lemma 3.3.4 shows that associativity of the product on M_F for any F follows from equality

$$A_*\begin{pmatrix} xyz \\ xyzyz \end{pmatrix} \begin{pmatrix} xyz \\ x \end{pmatrix} (a) \circ_{YZ} A_*\begin{pmatrix} yz \\ yzz \end{pmatrix} A^*\begin{pmatrix} yz \\ y \end{pmatrix} (b) = A_*\begin{pmatrix} xyz \\ xyzz \end{pmatrix} A^*\begin{pmatrix} xyz \\ xy \end{pmatrix} (a \times b)$$

which expresses associativity of the product on $M_{F_{A_Z}} = A_Z$. So the product on M_F is associative.

Now concerning unit, if X is a G-set, and if $m \in M_F(X) = F(X)$, then

$$\varepsilon \times m = F\left(A_*\begin{pmatrix} x \\ xx \end{pmatrix} A^*\begin{pmatrix} x \\ \bullet \end{pmatrix}(\varepsilon)\right)(m) = F(1_{A(X^2)})(m) = m$$

and this completes the proof of lemma 3.3.2. ∎

The previous lemmas lead to the following theorem:

Theorem 3.3.5: The constructions which to an A-module M associate the representation F_M of $\mathcal{C}_A$, and to the representation F of $\mathcal{C}_A$ associate the A-module M_F are mutual inverse equivalences of categories from A-Mod to $Funct_R(\mathcal{C}_A)$.

Proof: I have already proved that for any A-module N, the A-module M_{F_N} identifies with N. I need to prove that the construction $E \mapsto M_E$ is functorial in E, and that for any representation E of $\mathcal{C}_A$, the representation F_{M_E} is naturally isomorphic to E.

If $\theta : E \to E'$ is a morphism of representations of $\mathcal{C}_A$, i.e. a natural transformation of functors, then θ is determined by giving, for any G-set X, a morphism of R-modules θ_X from $E(X)$ to $E'(X)$. If Y is a G-set, and if $a \in A(YX) = \mathrm{Hom}_{\mathcal{C}_A}(X, Y)$, then the square

$$\begin{array}{ccc} E(X) & \xrightarrow{E(a)} & E(Y) \\ \theta_X \downarrow & & \downarrow \theta_Y \\ E'(X) & \xrightarrow[E'(a)]{} & E'(Y) \end{array}$$

must be commutative.

By definition, I have $M_E(X) = E(X)$, and to prove that $E \mapsto M_E$ is functorial in E, I must check that the morphisms θ_X define a morphism of A-modules from M_E to $M_{E'}$. If $f : X \to Y$ is a morphism of G-sets, then the square

$$\begin{array}{ccc} M_E(X) & \xrightarrow{M_{E,*}(f) = E(f_*)} & M_E(Y) \\ \theta_X \downarrow & & \downarrow \theta_Y \\ M_{E'}(X) & \xrightarrow[M_{E',*}(f) = E'(f_*)]{} & M_{E'}(Y) \end{array}$$

is the above square for $a = f_*$. Similarly, if $f' : Y \to X$ is a morphism of G-sets, then the square

$$\begin{array}{ccc} M_E(X) & \xrightarrow{M_E^*(f') = E(f'^*)} & M_E(Y) \\ \theta_X \downarrow & & \downarrow \theta_Y \\ M_{E'}(X) & \xrightarrow[M_{E'}^*(f') = E'(f'^*)]{} & M_{E'}(Y) \end{array}$$

is the same square for $a = f'^*$. Hence the maps θ_X induce a morphism of Mackey functors from M_E to $M_{E'}$.
Moreover, for any G-set Y, for all $a \in A(Y)$ and all $m \in M_E(X)$, I have

$$a \times \theta_X(m) = E'\left(A_*\begin{pmatrix} yx \\ yxx \end{pmatrix} A^*\begin{pmatrix} yx \\ y \end{pmatrix}(a)\right)\theta_X(m) = \dots$$

$$\dots = \theta_{YX} E\left(A_*\begin{pmatrix} yx \\ yxx \end{pmatrix} A^*\begin{pmatrix} yx \\ y \end{pmatrix}(a)\right) = \theta_{YX}(a \times m)$$

so θ is a morphism of A-modules.

Finally, by definition, for any E and any X I have

$$F_{M_E}(X) = M_E(X) = E(X)$$

Moreover, if $a \in A(YX) = \mathrm{Hom}_{\mathcal{C}_A}(X, Y)$, and if $m \in E(X)$, then

$$F_{M_E}(a)(m) = M_{E,*}\begin{pmatrix} yx \\ y \end{pmatrix} M_E^*\begin{pmatrix} yx \\ yxx \end{pmatrix}(a \times m) = \dots$$

$$\dots = E\left(\begin{pmatrix} yx \\ y \end{pmatrix}_*\right) E\left(\begin{pmatrix} yx \\ yxx \end{pmatrix}^*\right) E\left[A_*\begin{pmatrix} yx_1x_2 \\ yx_1x_2x_2 \end{pmatrix} A^*\begin{pmatrix} yx_1x_2 \\ yx_1 \end{pmatrix}(a)\right](m) = \dots$$

$$\dots = E\left(\begin{pmatrix} yx \\ y \end{pmatrix}_* \circ_{YX} \begin{pmatrix} yx \\ yxx \end{pmatrix}^* \circ_{YX^2} A_*\begin{pmatrix} yx_1x_2 \\ yx_1x_2x_2 \end{pmatrix} A^*\begin{pmatrix} yx_1x_2 \\ yx_1 \end{pmatrix}(a)\right)(m)$$

I set

$$u = \begin{pmatrix} yx \\ y \end{pmatrix}_* = A_*\begin{pmatrix} yx \\ yyx \end{pmatrix} A^*\begin{pmatrix} yx \\ \bullet \end{pmatrix}(\varepsilon) \in A(Y^2X)$$

$$v = \begin{pmatrix} yx \\ yxx \end{pmatrix}^* = A_*\begin{pmatrix} yx \\ yxyxx \end{pmatrix} A^*\begin{pmatrix} yx \\ \bullet \end{pmatrix}(\varepsilon) \in A(YXYX^2)$$

and

$$w = A_*\begin{pmatrix} yx_1x_2 \\ yx_1x_2x_2 \end{pmatrix} A^*\begin{pmatrix} yx_1x_2 \\ yx_1 \end{pmatrix}(a) \in A(YX^3)$$

The elements u, v, and w do not depend on F, and moreover

$$F_{M_E}(a)(m) = E\big(u \circ_{YX} v \circ_{YX^2} w\big)(m) \tag{3.3}$$

If I take for E any functor F_N associated to an A-module N, then I know that $M_E \simeq N$, and the previous equality is

$$F_N(a)(m) = F_N\big(u \circ_{YX} v \circ_{YX^2} w\big)(m)$$

Then for any N, I have

$$F_N(a) = F_N\big(u \circ_{YX} v \circ_{YX^2} w\big)$$

and lemma 3.3.4 proves that

$$u \circ_{YX} v \circ_{YX^2} w = a$$

Changing equality (3.3) according to this proves that for any functor E

$$F_{M_E}(a)(m) = E(a)(m)$$

and the theorem follows. ∎

Chapter 4

The algebra associated to a Green functor

4.1 The evaluation functors

Theorem 3.3.5 states an equivalence between A-modules and representations of $\mathcal{C}_A$, so it is possible to use the general results on representations of categories (see [3]). In particular

Proposition 4.1.1: Let A be a Green functor. If X is a G-set, the functor E_X of evaluation at X (defined by $E_X(F) = F(X)$) from $Funct_R(\mathcal{C}_A)$ to $\mathrm{End}_{\mathcal{C}_A}(X)$-Mod has a left adjoint $L_{X,-} : V \mapsto L_{X,V}$ defined by

$$L_{X,V}(Y) = \mathrm{Hom}_{\mathcal{C}_A}(X,Y) \otimes_{\mathrm{End}_{\mathcal{C}_A}(X)} V = A(YX) \otimes_{A(X^2)} V$$

If $f \in \mathrm{Hom}_{\mathcal{C}_A}(Y,Z) = A(ZY)$, then the image under $L_{X,V}(f)$ of the element $a \otimes v$ of $L_{X,V}(Y)$ is equal to

$$L_{X,V}(f)(a \otimes v) = (f \circ_Y a) \otimes v \in L_{X,V}(Z)$$

It follows that $E_X \circ L_{X,-}$ is the identity functor on $A(X^2)$-Mod, which thereby may be identified to a full subcategory of A-Mod.

Remark: There are two different products on $A(X^2)$: one is the product ".", the other is the product $\circ_X$. The algebra structure on $A(X^2)$ in the proposition corresponds of course to the second one.

This adjunction provides in particular a co-unit η_X, which for any functor F on $\mathcal{C}_A$ induces a morphism $\eta_{X,F} : L_{X,F(X)} \to F$ defined for a G-set Y by

$$a \otimes v \in L_{X,F(X)}(Y) \mapsto \eta_{X,F,Y}(a \otimes v) = F(a)(v) \in F(Y)$$

This definition makes sense because if $a \in A(YX) = \mathrm{Hom}_{\mathcal{C}_A}(X,Y)$, then $F(a)$ is a morphism of R-modules from $F(X)$ to $F(Y)$, and so $F(a)(v) \in F(Y)$.

4.2 Evaluation and equivalence

Now a natural question is to know if there are G-sets Ω such that E_Ω is an equivalence of categories between $Funct_R(\mathcal{C}_A)$ and $A(\Omega^2)$-**Mod**. To answer that question, I need the following definitions:

Definitions: *1) Let Ω be a G-set. If n is a non-negative integer, I denote by $n\Omega$ the G-set obtained by taking the disjoint union of n copies of Ω, and I say that $n\Omega$ is a multiple of Ω.*
2) If X and Y are objects of $\mathcal{C}_A$, I say that X is a direct summand of Y in $\mathcal{C}_A$ (or that X divides Y in $\mathcal{C}_A$) if there exists a split monomorphism from X to Y in $\mathcal{C}_A$, that is if there exists an element $\alpha \in A(YX)$ and an element $\beta \in A(XY)$ such that $\beta \circ_Y \alpha = 1_{A(X^2)}$.

In particular, if i is an injective morphism from the G-set X into the G-set Y, then lemmas 3.2.3 and 3.3.1 show that $i^* \circ_Y i_* = 1_{A(X^2)}$. Indeed the square

$$\begin{array}{ccc} X & \xrightarrow{Id} & X \\ {\scriptstyle Id}\downarrow & & \downarrow{\scriptstyle i} \\ X & \xrightarrow[i]{} & Y \end{array}$$

is cartesian, and it follows that

$$i^* \circ_Y i_* = Id_* \circ_X Id^* = 1_{A(X^2)} \circ_X 1_{A(X^2)} = 1_{A(X^2)}$$

Thus injections in G-**set** are mapped by the functor I_* to split monomorphisms.

Proposition 4.2.1: Let A be a Green functor over R for the group G, and Ω be a G-set. The following conditions are equivalent:

1. **The evaluation functor E_Ω from $Funct_R(\mathcal{C}_A)$ to $A(\Omega^2)$-Mod is an equivalence of categories.**
2. **For any G-set X, there exists an integer n such that X is a direct summand of $n\Omega$ in $\mathcal{C}_A$.**

Proof: If E_Ω is an equivalence of categories, then the inverse equivalence is equal to its adjoint $L_{\Omega,-}$. I already know that $E_\Omega \circ L_{\Omega,-}$ is the identity functor. Now say that E_Ω is an equivalence of categories is equivalent to say that the co-unit η_Ω is an isomorphism. This means that for any functor F, the morphism $\eta_{\Omega,F}$ is an isomorphism. In particular, if X is a G-set, and if F is the functor F_{A_X} associated by theorem 3.3.5 to the A-module A_X, then the morphism

$$\eta_{\Omega,F_{A_X},X} : a \otimes v \in A(X\Omega) \otimes_{A(\Omega^2)} A_X(\Omega) \mapsto F_{A_X}(a)(v) \in F_{A_X}(X) = A_X(X)$$

is an isomorphism. But $A_X(X) = A(X^2)$ and

$$F_{A_X}(a)(v) = a \circ_\Omega v$$

so the morphism $\eta_{\Omega,A_X,X}$ is deduced from the composition

$$A(X\Omega)\otimes A(\Omega X)\rightarrow A(X^2)$$

In particular, this morphism must be surjective, so there exists an integer n, and elements $\alpha_i\in A(\Omega X)$ and $\beta_i\in A(X\Omega)$, for $1\leq i\leq n$, such that

$$\sum_{i=1}^{n}\beta_i\circ_\Omega\alpha_i=1_{A(X^2)} \tag{4.1}$$

Now the sum of α_i's (resp. the sum of β_i's) is an element α of $A\big((n\Omega)X\big)$ (resp. an element β of $A\big(X(n\Omega)\big)$), and equality (4.1) may be written

$$\beta\circ_{n\Omega}\alpha=1_{A(X^2)}$$

Thus X is a direct summand of $n\Omega$ in $\mathcal{C}_A$.

Conversely, if hypothesis 2 of the proposition holds, and if X is a G-set, then there exists an integer n and elements $\alpha\in A\big((n\Omega)X\big)$, and $\beta\in A\big(X(n\Omega)\big)$ such that $\beta\circ_{n\Omega}\alpha=1_{A(X^2)}$. So there exists elements $\alpha_i\in A(\Omega X)$ and $\beta_i\in A(X\Omega)$, for $1\leq i\leq n$, such that equality (4.1) holds. Then if F is a functor on $\mathcal{C}_A$, let $\Phi=\eta_{\Omega,F,X}$. I denote by Θ the morphism from $F(X)$ to $A(X\Omega)\otimes_{A(\Omega^2)}F(X)$ defined by

$$\Theta(u)=\sum_{i=1}^{n}\beta_i\otimes F(\alpha_i)(u)$$

With those notations, I have

$$\Theta\Phi(a\otimes v)=\sum_{i=1}^{n}\beta_i\otimes F(\alpha_i)F(a)(v)=\sum_{i=1}^{n}\beta_i\otimes F(\alpha_i\circ_X a)(v)$$

But $v\in F(\Omega)$, and the action of element $\lambda\in A(\Omega^2)$ on $F(\Omega)$ is defined by $\lambda.v=F(\lambda)(v)$. As $\alpha_i\circ_X a$ is in $A(\Omega^2)$, I have also

$$\Theta\Phi(a\otimes v)=\sum_{i=1}^{n}\beta_i\otimes(\alpha_i\circ_X a).v=\ldots$$

$$\ldots=\sum_{i=1}^{n}\Big(\beta_i\circ_\Omega(\alpha_i\circ_X a)\Big)\otimes v=\left(\sum_{i=1}^{n}(\beta_i\circ_\Omega\alpha_i)\circ_X a\right)\otimes v=a\otimes v$$

and this proves that $\Theta\Phi$ is the identity.
Conversely

$$\Phi\Theta(u)=\sum_{i=1}^{n}F(\beta_i)F(\alpha_i)(u)=\sum_{i=1}^{n}F(\beta_i\circ_\Omega\alpha_i)(u)=F\left(\sum_{i=1}^{n}\beta_i\circ_\Omega\alpha_i\right)(u)=u$$

Thus Θ and Φ are mutual inverse isomorphisms. As the G-set X was arbitrary, this proves that η_Ω is an isomorphism, and E_Ω is an equivalence of categories. This completes the proof of the proposition. ■

4.3 The algebra $A(\Omega^2)$

Condition 2 of proposition 4.2.1 holds in particular if I take for Ω the disjoint union of G/H, for all subgroups H of G: indeed, if X is a G-set, let E a system of representatives of orbits $G\backslash X$, viewed as a G-set with trivial G-action. Let i be the map from X to $\Omega\times E$ defined as follows: if $x\in X$, then there exists a unique x_0 in $G.x\cap E$, and an element $g\in G$ such that $x=g.x_0$. Then denoting by G_{x_0} the stabilizer of x_0, the class $gG_{x_0}\in\Omega$ does not depend on the choice of g, and I set $i(x)=(gG_{x_0},x_0)$. The map i is injective: if $i(x)=i(x')$, then $x_0=x'_0$ and $gG_{x_0}=g'G_{x'_0}$, so $\{x\}=gG_{x_0}.x_0=g'G_{x'_0}.x'_0=\{x'\}$. Moreover, it is a morphism of G-sets: it is clear that $i(h.x)=(hgG_{x_0},x_0)=h.i(x)$.
As G acts trivially on E, I see that $\Omega\times E\simeq|E|\Omega$, and then any G-set is a subset of a disjoint union of copies of Ω. Hence, in $\mathcal{C}_A$, any G-set X is a direct summand of a multiple of Ω.
Those remarks show that the evaluation functor E_Ω is an equivalence of categories from $Funct_R(\mathcal{C}_A)$ to $A(\Omega^2)$-**Mod**. If M is an A-module, theorem 3.3.5 gives $M(\Omega)$ an $A(\Omega^2)$-module structure, defined for $\alpha\in A(\Omega^2)$ and $m\in M(\Omega)$ by

$$\alpha.m=\alpha\circ_\Omega m=A_*\begin{pmatrix}\omega_1\omega_2\\ \omega_1\end{pmatrix}A^*\begin{pmatrix}\omega_1\omega_2\\ \omega_1\omega_2\omega_2\end{pmatrix}(\alpha\times m)$$

Finally, I have proved the first assertion of the following theorem:

Theorem 4.3.1: Let A be a Green functor for the group G. Let

$$\Omega=\coprod_{H\subseteq G}G/H$$

Then the functor which maps the A-module M to the $A(\Omega^2)$-module $M(\Omega)$ is an equivalence of categories from A-Mod to $A(\Omega^2)$-Mod.
The inverse equivalence maps the $A(\Omega^2)$-module V to the A-module L_V defined by

$$L_V(X)=A(X\Omega)\otimes_{A(\Omega^2)}V$$

If $f:X\to Y$ is a morphism of G-sets, if $a\in A(X\Omega)$, $b\in A(Y\Omega)$, then

$$L_{V,*}(f)(a\otimes v)=(f_*\circ_X a)\otimes v=A_*\begin{pmatrix}x\omega\\ f(x)\omega\end{pmatrix}(a)\otimes v$$

$$L_V^*(f)(b\otimes v)=(f^*\circ_X b)\otimes v=A^*\begin{pmatrix}x\omega\\ f(x)\omega\end{pmatrix}(b)\otimes v$$

Finally, if Z is a G-set, and if $c\in A(Z)$, then

$$c\times(a\otimes v)=(c\times a)\otimes v$$

Proof: To complete the proof, is suffices to keep track of identifications of theorem 3.3.5, which prove that

$$L_{V,*}(f)(a\times v)=(f_*\circ_X a)\otimes v\qquad L_V^*(f)(b\times v)=(f^*\circ_X b)\otimes v$$

But I have seen that for any A-module M and any $m \in M(X)$ I have

$$f_* \circ_X m = M_*(f)(m)$$

It suffices now to take $M = A_\Omega$ to conclude that

$$f_* \circ_X a = A_{\Omega,*}(f)(a) = A_* \begin{pmatrix} x\omega \\ f(x)\omega \end{pmatrix} (a)$$

and a similar argument proves the corresponding equality for f^*.
Finally, theorem 3.3.5 shows that

$$c \times (a \otimes v) = \left(A_* \begin{pmatrix} zx \\ zxx \end{pmatrix} A^* \begin{pmatrix} zx \\ z \end{pmatrix} (c) \circ_X a \right) \otimes v$$

and by lemma 3.3.3

$$A_* \begin{pmatrix} zx \\ zxx \end{pmatrix} A^* \begin{pmatrix} zx \\ z \end{pmatrix} (c) \circ_X a = c \times a$$

4.4 Presentation by generators and relations

Theorem 4.3.1 shows that the category A-**Mod** of modules over the Green functor A is equivalent to the category of modules over the algebra $A(\Omega^2)$. The following proposition, where $\varepsilon_K = A^* \binom{gK}{\bullet} (\varepsilon)$ denotes the unit of $A(K) = A(G/K)$, gives the structure of this algebra:

Proposition 4.4.1: Let A be a Green functor for the group G, and let $\mu(A)$ the algebra over R defined by the following generators and relations:

- **The generators of $\mu(A)$ are:**
  - **The elements t_K^H and r_K^H, for $K \subseteq H \subseteq G$.**
  - **The elements $c_{x,H}$ for $x \in G$ and $H \subseteq G$.**
  - **The elements $\lambda_{K,a}$ for $K \subseteq G$ and $a \in A(K)$.**
- **The relations of $\mu(A)$ are:**
  - **The relations of the Mackey algebra for r_K^H, t_K^H, and $c_{x,H}$, i.e.**

$$t_K^H t_L^K = t_L^H, \quad r_L^K r_K^H = r_L^H \quad \forall L \subseteq K \subseteq H \tag{1}$$

$$c_{y,{}^xH} c_{x,H} = c_{yx,H} \quad \forall x, y, H \tag{2}$$

$$t_H^H = r_H^H = c_{h,H} \quad \forall h \in H \tag{3}$$

$$c_{x,H} t_K^H = t_{{}^xK}^{{}^xH} c_{x,K}, \quad c_{x,K} r_K^H = r_{{}^xK}^{{}^xH} c_{x,H} \quad \forall x, K, H \tag{4}$$

$$\sum_H t_H^H = \sum_H r_H^H = 1 \tag{5}$$

$$r_K^H t_L^H = \sum_{x \in K\backslash H/L} t_{K\cap {}^xL}^K c_{x,K^x\cap L} r_{K^x\cap L}^L \quad \forall K \subseteq H \supseteq L \tag{6}$$

the other products of r_H^K, t_H^K and $c_{g,H}$ being zero.

– **The additional following relations:**

$$\lambda_{K,a}+\lambda_{K,a'}=\lambda_{K,a+a'},\quad \lambda_{K,a}\lambda_{K,a'}=\lambda_{K,aa'}\quad \forall a,a'\in A(K),\ \forall K\subseteq G \quad (7)$$

$$\lambda_{K,z\varepsilon_K}=zt_K^K\quad \forall K\subseteq G,\quad z\in R \quad (8)$$

$$r_K^H\lambda_{H,a}=\lambda_{K,r_K^H(a)}r_K^H\quad \forall a\in A(H),\ \forall K\subseteq H\subseteq G \quad (9)$$

$$\lambda_{H,a}t_K^H=t_K^H\lambda_{K,r_K^H(a)}\quad \forall a\in A(H),\ \forall K\subseteq H\subseteq G \quad (10)$$

$$t_K^H\lambda_{K,a}r_K^H=\lambda_{H,t_K^H(a)}\quad \forall a\in A(K),\ \forall K\subseteq H\subseteq G \quad (11)$$

$$\lambda_{{}^xH,c_{x,H}(a)}c_{x,H}=c_{x,H}\lambda_{H,a}\quad \forall x\in G,\ \forall a\in A(H),\ \forall H\subseteq G \quad (12)$$

Then the algebra $A(\Omega^2)$ is isomorphic to $\mu(A)$.

Proof: First it is clear that any non-zero product of generators of $\mu(A)$ involving only the generators r_K^H, t_K^H, and $c_{x,H}$ can be put in the form $t_V^U c_{g,V} r_{V^g}^W$, for suitable subgroups U, V and W of G and an element $g\in G$: relations (4) and (6) allow to put the generators t_K^H on the left and the generators r_K^H on the right, and relation (1) allows to reduce the products of t_K^H's and the products of r_K^H's. On the other hand, the product $r_K^H\lambda_{H',a}$ is zero if $H'\neq H$: indeed $r_K^H=r_K^Hr_H^H$ by relation (1) and

$$\lambda_{H',a}=\lambda_{H',\varepsilon_{H'}}\lambda_{H',a}=t_{H'}^{H'}\lambda_{H',a}$$

by relations (7) and (8), and $r_H^Ht_{H'}^{H'}=0$ if $H\neq H'$. Then

$$r_K^H\lambda_{H',a}=r_K^Hr_H^Ht_{H'}^{H'}\lambda_{H',a}=0$$

Now relation (9) allows to put the r_K^H's on the right of the $\lambda_{H,a}$'s. Similarly, the product $c_{x,H}\lambda_{H',a}$ is zero if $H'\neq H$, since $c_{x,H}=c_{x,H}c_{1,H}=c_{x,H}t_H^H$ by relations (2) and (3). Now relation (12) allows to put the $c_{x,H}$'s on the right of the $\lambda_{H,a}$'s. Hence any non-zero product of generators of $\mu(A)$ obtained by multiplying a product of generators r_K^H, t_K^H, and $c_{x,H}$ on the right by a generator $\lambda_{H,a}$ can be put in the following form

$$\beta=t_V^U\lambda_{V,a}c_{g,V^g}r_{V^g}^W$$

for suitable subgroups U, V, W and an element g of G, and an element a of $A(V)$.

Now I claim that those elements generate $\mu(A)$ as an R-module: to see this, it suffices to check that the R-submodule M of $\mu(A)$ they generate contains the generators of $\mu(A)$, and is closed for the product. But

$$t_K^H=t_K^Ht_K^K=t_K^H\lambda_{K,\varepsilon_K}c_{1,K}\sum_L r_L^L=t_K^H\lambda_{K,\varepsilon_K}c_{1,K}r_K^K$$

$$r_K^H=r_K^Kr_K^H=(\sum_L t_L^L)\lambda_{K,\varepsilon_K}c_{1,K}r_K^H=t_K^K\lambda_{K,\varepsilon_K}c_{1,K}r_K^H$$

$$c_{x,H}=(\sum_L t_L^L)c_{x,H}(\sum_N r_N^N)=t_{{}^xH}^{{}^xH}c_{x,H}r_H^H=t_{{}^xH}^{{}^xH}\lambda_{{}^xH,\varepsilon_{({}^xH)}}c_{x,H}r_H^H$$

$$\lambda_{H,a}=\lambda_{H,\varepsilon_H}\lambda_{H,a}\lambda_{H,\varepsilon_H}=t_H^H\lambda_{H,a}r_H^H=t_H^H\lambda_{H,a}c_{1,H}r_H^H$$

So it remains to check that M closed for the product. First I observe that setting

$$c_g = \sum_V c_{g,V}$$

relations (2) and (3) show that $g \mapsto c_g$ is a morphism from G to the group of invertible elements of $\mu(A)$, i.e. that $\mu(A)$ is an interior G-algebra (in Puig's sense [11]). With this notation, I have the following relations:

$$c_g t_K^H = t_{{}^gK}^{{}^gH} c_g \quad c_g r_K^H = r_{{}^gK}^{{}^gH} c_g \quad c_g \lambda_{H,a} = \lambda_{{}^gH,{}^ga} c_g \quad c_h t_H^H = t_H^H = r_H^H = r_H^H c_h \text{ if } h \in H$$

and I can rewrite β in the form

$$\beta = t_V^U \lambda_{V,a} c_g r_{V^g}^W = t_{U\cap {}^gW}^U t_V^{U\cap {}^gW} \lambda_{V,a} c_g r_{V^g}^{U^g\cap W} r_{U^g\cap W}^W = \dots$$

$$\dots = t_{U\cap {}^gW}^U t_V^{U\cap {}^gW} \lambda_{V,a} r_V^{U\cap {}^gW} c_g r_{U^g\cap W}^W = t_{U\cap {}^gW}^U \lambda_{U\cap {}^gW, t_V^{U\cap {}^gW}(a)} c_g r_{U^g\cap W}^W$$

so any element of the form β is one of the following

$$\beta_{U,W,g,b} = t_{U\cap {}^gW}^U \lambda_{U\cap {}^gW,b} c_g r_{U^g\cap W}^W$$

where U and W are subgroups of G and g is an element of G, and b an element of $A(U \cap {}^gW)$.

Now let P be the product

$$P = \beta_{S,T,x,a}\beta_{U,V,y,b}$$

for subgroups S, T, U, V and elements x and y of G, and elements $a \in A(S \cap {}^xT)$ and $b \in A(U \cap {}^yT)$.

This product can also be written as

$$P = t_{S\cap {}^xT}^S \lambda_{S\cap {}^xT,a} c_x r_{S^x\cap T}^T t_{U\cap {}^yV}^U \lambda_{U\cap {}^yV,b} c_y r_{U^y\cap V}^V$$

It is zero if $U \neq T$, and if $U = T$, then by relation (6)

$$r_{S^x\cap T}^T t_{U\cap {}^yV}^U = \sum_{z\in S^x\cap T\backslash T/T\cap {}^yV} t_{S^x\cap T\cap {}^{zy}V}^{S^x\cap T} c_z r_{S^{xz}\cap T\cap {}^yV}^{T\cap {}^yV}$$

Relation (4) allows to exchange c_x with $t_{S^x\cap T\cap {}^{zy}V}^{S^x\cap T}$ so P can be written

$$P = \sum_{z\in S^x\cap T\backslash T/T\cap {}^yV} t_{S\cap {}^xT}^S \lambda_{S\cap {}^xT,a} t_{S\cap {}^xT\cap {}^{xzy}V}^{S\cap {}^xT} c_{xz} r_{S^{xz}\cap T\cap {}^yV}^{T\cap {}^yV} \lambda_{T\cap {}^yV,b} c_y r_{T^y\cap V}^V$$

Now relation (10) allows again to exchange $\lambda_{S\cap {}^xT,a}$ and $t_{S\cap {}^xT\cap {}^{xzy}V}^{S\cap {}^xT}$. Relations (9) and (4) allow to exchange $r_{S^{xz}\cap T\cap {}^yV}^{T\cap {}^yV}$ and $\lambda_{T\cap {}^yV,b} c_y$. The use of relations (1) gives

$$P = \sum_{z\in S^x\cap T\backslash T/T\cap {}^yV} t_{S\cap {}^xT\cap {}^{xzy}V}^S \lambda_{S\cap {}^xT\cap {}^{xzy}V, r_{S\cap {}^xT\cap {}^{xzy}V}^{S\cap {}^xT} a} c_{xz} \lambda_{S^{xz}\cap T\cap {}^yV, r_{S^{xz}\cap T\cap {}^yV}^{T\cap {}^yV} b} c_y r_{S^{xzy}\cap T^y\cap V}^V$$

Now relation (12) allows to exchange c_{xz} and $\lambda_{S^{xz}\cap T\cap {}^yV, r_{S^{xz}\cap T\cap {}^yV}^{T\cap {}^yV} b}$, and it follows that

$$P = \sum_{z\in S^x\cap T\backslash T/T\cap {}^yV} t_{S\cap {}^xT\cap {}^{xzy}V}^S \lambda_{S\cap {}^xT\cap {}^{xzy}V, r_{S\cap {}^xT\cap {}^{xzy}V}^{S\cap {}^xT} a} \lambda_{S\cap {}^xT\cap {}^{xzy}V, r_{S\cap {}^xT\cap {}^{xzy}V}^{T\cap {}^yV} ({}^{xz}b)} c_{xz} c_y r_{S^{xzy}\cap T^y\cap V}^V$$

Finally, relations (7) and (2) allow to group together the terms in λ and those in c, and then

$$P = \sum_{z\in S^x\cap T\backslash T/T\cap^y V} t^S_{S\cap^x T\cap^{xzy}V}\lambda_{S\cap^x T\cap^{xzy}V,\, r^{S\cap^x T}_{S\cap^x T\cap^{xzy}V}a.r^{T\cap^y V}_{S\cap^x T\cap^{xzy}V}{}^{xz}b}\, c_{xzy}\, r^V_{S^{xzy}\cap T^y\cap V}$$

So P is a sum of terms of the form β: more precisely, relation (11) gives finally

$$\beta_{S,T,x,a}\beta_{U,V,y,b} = \begin{cases} 0 \text{ if } U \neq T \\ \displaystyle\sum_{z\in S^x\cap T\backslash T/T\cap^y V} \beta_{S,V,xzy,t^{S\cap^{xzy}V}_{S\cap^x T\cap^{xzy}V}(r^{S\cap^x T}_{S\cap^x T\cap^{xzy}V}a.r^{T\cap^y V}_{S\cap^x T\cap^{xzy}V}{}^{xz}b)} \text{ otherwise}\end{cases} \tag{M}$$

It follows that the elements $\beta_{S,T,x,a}$, for S and T subgroups of G, for $x \in G$ and $a \in A(S\cap {}^xT)$, generate $\mu(A)$ as R-module. This set of generators is actually redundant: if $u \in U$ and $w \in W$, then ${}^ub \in A(U\cap {}^{ug}W) = A(U\cap {}^{ugw}W)$, and

$$\beta_{U,W,ugw,{}^ub} = t^U_{U\cap^{ug}W}\lambda_{U\cap^{ug}W,{}^ub}c_{ugw}r^W_{U^{gw}\cap W}$$

Since

$$c_{ugw}r^W_{U^{gw}\cap W} = c_{ug}c_w r^W_{U^{gw}\cap W} = \dots$$

$$\dots = c_{ug}r^W_{U^g\cap W}c_w = c_{ug}r^W_{U^g\cap W} = c_u c_g r^W_{U^g\cap W}$$

I have also

$$\beta_{U,W,ugw,{}^ub} = t^U_{U\cap^{ug}W}\lambda_{U\cap^{ug}W,{}^ub}c_u c_g r^W_{U^g\cap W} = \dots$$

$$\dots = t^U_{U\cap^{ug}W}c_u\lambda_{U\cap^g W,b}c_g r^W_{U^g\cap W} = c_u t^U_{U\cap^g W}\lambda_{U\cap^g W,b}c_g r^W_{U^g\cap W}$$

and finally

$$\beta_{U,W,ugw,{}^ub} = \beta_{U,W,g,b} \tag{N}$$

This equality shows that elements $\beta_{U,W,g,b}$, as U and W run through the set of subgroups of G, as g runs through a set of representatives of double classes $U\backslash G/W$, and b runs through $A(U\cap {}^gW)$, generate $\mu(A)$. Actually

Proposition 4.4.2: The algebra $\mu(A)$ is isomorphic to the algebra $\nu(A)$ defined by the following generators and relations:

- **The generators of $\nu(A)$ are the elements $\beta_{S,T,x,a}$ where S and T are subgroups of G, where $x \in G$ and $a \in A(S\cap {}^xT)$.**

- **The relations of $\nu(A)$ are the following:**

$$\beta_{U,W,ugw,{}^ub} = \beta_{U,W,g,b} \quad \textbf{for} \quad u\in U,\ w\in W \tag{N}$$

$$\beta_{S,T,x,a}\beta_{U,V,y,b} = \delta_{U,T}\sum_{z\in S^x\cap T\backslash T/T\cap^y V} \beta_{S,V,xzy,t^{S\cap^{xzy}V}_{S\cap^x T\cap^{xzy}V}(r^{S\cap^x T}_{S\cap^x T\cap^{xzy}V}a.r^{T\cap^y V}_{S\cap^x T\cap^{xzy}V}{}^{xz}b)} \tag{M}$$

where $\delta_{U,T} = 0$ if $U \neq T$ and $\delta_{U,T} = 1$ if $U = T$.

$$z\beta_{S,T,x,a} + z'\beta_{S,T,x,a'} = \beta_{S,T,x,za+z'a'} \quad \textbf{for} \quad z,\ z' \in R \tag{L}$$

Proof: To prove this proposition, it suffices to show that with these relations, and setting

$$t_K^H = \beta_{H,K,1,\varepsilon_K}$$
$$r_K^H = \beta_{K,H,1,\varepsilon_K}$$
$$c_{x,H} = \beta_{{}^xH,H,x,\varepsilon_{{}^xH}}$$
$$\lambda_{H,a} = \beta_{H,H,1,a}$$

relations (1) to (12) hold in $\nu(A)$. But

$$t_K^H t_L^K = \beta_{H,K,1,\varepsilon_K}\beta_{K,L,1,\varepsilon_L} = \sum_{z=1\in K\backslash K/K} \beta_{H,L,1,r_L^K(\varepsilon_K).r_L^L(\varepsilon_L)} = t_L^H$$

$$r_L^K r_K^H = \beta_{L,K,1,\varepsilon_L}\beta_{K,H,1,\varepsilon_K} = \sum_{z=1\in K\backslash K/K} \beta_{L,H,1,r_L^L(\varepsilon_L).r_L^K(\varepsilon_K)} = r_L^H$$

So relation (1) holds in $\nu(A)$. Similarly

$$c_{y,{}^xH}c_{x,H} = \beta_{{}^{yx}H,{}^xH,y,\varepsilon_{{}^{yx}H}}\beta_{{}^xH,H,x,\varepsilon_{({}^xH)}} = \sum_{z=1\in {}^xH\backslash {}^xH/{}^xH} \beta_{{}^{yx}H,H,yx,\varepsilon_{{}^{yx}H}.{}^y\varepsilon_{{}^xH}} = c_{yx,H}$$

whence relation (2). Relation (3) follows from

$$t_H^H = \beta_{H,H,1,\varepsilon_H} = r_H^H$$

and from

$$c_{h,H} = \beta_{H,H,h,\varepsilon_H} = \beta_{H,H,1,(\varepsilon_H)^h} = t_H^H$$

Relation (4) follows from

$$c_{x,H}t_K^H = \beta_{{}^xH,H,x,\varepsilon_{({}^xH)}}\beta_{H,K,1,\varepsilon_K} = \sum_{z=1\in H\backslash H/K} \beta_{{}^xH,K,x,{}^x\varepsilon_K}$$

whereas

$$t_{{}^xK}^{{}^xH}c_{x,K} = \beta_{{}^xH,{}^xK,1,\varepsilon_{{}^xK}}\beta_{{}^xK,K,x,\varepsilon_{{}^xK}} = \sum_{z=1\in {}^xK\backslash {}^xK/{}^xK} \beta_{{}^xH,K,x,{}^x\varepsilon_K}$$

Similarly

$$c_{x,K}r_K^H = \beta_{{}^xK,K,x,\varepsilon_{({}^xK)}}\beta_{K,H,1,\varepsilon_K} = \sum_{z=1\in K\backslash K/K} \beta_{{}^xK,H,x,{}^x\varepsilon_K}$$

whereas

$$r_{{}^xK}^{{}^xH}c_{x,H} = \beta_{{}^xK,{}^xH,1,\varepsilon_{{}^xK}}\beta_{{}^xH,H,x,\varepsilon_{{}^xH}} = \sum_{z=1\in {}^xK\backslash {}^xH/{}^xH} \beta_{{}^xK,H,x,\varepsilon_{{}^xK}}$$

To check relation (5), I set $e = \sum_L t_L^L$. Then $e = \sum_L r_L^L$ and moreover

$$\beta_{S,T,x,a}e = \sum_L \beta_{S,T,x,a}\beta_{L,L,1,\varepsilon_L} = \ldots$$

$$\ldots = \beta_{S,T,x,a}\beta_{T,T,1,\varepsilon_T} = \sum_{z=1\in(S\cap{}^xT)\backslash T/T} \beta_{S,T,x,a.r_{S\cap{}^xT}^{{}^xT}({}^x\varepsilon_T)} = \beta_{S,T,x,a}$$

Similarly

$$e\beta_{S,T,x,a} = \sum_L \beta_{L,L,1,\varepsilon_L}\beta_{S,T,x,a} = \beta_{S,S,1,\varepsilon_S}\beta_{S,T,x,a} = \sum_{z=1\in S\backslash S/(S\cap{}^xT)} \beta_{S,T,x,r_{S\cap{}^xT}^S(\varepsilon_S).a} = \beta_{S,T,x,a}$$

so e is the unit of $\nu(A)$.

Finally for the Mackey axiom

$$r_K^H t_L^H = \beta_{K,H,1,\varepsilon_K}\beta_{H,L,1,\varepsilon_L} = \sum_{x\in K\backslash H/L} \beta_{K,L,x,r^K_{K\cap {}^xL}(\varepsilon_K).r^{{}^xL}_{K\cap {}^xL}({}^x\varepsilon_L)}$$

Moreover

$$t_{K\cap {}^xL} c_{x,K^x\cap L} r^L_{K^x\cap L} = \beta_{K,K\cap {}^xL,1,\varepsilon_{K\cap {}^xL}}\beta_{K\cap {}^xL,L,x,{}^x\varepsilon_{K^x\cap L}} = \dots$$

$$\dots = \sum_{z=1\in(K\cap {}^xL)\backslash(K\cap {}^xL)/(K\cap {}^xL)} \beta_{K,L,x,\varepsilon_{K\cap {}^xL}}$$

The other products of t_K^H, r_K^H, and $c_{x,H}$ vanish, because

$$\beta_{S,T,x,a}\beta_{U,V,y,b} = 0 \quad \text{if} \quad U \neq T$$

Relation $\lambda_{K,a} + \lambda_{K,a'} = \lambda_{K,a+a'}$ is a trivial consequence of relations (L). Moreover

$$\lambda_{K,a}\lambda_{K,a'} = \beta_{K,K,1,a}\beta_{K,K,1,a'} = \sum_{z=1\in K\backslash K/K} \beta_{K,K,1,a.a'} = \lambda_{K,a.a'}$$

whence relation (7). Now equality

$$\lambda_{K,\varepsilon_K} = \beta_{K,K,1,\varepsilon_K} = t_K^K$$

and relations (L) prove (8). Moreover

$$r_K^H\lambda_{H,a} = \beta_{K,H,1,\varepsilon_K}\beta_{H,H,1,a} = \sum_{z=1\in K\backslash H/H} \beta_{K,H,1,r_K^H(a)} = \beta_{K,H,1,r_K^H(a)}$$

whereas

$$\lambda_{K,r_K^H(a)} r_K^H = \beta_{K,K,1,r_K^H(a)}\beta_{K,H,1,\varepsilon_K} = \sum_{z=1\in K\backslash K/K} \beta_{K,H,1,r_K^H(a)} = \beta_{K,H,1,r_K^H(a)}$$

proving (9). Similarly

$$\lambda_{H,a} t_K^H = \beta_{H,H,1,a}\beta_{H,K,1,\varepsilon_K} = \sum_{z=1\in H\backslash H/K} \beta_{H,K,1,r_K^H(a)} = \beta_{H,K,1,r_K^H(a)}$$

whereas

$$t_H^K\lambda_{K,r_K^H(a)} = \beta_{H,K,1,\varepsilon_K}\beta_{K,K,1,\varepsilon_K} = \sum_{z=1\in K\backslash K/K} \beta_{H,K,1,r_K^H(a)} = \beta_{H,K,1,r_K^H(a)}$$

which proves (10).

For relation (11), I write

$$t_K^H\lambda_{K,a} = \beta_{H,K,1,\varepsilon_K}\beta_{K,K,1,a} = \sum_{z=1\in K\backslash K/K} \beta_{H,K,1,a} = \beta_{H,K,1,a}$$

so that

$$t_K^H\lambda_{K,a} r_K^H = \beta_{H,K,1,a}\beta_{K,H,1,\varepsilon_K} = \sum_{z=1\in K\backslash K/K} \beta_{H,H,1,t_K^H(a.\varepsilon_K)} = \lambda_{H,t_K^H(a)}$$

Finally, for relation (12)

$$\lambda_{^xH,^xa}c_{x,H} = \beta_{^xH,^xH,1,^xa}\beta_{^xH,H,x,\varepsilon_{^xH}} = \sum_{z=1\in {^xH}\backslash {^xH}/{^xH}} \beta_{^xH,H,x,^xa} = \beta_{^xH,H,x,^xa}$$

whereas

$$c_{x,H}\lambda_{H,a} = \beta_{^xH,H,x,\varepsilon_{^xH}}\beta_{H,H,1,a} = \sum_{z=1\in H\backslash H/H} \beta_{^xH,H,x,^xa} = \beta_{^xH,H,x,^xa}$$

This proves (12), and proposition 4.4.2. ∎

To complete the proof of proposition 4.4.1, I observe that relations (N) and relation (7) and (8) show that the map $b \in A(U \cap {^gW}) \mapsto \beta_{U,W,g,b} \in \mu(A)$ is R-linear, and induces a surjective morphism of R-modules

$$\bigoplus_{\substack{U,W\subseteq G\\ g\in U\backslash G/W}} A(U \cap {^gW}) \to \mu(A)$$

Moreover, if $\Omega = \coprod_{H\subseteq G} G/H$, then

$$\Omega^2 \simeq \coprod_{\substack{H,K\subseteq G\\ x\in H\backslash G/K}} G/(H\cap {^xK})$$

Thus

$$A(\Omega^2) \simeq \bigoplus_{\substack{H,K\subseteq G\\ x\in H\backslash G/K}} A(H\cap {^xK}) \qquad (I)$$

and the previous morphism induces a surjective R-module homomorphism from $A(\Omega^2)$ to $\mu(A)$.

In order to build a morphism from $\mu(A)$ to $A(\Omega^2)$, and to prove that it is a morphism of algebras, I must state precisely the above isomorphism (I). If H and K are subgroups of G, and $x \in H\backslash G/K$, I denote by $i_{H,x,K}$ the injection from $G/(H\cap {^xK})$ into Ω^2 defined by

$$i_{H,x,K}\big(g(H\cap {^xK})\big) = (gH, gxK)$$

and I denote by $\phi_{H,x,K} = A^*(i_{H,x,K})$ the associated morphism from $A(H\cap {^xK})$ to $A(\Omega^2)$. If S and T are subgroups of G, if $x \in G$ and if $a \in A(S\cap {^xT})$, I set

$$\gamma_{S,T,x,a} = \phi_{S,x,T}(a) = A_*(i_{S,x,T})(a)$$

It is an element of $A(\Omega^2)$. Let Q be the product

$$Q = \gamma_{S,T,x,a} \circ_\Omega \gamma_{U,V,y,b}$$

By definition of the product $\circ_\Omega$, I have

$$Q = A_*\begin{pmatrix}\omega_1\omega_2\omega_3\\ \omega_1\omega_3\end{pmatrix} A^*\begin{pmatrix}\omega_1\omega_2\omega_3\\ \omega_1\omega_2\omega_2\omega_3\end{pmatrix}(\gamma_{S,T,x,a} \times \gamma_{U,V,y,b})$$

Moreover

$$\gamma_{S,T,x,a} \times \gamma_{U,V,y,b} = A_*\left(i_{S,x,T} \times i_{U,y,V}\right)(a \times b)$$

Let Z be the pullback of Ω^3 and $G/(S\cap {}^xT)\times G/(U\times {}^yV)$ over the maps $\begin{pmatrix}\omega_1\omega_2\omega_3\\ \omega_1\omega_2\omega_2\omega_3\end{pmatrix}$ and $i_{S,x,T}\times i_{U,y,V}$: the set Z is the set of couples

$$\big(g(S\cap {}^xT), h(U\cap {}^yV)\big)\in G/(S\cap {}^xT)\times G/(U\times {}^yV)$$

such that $gxT=hU$. Denoting by i the injection from Z into $G/(S\cap {}^xT)\times G/(U\times {}^yV)$, and j the map from Z to Ω^3 defined by

$$j\left(\big(g(S\cap {}^xT), h(U\cap {}^yV)\big)\right)=(gS, gxT, hyV)$$

I have a cartesian square

$$\begin{array}{ccc}
Z & \xrightarrow{\quad i\quad} & G/(S\cap {}^xT)\times G/(U\times {}^yV)\\
{\scriptstyle j}\downarrow & & \downarrow{\scriptstyle i_{S,x,T}\times i_{U,y,V}}\\
\Omega^3 & \xrightarrow[\begin{pmatrix}\omega_1\omega_2\omega_3\\ \omega_1\omega_2\omega_2\omega_3\end{pmatrix}]{} & \Omega^4
\end{array}$$

and the product Q can also be written

$$Q=A_*\begin{pmatrix}\omega_1\omega_2\omega_3\\ \omega_1\omega_3\end{pmatrix}A_*(j)A^*(i)(a\times b)$$

But equality $gxT=hU$ implies $T=U$. Hence if $T\neq U$, the set Z is empty, and Q is zero. And if $T=U$, then the map

$$\big(g(S\cap {}^xT), h(U\cap {}^yV)\big)\in Z\mapsto x^{-1}g^{-1}h\in (S^x\cap T)\backslash T/(T\cap {}^yV)$$

is well defined, and induces a bijection

$$G\backslash Z\simeq (S^x\cap T)\backslash T/(T\cap {}^yV)$$

The orbit corresponding to $z\in (S^x\cap T)\backslash T/(T\cap {}^yV)$ is the orbit of the couple

$$\big((S\cap {}^xT), xz(T\cap {}^yV)\big)$$

the stabilizer of which in G is the group $S\cap {}^xT\cap {}^{xzy}V$. So the set Z identifies with

$$Z\simeq \coprod_{z\in (S^x\cap T)\backslash T/(T\cap {}^yV)} G/(S\cap {}^xT\cap {}^{xzy}V)$$

so that

$$A(Z)\simeq \bigoplus_{z\in (S^x\cap T)\backslash T/(T\cap {}^yV)} A(S\cap {}^xT\cap {}^{xzy}V)$$

Now the product $G/(S\cap {}^xT)\times G/(U\times {}^yV)$ identifies with

$$\coprod_{z\in (S^x\cap T)\backslash G/(T\cap {}^yV)} G/(S\cap {}^xT\cap {}^{xzy}V)$$

and then

$$A\big(G/(S\cap {}^xT)\times G/(U\times {}^yV)\big)=\oplus_{z\in (S^x\cap T)\backslash G/(T\cap {}^yV)}A(S\cap {}^xT\cap {}^{xzy}V)$$

By construction of the product $\times$, I know that

$$a \times b = \bigoplus_{z \in (S^x \cap T) \backslash G/(T \cap {}^y V)} r^{S \cap {}^x T}_{S \cap {}^x T \cap {}^{xzy} V}(a) . r^{{}^x T \cap {}^{xzy} V}_{S \cap {}^x T \cap {}^{xzy} V}({}^{xz} b)$$

Up to those identifications, the map $A_*(i)$ is the projection on the components corresponding to the elements z of $(S^x \cap T) \backslash T/(T \cap {}^y V)$. Now the map $\binom{\omega_1 \omega_2 \omega_3}{\omega_1 \omega_3} j$ sends the element $g(S \cap {}^x T \cap {}^{xzy} V)$ to

$$\binom{\omega_1 \omega_2 \omega_3}{\omega_1 \omega_3} j\Big(g(S \cap {}^x T \cap {}^{xzy} V)\Big) = \binom{\omega_1 \omega_2 \omega_3}{\omega_1 \omega_3} (gS, gxT, gxzyV) = (gS, gxzyV)$$

This shows that it factors as the projection

$$\pi^{S \cap {}^{xzy} V}_{S \cap {}^x T \cap {}^{xzy} V} : G/(S \cap {}^x T \cap {}^{xzy} V) \to G/(S \cap {}^{xzy} V)$$

followed by $i_{S,xzy,V}$. Finally, I get

$$Q = \sum_{z \in (S^x \cap T) \backslash T/(T \cap {}^y V)} A_*(i_{S,xzy,V}) \Big(t^{S \cap {}^{xzy} V}_{S \cap {}^x T \cap {}^{xzy} V} \Big(r^{S \cap {}^x T}_{S \cap {}^x T \cap {}^{xzy} V}(a) . r^{{}^x T \cap {}^{xzy} V}_{S \cap {}^x T \cap {}^{xzy} V}({}^{xz} b) \Big) \Big)$$

and

$$\gamma_{S,T,x,a} \circ_\Omega \gamma_{U,V,y,b} = \delta_{U,T} \sum_{z \in (S^x \cap T) \backslash T/(T \cap {}^y V)} \gamma_{S,V,xzy,t^{S \cap {}^{xzy} V}_{S \cap {}^x T \cap {}^{xzy} V}(r^{S \cap {}^x T}_{S \cap {}^x T \cap {}^{xzy} V}(a) . r^{{}^x T \cap {}^{xzy} V}_{S \cap {}^x T \cap {}^{xzy} V}({}^{xz} b))} \qquad (M')$$

The analogy between those formulae and the formulae (M) giving the product in $\mu(A)$ shows that the map

$$\gamma_{S,T,x,a} \in A(\Omega^2) \mapsto \beta_{S,T,x,a} \in \mu(A)$$

induces a surjective algebra homomorphism from $A(\Omega^2)$ to $\mu(A)$. Moreover, it is easy to see that the identification relations

$$\gamma_{S,T,sxt,{}^s b} = \gamma_{S,T,x,b} \text{ pour } s \in S,\ t \in T \qquad (N')$$

analogous to relations (N) hold: indeed, if α_s denotes the map

$$\alpha_s : g(S \cap {}^x T) \in G/(S \cap {}^x T) \to gs^{-1}(S \cap {}^{sx} T) \in G/(S \cap {}^{sx} T) = G/(S \cap {}^{sxt} T)$$

then

$$i_{S,sxt,T}\alpha_s\Big(g(S \cap {}^x T)\Big) = i_{S,sxt,T}\Big(gs^{-1}(S \cap {}^{sx} T)\Big) = (gs^{-1}S, gs^{-1}sxtT) = \dots$$

$$\dots = (gS, gxT) = i_{S,x,T}\Big(g(S \cap {}^x T)\Big)$$

So $i_{S,sxt,T}\alpha_s = i_{S,x,T}$, and taking image under A_* gives relation (N').

Now I can define a morphism from $\nu(A)$ to $A(\Omega^2)$ by mapping $\beta_{S,T,x,a}$ to $\gamma_{S,T,x,a}$, and this is equivalent to define a morphism from $\mu(A)$ to $A(\Omega^2)$ by

$$t^H_K \mapsto t'^H_K = \gamma_{H,K,1,\varepsilon_K}$$

$$r^H_K \mapsto r'^H_K = \gamma_{K,H,1,\varepsilon_K}$$

$$c_{x,H} \mapsto c'_{x,H} = \gamma_{{}^x H,H,x,\varepsilon({}^x H)}$$

$$\lambda_{H,a} \mapsto \lambda'_{H,a} = \gamma_{H,H,1,a}$$

Proposition 4.4.2 shows indeed that since relations (M') and (N') hold, and since the map $a \mapsto \lambda'_{H,a} = A_*(i_{H,1,H})(a)$ is R-linear, then relations (1) to (12) hold for t'^H_K, r'^H_K, $c'_{x,H}$, $\lambda'_{H,a}$.

Those morphisms between $\mu(A)$ and $A(\Omega^2)$ are clearly mutual inverse, and this completes the proof of proposition 4.4.1. ∎

4.5 Examples

4.5.1 The Mackey algebra

In the case when A is the Burnside functor b, then the algebra $b(\Omega^2)$ is isomorphic to the Mackey algebra $\mu(G)$: the relations of Mackey algebra hold in $b(\Omega^2)$, so there is a natural morphism Φ from $\mu(G)$ to $b(\Omega^2)$, mapping the generators of the Mackey algebra to the generators of $b(\Omega^2)$ having the same name. Conversely, there is a morphism Θ from $b(\Omega^2)$ to $\mu(G)$ defined by

$$\Theta(t_K^G) = t_K^G \quad \Theta(r_K^G) = r_K^G \quad \Theta(c_{x,H}) = c_{x,H}$$

The image of $\lambda_{H,a}$ for $a \in b(H)$ is defined by linearity from

$$\Theta(\lambda_{H,H/K}) = t_K^H r_K^H$$

This definition is made in order to satisfy relation (12): indeed, as $H/K = t_K^H(K/K)$ in $b(H)$, I have the following equality in $b(\Omega^2)$

$$\lambda_{H,H/K} = \lambda_{H,t_K^H(\varepsilon_K)} = t_K^H \lambda_{K,\varepsilon_K} r_K^H = t_K^H t_K^K r_K^H = t_K^H r_K^H$$

It is not difficult then to check relations (7) to (12). So it is clear that Φ and Θ are mutual inverse algebra isomorphisms.

Proposition 4.5.1: The algebra $b(\Omega^2)$ is isomorphic to the Mackey algebra.

Remarks: 1) The identification of $b(X)$ given in propositions 2.4.2 and 2.4.3 is a way to recover the construction of Lindner ([9]) (see also the article of Thévenaz and Webb [15]): if X and Y are G-sets, then an element of $b(XY)$ is the difference of two G-sets over XY. Now a G-set V over XY is determined by a morphism f from V to X and a morphism g from V to Y. If W is the G-set over YZ associated to the morphism $h : W \to Y$ and to the morphism $k : W \to Z$, then by definition of composition in $\mathcal{C}_b$, I have

$$V \circ_Y W = b_* \begin{pmatrix} xyz \\ xz \end{pmatrix} b^* \begin{pmatrix} xyz \\ xyyz \end{pmatrix} (V \times W)$$

where $V \times W$ is the product V and W, viewed as a G-set over XY^2Z by the morphism $l(v,w) = \big(f(v), g(v), h(w), k(w)\big)$. Proposition 2.4.2 says how to compute $b^*\begin{pmatrix} xyz \\ xyyz \end{pmatrix}(V \times W)$: I must fill the cartesian square

$$\begin{array}{ccc} U & \xrightarrow{\;r\;} & V \times W \\ {\scriptstyle s}\downarrow & & \downarrow{\scriptstyle l} \\ XYZ & \xrightarrow[\begin{pmatrix} xyz \\ xyyz \end{pmatrix}]{} & XY^2Z \end{array}$$

Now U clearly identifies with the pull-back

$$U = \{(v,w) \in V \times W \mid g(v) = h(w)\}$$

with moreover $r(v,w) = \big(f(v), g(v), h(w)\big)$ and $s(v,w) = (v,w) \in V \times W$. With those identifications

$$b^*\begin{pmatrix} xyz \\ xyyz \end{pmatrix}(V \times W) = (U, r)$$

and proposition 2.4.2 now gives

$$b_*\begin{pmatrix} xyz \\ xz \end{pmatrix} b^*\begin{pmatrix} xyz \\ xyyz \end{pmatrix}(V \times W) = (U, \begin{pmatrix} xyz \\ yz \end{pmatrix} r)$$

In other words, the product $V \circ_Y W$ is obtained from the diagram

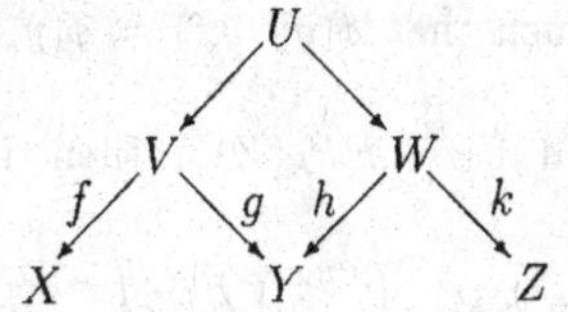

where the top square is cartesian.

2) An easy counting argument on the ranks over R in the isomorphism

$$\mu(G) \simeq b(\Omega^2) \simeq \bigoplus_{\substack{H,K \subseteq G \\ x \in H\backslash G/K}} b(H \cap {}^xK)$$

shows that the elements $t_L^H c_x r_{L^x}^K$, for H and K subgroups of G, for $x \in H\backslash G/K$, and L subgroup of $H \cap {}^xK$ up to conjugation by $H \cap {}^xK$ (i.e. for a well defined element $(H \cap {}^xK)/L$ of $b(H \cap {}^xL)$), are a basis over R of the Mackey algebra: it is proposition (3.2) of Thévenaz and Webb ([15]).

4.5.2 The Yoshida algebra

The second example of an algebra $A(\Omega^2)$ is the case when A is the fixed points functor FP_R (see [14]), defined as follows: for any subgroup H of G, the module $FP_R(H)$ is equal to R. The restriction maps r_K^H are all identity, and the transfers t_K^H are multiplication by $[H : K]$. It follows that denoting by $[X]$ the permutation R-module associated to the G-set X, I have for any G-set X

$$FP_R(X) = \mathrm{Hom}_G([X], R)$$

If $f : X \to Y$ is a morphism of G-sets, then $[f]$ is a morphism from $[X]$ in $[Y]$, giving by composition the map

$$FP_R^*(f) : FP_R(Y) \to FP_R(X)$$

The map $FP_{R,*}(f)$ is defined by identifying X to a basis of $[X]$ and Y to a basis of $[Y]$, and setting for $\phi \in FP_R(X) = \mathrm{Hom}_G([X], R)$ and $y \in Y$

$$FP_{R,*}(f)(\phi)(y) = \sum_{x \in f^{-1}(y)} \phi(x)$$

The Green functor structure of FP_R is obvious: if $\phi \in FP_R(X) = \mathrm{Hom}_G([X], R)$ and if $\psi \in FP_R(Y) = \mathrm{Hom}_G([Y], R)$, then $\phi \times \psi$ is the element of $\mathrm{Hom}_G([X \times Y], R)$ defined by

$$(\phi \times \psi)\big((x, y)\big) = \phi(x)\psi(y)$$

If X and Y are G-sets, the module $FP_R(YX)$ identifies with the G-homomorphisms from $[X]$ to $[Y]$ by the map sending $\phi \in \mathrm{Hom}_G([YX], R)$ to the map

$$x \in X \mapsto \sum_{y \in Y} \phi(y, x)y$$

In other words, the element ϕ is determined by its matrix $\big(\phi(y, x)\big)_{y \in Y, x \in X}$, which must be G-invariant (i.e. such that $\phi(gy, gx) = \phi(y, x)$ for all $g \in G$, $y \in Y$ and $x \in X$).

If Z is another G-set, and if $\psi \in FP_R(ZY)$, then the composite $\psi \circ_Y \phi$ is defined by

$$\psi \circ_Y \phi = FP_{R,*}\begin{pmatrix} zyx \\ zx \end{pmatrix} FP_R^*\begin{pmatrix} zyx \\ zyyx \end{pmatrix}(\psi \times \phi)$$

Here $\psi \times \phi$ is the morphism from $[ZY^2X]$ to R defined by

$$(\psi \times \phi)(z, y_1, y_2, x) = \psi(z, y_1)\phi(y_2, x)$$

so $A^*\left(\begin{smallmatrix} zyx \\ zyyx \end{smallmatrix}\right)(\psi \times \phi)$ is the morphism from $[ZYX]$ to R defined by

$$A^*\begin{pmatrix} zyx \\ zyyx \end{pmatrix}(\psi \times \phi)(z, y, x) = \psi(z, y)\phi(y, x)$$

Finally, the element $\psi \circ_Y \phi$ of $FP_R(ZX)$ is defined by

$$(\psi \circ_Y \phi)(z, x) = \sum_{(z,y,x) \in \left(\begin{smallmatrix} zyx \\ zx \end{smallmatrix}\right)^{-1}(z,x)} \psi(z, y)\phi(y, x) = \sum_{y \in Y} \psi(z, y)\phi(y, x)$$

so the product $\circ_Y$ corresponds to the matrix product. It follows that for any X, the algebra $A(X^2)$ is the algebra of endomorphisms of the RG-module $[X]$. In particular

Proposition 4.5.2: If $A = FP_R$, then the algebra $A(\Omega^2)$ is the Hecke algebra

$$\mathrm{End}_G\left(\bigoplus_{H \subseteq G} \mathrm{Ind}_H^G R\right)$$

This algebra was defined by Yoshida in [17]. The modules over FP_R are the cohomological Mackey functors (see [15] prop. 16.3), i.e. the Mackey functors such that

$$t_K^H r_K^H(m) = [H : K]m$$

for any $K \subseteq H$ and any $m \in M(H)$.

It is also clear that the functor $X \mapsto [X]$ is an equivalence of categories from C_{FP_R} on the full subcategory of the category of RG-modules formed by finitely generated permutation modules: this functor is indeed fully faithful by definition of C_{FP_R}, and essentially surjective, since any finitely generated permutation module is the module associated to a (finite) G-set. Theorem 3.3.5 now says that cohomological Mackey functors are exactly the additive functors on that category: this is a theorem of Yoshida ([17]).

The functor FP_R is also the functor $H^0(-,R) = \mathrm{Ext}^0_{RG}(R,R)$. It is also a subfunctor of the Green functor $H^{\oplus}(-,R) = \oplus_i H^i(-,R)$ (the standard notation is $H^*(-,R)$, but I prefer not to use it here). A similar argument proves the

Proposition 4.5.3: If A is the functor $H^{\oplus}(-,R) = \oplus_i H^i(-,R)$, then

$$A(\Omega^2) \simeq \mathrm{Ext}^{\oplus}_{RG}\left(\bigoplus_{H\subseteq G} \mathrm{Ind}_H^G R, \bigoplus_{H\subseteq G} \mathrm{Ind}_H^G R\right)$$

Chapter 5

Morita equivalence and relative projectivity

By its very definition, the algebra $A(\Omega^2)$ is only defined up to Morita equivalence: it has indeed been chosen so that the category $A(\Omega^2)$-**Mod** is equivalent to A-**Mod**. A natural question is then to know if it is possible to replace Ω by another G-set.

5.1 Morita equivalence of algebras $A(X^2)$

The next proposition gives a way to know if the algebras $A(X^2)$ and $A(Y^2)$ are Morita-equivalent:

Proposition 5.1.1: Let A be a Green functor for G, and let X and Y be G-sets. Then $A(YX) = \mathrm{Hom}_{\mathcal{C}_A}(X,Y)$ is an $A(Y^2)$-module-$A(X^2)$, and $A(XY)$ is an $A(X^2)$-module-$A(Y^2)$. The following conditions are equivalent:

1. **The bimodules $A(YX)$ and $A(XY)$ induce mutual inverse Morita equivalences between $A(X^2)$ and $A(Y^2)$.**

2. **There exists integers n and m such that X divides nY and Y divides mX in $\mathcal{C}_A$.**

Proof: First it is clear that composition induces bimodule morphisms

$$\Phi : A(XY) \otimes_{A(Y^2)} A(YX) \to A(X^2) : \alpha \otimes \beta \mapsto \alpha \circ_Y \beta$$

and

$$\Theta : A(YX) \otimes_{A(X^2)} A(XY) \to A(Y^2) : \beta \otimes \alpha \mapsto \beta \circ_X \alpha$$

Those morphisms are balanced in the sense of Curtis-Reiner ([4]): as composition is associative, I have

$$\Phi(\alpha \otimes \beta) \circ_X \alpha' = \alpha \circ_Y \Theta(\beta \otimes \alpha') = \alpha \circ_Y \beta \circ_X \alpha'$$

and similarly

$$\Theta(\beta \otimes \alpha) \circ_Y \beta' = \beta \circ_X \Phi(\alpha \otimes \beta') = \beta \circ_X \alpha \circ_T \beta'$$

They form a Morita context (see [4] (3.54)). As the image of Φ (resp. Θ) is a sub-bimodule of the corresponding algebra, i.e. a two sided ideal, the map Φ (resp. Θ) is surjective if and only if its image contains the unit. Thus Φ and Θ are surjective if and only if there exists elements $\alpha_i \in A(XY)$ and $\beta_i \in A(YX)$, for $1 \leq i \leq n$ such that

$$\sum_{i=1}^{n} \alpha_i \circ_Y \beta_i = 1_{A(X^2)}$$

and elements $\beta'_j \in A(YX)$ and $\alpha'_j \in A(XY)$ for $1 \leq j \leq m$ such that

$$\sum_{j=1}^{m} \beta'_j \circ_X \alpha'_j = 1_{A(Y^2)}$$

The first equality means precisely that X divides nY in $\mathcal{C}_A$, and the second that Y divides mX, and the proposition follows. ∎

Definition: *Let X and Y be G-sets. I will say that X and Y have the same stabilizers if*

$$\forall H \subseteq G,\ \exists x \in X,\ G_x = H \quad \Leftrightarrow \quad \exists y \in Y,\ G_y = H$$

This is equivalent to say that if

$$X \simeq \coprod_{H \subseteq G} n_H(G/H) \qquad\qquad Y \simeq \coprod_{H \subseteq G} m_H(G/H)$$

then $n_H \neq 0$ if and only if $m_H \neq 0$. Now if m is an integer greater or equal to all quotients m_H/n_H (for $n_H \neq 0$), I have $mn_H \geq m_H$, and Y is isomorphic to a sub-G-set of mX. Now by lemma 3.2.3, the set Y divides mX in $\mathcal{C}_A$. Similarly, if n is an integer greater or equal to all n_H/m_H (for $m_H \neq 0$), then X divides nY in $\mathcal{C}_A$. Thus

Proposition 5.1.2: If X and Y are G-sets having the same stabilizers, then $A(X^2)$ and $A(Y^2)$ are Morita-equivalent.

5.2 Relative projectivity

Let X be a candidate G-set to replace Ω. Then in particular, there must exist an integer n such that $\bullet$ divides nX in $\mathcal{C}_A$. The next proposition describes this situation:

Proposition 5.2.1: Let A be a Green functor for G, and let X be a G-set. The following conditions are equivalent:

1. **There exists an integer n such that $\bullet$ divides nX in $\mathcal{C}_A$.**
2. **The set $\bullet$ divides X in $\mathcal{C}_A$.**
3. **The map $A_*\binom{x}{\bullet}$ is surjective.**
4. **The image of the map $A_*\binom{x}{\bullet}$ contains ε.**

Poof: The proof of this proposition uses the next lemma:

Lemma 5.2.2: Let A be a Green functor for G, and let $f : X \to Y$ be a morphism of G-sets. Then $A^*(f)$ is a morphism of rings (with unit) from $\big(A(Y),.\big)$ to $\big(A(X),.\big)$, and the image of $A(X)$ under $A_*(f)$ is a two-sided ideal of $\big(A(Y),.\big)$.

Proof: The product "." can be recovered from the product $\times$ by the formula

$$a.b = A^*\begin{pmatrix} x \\ xx \end{pmatrix}(a \times b)$$

for $a, b \in A(X)$. Now for $a, b \in A(Y)$, I have

$$A^*(f)(a).A^*(f)(b) = A^*\begin{pmatrix} x \\ xx \end{pmatrix}\left(A^*\begin{pmatrix} x \\ f(x) \end{pmatrix}(a) \times A^*\begin{pmatrix} x \\ f(x) \end{pmatrix}(b)\right) = \ldots$$

$$\ldots = A^*\begin{pmatrix} x \\ xx \end{pmatrix} A^*\begin{pmatrix} x_1x_2 \\ f(x_1)f(x_2) \end{pmatrix}(a \times b) = A^*\begin{pmatrix} x \\ f(x)f(x) \end{pmatrix}(a \times b) = \ldots$$

$$\ldots = A^*\begin{pmatrix} x \\ f(x) \end{pmatrix} A^*\begin{pmatrix} y \\ yy \end{pmatrix}(a \times b) = A^*(f)(a.b)$$

and the first assertion follows, since moreover

$$A^*(f)(\varepsilon_Y) = A^*\begin{pmatrix} x \\ f(x) \end{pmatrix} A^*\begin{pmatrix} y \\ \bullet \end{pmatrix}(\varepsilon) = A^*\begin{pmatrix} x \\ \bullet \end{pmatrix}(\varepsilon) = \varepsilon_X$$

Remark: It is actually a reformulation of assertion iii) of proposition 1.8.3.

For the second assertion, if $a \in A(X)$ and $b \in A(Y)$, then

$$b.A_*(f)(a) = A^*\begin{pmatrix} y \\ yy \end{pmatrix}\big(b \times A_*(f)(a)\big) = A^*\begin{pmatrix} y \\ yy \end{pmatrix} A_*\begin{pmatrix} yx \\ yf(x) \end{pmatrix}(b \times a)$$

As the square

$$\begin{array}{ccc} X & \xrightarrow{\begin{pmatrix} x \\ f(x)x \end{pmatrix}} & YX \\ f\big\downarrow & & \big\downarrow\begin{pmatrix} yx \\ yf(x) \end{pmatrix} \\ Y & \xrightarrow[\begin{pmatrix} y \\ yy \end{pmatrix}]{} & Y^2 \end{array}$$

is cartesian, I have also

$$b.A_*(f)(a) = A_*(f)A^*\begin{pmatrix} x \\ f(x)x \end{pmatrix}(b \times a)$$

A similar argument shows that

$$A_*(f)(a).b = A_*(f)A^*\begin{pmatrix} x \\ xf(x) \end{pmatrix}(a \times b)$$

so the image of $A_*(f)$ is a two-sided ideal of $A(Y)$. ■

Proof of proposition 5.2.1: It is clear that 2) implies 1). Now if 1) holds, then there exists elements $\alpha_i \in A(X\bullet) = A(X)$ and $\beta_i \in A(\bullet X) = A(X)$, for $1 \leq i \leq n$, such that

$$\sum_{i=1}^{n} \beta_i \circ_X \alpha_i = 1_{A(\bullet^2)}$$

But $A(\bullet^2) = A(\bullet)$ and

$$1_{A(\bullet^2)} = A_*\begin{pmatrix} \bullet \\ \bullet\bullet \end{pmatrix} A^*\begin{pmatrix} \bullet \\ \bullet \end{pmatrix}(\varepsilon) = \varepsilon$$

Moreover

$$\beta_i \circ_X \alpha_i = A_*\begin{pmatrix} \bullet x \bullet \\ \bullet \end{pmatrix} A^*\begin{pmatrix} \bullet x \bullet \\ \bullet x x \bullet \end{pmatrix}(\beta_i \times \alpha_i) = A_*\begin{pmatrix} x \\ \bullet \end{pmatrix}\begin{pmatrix} x \\ xx \end{pmatrix}(\beta_i \times \alpha_i) = A_*\begin{pmatrix} x \\ \bullet \end{pmatrix}(b_i.a_i)$$

This proves that

$$\epsilon = A_*\begin{pmatrix} x \\ \bullet \end{pmatrix}\left(\sum_{i=1}^{n} \beta_i.\alpha_i\right)$$

so 1) implies 4).

By the previous lemma, the image of $A_*\begin{pmatrix} x \\ \bullet \end{pmatrix}$ is a two-sided ideal. So 4) implies 3). Finally if 3) holds, let $\alpha \in A(X)$ such that $A_*\begin{pmatrix} x \\ \bullet \end{pmatrix}(\alpha) = \varepsilon$. I can view α as an element of $A(X\bullet)$ and ε_X as an element of $A(\bullet X)$. Then by the previous computation

$$\varepsilon_X \circ_X \alpha = A_*\begin{pmatrix} x \\ \bullet \end{pmatrix}(\epsilon_X.\alpha) = A_*\begin{pmatrix} x \\ \bullet \end{pmatrix}(\alpha) = \varepsilon$$

and this proves that $\bullet$ divides X in $\mathcal{C}_A$. ■

Definition: *Let A be a Green functor for the group G. If X is a G-set having the equivalent properties of proposition 5.2.1, the functor A is said to be projective relative to X.*

This definition coincides with the usual one (see for example Webb [16]). In the case $X = G/H$, the map $A_*\begin{pmatrix} x \\ \bullet \end{pmatrix}$ is the transfer $t_H^G : A(H) \rightarrow A(G) = A(\bullet)$, and this is the classical notion of relative H-projectivity.

Proposition 5.2.3: Let A be a Green functor for G, and let X and Y be G-sets.

1. **If A is projective relative to X, and if $\mathrm{Hom}_{G\text{-}set}(X, Y) \neq \emptyset$, then A is projective relative to Y.**

2. **The functor A is projective relative to $X \times Y$ if and only if it is projective relative to X and relative to Y.**

3. **If $A(\bullet)$ has a unique maximal two-sided ideal (for instance if it is a local ring), then A is projective relative to $X \coprod Y$ if and only if it is projective relative to X or relative to Y.**
   In this case, if H is a minimal subgroup of G such that A is projective relative to G/H, then A is projective relative to X if and only if $X^H \neq \emptyset$. In particular, the subgroup H is unique up to G-conjugation.

Proof: 1) Let $f : X \to Y$ be a morphism of G-sets. Assertion 1) follows from the equality

$$A_*\binom{x}{\bullet} = A_*\binom{y}{\bullet} A_*(f)$$

2) If A is projective relative to $X \times Y$, as there exists morphisms of G-sets from $X \times Y$ to X and Y, assertion 1) implies that A is projective relative to X and relative to Y. Conversely, if A is projective relative to X and relative to Y, let $\alpha \in A(X)$ and $\beta \in A(Y)$ such that

$$A_*\binom{x}{\bullet}(\alpha) = \varepsilon = A_*\binom{y}{\bullet}(\beta)$$

Then

$$A_*\binom{xy}{\bullet}(\alpha \times \beta) = A_*\binom{xy}{\bullet\bullet}(\alpha \times \beta) = A_*\binom{x}{\bullet}(\alpha) \times A_*\binom{y}{\bullet}(\beta) = \varepsilon \times \varepsilon = \varepsilon$$

and this proves that A is projective relative to $X \times Y$.
3) Let $Z = X \coprod Y$, and $\imath_X$ and $\imath_Y$ the injections from X and Y into Z. It follows from assertion 1) that if A is projective relative to X or relative to Y, then A is projective relative to Z. Conversely, as $A(Z) = A_*(i_X)\big(A(X)\big) \oplus A_*(i_Y)\big(A(Y)\big)$, and as

$$A_*\binom{z}{\bullet} A_*(i_X) = A_*\binom{x}{\bullet}$$

I see that if A is projective relative to Z, then

$$A(\bullet) = A_*\binom{z}{\bullet}\big(A(Z)\big) = A_*\binom{x}{\bullet}\big(A(X)\big) + A_*\binom{y}{\bullet}\big(A(Y)\big)$$

Now $A(Z)$ is the sum of two of its two-sided ideals. If $A(\bullet)$ has a unique maximal two-sided ideal, this forces one of them to be the whole of $A(Z)$, thus A is projective relative to X or relative to Y.
Let H be a minimal subgroup such that A is projective relative to G/H: such a subgroup certainly exists, since A is trivially projective relative to $G/G = \bullet$. Since X^H identifies with $\mathrm{Hom}_{G\text{-}set}(G/H, X)$, it follows from assertion 1) that if $X^H \neq \emptyset$, then A is projective relative to X.
Conversely, if A is projective relative to X, it follows from assertion 2) that A is projective relative to $(G/H) \times X$. Now the first part of assertion 3) shows that there exists $x \in X$ such that A is projective relative to $(G/H) \times (G/G_x)$. This product is isomorphic to the disjoint union of sets $G/(H \cap {}^gG_x) = G/(H \cap G_{gx})$, for suitable elements $g \in G$. Then there exists g such that A is projective relative to $G/(H \cap G_{gx})$, and now the minimality of H implies that $H \subseteq G_{gx}$, i.e. $gx \in X^H$. ■

5.3 Cartesian product in C_A

5.3.1 Definition

Let A be a Green functor, and U be a G-set. On the category A-**Mod**, I have the functor $M \mapsto M_U$ (see lemma 3.1.1). As

$$(A_X)_U(Z) = A_X(ZU) = A(ZUX) = A_{UX}(Z)$$

it follows that $(A_X)_U \simeq A_{UX}$, and that the above functor preserves the subcategory of A-**Mod** formed by the modules A_X. This subcategory is isomorphic to $\mathcal{C}_A$, so it is possible to define the functor $X \mapsto UX$ on $\mathcal{C}_A$:

Lemma 5.3.1: The correspondence

$$X \mapsto UX$$

$$\alpha \in \mathrm{Hom}_{\mathcal{C}_A}(X,Y) \mapsto 1_{A((UY)^2)} \circ_Y \alpha = A_* \begin{pmatrix} uyx \\ uyux \end{pmatrix} A^* \begin{pmatrix} uyx \\ yx \end{pmatrix} (\alpha) \in \mathrm{Hom}_{\mathcal{C}_A}(UX, UY)$$

is a functor $U \times -$ from $\mathcal{C}_A$ to $\mathcal{C}_A$.

Proof: It suffices to make the isomorphisms of proposition 3.1.3 explicit: an element α of $A(YX)$ defines a morphism from A_Y to A_X, the evaluation of which at a G-set Z is given by

$$\beta \in A_Y(Z) = A(ZY) \mapsto \beta \circ_Y \alpha \in A(ZX) = A_X(Z)$$

This morphism from A_Y to A_X defines in turn a morphism from $A_{UY} = (A_Y)_U$ to $A_{UX} = (A_X)_U$, evaluating at Z as

$$\beta \in A_{UY}(Z) = A(ZUY) \mapsto \beta \circ_Y \alpha \in A(ZUX) = A_{UX}(Z)$$

Now this morphism is determined by the image of the element $1_{A((UY)^2)} \in A_{UY}(UY)$, i.e. by the element

$$1_{A((UY)^2)} \circ_Y \alpha$$

of $A(UYUX)$. Now it suffices to write

$$1_{A((UY)^2)} \circ_Y \alpha = A_* \begin{pmatrix} u_1y_1u_2y_2x \\ u_1y_1u_2x \end{pmatrix} A^* \begin{pmatrix} u_1y_1u_2y_2x \\ u_1y_1u_2y_2y_2x \end{pmatrix} \left[A_* \begin{pmatrix} uy \\ uyuy \end{pmatrix} A^* \begin{pmatrix} uy \\ \bullet \end{pmatrix} (\varepsilon) \times \alpha \right]$$

The product inside hooks is equal to

$$A_* \begin{pmatrix} uy_1y_2x \\ uy_1uy_1y_2x \end{pmatrix} A^* \begin{pmatrix} uy_1y_2x \\ y_2x \end{pmatrix} (\alpha)$$

As the square

$$\begin{array}{ccc} UYX & \xrightarrow{\begin{pmatrix} uyx \\ uyyx \end{pmatrix}} & UY^2X \\ {\scriptstyle \begin{pmatrix} uyx \\ uyuyx \end{pmatrix}} \downarrow & & \downarrow {\scriptstyle \begin{pmatrix} uy_1y_2x \\ uy_1uy_1y_2x \end{pmatrix}} \\ (UY)^2X & \xrightarrow[\begin{pmatrix} u_1y_1u_2y_2x \\ u_1y_1u_2y_2u_2y_2x \end{pmatrix}]{} & UYUY^2X \end{array}$$

is cartesian, I have

$$A^* \begin{pmatrix} u_1y_1u_2y_2x \\ u_1y_1u_2y_2y_2x \end{pmatrix} A_* \begin{pmatrix} uy_1y_2x \\ uy_1uy_1y_2x \end{pmatrix} = A_* \begin{pmatrix} uyx \\ uyuyx \end{pmatrix} A^* \begin{pmatrix} uyx \\ uyyx \end{pmatrix}$$

and then

$$1_{A((UY)^2)} \circ_Y \alpha = A_*\begin{pmatrix} u_1y_1u_2y_2x \\ u_1y_1u_2x \end{pmatrix} A_*\begin{pmatrix} uyx \\ uyuyx \end{pmatrix} A^*\begin{pmatrix} uyx \\ uyyx \end{pmatrix} A^*\begin{pmatrix} uy_1y_2x \\ y_2x \end{pmatrix}(\alpha) = \dots$$

$$\dots = A_*\begin{pmatrix} uyx \\ uyux \end{pmatrix} A^*\begin{pmatrix} uyx \\ yx \end{pmatrix}(\alpha)$$

proving the lemma. ■

In other words, this lemma means that if X, Y and Z are G-sets, if $\alpha \in A(YX)$ and $\beta \in A(ZY)$, then

$$\left[1_{A((UZ)^2)} \circ_Z \beta\right] \circ_{UY} \left[1_{A((UY)^2)} \circ_Y \alpha\right] = 1_{A((UZ)^2)} \circ_Z (\beta \circ_Y \alpha)$$

This equality can be "proved" by the following argument: let

$$P = \left[1_{A((UZ)^2)} \circ_Z \beta\right] \circ_{UY} \left[1_{A((UY)^2)} \circ_Y \alpha\right]$$

As the product $\circ$ is associative, I also have

$$P = \left(\left[1_{A((UZ)^2)} \circ_Z \beta\right] \circ_{UY} 1_{A((UY)^2)}\right) \circ_Y \alpha$$

Observing now that $1_{A((UY)^2)}$ is a unit for the product $\circ_{UY}$, it follows that

$$P = \left[1_{A((UZ)^2)} \circ_Z \beta\right] \circ_Y \alpha$$

and using again associativity of the product $\circ$, I get

$$P = 1_{A((UZ)^2)} \circ_Z (\beta \circ_Y \alpha)$$

The only trouble with that argument is that I never proved that the product $\circ$ is associative in those conditions. It is the reason for the following lemma

Lemma 5.3.2: Let X, Y, Z, T and U be G-sets. Let $a \in A(XY)$, $b \in A(YUZ)$, and $c \in A(ZT)$. Then

$$(a \circ_Y b) \circ_Z c = a \circ_Y (b \circ_Z c) \in A(XUT)$$

Proof: By definition

$$a \circ_Y b = A_*\begin{pmatrix} xyuz \\ xuz \end{pmatrix} A^*\begin{pmatrix} xyuz \\ xyyuz \end{pmatrix}(a \times b)$$

So

$$(a \circ_Y b) \circ_Z c = A_*\begin{pmatrix} xuzt \\ xut \end{pmatrix} A^*\begin{pmatrix} xuzt \\ xuzzt \end{pmatrix}\left[A_*\begin{pmatrix} xyuz \\ xuz \end{pmatrix} A^*\begin{pmatrix} xyuz \\ xyyuz \end{pmatrix}(a \times b) \times c\right]$$

The product inside hooks is equal to

$$A_*\begin{pmatrix} xyuz_1z_2t \\ xuz_1z_2t \end{pmatrix} A^*\begin{pmatrix} xyuz_1z_2t \\ xyyuz_1z_2t \end{pmatrix}(a \times b \times c)$$

As the square

$$\begin{array}{ccc}
XYUZT & \xrightarrow{\begin{pmatrix} xyuzt \\ xyuzzt \end{pmatrix}} & XYUZ^2T \\
{\scriptstyle \begin{pmatrix} xyuzt \\ xuzt \end{pmatrix}} \Big\downarrow & & \Big\downarrow {\scriptstyle \begin{pmatrix} xyuz_1z_2t \\ xuz_1z_2t \end{pmatrix}} \\
XUZT & \xrightarrow[\begin{pmatrix} xuzt \\ xuzzt \end{pmatrix}]{} & XUZ^2T
\end{array}$$

is cartesian, it follows that

$$A^*\begin{pmatrix} xuzt \\ xuzzt \end{pmatrix} A_*\begin{pmatrix} xyuz_1z_2t \\ xuz_1z_2t \end{pmatrix} = A_*\begin{pmatrix} xyuzt \\ xuzt \end{pmatrix} A^*\begin{pmatrix} xyuzt \\ xyuzzt \end{pmatrix}$$

and then

$$(a \circ_Y b) \circ_Z c = A_*\begin{pmatrix} xuzt \\ xut \end{pmatrix} A_*\begin{pmatrix} xyuzt \\ xuzt \end{pmatrix} A^*\begin{pmatrix} xyuzt \\ xyuzzt \end{pmatrix} A^*\begin{pmatrix} xyuz_1z_2t \\ xyyuz_1z_2t \end{pmatrix} = \dots$$

$$\dots = A_*\begin{pmatrix} xyuzt \\ xut \end{pmatrix} A^*\begin{pmatrix} xyuzt \\ xyyuzzt \end{pmatrix} (a \times b \times c)$$

On the other hand

$$a \circ_Y (b \circ_Z c) = A_*\begin{pmatrix} xyut \\ xut \end{pmatrix} A^*\begin{pmatrix} xyut \\ xyyut \end{pmatrix} \left[a \times A_*\begin{pmatrix} yuzt \\ yut \end{pmatrix} A^*\begin{pmatrix} yuzt \\ yuzzt \end{pmatrix} \right]$$

The product inside hooks is equal to

$$A_*\begin{pmatrix} xy_1y_2uzt \\ xy_1y_2ut \end{pmatrix} A^*\begin{pmatrix} xy_1y_2uzt \\ xy_1y_2uzzt \end{pmatrix} (a \times b \times c)$$

As the square

$$\begin{array}{ccc}
XYUZT & \xrightarrow{\begin{pmatrix} xyuzt \\ xyyuzt \end{pmatrix}} & XY^2UZT \\
{\scriptstyle \begin{pmatrix} xyuzt \\ xyut \end{pmatrix}} \Big\downarrow & & \Big\downarrow {\scriptstyle \begin{pmatrix} xy_1y_2uzt \\ xy_1y_2ut \end{pmatrix}} \\
XYUT & \xrightarrow[\begin{pmatrix} xyut \\ xyyut \end{pmatrix}]{} & XY^2UT
\end{array}$$

is cartesian, I have

$$A^*\begin{pmatrix} xyut \\ xyyut \end{pmatrix} A_*\begin{pmatrix} xy_1y_2uzt \\ xy_1y_2ut \end{pmatrix} = A_*\begin{pmatrix} xyuzt \\ xyut \end{pmatrix} A^*\begin{pmatrix} xyuzt \\ xyyuzt \end{pmatrix}$$

and then

$$a \circ_Y (b \circ_Z c) = A_*\begin{pmatrix} xyut \\ xut \end{pmatrix} A_*\begin{pmatrix} xyuzt \\ xyut \end{pmatrix} A^*\begin{pmatrix} xyuzt \\ xyyuzt \end{pmatrix} A^*\begin{pmatrix} xy_1y_2uzt \\ xy_1y_2uzzt \end{pmatrix} (a \times b \times c) = \dots$$

$$\dots = A_*\begin{pmatrix} xyuzt \\ xut \end{pmatrix} A^*\begin{pmatrix} xyuzt \\ xyyuzzt \end{pmatrix} (a \times b \times c)$$

proving the lemma. ∎

5.3.2 Adjunction

The previous lemma leads to a product on the right $-\times U$:

Proposition 5.3.3: Let A be a Green functor for G, and U be a G-set. The correspondence

$$X \mapsto XU$$

$$\alpha \in A(YX) \mapsto \alpha \circ_X 1_{A((XU)^2)} \in A(YUXU)$$

is a functor $-\times U$ from $\mathcal{C}_A$ to $\mathcal{C}_A$, which is right adjoint to the functor $U\times -$.

Proof: Say that $-\times U$ is a functor is equivalent to say that if X, Y, and Z are G-sets, if $\alpha \in A(YX)$ and $\beta \in A(ZY)$, then

$$\left[\beta \circ_Y 1_{A((YU)^2)}\right] \circ_{YU} \left[\alpha \circ_X 1_{A((XU)^2)}\right] = (\beta \circ_X \alpha) \circ_X 1_{A((XU)^2)}$$

This follows from lemma 5.3.2 and from the fact that $1_{A((YU)^2)}$ is a unit for the product $\circ_{YU}$. It is clear moreover that

$$1_{A(X^2)} \times U = 1_{A(X^2)} \circ_X 1_{A((XU)^2)} = 1_{A((XU)^2)}$$

so $-\times U$ is a functor.

To prove that this functor is right adjoint to the functor $U\times -$, I observe that

$$\mathrm{Hom}_{\mathcal{C}_A}(X, YU) = A(YUX) = \mathrm{Hom}_{\mathcal{C}_A}(UX, Y)$$

and it suffices to check the functoriality of this equality with respect to X and Y. So let Z and T be G-sets, and $f \in A(XZ)$ and $g \in A(TY)$. I must check that the squares

$$\begin{array}{ccc}
\mathrm{Hom}_{\mathcal{C}_A}(X, YU) = A(YUX) & \xrightarrow{Id} & A(YUX) = \mathrm{Hom}_{\mathcal{C}_A}(UX, Y) \\
\downarrow & & \downarrow \\
\mathrm{Hom}_{\mathcal{C}_A}(Z, TU) = A(TUZ) & \xrightarrow[Id]{} & A(TUZ) = \mathrm{Hom}_{\mathcal{C}_A}(UZ, T)
\end{array}$$

are commutative. That is for $h \in A(YUX)$, check that

$$(g\times U) \circ_{YU} h \circ_X f = g \circ_Y h \circ_{UX} (U \times f)$$

But

$$(g\times U)\circ_{YU} h \circ_X f = (g\circ_Y 1_{A((YU)^2)})\circ_{YU} h\circ_X f = g\circ_Y \left(1_{A((YU)^2)}\circ_{YU} h\circ_X f\right) = g\circ_Y (h\circ_X f)$$

whereas

$$g\circ_Y h\circ_{UX}(U\times f) = g\circ_Y h\circ_{UX}(1_{A((UX)^2)}\circ_X f) = \left(g\circ_Y h\circ_{UX} 1_{A((UX)^2)}\right)\circ_X f) = (g\circ_Y h)\circ_X f$$

Now equality follows from lemma 5.3.2. ∎

Remark: By a similar computation as in lemma 5.3.1, it is possible to show that if $\alpha \in A(YX)$, then

$$\alpha \times U = A_*\begin{pmatrix} yxu \\ yuxu \end{pmatrix} A^*\begin{pmatrix} yxu \\ yx \end{pmatrix}(\alpha)$$

Notation: *If X is a G-set, I denote by $\sigma_{U,X}$ the natural bijection from UX to XU switching the components, and by $\theta_{U,X} = (\sigma_{U,X})_*$ the associated morphism from UX to XU in $\mathcal{C}_A$.*
It follows from lemma 3.2.3 that

$$\theta_{U,X} \circ_{UX} \theta_{X,U} = (\sigma_{U,X})_* \circ_{UX} (\sigma_{X,U})_* = (\sigma_{U,X} \circ \sigma_{X,U})_* = Id_* = 1_{A((UX)^2)}$$

so $\theta_{U,X}$ is an isomorphism.

Proposition 5.3.4: If X and Y are G-sets, and if $\alpha \in A(YX)$, then the square

$$\begin{array}{ccc} UX & \xrightarrow{\theta_{U,X}} & XU \\ {\scriptstyle U\times\alpha}\big\downarrow & & \big\downarrow{\scriptstyle \alpha\times U} \\ UY & \xrightarrow[\theta_{U,Y}]{} & YU \end{array}$$

is commutative in $\mathcal{C}_A$, and the morphisms $\theta_{U,X}$ define an isomorphism from the functor $U \times -$ to the functor $- \times U$.

Proof: By lemma 5.3.2, I have

$$\theta_{U,Y} \circ_{UY} (U \times \alpha) = \theta_{U,Y} \circ_{UY} (1_{A((UY)^2)} \circ_Y \alpha) = \theta_{U,Y} \circ_Y \alpha$$

Whereas

$$(\alpha \times U) \circ_{XU} \theta_{U,X} = (\alpha \circ_X 1_{A((XU)^2)}) \circ_{XU} \theta_{U,X} = \alpha \circ_X \theta_{U,X}$$

This product is equal to

$$\alpha \circ_X \theta_{U,X} = \alpha \circ_X A_*\begin{pmatrix} ux \\ xuux \end{pmatrix} A^*\begin{pmatrix} ux \\ \bullet \end{pmatrix}(\varepsilon) = \ldots$$

$$\ldots = A_*\begin{pmatrix} yx_1u_1u_2x_2 \\ yu_1u_2x_2 \end{pmatrix} A^*\begin{pmatrix} yx_1u_1u_2x_2 \\ yx_1x_1u_1u_2x_2 \end{pmatrix} \left[\alpha \times A_*\begin{pmatrix} ux \\ xuux \end{pmatrix} A^*\begin{pmatrix} ux \\ \bullet \end{pmatrix}(\varepsilon)\right]$$

The product inside hooks is

$$A_*\begin{pmatrix} yx_1ux_2 \\ ux_1x_2uux_2 \end{pmatrix} A^*\begin{pmatrix} yx_1ux_2 \\ yx_1 \end{pmatrix}(\alpha)$$

As the square

$$\begin{array}{ccc} YXU & \xrightarrow{\begin{pmatrix} yxu \\ yxux \end{pmatrix}} & YXUX \\ {\scriptstyle \begin{pmatrix} yxu \\ yxuux \end{pmatrix}}\big\downarrow & & \big\downarrow{\scriptstyle \begin{pmatrix} yx_1ux_2 \\ yx_1x_2uux_2 \end{pmatrix}} \\ YXU^2X & \xrightarrow[\begin{pmatrix} yx_1u_1u_2x_2 \\ yx_1x_1u_1u_2x_2 \end{pmatrix}]{} & YX^2U^2X \end{array}$$

is cartesian, I have

$$A^*\begin{pmatrix} yx_1u_1u_2x_2 \\ yx_1x_1u_1u_2x_2 \end{pmatrix} A_*\begin{pmatrix} yx_1ux_2 \\ ux_1x_2uux_2 \end{pmatrix} = A_*\begin{pmatrix} yxu \\ yxuux \end{pmatrix} A^*\begin{pmatrix} yxu \\ yxux \end{pmatrix}$$

and then

$$\alpha \circ_X \theta_{U,X} = A_*\begin{pmatrix} yx_1u_1u_2x_2 \\ yu_1u_2x_2 \end{pmatrix} A_*\begin{pmatrix} yxu \\ yxuux \end{pmatrix} A^*\begin{pmatrix} yxu \\ yxux \end{pmatrix} A^*\begin{pmatrix} yx_1ux_2 \\ yx_1 \end{pmatrix}(\alpha)$$

Finally

$$\alpha \circ_X \theta_{U,X} = A_*\begin{pmatrix} yxu \\ yuux \end{pmatrix} A^*\begin{pmatrix} yxu \\ yx \end{pmatrix}(\alpha)$$

On the other hand

$$\theta_{U,Y} \circ_Y \alpha = A_*\begin{pmatrix} uy \\ yuuy \end{pmatrix} A^*\begin{pmatrix} yxu \\ yx \end{pmatrix}(\varepsilon) \circ_Y \alpha = \ldots$$

$$\ldots = A_*\begin{pmatrix} y_1u_1u_2y_2x \\ y_1u_1u_2x \end{pmatrix} A^*\begin{pmatrix} y_1u_1u_2y_2x \\ y_1u_1u_2y_2y_2x \end{pmatrix} A_*\begin{pmatrix} uy_1y_2x \\ y_1uuy_1y_2x \end{pmatrix} A^*\begin{pmatrix} uy_1y_2x \\ y_2x \end{pmatrix}(\alpha)$$

As the square

$$\begin{array}{ccc} YXU & \xrightarrow{\begin{pmatrix} yxu \\ uyyx \end{pmatrix}} & UY^2X \\ {\scriptstyle\begin{pmatrix} yxu \\ yuuyx \end{pmatrix}}\big\downarrow & & \big\downarrow{\scriptstyle\begin{pmatrix} uy_1y_2x \\ y_1uuy_1y_2x \end{pmatrix}} \\ YU^2YX & \xrightarrow[\begin{pmatrix} y_1u_1u_2y_2x \\ y_1u_1u_2y_2y_2x \end{pmatrix}]{} & YU^2Y^2X \end{array}$$

is cartesian, I have

$$A^*\begin{pmatrix} y_1u_1u_2y_2x \\ y_1u_1u_2y_2y_2x \end{pmatrix} A_*\begin{pmatrix} uy_1y_2x \\ y_1uuy_1y_2x \end{pmatrix} = A_*\begin{pmatrix} yxu \\ yuuyx \end{pmatrix} A^*\begin{pmatrix} yxu \\ uyyx \end{pmatrix}$$

so

$$\theta_{U,Y} \circ_Y \alpha = A_*\begin{pmatrix} y_1u_1u_2y_2x \\ y_1u_1u_2x \end{pmatrix} A_*\begin{pmatrix} yxu \\ yuuyx \end{pmatrix} A^*\begin{pmatrix} yxu \\ uyyx \end{pmatrix} A^*\begin{pmatrix} uy_1y_2x \\ y_2x \end{pmatrix}(\alpha)$$

or

$$\theta_{U,Y} \circ_Y \alpha = A_*\begin{pmatrix} yxu \\ yuux \end{pmatrix} A^*\begin{pmatrix} yxu \\ yx \end{pmatrix}(\alpha)$$

and this proves the proposition. ■

5.3.3 Cartesian product in $\mathcal{C}_A \times \mathcal{C}_A$

In view of the previous paragraphs, one may try to define a cartesian product functor on $\mathcal{C}_A \times \mathcal{C}_A$, mapping two G-sets X and Y to $X \times Y$, and morphisms $f \in A(X'X)$ and $g \in A(Y'Y)$ to the morphism (denoted by $f \otimes g$ to avoid confusion) defined by

$$f \otimes g = (f \times Y') \circ_{XY'} (X \times g) \in A(X'Y'XY)$$

However, this definition is generally not functorial in X and Y:

Proposition 5.3.5: Let A be a Green functor for the group G. The following conditions are equivalent:

1. **The above definitions turn the cartesian product into a functor from $\mathcal{C}_A \times \mathcal{C}_A$ to $\mathcal{C}_A$.**

2. **For any G-sets X and Y, and any $\alpha \in A(X)$ and $\beta \in A(Y)$,**

$$\beta \times \alpha = A_* \begin{pmatrix} xy \\ yx \end{pmatrix} (\alpha \times \beta)$$

Remark: I will see later that this is equivalent to say that A is *commutative* (i.e. that for any G-set X, the ring $(A(X), .)$ is commutative).

Proof: Say that the cartesian product is functorial is equivalent to say that if $f \in A(X'X)$, if $g \in A(Y'Y)$, if $f' \in A(X"X')$ and $g' \in A(Y"Y')$ are morphisms in $\mathcal{C}_A$, then

$$(f' \otimes g') \circ_{X'Y'} (f \otimes g) = (f' \circ_{X'} f) \otimes (g' \circ_{Y'} g)$$

The case $f' = 1_{A(X'^2)}$ and $g = 1_{A(Y^2)}$ now gives

$$(X' \times g') \circ_{X'Y'} (f \times Y) = f \otimes g' = (f \times Y") \circ_{XY"} (X \times g')$$

In other words, for any X, Y, X', Y', and any $f \in A(X'X)$, $g \in A(Y'Y)$, I must have

$$(X' \times g) \circ_{X'Y} (f \times Y) = (f \times Y') \circ_{XY'} (X \times g) \qquad (C)$$

The left hand side P is equal to

$$P = A_* \begin{pmatrix} x'_1y'x'_2y_1xy_2 \\ x'_1y'xy_2 \end{pmatrix} A^* \begin{pmatrix} x'_1y'x'_2y_1xy_2 \\ x'_1y'x'_2y_1x'_2y_1xy_2 \end{pmatrix} [(X' \times g) \times (f \times Y)]$$

The product inside hooks is equal to

$$A_* \begin{pmatrix} x'_1y'y_1x'_2xy_2 \\ x'_1y'x'_1y_1x'_2y_2xy_2 \end{pmatrix} A^* \begin{pmatrix} x'_1y'y_1x'_2xy_2 \\ y'y_1x'_2x \end{pmatrix} (g \times f)$$

As the square

$$\begin{array}{ccc} X'Y'XY & \xrightarrow{\begin{pmatrix} x'y'xy \\ x'y'yx'xy \end{pmatrix}} & X'Y'YX'XY \\ \begin{pmatrix} x'y'xy \\ x'y'x'yxy \end{pmatrix} \Big\downarrow & & \Big\downarrow \begin{pmatrix} x'_1y'y_1x'_2xy_2 \\ x'_1y'x'_1y_1x'_2y_2xy_2 \end{pmatrix} \\ X'Y'X'YXY & \xrightarrow[\begin{pmatrix} x'_1y'x'_2y_1xy_2 \\ x'_1y'x'_2y_1x'_2y_1xy_2 \end{pmatrix}]{} & X'Y'(X'Y)^2XY \end{array}$$

is cartesian, I have

$$A^* \begin{pmatrix} x'_1y'x'_2y_1xy_2 \\ x'_1y'x'_2y_1x'_2y_1xy_2 \end{pmatrix} A_* \begin{pmatrix} x'_1y'y_1x'_2xy_2 \\ x'_1y'x'_1y_1x'_2y_2xy_2 \end{pmatrix} = A_* \begin{pmatrix} x'y'xy \\ x'y'x'yxy \end{pmatrix} A^* \begin{pmatrix} x'y'xy \\ x'y'yx'xy \end{pmatrix}$$

so

$$P = A_* \begin{pmatrix} x'_1y'x'_2y_1xy_2 \\ x'_1y'xy_2 \end{pmatrix} A_* \begin{pmatrix} x'y'xy \\ x'y'x'yxy \end{pmatrix} A^* \begin{pmatrix} x'y'xy \\ x'y'yx'xy \end{pmatrix} A^* \begin{pmatrix} x'_1y'y_1x'_2xy_2 \\ y'y_1x'_2x \end{pmatrix} (g \times f) = \ldots$$

$$\dots = A_*\begin{pmatrix} x'y'xy \\ x'y'xy \end{pmatrix} A^*\begin{pmatrix} x'y'xy \\ y'yx'x \end{pmatrix} (g \times f) = A^*\begin{pmatrix} x'y'xy \\ y'yx'x \end{pmatrix} (g \times f)$$

The right hand side of (C) is

$$Q = A_*\begin{pmatrix} x'y_1x_1y'x_2y_2 \\ x'y_1x_2y_2 \end{pmatrix} A^*\begin{pmatrix} x'y_1x_1y'x_2y_2 \\ x'y_1x_1y'x_1y'x_2y_2 \end{pmatrix} [(f \times Y') \times (X \times g)]$$

The product inside hooks is equal to

$$A_*\begin{pmatrix} x'x_1y_1'x_2y_2'y \\ x'y_1'x_1y_1'x_2y_2'x_2y \end{pmatrix} A^*\begin{pmatrix} x'x_1y_1'x_2y_2'y \\ x'x_1y_2'y \end{pmatrix} (f \times g)$$

As the square

$$\begin{array}{ccc} X'XY'Y & \xrightarrow{\begin{pmatrix} x'xy'y \\ x'xy'xy'y \end{pmatrix}} & X'XY'XY'Y \\ {\scriptstyle\begin{pmatrix} x'xy'y \\ x'y'xy'xy \end{pmatrix}}\Big\downarrow & & \Big\downarrow{\scriptstyle\begin{pmatrix} x'x_1y_1'x_2y_2'y \\ x'y_1'x_1y_1'x_2y_2'x_2y \end{pmatrix}} \\ X'YXY'XY & \xrightarrow[\begin{pmatrix} x'y_1x_1y'x_2y_2 \\ x'y_1x_1y'x_1y'x_2y_2 \end{pmatrix}]{} & X'Y(XY')^2XY \end{array}$$

is cartesian, I have

$$A^*\begin{pmatrix} x'y_1x_1y'x_2y_2 \\ x'y_1x_1y'x_1y'x_2y_2 \end{pmatrix} A_*\begin{pmatrix} x'x_1y_1'x_2y_2'y \\ x'y_1'x_1y_1'x_2y_2'x_2y \end{pmatrix} = A_*\begin{pmatrix} x'xy'y \\ x'y'xy'xy \end{pmatrix} A^*\begin{pmatrix} x'xy'y \\ x'xy'xy'y \end{pmatrix}$$

and then

$$Q = A_*\begin{pmatrix} x'y_1x_1y'x_2y_2 \\ x'y_1x_2y_2 \end{pmatrix} A_*\begin{pmatrix} x'xy'y \\ x'y'xy'xy \end{pmatrix} A^*\begin{pmatrix} x'xy'y \\ x'xy'xy'y \end{pmatrix} A^*\begin{pmatrix} x'x_1y_1'x_2y_2'y \\ x'x_1y_2'y \end{pmatrix} (f \times g) = \dots$$

$$\dots = A_*\begin{pmatrix} x'xy'y \\ x'y'xy \end{pmatrix} A^*\begin{pmatrix} x'xy'y \\ x'xy'y \end{pmatrix} (f \times g) = A_*\begin{pmatrix} x'xy'y \\ x'y'xy \end{pmatrix} (f \times g)$$

Finally, equation (C) can be written

$$A^*\begin{pmatrix} x'y'xy \\ y'yx'x \end{pmatrix} (g \times f) = A_*\begin{pmatrix} x'xy'y \\ x'y'xy \end{pmatrix} (f \times g)$$

Taking its image under $A_*\left(\begin{smallmatrix} x'y'xy \\ y'yx'x \end{smallmatrix}\right)$, which is the inverse isomorphism of $A^*\left(\begin{smallmatrix} x'y'xy \\ y'yx'x \end{smallmatrix}\right)$, I get

$$g \times f = A_*\begin{pmatrix} x'xy'y \\ y'yx'x \end{pmatrix} (f \times g)$$

and the case $X' = Y' = \bullet$ is assertion 2).

Conversely, assertion 2) implies the previous equality, so it implies equality (C). Then

$$(f' \otimes g') \circ_{X'Y'} (f \otimes g) = (f' \times Y'') \circ_{X'Y''} (X' \times g') \circ_{X'Y'} (f \times Y') \circ_{XY'} (X \times g) = \dots$$

$$\dots = (f' \times Y'') \circ_{X'Y''} (f \times Y'') \circ_{XY''} (X \times g') \circ_{XY'} (X \times g) = (f' \circ_{X'} f) \otimes (g' \circ_{Y'} g)$$

since $X \times -$ and $- \times Y''$ are functors. Thus 2) implies 1), and this proves the proposition. ∎

5.4 Morita equivalence and relative projectivity

The existence of a cartesian product functor in $\mathcal{C}_A$ has the following consequence:

Proposition 5.4.1: Let A be a Green functor for the group G, and Y be G-set. Let

$$\Omega_Y = \coprod_{\substack{H \subseteq G \\ Y^H \neq \emptyset}} G/H$$

The following conditions are equivalent:

1. **The evaluation functor $M \mapsto M(\Omega_Y)$ is an equivalence of categories from A-Mod to $A(\Omega_Y^2)$-Mod.**
2. **The functor A is projective relative to Y.**

Proof: If 2) holds, I know that $\bullet$ divides Y in $\mathcal{C}_A$. Taking images by the functor $\Omega \times -$, I see that Ω divides $\Omega \times Y$ in $\mathcal{C}_A$. I also know that any G-set divides a multiple of Ω in $\mathcal{C}_A$. So any G-set divides a multiple of $\Omega \times Y$ in $\mathcal{C}_A$..

But if $(\omega, y) \in \Omega \times Y$, then its stabilizer $G_{(\omega,y)}$ in G is contained in G_y, so $Y^{G_{(\omega,y)}} \neq \emptyset$, and then $G_{(\omega,y)}$ is the stabilizer in G of the element $G_{(\omega,y)}$ of Ω_Y. Conversely, if H is a subgroup of G such that $Y^H \neq \emptyset$, and if $y \in Y^H$, then H is the stabilizer of the element (H, y) of $\Omega \times Y$. It follows that $\Omega \times Y$ and Ω_Y have the same stabilizers, so $\Omega \times Y$ is a subset of a multiple of Ω_Y, hence it divides a multiple of Ω_Y in $\mathcal{C}_A$.

Now any G-set divides a multiple of Ω_Y in $\mathcal{C}_A$, and proposition 4.2.1 shows that the evaluation functor $M \mapsto M(\Omega_Y)$ is an equivalence of categories from A-**Mod** to $A(\Omega_Y^2)$-**Mod**.

Conversely, if this functor is an equivalence of categories, then any G-set divides a multiple of Ω_Y in $\mathcal{C}_A$. As Ω_Y and $\Omega \times Y$ have the same stabilizers, any G-set divides a multiple of $\Omega \times Y$. In particular $\bullet$ divides such a multiple, and A is projective relative to $\Omega \times Y$.

But as $\bullet \simeq G/G$ is a subset of Ω, it divides Ω in $\mathcal{C}_A$, so A is projective relative to Ω. Now proposition 5.2.3.2) shows that A is projective relative to $\Omega \times Y$ if and only if it is projective relative to Y, and this completes the proof of the proposition. ■

Remark: There is a natural inclusion i from Ω_Y into Ω, so an inclusion $f = A_*(i \times i)$ from $A(\Omega_Y^2)$ to $A(\Omega^2)$. If $x \in G$, then x defines an endomorphism a_x of the G-set Ω by

$$a_x(g.H) = gHx^{-1} = gx^{-1}.{}^xH$$

and this endomorphism stabilizes Ω_Y (as a set). Now the map $x \mapsto (a_x)_*$ turn the algebras $A(\Omega^2)$ and $A(\Omega_Y^2)$ into interior G-algebras, and f is compatible with those structures. The following lemma shows that f is actually an embedding of interior algebras (see Puig [11]):

Lemma 5.4.2: Let i be an injective morphism of G-sets from X to Y. Then $A_*(i \times i)$ is an embedding from $A(X^2)$ into $A(Y^2)$.

Proof: If i is an injective morphism from X to Y, then i_* is a split monomorphism from X to Y in $\mathcal{C}_A$, since $i^* \circ i_* = 1_{A(X^2)}$. Then it is clear that the map

$$f \in A(X^2) = \mathrm{End}_{\mathcal{C}_A}(X) \mapsto i_* \circ f \circ i^* \in A(Y^2) = \mathrm{End}_{\mathcal{C}_A}(Y)$$

is an embedding of algebras. So to prove the lemma, it suffices to prove that for any $f \in A(X^2)$

$$i_* \circ_X f \circ_X i^* = A_*(i \times i)(f)$$

But $i_* = A_*\begin{pmatrix} x \\ i(x)x \end{pmatrix} A^*\begin{pmatrix} x \\ \bullet \end{pmatrix}(\varepsilon)$, so

$$i_* \circ_X f = A_*\begin{pmatrix} yx_1x_2 \\ yx_2 \end{pmatrix} A^*\begin{pmatrix} yx_1x_2 \\ yx_1x_1x_2 \end{pmatrix}\left(A_*\begin{pmatrix} x \\ i(x)x \end{pmatrix} A^*\begin{pmatrix} x \\ \bullet \end{pmatrix}(\varepsilon) \times f\right) = \ldots$$

$$\ldots = A_*\begin{pmatrix} yx_1x_2 \\ yx_2 \end{pmatrix} A^*\begin{pmatrix} yx_1x_2 \\ yx_1x_1x_2 \end{pmatrix} A_*\begin{pmatrix} x_1x_2x_3 \\ i(x_1)x_1x_2x_3 \end{pmatrix} A^*\begin{pmatrix} x_1x_2x_3 \\ x_2x_3 \end{pmatrix}(f)$$

As the square

$$\begin{array}{ccc} X^2 & \xrightarrow{\begin{pmatrix} x_1x_2 \\ x_1x_1x_2 \end{pmatrix}} & X^3 \\ {\scriptstyle\begin{pmatrix} x_1x_2 \\ i(x_1)x_1x_2 \end{pmatrix}}\downarrow & & \downarrow{\scriptstyle\begin{pmatrix} x_1x_2x_3 \\ i(x_1)x_1x_2x_3 \end{pmatrix}} \\ YX^2 & \xrightarrow[\begin{pmatrix} yx_1x_2 \\ yx_1x_1x_2 \end{pmatrix}]{} & YX^3 \end{array}$$

is cartesian, I have

$$A^*\begin{pmatrix} yx_1x_2 \\ yx_1x_1x_2 \end{pmatrix} A_*\begin{pmatrix} x_1x_2x_3 \\ i(x_1)x_1x_2x_3 \end{pmatrix} = A_*\begin{pmatrix} x_1x_2 \\ i(x_1)x_1x_2 \end{pmatrix} A^*\begin{pmatrix} x_1x_2 \\ x_1x_1x_2 \end{pmatrix}$$

whence

$$i_* \circ_X f = A_*\begin{pmatrix} yx_1x_2 \\ yx_2 \end{pmatrix} A_*\begin{pmatrix} x_1x_2 \\ i(x_1)x_1x_2 \end{pmatrix} A^*\begin{pmatrix} x_1x_2 \\ x_1x_1x_2 \end{pmatrix} A^*\begin{pmatrix} x_1x_2x_3 \\ x_2x_3 \end{pmatrix}(f) = \ldots$$

$$\ldots = A_*\begin{pmatrix} x_1x_2 \\ i(x_1)x_2 \end{pmatrix} A^*\begin{pmatrix} x_1x_2 \\ x_1x_2 \end{pmatrix}(f) = A_*\begin{pmatrix} x_1x_2 \\ i(x_1)x_2 \end{pmatrix}(f)$$

Then

$$i_* \circ_X f \circ_X i^* = A_*\begin{pmatrix} y_1xy_2 \\ y_1y_2 \end{pmatrix} A^*\begin{pmatrix} y_1xy_2 \\ y_1xxy_2 \end{pmatrix}\left[A_*\begin{pmatrix} x_1x_2 \\ i(x_1)x_2 \end{pmatrix}(f) \times A_*\begin{pmatrix} x \\ xi(x) \end{pmatrix} A^*\begin{pmatrix} x \\ \bullet \end{pmatrix}(\varepsilon)\right]$$

The product inside hooks is

$$A_*\begin{pmatrix} x_1x_2x_3 \\ i(x_1)x_2x_3i(x_3) \end{pmatrix} A^*\begin{pmatrix} x_1x_2x_3 \\ x_1x_2 \end{pmatrix}(f)$$

But the square

$$\begin{array}{ccc} X^2 & \xrightarrow{\begin{pmatrix} x_1x_2 \\ x_1x_2x_2 \end{pmatrix}} & X^3 \\ \begin{pmatrix} x_1x_2 \\ i(x_1)x_2i(x_2) \end{pmatrix}\Big\downarrow & & \Big\downarrow\begin{pmatrix} x_1x_2x_3 \\ i(x_1)x_2x_3i(x_3) \end{pmatrix} \\ YXY & \xrightarrow[\begin{pmatrix} y_1xy_2 \\ y_1xxy_2 \end{pmatrix}]{} & YX^2Y \end{array}$$

is cartesian. So

$$A^*\begin{pmatrix} y_1xy_2 \\ y_1xxy_2 \end{pmatrix} A_*\begin{pmatrix} x_1x_2x_3 \\ i(x_1)x_2x_3i(x_3) \end{pmatrix} = A_*\begin{pmatrix} x_1x_2 \\ i(x_1)x_2i(x_2) \end{pmatrix} A^*\begin{pmatrix} x_1x_2 \\ x_1x_2x_2 \end{pmatrix}$$

and finally

$$i_*\circ_X f\circ_X i^* = A_*\begin{pmatrix} y_1xy_2 \\ y_1y_2 \end{pmatrix} A_*\begin{pmatrix} x_1x_2 \\ i(x_1)x_2i(x_2) \end{pmatrix} A^*\begin{pmatrix} x_1x_2 \\ x_1x_2x_2 \end{pmatrix} A^*\begin{pmatrix} x_1x_2x_3 \\ x_1x_2 \end{pmatrix}(f) = \ldots$$

$$\ldots = A_*\begin{pmatrix} x_1x_2 \\ i(x_1)i(x_2) \end{pmatrix} A^*\begin{pmatrix} x_1x_2 \\ x_1x_2 \end{pmatrix}(f) = A_*(i\times i)(f)$$

which proves the lemma. ■

Remark: If A is projective relative to Y, proposition 5.4.1 means moreover that

$$A(\Omega^2)f(1_{A(\Omega_Y^2)})A(\Omega^2) = A(\Omega^2)$$

so the algebra $A(\Omega_Y^2)$ is a kind of "source algebra" of $A(\Omega^2)$ (cf [11]). The next section is a refinement of this analogy.

5.5 Progenerators

5.5.1 Finitely generated modules

Proposition 4.2.1 provides a way to define the notion of a finitely generated module over a Green functor A: an A-module M is *finitely generated (or of finite type)* if $M(\Omega)$ is a finitely generated $A(\Omega^2)$-module. This notion depends only on the category A-**Mod**, by the following argument (Benson): a module M is finitely generated if and only for any set I, the map

$$(\phi_i)_{i\in I} \in \bigoplus_I \mathrm{Hom}_A(M,M) \mapsto \oplus_i\phi_i \in \mathrm{Hom}_A(M, \bigoplus_I M)$$

is an isomorphism.

The following lemma shows that this definition actually does not depend on Ω:

Lemma 5.5.1: Let A be a Green functor for the group G, and M be an A-module. The following conditions are equivalent:

1. **The module $M(\Omega)$ is a finitely generated $A(\Omega^2)$-module.**

2. **There exist a (finite) G-set X such that M is a quotient of A_X.**

Remark: The word "finite" is unnecessary, since the module A_X is not defined for an infinite X.

Proof: Suppose 1) holds. It means that $M(\Omega)$ is a quotient of a direct sum of a finite number of copies of $A(\Omega^2)$. Taking inverse images under the equivalence of proposition 4.2.1 shows that M is a quotient of a direct sum of a finite number, say n, of copies of A_Ω. As $nA_\Omega \simeq A_{n\Omega}$, condition 2) holds for $X = n\Omega$.

Conversely, if 2) holds, then as X divides a multiple $n\Omega$ of Ω in $\mathcal{C}_A$, the module A_X is a direct summand of $A_{n\Omega} \simeq nA_\Omega$. Now $M(\Omega)$ is a quotient of $nA_\Omega(\Omega) = nA(\Omega^2)$, so 1) holds. ■

5.5.2 Idempotents and progenerators

Let A be a Green functor for the group G. By definition, an A-module P will be called a *progenerator* (of A-**Mod**) if it is projective, of finite type, and if any A-module M is a quotient of a direct sum of copies of P. Proposition 4.3.1 shows that this is equivalent to say that $P(\Omega)$ is a progenerator of $A(\Omega^2)$-**Mod**.

Since $A_\Omega(\Omega) = A(\Omega^2)$ is a free module over $A(\Omega^2)$, it follows that A_Ω is a progenerator. More generally, proposition 3.1.3 shows that if Y is a G-set, then A_Y is a projective A-module, and proposition 4.2.1 shows that A_Y is a progenerator if and only if any G-set X divides a multiple of Y in $\mathcal{C}_A$. A natural question is now to know if any direct summands of A_Y are also progenerators.

As the algebra of endomorphisms of A_Y is $A(Y^2)$, it is equivalent to consider idempotents of this algebra. I will use the following notation:

Notation: *If j is an idempotent of $A(Y^2)$, I will denote by $A_Y.j$ the associated direct summand of A_Y.*

By definition, if X is a G-set, then

$$(A_Y.j)(X) = A(XY) \circ_Y j$$

This is the set of morphisms (in $\mathcal{C}_A$) from Y to X which factor through j. The module $A_Y.j$ is projective, and for any A-module M, I have

$$\mathrm{Hom}_{\mathcal{C}_A}(A_Y.j, M) \simeq j \circ_Y M(Y)$$

The question is now to find couples (Y, j) such that $A_Y.j$ is a progenerator.

Proposition 5.5.2: Let A be a Green functor for G. Let Y be a G-set, and j be an idempotent of $A(Y^2)$. The following conditions are equivalent:

1. **The module $A_Y.j$ is a progenerator.**

2. **Any G-set X divides a multiple of Y in $\mathcal{C}_A$, and the two-sided ideal of $A(Y^2)$ generated by j is equal to $A(Y^2)$.**

Proof: If $A_Y.j$ is a progenerator, as it is a direct summand of A_Y, the module A_Y is also a progenerator. Now by proposition 4.2.1, any X divides a multiple of Y. Moreover, the module A_Y is a quotient of a direct sum of copies of $A_Y.j$, so it is a direct summand, which means that the composition morphism

$$\Big(A(Y^2)\circ_Y j\Big)\otimes\Big(j\circ_Y A(Y^2)\Big)\to A(Y^2):\ \alpha\otimes\beta\mapsto\alpha\circ_Y\beta$$

is surjective. As the image of this morphism is $A(Y^2)\circ_Y j\circ_Y A(Y^2)$, it follows that the two-sided ideal of $A(Y^2)$ generated by j is the whole of $A(Y^2)$.

Conversely, if 2) holds, then A_Y is a progenerator (proposition 4.2.1), and equality

$$A(Y^2)\circ_Y j\circ_Y A(Y^2)=A(Y^2)$$

proves that A_Y is a direct summand of a sum of copies of $A_Y.j$. So $A_Y.j$ is a progenerator, and this completes the proof of the proposition. ∎

Lemma 5.5.3: Let A be a Green functor for G, and X be a G-set. Then $A_*\begin{pmatrix}x\\xx\end{pmatrix}$ is a morphism of rings (with unit) from $\big(A(X),.\big)$ to $\big(A(X^2),\circ_X\big)$.

Proof: Let a and b be elements of $A(X)$. Then

$$A_*\begin{pmatrix}x\\xx\end{pmatrix}(a)\circ_X A_*\begin{pmatrix}x\\xx\end{pmatrix}(b)=\ldots$$

$$\ldots=A_*\begin{pmatrix}x_1x_2x_3\\x_1x_3\end{pmatrix}A^*\begin{pmatrix}x_1x_2x_3\\x_1x_2x_2x_3\end{pmatrix}\left[A_*\begin{pmatrix}x\\xx\end{pmatrix}(a)\times A_*\begin{pmatrix}x\\xx\end{pmatrix}(b)\right]$$

The product inside hooks is equal to

$$A_*\begin{pmatrix}x_1x_2\\x_1x_1x_2x_2\end{pmatrix}(a\times b)$$

and the square

$$\begin{array}{ccc}X & \xrightarrow{\begin{pmatrix}x\\xx\end{pmatrix}} & X^2\\ {\scriptstyle\begin{pmatrix}x\\xxx\end{pmatrix}}\Big\downarrow & & \Big\downarrow{\scriptstyle\begin{pmatrix}x_1x_2\\x_1x_1x_2x_2\end{pmatrix}}\\ X^3 & \xrightarrow[\begin{pmatrix}x_1x_2x_3\\x_1x_2x_2x_3\end{pmatrix}]{} & X^4\end{array}$$

is cartesian. So

$$A^*\begin{pmatrix}x_1x_2x_3\\x_1x_2x_2x_3\end{pmatrix}A_*\begin{pmatrix}x_1x_2\\x_1x_1x_2x_2\end{pmatrix}=A_*\begin{pmatrix}x\\xxx\end{pmatrix}A^*\begin{pmatrix}x\\xx\end{pmatrix}$$

and

$$A_*\begin{pmatrix}x\\xx\end{pmatrix}(a)\circ_X A_*\begin{pmatrix}x\\xx\end{pmatrix}(b)=A_*\begin{pmatrix}x_1x_2x_3\\x_1x_3\end{pmatrix}A_*\begin{pmatrix}x\\xxx\end{pmatrix}A^*\begin{pmatrix}x\\xx\end{pmatrix}(a\times b)=A_*\begin{pmatrix}x\\xx\end{pmatrix}(a.b)$$

This proves the lemma, since moreover $A_*\begin{pmatrix}x\\xx\end{pmatrix}(\varepsilon_X)=1_{A(X^2)}$. ∎

Notation: *If X is a G-set, and if $a \in A(X)$, I will set*

$$\tilde{a} = A_* \begin{pmatrix} x \\ xx \end{pmatrix} (a)$$

If Y is a G-set, and if j is an idempotent of $\big(A(Y), .\big)$, then $\tilde{j}$ is an idempotent of $A(Y^2)$. So $\Omega \times \tilde{j}$ is an idempotent of $A(\Omega \times Y)$. Then:

Proposition 5.5.4: With those notations, if $\epsilon \in A_* \left(\begin{smallmatrix} y \\ \bullet \end{smallmatrix}\right) \big(A(Y).j.A(Y)\big)$, the module $A_{\Omega \times Y}.(\Omega \times \tilde{j})$ is a progenerator.

Proof: The hypothesis implies that A is projective relative to Y, and that there exists elements α_i and β_i in $A(Y)$, for $1 \leq i \leq n$, such that

$$\varepsilon = A_* \begin{pmatrix} y \\ \bullet \end{pmatrix} (\sum_{i=1}^{n} \alpha_i.j.\beta_i) \tag{5.1}$$

This expression can be transformed thanks to the following lemma:

Lemma 5.5.5: If $\alpha \in A(Y)$, let $\varepsilon \times \alpha$ (resp. $\alpha \times \varepsilon$) be the element α of $A(Y)$, viewed as $\mathrm{Hom}_{\mathcal{C}_A}(Y, \bullet)$ (resp. as $\mathrm{Hom}_{\mathcal{C}_A}(\bullet, Y)$). Then for α and β in $A(Y)$

$$(\varepsilon \times \alpha) \circ_Y (\beta \times \varepsilon) = A_* \begin{pmatrix} y \\ \bullet \end{pmatrix} (\alpha.\beta)$$

$$\tilde{\beta} \circ_Y (\alpha \times \varepsilon) = (\beta.\alpha) \times \varepsilon$$

$$(\varepsilon \times \alpha) \circ_Y \tilde{\beta} = \varepsilon \times .(\alpha.\beta)$$

Proof: By definition

$$(\varepsilon \times \alpha) \circ_Y (\beta \times \varepsilon) = A_* \begin{pmatrix} y \\ \bullet \end{pmatrix} A^* \begin{pmatrix} y \\ yy \end{pmatrix} (\alpha \times \beta) = A_* \begin{pmatrix} y \\ \bullet \end{pmatrix} (\alpha.\beta)$$

and the first equality follows. For the second

$$\tilde{\beta} \circ_Y (\alpha \times \varepsilon) = A_* \begin{pmatrix} y_1 y_2 \\ y_1 \end{pmatrix} A^* \begin{pmatrix} y_1 y_2 \\ y_1 y_2 y_2 \end{pmatrix} \left[A_* \begin{pmatrix} y \\ yy \end{pmatrix} (\beta) \times \alpha \right]$$

The product inside hooks is

$$A_* \begin{pmatrix} y_1 y_2 \\ y_1 y_1 y_2 \end{pmatrix} (\beta \times \alpha)$$

As the square

$$\begin{array}{ccc} Y & \xrightarrow{\begin{pmatrix} y \\ yy \end{pmatrix}} & Y^2 \\ {\scriptstyle \begin{pmatrix} y \\ yy \end{pmatrix}} \downarrow & & \downarrow {\scriptstyle \begin{pmatrix} y_1 y_2 \\ y_1 y_1 y_2 \end{pmatrix}} \\ Y^2 & \xrightarrow[\begin{pmatrix} y_1 y_2 \\ y_1 y_2 y_2 \end{pmatrix}]{} & Y^3 \end{array}$$

is cartesian, I have

$$A^*\begin{pmatrix} y_1y_2 \\ y_1y_2y_2 \end{pmatrix} A_*\begin{pmatrix} y_1y_2 \\ y_1y_1y_2 \end{pmatrix} = A_*\begin{pmatrix} y \\ yy \end{pmatrix} A^*\begin{pmatrix} y \\ yy \end{pmatrix}$$

and so

$$\tilde{\beta} \circ_Y (\alpha\times\varepsilon) = A_*\begin{pmatrix} y_1y_2 \\ y_1 \end{pmatrix} A_*\begin{pmatrix} y \\ yy \end{pmatrix} A^*\begin{pmatrix} y \\ yy \end{pmatrix} (\beta\times\alpha) = A_*\begin{pmatrix} y \\ y \end{pmatrix} A^*\begin{pmatrix} y \\ yy \end{pmatrix} (\beta\times\alpha) = \alpha.\beta$$

which is the element $(\beta.\alpha)\times\varepsilon$. A similar computation proves the last assertion. ■

Proof of proposition: It follows from lemma that

$$(\varepsilon\times\alpha)\circ_Y \tilde{j} \circ_Y (\beta\times\varepsilon) = A_*\begin{pmatrix} y \\ \bullet \end{pmatrix} (\alpha.\beta)$$

and equality (5.1) can also be written

$$\varepsilon = \sum_{i=1}^{n} (\varepsilon\times\alpha_i)\circ_Y \tilde{j} \circ_Y (\beta_i\times\varepsilon)$$

Taking the image of this equality under the functor $\Omega\times -$ gives

$$1_{A(\Omega^2)} = \sum_{i=1}^{n} \big(\Omega\times(\alpha_i\times\varepsilon)\big) \circ_{\Omega\times Y} (\Omega\times\tilde{j}) \circ_{\Omega\times Y} \big(\Omega\times(\beta_i\times\varepsilon)\big)$$

This proves that the unit of A_Ω factors through $A_{\Omega\times Y}(\Omega\times\tilde{j})$, so the latter is a progenerator. ■

Proposition 5.5.6: If H is a subgroup of G, let

$$\Omega_H = \coprod_{K\subseteq H} G/K$$

I denote by π the map from Ω_H to G/H defined by $\pi(xK) = xH$. If j is an idempotent of $A(H)$, I denote by $J = A^*(\widetilde{\pi})(j)$ the associated idempotent of $A(\Omega_H^2)$. Then:

1. **If $\varepsilon \in t_H^G\big(A(H).j.A(H)\big)$, then $A_{\Omega_H}.J$ is a progenerator.**
2. **If $A(\bullet)$ has a unique maximal two-sided ideal, if H is minimal such that A is projective relative to G/H, and if $A_{\Omega_H}.J$ is a progenerator, then $\varepsilon \in t_H^G\big(A(H).j.A(H)\big)$.**

Proof: The inclusion from Ω_H into Ω induces an embedding from $A(\Omega_H^2)$ into $A(\Omega^2)$, which can easily be described: indeed, the isomorphism of proposition 4.4.1

$$\mu(A) \simeq A(\Omega^2) = \bigoplus_{\substack{K,L\subseteq G \\ x\in K\backslash G/L}} A(K\cap {}^xL)$$

provides an identification of the element $a \in A(K \cap {}^xL)$ with

$$t_{K\cap^x L}^K \lambda_{K\cap^x L,a} c_x r_{K^x\cap L}^L$$

Now the image of $A(\Omega_H^2)$ identifies with

$$A(\Omega_H^2) \simeq \bigoplus_{\substack{K,L\subseteq H\\ x\in K\backslash G/L}} A(K\cap {}^xL)$$

and the embedding is just an inclusion of components. Moreover, by definition of π

$$A^*(\pi)(j) = \bigoplus_{L\subseteq H} r_L^H(j) \in A(\Omega_H) = \bigoplus_{L\subseteq H} A(L)$$

It follows that the element $J = \widetilde{A^*(\pi)}(j)$ is equal to the element

$$J = \sum_{L\subseteq H} t_L^L \lambda_{L,r_L^H(j)} c_1 r_L^L$$

corresponding to the double class $L.1.L$. If

$$A(\Omega^2) J A(\Omega^2) = A(\Omega^2)$$

then the identity morphism of Ω (in $\mathcal{C}_A$) factors through $A_{\Omega_H}.J$, so the latter is a progenerator.

But the unit $1_{A(\Omega^2)}$ is equal to

$$1_{A(\Omega^2)} = \sum_{K\subseteq G} t_K^K \lambda_{K,\varepsilon_K} c_1 r_K^K$$

If $\varepsilon \in t_H^G\big(A(H).j.A(H)\big)$, then there exists elements α_i and β_i of $A(H)$, for $1 \le i \le n$, such that

$$\varepsilon = t_H^G(\sum_{i=1}^n \alpha_i.j.\beta_i)$$

Now for any subgroup K of G

$$\varepsilon_K = r_K^G(\varepsilon) = \sum_{\substack{i=1\\ x\in K\backslash G/H}}^n t_{K\cap^x H}^K\Big({}^x r_{K^x\cap H}^H(\alpha_i).{}^x r_{K^x\cap H}^H(j).{}^x r_{K^x\cap H}^H(\beta_i)\Big)$$

Now setting $K_x = K^x \cap H$, replacing this expression of ε_K in $1_{A(\Omega^2)}$, and using relations of proposition 4.4.1, I get

$$1_{A(\Omega^2)} = \sum_{\substack{i=1\\ x\in K\backslash G/H}}^n t_{{}^xK_x}^K \lambda_{{}^xK_x,{}^x r_{K_x}^H(\alpha_i)} c_x r_{K_x}^{K_x} . t_{K_x}^{K_x} \lambda_{K_x,r_{K_x}^H(j)} r_{K_x}^{K_x} . t_{K_x}^{K_x} \lambda_{K_x,r_{K_x}^H(\beta_i)} c_x^{-1} r_{{}^xK_x}^K$$

Setting finally

$$u_{i,K,x} = t_{{}^xK_x}^K \lambda_{{}^xK_x,{}^x r_{K_x}^H(\alpha_i)} c_x r_{K_x}^{K_x}$$

$$v_{i,K,x} = t_{K_x}^{K_x} \lambda_{K_x,r_{K_x}^H(\beta_i)} c_x^{-1} r_{{}^xK_x}^K$$

I have the following equality

$$1_{A(\Omega^2)} = \sum_{\substack{i=1 \\ K\subseteq G \\ x\in K\backslash G/H}}^{n} u_{i,K,x} J v_{i,K,x}$$

proving that $1_{A(\Omega^2)}$ is in the two-sided ideal generated by J, so that $A_{\Omega_H}.J$ is a progenerator.

Conversely, if $A_{\Omega_H}.J$ is a progenerator, then the identity of $A(\Omega^2)$ factors through $A_{\Omega_H}.J$, and then

$$A(\Omega^2)JA(\Omega^2) = A(\Omega^2)$$

Now let α_i and β_i, for $1 \leq i \leq n$, such that

$$1_{A(\Omega^2)} = \sum_{1\leq i\leq n} \alpha_i J \beta_i$$

Then

$$t_G^G r_G^G = t_G^G \lambda_{G,\varepsilon} r_G^G = \sum_{1\leq i\leq n} t_G^G \alpha_i J \beta_i r_G^G$$

The elements $t_G^G \alpha_i$ are linear combinations of elements of the form

$$t_K^G \lambda_{K,a} c_x r_{K^x}^{K^x}$$

for suitable subgroups K, and elements $a \in A(K)$ and $x \in G$. Moreover

$$t_K^G \lambda_{K,a} c_x r_{K^x}^{K^x} = c_x t_{K^x}^G \lambda_{K^x,a^x} r_{K^x}^{K^x} = t_{K^x}^G \lambda_{K^x,a^x} r_{K^x}^{K^x}$$

and the elements $t_G^G \alpha_i$ are linear combinations of elements of the form

$$t_K^G \lambda_{K,a} r_K^K$$

Similarly, the elements $\beta_i r_G^G$ are linear combinations of elements of the form

$$t_K^K \lambda_{K,a} r_G^G$$

Moreover, as

$$J = \sum_{L\subseteq H} t_L^L \lambda_{L,r_L^H(j)} r_L^L$$

It follows that

$$t_K^G \lambda_{K,a} r_K^K J r_{K'}^{K'} \lambda_{K',a'} r_{K'}^G$$

is zero is $K \neq K'$, or if $K = K' \not\subseteq H$. Finally, there exists element a_U and b_U of $A(U)$, for $U \subseteq H$, such that

$$t_G^G \lambda_{G,\varepsilon} r_G^G = \sum_{U\subseteq H} t_U^G \lambda_{U,a_U} r_U^U t_U^U \lambda_{U,r_U^H(j)} r_U^U t_U^U \lambda_{U,b_U} r_U^G = \ldots$$

$$\ldots = \sum_{U\subseteq H} t_U^G \lambda_{U,a_U.r_U^H(j).b_U} r_U^G = t_G^G \lambda_{G,\sum_{U\subseteq H} t_U^G(a_U.r_U^H(j).b_U)} r_G^G$$

Now proposition 4.4.1 shows that

$$\varepsilon = \sum_{U \subseteq H} t_U^G(a_U.r_U^H(j).b_U)$$

But the images $t_U^G\big(A(U).r_U^H(j).A(U)\big)$ are two-sided ideals of $A(G) = A(\bullet)$. If $A(\bullet)$ has a unique maximal two-sided ideal, then there exists a subgroup H of G such that

$$\varepsilon \in t_U^G\big(A(U).r_U^H(j).A(U)\big)$$

Now the minimality of H implies $U = H$, and

$$\varepsilon \in t_H^G\big(A(H).j.A(H)\big)$$

which proves the proposition. ∎

Remark: If $A(\bullet)$ is a local ring, then there exists elements a and b of $A(H)$ such that

$$\varepsilon = t_H^G(a.j.b)$$

If Rosenberg's lemma holds in $A(H)$ (for instance if R is a complete local noetherian ring, and if $A(H)$ is of finite type over R), and if j is primitive, then this condition entirely determines j up to conjugation by an invertible element of $A(H)$ (the argument of Puig in [11] can be applied here without change).

Chapter 6

Construction of Green functors

6.1 The functors $\mathcal{H}(M,M)$

If M is a Mackey functor, I know that $\mathcal{H}(M,M)$ is a Green functor. To describe its product $\times$, I observe that the identity morphism from $\mathcal{H}(M,M)$ to itself must turn M into an $\mathcal{H}(M,M)$-module. Let X and Y be G-sets. If $f \in \mathcal{H}(M,M)(X) = \mathrm{Hom}_{Mack(G)}(M, M_X)$, and if $m \in M(Y)$, I must find an element $f \times m$ of $M(X \times Y)$. But f is determined by morphisms

$$f_Z : M(Z) \rightarrow M_X(Z) = M(ZX)$$

In particular, I have the element $f_Y(m)$ in $M(Y \times X)$, so the element $M_*\binom{yx}{xy}\big(f_Y(m)\big)$ in $M(XY)$. This is the element I am looking for:

Proposition 6.1.1: Let M be a Mackey functor, and let X and Y be G-sets. The Green functor structure of $\mathcal{H}(M,M)$ and the $\mathcal{H}(M,M)$-module structure of M are given by the following products:

- **If $f \in \mathcal{H}(M,M)(X)$ and $g \in \mathcal{H}(M,M)(Y)$, then $f \times g$ is the morphism from M to M_{XY} defined for a G-set Z and $m \in M(Z)$ by**

$$(f \times g)_Z(m) = M_*\begin{pmatrix} zyx \\ zxy \end{pmatrix} \circ f_{ZY} \circ g_Z(m)$$

- **If $f \in \mathcal{H}(M,M)(X) = \mathrm{Hom}_{Mack(G)}(M, M_X)$ and $m \in M(Y)$, then $f \times m$ is the element of $M(XY)$ defined by**

$$f \times m = M_*\begin{pmatrix} yx \\ xy \end{pmatrix} \circ f_Y(m)$$

In particular, if moreover M is an A-module for the Green functor A, then the image of the element $a \in A(X)$ under the morphism from A to $\mathcal{H}(M,M)$ is the morphism from M to M_X defined by

$$m \in M(Y) \mapsto M_*\begin{pmatrix} yx \\ xy \end{pmatrix}(a \times m)$$

Proof: Let A be a Green functor, and let M be an A-module. In the case when $X = G/H$ and $Y = G/K$ for subgroups H and K of G, I know that the product of $a \in A(X)$ and $m \in M(Y)$ is the element of $M(G/H \times G/K)$ defined by

$$a \times m = \bigoplus_{x \in H\backslash G/K} r^H_{H\cap {}^xK}(a).r^{{}^xK}_{H\cap {}^xK}({}^xm)$$

up to identification of $G/H \times G/K$ with $\coprod_{x\in H\backslash G/K} G/(H \cap {}^xK)$ by the inverse of the map $g.(H \cap {}^xK) \mapsto (gH, gxK)$.

On the other hand, the element a defines an endomorphism ϕ_a of $\mathrm{Res}^G_H M$ by

$$m \in M(L) \mapsto \phi_{a,L}(m) = r^H_L(a).m \quad \text{for} \quad L \subseteq H$$

By adjunction, this element defines a morphism from M to $M_{G/H} = \mathrm{Ind}^G_H \mathrm{Res}^G_H M$ given by

$$m \in M(K) \mapsto \bigoplus_{x\in H\backslash G/K} \phi_{a,H\cap {}^xK}\Big(r^{{}^xK}_{H\cap {}^xK}({}^xm)\Big) = \bigoplus_{x\in H\backslash G/K} r^H_{H\cap {}^xK}(a).r^{{}^xK}_{H\cap {}^xK}({}^xm)$$

up to identification of $G/K \times G/H$ with $\coprod_{x\in H\backslash G/K} G/(H \cap {}^xK)$ by the inverse of the map $g.(H \cap^x K) \mapsto (gxK, gH)$.

It follows that a does map to the morphism from M to M_X defined by

$$m \in M(Y) \mapsto M_* \begin{pmatrix} xy \\ yx \end{pmatrix} (a \times m)$$

In the case $A = \mathcal{H}(M,M)$, this formula gives for $f \in \mathcal{H}(M,M)(X)$ and $m \in M(Y)$

$$f_Y(m) = M_* \begin{pmatrix} xy \\ yx \end{pmatrix} (f \times m)$$

Taking the image under $M_* \binom{yx}{xy}$ gives

$$f \times m = M_* \begin{pmatrix} yx \\ xy \end{pmatrix} f_Y(m)$$

This formula gives a way to compute $f \times g$, for $f \in \mathcal{H}(M,M)(X)$ and $g \in \mathcal{H}(M,M)(Y)$: indeed, if $m \in M(Z)$, I have

$$f \times (g \times m) = f \times \left[M_* \begin{pmatrix} zy \\ yz \end{pmatrix} \circ g_Z(m) \right] = M_* \begin{pmatrix} yzx \\ xyz \end{pmatrix} \circ f_{YZ} \circ M_* \begin{pmatrix} zy \\ yz \end{pmatrix} \circ g_Z(m)$$

But f is a morphism of Mackey functors. So

$$f_{YZ} \circ M_* \begin{pmatrix} zy \\ yz \end{pmatrix} = M_* \begin{pmatrix} zyx \\ yzx \end{pmatrix} \circ f_{ZY}$$

and then

$$f \times (g \times m) = M_* \begin{pmatrix} yzx \\ xyz \end{pmatrix} M_* \begin{pmatrix} zyx \\ yzx \end{pmatrix} \circ f_{ZY} \circ g_Z(m) = M_* \begin{pmatrix} zyx \\ xyz \end{pmatrix} \circ f_{ZY} \circ g_Z(m) = \ldots$$

$$\dots = M_*\begin{pmatrix} zxy \\ xyz \end{pmatrix} \circ \left(M_*\begin{pmatrix} zyx \\ zxy \end{pmatrix} \circ f_{ZY} \circ g_Z\right)(m)$$

This expression must be equal to

$$(f \times g) \times m = M_*\begin{pmatrix} zxy \\ xyz \end{pmatrix} \circ (f \times g)_Z(m)$$

and it follows that

$$(f \times g)_Z = M_*\begin{pmatrix} zyx \\ zxy \end{pmatrix} \circ f_{ZY} \circ g_Z$$

which proves the proposition. ∎

6.1.1 The product $\hat{\circ}$

The previous proposition leads to the following definition:

Definition: *Let M, N and P be Mackey functors. If X and Y are G-sets, if $a \in \mathcal{H}(M,N)(X)$ and $b \in \mathcal{H}(N,P)(Y)$, I denote by $b \,\hat{\circ}\, a$ the element of $\mathcal{H}(M,P)(Y \times X)$ defined on a G-set U by*

$$(b \,\hat{\circ}\, a)_U = P_*\begin{pmatrix} uxy \\ uyx \end{pmatrix} \circ b_{UX} \circ a_U : M(U) \to M(UYX) = M_{YX}(U)$$

Actually, I will only know that $b \,\hat{\circ}\, a$ is in $\mathcal{H}(M,P)(YX)$ after proving the next proposition:

Proposition 6.1.2: The product $\hat{\circ}$ defines a bilinear morphism from $\mathcal{H}(N,P)$, $\mathcal{H}(M,N)$ to $\mathcal{H}(M,P)$, which is moreover associative.

Proof: First it is clear that $b \,\hat{\circ}\, a$ is a morphism of Mackey functors from M to P_{YX}: indeed, if $f : U \to V$ is a morphism of G-sets, then the following diagram is commutative:

$$\begin{array}{ccccccc}
M(U) & \xrightarrow{a_U} & N(UX) & \xrightarrow{b_{UX}} & P(UXY) & \xrightarrow{P_*\begin{pmatrix} uxy \\ uyx \end{pmatrix}} & P(UYX) \\
{\scriptstyle M_*(f)}\downarrow & & {\scriptstyle N_*(fI_X)}\downarrow & & {\scriptstyle P_*(fI_XI_Y)}\downarrow & & \downarrow{\scriptstyle P_*(fI_YI_X)} \\
M(V) & \xrightarrow[a_V]{} & N(VX) & \xrightarrow[b_{VX}]{} & P(VXY) & \xrightarrow[P_*\begin{pmatrix} vxy \\ vyx \end{pmatrix}]{} & P(VYX)
\end{array}$$

The two squares on the left are commutative because a and b arc morphisms of Mackey functors. The right square is commutative because its the image by P_* of a commutative square. It follows that the products along the edgés are equal, so

$$P_{YX,*}(f) \circ (b \,\hat{\circ}\, a)_U = (b \,\hat{\circ}\, a)_V \circ M_*(f)$$

The diagram obtained with M^*, N^* and P^*, and by reversing the vertical arrows, is also commutative, for the same reasons (and also because $P_*\begin{pmatrix} uxy \\ uyx \end{pmatrix} = P^*\begin{pmatrix} uyx \\ uxy \end{pmatrix}$). Thus $b \,\hat{\circ}\, a$ is a morphism of Mackey functors.

The product $\hat{\circ}$ is moreover bifunctorial in X and Y: if $\alpha : X \to X'$ and $\beta : Y \to Y'$ are morphisms of G-sets, then the following diagram, where I set

$$\mathcal{H} = \mathcal{H}(M,N) \qquad \mathcal{K} = \mathcal{H}(N,P)$$

is commutative:

$$\begin{array}{ccccccc}
M(U) & \xrightarrow{a_U} & N(UX) & \xrightarrow{b_{UX}} & P(UXY) & \xrightarrow{P_*\begin{pmatrix} uxy \\ uyx \end{pmatrix}} & P(UYX) \\
{\scriptstyle Id}\downarrow & (1) & \downarrow{\scriptstyle N_*(I_U\alpha)} & (2) & \downarrow{\scriptstyle P_*(I_U\alpha I_Y)} & (3) & \downarrow{\scriptstyle P_*(I_U I_Y \alpha)} \\
M(U) & \xrightarrow{\mathcal{H}_*(\alpha)(a)_U} & N(UX') & \xrightarrow{b_{UX'}} & P(UX'Y) & \xrightarrow{P_*\begin{pmatrix} ux'y \\ uyx' \end{pmatrix}} & P(UYX') \\
{\scriptstyle Id}\downarrow & (4) & \downarrow{\scriptstyle Id} & (5) & \downarrow{\scriptstyle P_*(I_U I_{X'}\beta)} & (6) & \downarrow{\scriptstyle P_*(I_U \beta I_{X'})} \\
M(U) & \xrightarrow{\mathcal{H}_*(\alpha)(a)_U} & N(UX') & \xrightarrow{\mathcal{K}_*(\beta)(b)_{UX'}} & P(UX'Y') & \xrightarrow{P_*\begin{pmatrix} ux'y' \\ uy'x' \end{pmatrix}} & P(UY'X')
\end{array}$$

The squares (1) and (5) are commutative by definition of $\mathcal{H}_*(\alpha)$ and of $\mathcal{K}_*(\beta)$. The square (2) is commutative because b is a morphism of Mackey functors from N to P_Y. The squares (3) and (6) are commutative because they are the image under P_* of a commutative square. Finally the square (4) is trivially commutative.

The two possible products from $M(U)$ to $P(UY'X')$ along the edges of that diagram give the equality

$$\mathcal{H}(N,P)_*(\beta)(b) \,\hat{\circ}\, \mathcal{H}(M,N)_*(\alpha)(a) = \mathcal{H}(M,P)_*(\beta \times \alpha)(b \,\hat{\circ}\, a)$$

which proves that the product $\hat{\circ}$ is covariant.

If now $\alpha : X' \to X$ and $\beta : Y' \to Y$ are morphisms of G-sets, then the similar commutative diagram

$$\begin{array}{ccccccc}
M(U) & \xrightarrow{a_U} & N(UX) & \xrightarrow{b_{UX}} & P(UXY) & \xrightarrow{P_*\begin{pmatrix} uxy \\ uyx \end{pmatrix}} & P(UYX) \\
{\scriptstyle Id}\downarrow & & \downarrow{\scriptstyle N^*(I_U\alpha)} & & \downarrow{\scriptstyle P^*(I_U\alpha I_Y)} & & \downarrow{\scriptstyle P^*(I_U I_Y \alpha)} \\
M(U) & \xrightarrow{\mathcal{H}^*(\alpha)(a)_U} & N(UX') & \xrightarrow{b_{UX'}} & P(UX'Y) & \xrightarrow{P_*\begin{pmatrix} ux'y \\ uyx' \end{pmatrix}} & P(UYX') \\
{\scriptstyle Id}\downarrow & & \downarrow{\scriptstyle Id} & & \downarrow{\scriptstyle P^*(I_U I_{X'}\beta)} & & \downarrow{\scriptstyle P^*(I_U \beta I_{X'})} \\
M(U) & \xrightarrow{\mathcal{H}^*(\alpha)(a)_U} & N(UX') & \xrightarrow{\mathcal{K}^*(\beta)(b)_{UX'}} & P(UX'Y') & \xrightarrow{P_*\begin{pmatrix} ux'y' \\ uy'x' \end{pmatrix}} & P(UY'X')
\end{array}$$

shows that the product $\hat{\circ}$ is contravariant, hence that if defines a bilinear morphism from $\mathcal{H}(N,P)$, $\mathcal{H}(M,N)$ to $\mathcal{H}(M,P)$.

I must still show that the product $\hat{\circ}$ is associative: so let Q be a Mackey functor, let Z be a G-set, and let c be an element of $\mathcal{H}(P,Q)(Z)$. Then

$$\big(c\,\hat{\circ}\,(b\,\hat{\circ}\,a)\big)_U = Q_*\begin{pmatrix} uyxz \\ uzyx \end{pmatrix} \circ c_{UYX} \circ (b\,\hat{\circ}\,a)_U = Q_*\begin{pmatrix} uyxz \\ uzyx \end{pmatrix} \circ c_{UYX} \circ P_*\begin{pmatrix} uxy \\ uyx \end{pmatrix} \circ b_{UX} \circ a_U$$

Since c is a morphism of Mackey functors from P to Q_Z, I have

$$c_{UYX} \circ P_*\begin{pmatrix} uxy \\ uyx \end{pmatrix} = Q_*\begin{pmatrix} uxyz \\ uyxz \end{pmatrix} \circ c_{UXYZ}$$

and so

$$\big(c\,\hat{\circ}\,(b\,\hat{\circ}\,a)\big)_U = Q_*\begin{pmatrix} uyxz \\ uzyx \end{pmatrix} \circ Q_*\begin{pmatrix} uxyz \\ uyxz \end{pmatrix} \circ c_{UXYZ} \circ b_{UX} \circ a_U = \ldots$$

$$\ldots = Q_*\begin{pmatrix} uxyz \\ uzyx \end{pmatrix} \circ c_{UXYZ} \circ b_{UX} \circ a_U$$

On the other hand

$$\big((c\,\hat{\circ}\,b)\,\hat{\circ}\,a\big)_U = Q_*\begin{pmatrix} uxzy \\ uzyx \end{pmatrix} \circ (c\,\hat{\circ}\,b)_{UX} \circ a_U = \ldots$$

$$\ldots = Q_*\begin{pmatrix} uxzy \\ uzyx \end{pmatrix} \circ Q_*\begin{pmatrix} uxyz \\ uxzy \end{pmatrix} \circ c_{UXY} \circ b_{UX} \circ a_U = Q_*\begin{pmatrix} uxyz \\ uzyx \end{pmatrix} \circ c_{UXYZ} \circ b_{UX} \circ a_U$$

This shows that the product $\hat{\circ}$ is associative, and proves the proposition. ∎

6.2 The opposite functor of a Green functor

Definition: *Let A be a Green functor for the group G. I call opposite functor of A, and I denote by A^{op}, the Green functor equal to A as Mackey functor, and endowed with the product $\times^{op}$ defined for G-sets X and Y and elements $a \in A^{op}(X) = A(X)$ and $b \in A^{op}(Y)$ by*

$$a \times^{op} b = A_*\begin{pmatrix} yx \\ xy \end{pmatrix}(b \times a)$$

and the unit $\varepsilon \in A^{op}(\bullet) = A(\bullet)$

The reason for this definition is the following proposition:

Proposition 6.2.1: Let A be a Green functor for the group G. Then A^{op} is a Green functor. Moreover for any G-set X

$$(A^{op}(X), .^{op}) = (A(X), .)^{op}$$

Proof: I must show that the product $\times^{op}$ is bifunctorial, associative, and unitary. So let $f : X \to X'$ and $g : Y \to Y'$ be morphisms of G-sets. Then for $a \in A(X)$ and $b \in A(Y)$, I have

$$A_*^{op}(f)(a)\times^{op} A_*^{op}(g)(b) = A_*\begin{pmatrix} y'x' \\ x'y' \end{pmatrix}\big(A_*(g)(b)\times A_*(f)(a)\big) = A_*\begin{pmatrix} y'x' \\ x'y' \end{pmatrix} A_*(g\times f)(b\times a)$$

But

$$\begin{pmatrix} x'y' \\ y'x' \end{pmatrix}\circ(f\times g) = (g\times f)\circ\begin{pmatrix} xy \\ yx \end{pmatrix}$$

so

$$A_*(g\times f) = A_*\left(\begin{pmatrix} x'y' \\ y'x' \end{pmatrix}\circ(f\times g)\circ\begin{pmatrix} yx \\ xy \end{pmatrix}\right) = A_*\begin{pmatrix} x'y' \\ y'x' \end{pmatrix} A_*(f\times g) A_*\begin{pmatrix} yx \\ xy \end{pmatrix}$$

and

$$A_*^{op}(f)(a)\times^{op} A_*^{op}(g)(b) = A_*\begin{pmatrix} y'x' \\ x'y' \end{pmatrix} A_*\begin{pmatrix} x'y' \\ y'x' \end{pmatrix} A_*(f\times g) A_*\begin{pmatrix} yx \\ xy \end{pmatrix}(b\times a) = \dots$$

$$\dots = A_*(f\times g)(a\times^{op} b)$$

proving that $\times^{op}$ is covariant. Similarly, if $a' \in A(X')$ and $b' \in A(Y')$, then

$$A^{op*}(f)(a')\times^{op} A^{op*}(f)(b') = A_*\begin{pmatrix} yx \\ xy \end{pmatrix}\big(A^*(g)(b')\times A^*(f)(a')\big) = \dots$$

$$\dots = A_*\begin{pmatrix} yx \\ xy \end{pmatrix} A^*(g\times f)(b'\times a')$$

Since

$$A^*(g\times f) = A^*\begin{pmatrix} yx \\ xy \end{pmatrix} A^*(f\times g) A^*\begin{pmatrix} x'y' \\ y'x' \end{pmatrix}$$

and since $A^*\left(\begin{smallmatrix} x'y' \\ y'x' \end{smallmatrix}\right) = A_*\left(\begin{smallmatrix} y'x' \\ x'y' \end{smallmatrix}\right)$, I have

$$A^{op*}(f)(a')\times^{op} A^{op*}(f)(b') = A_*\begin{pmatrix} yx \\ xy \end{pmatrix} A^*\begin{pmatrix} yx \\ xy \end{pmatrix} A^*(f\times g) A_*\begin{pmatrix} y'x' \\ x'y' \end{pmatrix}(b'\times a') = \dots$$

$$\dots = A^*(f\times g)(a'\times^{op} b')$$

which proves that $\times^{op}$ is contravariant.

The product $\times^{op}$ is also associative: if X, Y, Z are G-sets, if $a \in A(X)$, $b \in A(Y)$, and $c \in A(Z)$, then

$$(a\times^{op} b)\times^{op} c = A_*\begin{pmatrix} zxy \\ xyz \end{pmatrix}\big(c\times(a\times^{op} b)\big) = A_*\begin{pmatrix} zxy \\ xyz \end{pmatrix}\left(c\times A_*\begin{pmatrix} yx \\ xy \end{pmatrix}(b\times a)\right) = \dots$$

$$\dots = A_*\begin{pmatrix} zxy \\ xyz \end{pmatrix} A_*\begin{pmatrix} zyx \\ zxy \end{pmatrix}(c\times b\times a) = A_*\begin{pmatrix} zyx \\ xyz \end{pmatrix}(c\times b\times a)$$

Similarly

$$a\times^{op}(b\times^{op} c) = A_*\begin{pmatrix} yzx \\ xyz \end{pmatrix}\big((b\times^{op} c)\times a\big) = A_*\begin{pmatrix} yzx \\ xyz \end{pmatrix}\left(A_*\begin{pmatrix} zy \\ yz \end{pmatrix}(c\times b)\times a\right) = \dots$$

$$\ldots = A_* \begin{pmatrix} yzx \\ xyz \end{pmatrix} A_* \begin{pmatrix} zyx \\ yzx \end{pmatrix} (c \times b \times a) = A_* \begin{pmatrix} zyx \\ xyz \end{pmatrix} (c \times b \times a)$$

which proves that $\times^{op}$ is associative. Finally if $a \in A(X)$, then

$$\varepsilon \times^{op} a = A_* \begin{pmatrix} x\bullet \\ \bullet x \end{pmatrix} (a \times \varepsilon) = A_* \begin{pmatrix} x \\ x \end{pmatrix} (a) = a$$

and it is clear similarly that $a \times^{op} \varepsilon = a$. So ε is a unit for A^{op}.

To complete the proof, I compute the product "$.^{op}$": if $a, b \in A(X)$, then

$$a.^{op}b = A^* \begin{pmatrix} x \\ xx \end{pmatrix} (a \times^{op} b) = A^* \begin{pmatrix} x \\ xx \end{pmatrix} A_* \begin{pmatrix} x_1x_2 \\ x_2x_1 \end{pmatrix} (b \times a)$$

As

$$A^* \begin{pmatrix} x \\ xx \end{pmatrix} A_* \begin{pmatrix} x_1x_2 \\ x_2x_1 \end{pmatrix} = A^* \begin{pmatrix} x \\ xx \end{pmatrix} A^* \begin{pmatrix} x_2x_1 \\ x_1x_2 \end{pmatrix}$$

and as

$$\begin{pmatrix} x_2x_1 \\ x_1x_2 \end{pmatrix} \circ \begin{pmatrix} x \\ xx \end{pmatrix} = \begin{pmatrix} x \\ xx \end{pmatrix}$$

I have

$$a.^{op}b = A^* \begin{pmatrix} x \\ xx \end{pmatrix} (b \times a) = b.a$$

and this proves the proposition. ∎

Remark: It follows that condition 2) of proposition 5.3.5 is equivalent to the commutativity of all the rings $(A(X), .)$, or to the commutativity of all the rings $A(H)$, for $H \subseteq G$.

6.2.1 Right modules

Definition: *Let A be a Green functor for G. A right module M for A, or module-A, is a Mackey functor for G, together with bilinear maps*

$$M(X) \times A(Y) \to M(X \times Y)$$

denoted by $(m, a) \mapsto m \times a$ for any G-sets X and Y, which are bifunctorial, associative, and unitary (in the obvious sense).
A morphism θ of modules-A from M to N is a morphism of Mackey functors from M to N, such that for any G-sets X and Y, and any $m \in M(X)$ and $a \in A(X)$

$$\theta_{X \times Y}(m \times a) = \theta_X(m) \times a$$

I can now speak of the category of modules-A, that I will denote by **Mod**-A. As in the case of ordinary rings:

Proposition 6.2.2: The category Mod-A is equivalent to the category A^{op}-Mod.

Proof: If M is a module-A, I turn it into an A^{op}-module by setting, for $a \in A(X)$ and $m \in M(Y)$

$$a \times^{op} m = M_*\begin{pmatrix} yx \\ xy \end{pmatrix}(m \times a)$$

Now the proof of proposition 6.2.1 can be carried word for word here, to prove that this product is bifunctorial, associative and unitary.
Conversely, if M is an A^{op}-module, I can turn it into a module-A by setting in the same conditions

$$m \times a = M_*\begin{pmatrix} xy \\ yx \end{pmatrix}(a \times^{op} m)$$

It is clear that those correspondences are functorial, and inverse to each other. The proposition follows.

6.2.2 The dual of an A-module

Definition: *If M is an A-module, the dual of M is the Mackey functor M° defined on the G-set X by*

$$M^\circ(X) = \mathrm{Hom}_R\big(M(X), R\big)$$

If $f : X \to Y$ is a morphism of G-sets, then

$$(M^\circ)_*(f) = {}^tM^*(f) \qquad (M^\circ)^*(f) = {}^tM_*(f)$$

where ${}^tl$ is the transpose map of l.

I define a right product on M° by setting, for G-sets X and Y, for $\phi \in M^\circ(X)$, for $a \in A(Y)$, and $m \in M(X \times Y)$

$$(\phi \times a)(m) = \phi(a \circ_Y^{op} m)$$

where by definition

$$a \circ_Y^{op} m = M_*\begin{pmatrix} xy \\ x \end{pmatrix} M^*\begin{pmatrix} xy \\ yxy \end{pmatrix}(a \times m)$$

Proposition 6.2.3: These definitions turn M° into a module-A.

Proof: First the product is bifunctorial: if $f : X \to X'$ and $g : Y \to Y'$ are morphisms of G-sets, if $\phi \in M^\circ(X)$, if $a \in A(Y)$, and $m' \in M(X' \times Y')$, then

$$\big((M^\circ)_*(f)(\phi) \times A_*(g)(a)\big)(m') = (M^\circ)_*(f)(\phi)\big(A_*(g)(a) \circ_{Y'}^{op} m'\big) = \dots$$

$$\dots = \phi M^*(f)\big(A_*(g)(a) \circ_{Y'}^{op} m'\big)$$

Moreover

$$A_*(g)(a) \circ_{Y'}^{op} m' = M_*\begin{pmatrix} x'y' \\ x' \end{pmatrix} M^*\begin{pmatrix} x'y' \\ y'x'y' \end{pmatrix}(A_*(g)(a) \times m') = \dots$$

$$\dots = M_*\begin{pmatrix} x'y' \\ x' \end{pmatrix} M^*\begin{pmatrix} x'y' \\ y'x'y' \end{pmatrix} M_*\begin{pmatrix} yx'y' \\ g(y)x'y' \end{pmatrix}(a \times m)$$

But the square

$$\begin{array}{ccc} X'Y & \xrightarrow{\begin{pmatrix} x'y \\ yx'g(y) \end{pmatrix}} & YX'Y' \\ {\scriptstyle \begin{pmatrix} x'y \\ x'g(y) \end{pmatrix}}\Big\downarrow & & \Big\downarrow{\scriptstyle \begin{pmatrix} yx'y' \\ g(y)x'y' \end{pmatrix}} \\ X'Y' & \xrightarrow[\begin{pmatrix} x'y' \\ y'x'y' \end{pmatrix}]{} & Y'X'Y' \end{array}$$

is cartesian. So

$$A_*(g)(a) \circ^{op}_{Y'} m' = M_*\begin{pmatrix} x'y' \\ x' \end{pmatrix} M_*\begin{pmatrix} x'y \\ x'g(y) \end{pmatrix} M^*\begin{pmatrix} x'y \\ yx'g(y) \end{pmatrix} (a \times m')$$

It follows that

$$\Big((M^o)_*(f)(\phi) \times A_*(g)(a)\Big)(m') = \ldots$$

$$\ldots = \phi M^*(f) M_*\begin{pmatrix} x'y' \\ x' \end{pmatrix} M_*\begin{pmatrix} x'y \\ x'g(y) \end{pmatrix} M^*\begin{pmatrix} x'y \\ yx'g(y) \end{pmatrix} (a \times m') = \ldots$$

$$\ldots = \phi M^*(f) M_*\begin{pmatrix} x'y \\ x' \end{pmatrix} M^*\begin{pmatrix} x'y \\ yx'g(y) \end{pmatrix} (a \times m')$$

The square

$$\begin{array}{ccc} XY & \xrightarrow{\begin{pmatrix} xy \\ f(x)y \end{pmatrix}} & X'Y \\ {\scriptstyle \begin{pmatrix} xy \\ x \end{pmatrix}}\Big\downarrow & & \Big\downarrow{\scriptstyle \begin{pmatrix} x'y \\ x' \end{pmatrix}} \\ X & \xrightarrow[f]{} & X' \end{array}$$

is cartesian. So

$$\Big((M^o)_*(f)(\phi) \times A_*(g)(a)\Big)(m') = \phi M_*\begin{pmatrix} xy \\ x \end{pmatrix} M^*\begin{pmatrix} xy \\ f(x)y \end{pmatrix} M^*\begin{pmatrix} x'y \\ yx'g(y) \end{pmatrix} (a \times m') = \ldots$$

$$\ldots = \phi M_*\begin{pmatrix} xy \\ x \end{pmatrix} M^*\begin{pmatrix} xy \\ yf(x)g(y) \end{pmatrix} (a \times m) = \ldots$$

$$\ldots = \phi M_*\begin{pmatrix} xy \\ x \end{pmatrix} M^*\begin{pmatrix} xy \\ yxy \end{pmatrix} M^*\begin{pmatrix} y_1xy_2 \\ y_1f(x)g(y_2) \end{pmatrix} (a \times m) = \ldots$$

$$\ldots = \phi M_*\begin{pmatrix} xy \\ x \end{pmatrix} M^*\begin{pmatrix} xy \\ yxy \end{pmatrix} \Big(a \times M^*(f \times g)(m')\Big) = \phi\Big(a \circ^{op}_Y M^*(f \times g)(m')\Big) = \ldots$$

$$\ldots = (\phi \times a) M^*(f \times g)(m') = (M^o)_*(f \times g)(\phi \times a)(m')$$

and this proves that the product is covariant.

If now $\phi' \in M^o(X')$, if $a' \in A(Y')$, and if $m \in M(X \times Y)$, then

$$\Big((M^o)^*(f)(\phi') \times A^*(g)(a')\Big)(m) = (M^o)^*(f)(\phi')\Big(A^*(g)(a') \circ^{op}_Y m\Big) = \ldots$$

$$\ldots = \phi' M_*(f)\Big(A^*(g)(a') \circ_Y^{op} m\Big)$$

Moreover

$$A^*(g)(a') \circ_Y^{op} m = M_*\begin{pmatrix} xy \\ x \end{pmatrix} M^*\begin{pmatrix} xy \\ yxy \end{pmatrix}\Big(A^*(g)(a') \times m\Big) = \ldots$$

$$\ldots = M_*\begin{pmatrix} xy \\ x \end{pmatrix} M^*\begin{pmatrix} xy \\ yxy \end{pmatrix} M^*\begin{pmatrix} y_1xy_2 \\ g(y_1)xy_2 \end{pmatrix}(a' \times m) = \ldots$$

$$\ldots = M_*\begin{pmatrix} xy \\ x \end{pmatrix} M^*\begin{pmatrix} xy \\ g(y)xy \end{pmatrix}(a' \times m)$$

So

$$\Big((M^o)^*(f)(\phi') \times A^*(g)(a')\Big)(m) = \phi' M_*(f) M_*\begin{pmatrix} xy \\ x \end{pmatrix} M^*\begin{pmatrix} xy \\ g(y)xy \end{pmatrix}(a' \times m) = \ldots$$

$$\ldots = \phi' M_*\begin{pmatrix} xy \\ f(x) \end{pmatrix} M^*\begin{pmatrix} xy \\ g(y)xy \end{pmatrix}(a' \times m)$$

On the other hand

$$(M^o)^*(f \times g)(\phi' \times a')(m) = (\phi' \times a')M_*(f \times g)(m) = \phi'\Big(a' \circ_{Y'}^{op} M_*(f \times g)(m)\Big)$$

But

$$a' \circ_{Y'}^{op} M_*(f \times g)(m) = M_*\begin{pmatrix} x'y' \\ x' \end{pmatrix} M^*\begin{pmatrix} x'y' \\ y'x'y' \end{pmatrix}\Big(a' \times M_*(f \times g)(m)\Big) = \ldots$$

$$\ldots = M_*\begin{pmatrix} x'y' \\ x' \end{pmatrix} M^*\begin{pmatrix} x'y' \\ y'x'y' \end{pmatrix} M_*\begin{pmatrix} y'xy \\ y'f(x)g(y) \end{pmatrix}(a' \times m)$$

As the square

$$\begin{array}{ccc} XY & \xrightarrow{\begin{pmatrix} xy \\ g(y)xy \end{pmatrix}} & Y'XY \\ {\scriptstyle f \times g}\downarrow & & \downarrow{\begin{pmatrix} y'xy \\ y'f(x)g(y) \end{pmatrix}} \\ X'Y' & \xrightarrow[\begin{pmatrix} x'y' \\ y'x'y' \end{pmatrix}]{} & Y'X'Y' \end{array}$$

is cartesian, this is also

$$a' \circ_{Y'}^{op} M_*(f \times g)(m) = M_*\begin{pmatrix} x'y' \\ x' \end{pmatrix} M_*(f \times g) M^*\begin{pmatrix} xy \\ g(y)xy \end{pmatrix}(a' \times m) = \ldots$$

$$\ldots = M_*\begin{pmatrix} xy \\ f(x) \end{pmatrix} M^*\begin{pmatrix} xy \\ g(y)xy \end{pmatrix}(a' \times m)$$

So I have

$$(M^o)^*(f \times g)(\phi' \times a')(m) = \phi' M_*\begin{pmatrix} xy \\ f(x) \end{pmatrix} M^*\begin{pmatrix} xy \\ g(y)xy \end{pmatrix}(a' \times m) = \ldots$$

$$\ldots = \Big((M^o)^*(f)(\phi') \times A^*(g)(a')\Big)(m)$$

which proves that the product is also contravariant.

I must still prove that the product is associative and unitary. So let X, Y, and Z be G-sets. If $\phi \in M^o(X)$, if $a \in A(Y)$, and $b \in A(Z)$, then for $m \in M(X \times Y \times Z)$, I have

$$\Big((\phi \times a) \times b\Big)(m) = (\phi \times a)(b \circ^{op}_Z m) = \phi\Big(a \circ^{op}_Y (b \circ^{op}_Z m)\Big)$$

But

$$a \circ^{op}_Y (b \circ^{op}_Z m) = M_*\begin{pmatrix} xy \\ x \end{pmatrix} M^*\begin{pmatrix} xy \\ yxy \end{pmatrix}\Big(a \times (b \circ^{op}_Z m)\Big)$$

Moreover

$$a \times (b \circ^{op}_Z m) = a \times M_*\begin{pmatrix} xyz \\ xy \end{pmatrix} M^*\begin{pmatrix} xyz \\ zxyz \end{pmatrix}(b \times m) = \ldots$$

$$\ldots = M_*\begin{pmatrix} y_1xy_2z \\ y_1xy_2 \end{pmatrix} M^*\begin{pmatrix} y_1xy_2z \\ y_1zxy_2z \end{pmatrix}(a \times b \times m)$$

As the square

$$\begin{array}{ccc} XYZ & \xrightarrow{\begin{pmatrix} xyz \\ yxyz \end{pmatrix}} & YXYZ \\ {\scriptstyle\begin{pmatrix} xyz \\ xy \end{pmatrix}}\Big\downarrow & & \Big\downarrow{\scriptstyle\begin{pmatrix} y_1xy_2z \\ y_1xy_2 \end{pmatrix}} \\ XY & \xrightarrow[\begin{pmatrix} xy \\ yxy \end{pmatrix}]{} & YXY \end{array}$$

is cartesian, I have

$$M^*\begin{pmatrix} xy \\ yxy \end{pmatrix} M_*\begin{pmatrix} y_1xy_2z \\ y_1xy_2 \end{pmatrix} = M_*\begin{pmatrix} xyz \\ xy \end{pmatrix} M^*\begin{pmatrix} xyz \\ yxyz \end{pmatrix}$$

and then

$$a\circ^{op}_Y(b\circ^{op}_Z m) = M_*\begin{pmatrix} xy \\ x \end{pmatrix} M_*\begin{pmatrix} xyz \\ xy \end{pmatrix} M^*\begin{pmatrix} xyz \\ yxyz \end{pmatrix} M^*\begin{pmatrix} y_1xy_2z \\ y_1zxy_2z \end{pmatrix}(a\times b\times m) = \ldots$$

$$\ldots = M_*\begin{pmatrix} xyz \\ x \end{pmatrix} M^*\begin{pmatrix} xyz \\ yzxyz \end{pmatrix}(a \times b \times m)$$

On the other hand

$$\Big(\phi \times (a \times b)\Big)(m) = \phi\Big((a \times b) \circ^{op}_{YZ} m\Big)$$

Since

$$(a \times b) \circ^{op}_{YZ} m = M_*\begin{pmatrix} xyz \\ x \end{pmatrix} M_*\begin{pmatrix} xyz \\ yzxyz \end{pmatrix}(a \times b \times m)$$

the product is associative.

Moreover, if $\phi \in M^o(X)$, and if $m \in M(X)$, then

$$(\phi \times \varepsilon_A)(m) = \phi(\varepsilon_A \circ^{op}_\bullet m)$$

As

$$\varepsilon_A \circ^{op}_\bullet m = M_*\begin{pmatrix} x\bullet \\ x \end{pmatrix} M^*\begin{pmatrix} x\bullet \\ \bullet x\bullet \end{pmatrix}(\varepsilon \times m) = m$$

the product is unitary, which completes the proof of the proposition. ∎

6.3 Tensor product of Green functors

If A and B are Green functors for the group G, then their tensor product $A\hat{\otimes}B$ can be given a structure of Green functor:

Definition: *If X and Y are G-sets, if (U,ϕ) is a G-set over X and (V,ψ) is a G-set over Y, if $a \in A(U)$, $b \in B(U)$, $c \in A(V)$ and $d \in B(V)$, I define the product of the element $[a\otimes b]_{(U,\phi)}$ of $A\hat{\otimes}B(X)$ by the element $[c\otimes d]_{(V,\psi)}$ of $A\hat{\otimes}B(Y)$ by setting*

$$[a\otimes b]_{(U,\phi)} \times [c\otimes d]_{(V,\psi)} = [(a\times c)\otimes(b\times d)]_{(U\times V,\phi\times\psi)} \in (A\hat{\otimes}B)(X\times Y)$$

If ε_A and ε_B are the units A and B, I set

$$\varepsilon = [\varepsilon_A\otimes\varepsilon_B]_{(\bullet,Id)} \in A\hat{\otimes}B(\bullet)$$

Proposition 6.3.1: Let A and B be Green functors for the group G. Then the product $\times$ turns $A\hat{\otimes}B$ into a Green functor, with unit ε.

Proof: First I must check that this product is well defined. So let $f:(U,\phi)\to(U',\phi')$ be a morphism of G-sets over X, let $a\in A(U)$ and $b'\in B(U')$. Then

$$[A_*(f)(a)\otimes b']_{(U',\phi')}\times[c\otimes d]_{(V,\psi)} = \Big[\big(A_*(f)(a)\times c\big)\otimes(b'\times d)\Big]_{(U'\times V,\phi'\times\psi)} = \ldots$$

$$\ldots = \Big[\big(A_*(f\times Id)(a\times c)\big)\otimes(b'\times d)\Big]_{(U'\times V,\phi'\times\psi)} = \ldots$$

$$\ldots = [(a\times c)\otimes B^*(f\times id)(b'\times d)]_{(U\times V,\phi\times\psi)} = \ldots$$

$$\ldots = \Big[(a\times c)\otimes\big(B^*(f)(b')\times d\big)\Big]_{(U\times V,\phi\times\psi)}$$

which does coincide with the definition of

$$[a\otimes B^*(f)(b')]_{(U,\phi)}\times[c\otimes d]_{(V,\psi)}$$

Similar arguments show that the product $\times$ on $A\hat{\otimes}B$ is well defined.

This product is moreover bifunctorial: if $f:X\to X'$ and $g:Y\to Y'$ are morphisms of G-sets, if $u=[a\otimes b]_{(U,\phi)}$ is an element of $A\hat{\otimes}B(X)$ and $v=[c\otimes d]_{(V,\psi)}$ an element of $A\hat{\otimes}B(Y)$, then

$$(A\hat{\otimes}B)_*(f)(u)\times(A\hat{\otimes}B)_*(g)(v) = [a\otimes b]_{(U,f\phi)}\times[c\otimes d]_{(V,g\psi)} = [(a\times c)\otimes(b\times d)]_{(U\times V,f\phi\times g\psi)}$$

On the other hand

$$(A\hat{\otimes}B)_*(f\times g)(u\times v) = [(a\times c)\otimes(b\times d)]_{\big(U\times V,(f\times g)(\phi\times\psi)\big)}$$

and equality $f\phi\times g\psi=(f\times g)(\phi\times\psi)$ proves that the product is covariant.

If now f is a morphism from X' to X and g is a morphism from Y' to Y, then to compute $(A\hat{\otimes}B)^*(f)(u)$ and $(A\hat{\otimes}B)^*(g)(v)$, I fill the cartesian squares

$$\begin{array}{ccc} P & \xrightarrow{\alpha} & U \\ {\scriptstyle\beta}\downarrow & & \downarrow{\scriptstyle\phi} \\ X' & \xrightarrow[f]{} & X \end{array} \qquad\qquad \begin{array}{ccc} Q & \xrightarrow{\gamma} & V \\ {\scriptstyle\delta}\downarrow & & \downarrow{\scriptstyle\psi} \\ Y' & \xrightarrow[g]{} & Y \end{array}$$

so that

$$(A\hat{\otimes}B)^*(f)(u) = [A^*(\alpha)(a)\otimes B^*(\alpha)(b)]_{(P,\beta)} \quad (A\hat{\otimes}B)^*(g)(v) = [A^*(\gamma)(c)\otimes B^*(\gamma)(d)]_{(Q,\delta)}$$

The product $(A\hat{\otimes}B)^*(f)(u) \times (A\hat{\otimes}B)^*(g)(v)$ is then equal to

$$\Big[\big(A^*(\alpha)(a) \times A^*(\gamma)(c)\big) \otimes \big(B^*(\alpha)(b) \times B^*(\gamma)(d)\big)\Big]_{(P\times Q,\beta\times\delta)}$$

which can also be written as

$$\Big[\big(A^*(\alpha\times\gamma)(a\times c)\big) \otimes \big(B^*(\alpha\times\gamma)(b\times d)\big)\Big]_{(P\times Q,\beta\times\delta)}$$

As the square

$$\begin{array}{ccc} P\times Q & \xrightarrow{\alpha\times\beta} & U\times V \\ {\scriptstyle\beta\times\delta}\downarrow & & \downarrow{\scriptstyle\phi\times\psi} \\ X'\times X & \xrightarrow[f\times g]{} & X\times Y \end{array}$$

is also cartesian, I have

$$(A\hat{\otimes}B)^*(f\times g)(u\times v) = \Big[\big(A^*(\alpha\times\gamma)(a\times c)\big) \otimes \big(B^*(\alpha\times\gamma)(b\times d)\big)\Big]_{(P\times Q,\beta\times\delta)}$$

which proves that the product $\times$ on $A\hat{\otimes}B$ is contravariant, hence bifunctorial.

It is clear from the definitions that the product $\times$ is associative. Finally, the element ε is a unit, since

$$[a\otimes b]_{(U,\phi)} \times [\varepsilon_A\otimes\varepsilon_B]_{(\bullet,Id)} = [(a\times\varepsilon_A)\otimes(b\times\varepsilon_B)]_{(U\times\bullet,\phi\times\bullet)} = [a\otimes b]_{(U,\phi)}$$

$$[\varepsilon_A\otimes\varepsilon_B]_{(\bullet,Id)} \times [a\otimes b]_{(U,\phi)} = [(\varepsilon_A\times a)\otimes(\varepsilon_B\times b)]_{(\bullet\times U,\bullet\times\phi)} = [a\otimes b]_{(U,\phi)}$$

and this completes the proof of the proposition. ■

The tensor product of Green functors is functorial:

Lemma 6.3.2: : If $f : A \to A'$ and $g : B \to B'$ are morphisms of Green functors for the group G, then $f\hat{\otimes}g : A\hat{\otimes}B \to A'\hat{\otimes}B'$ is a morphism of Green functors. If f and g are unitary, so is $f\hat{\otimes}g$.

Proof: The morphism f (resp. the morphism g) is determined by morphisms f_U (resp. g_U) from $A(U)$ to $A'(U)$ (resp. from $B(U)$ to $B'(U)$), for any G-set U. Moreover, the image of the element $[a\otimes b]_{(U,\phi)}$ of $(A\hat{\otimes}B)(X)$ under $(f\hat{\otimes}g)_X$ is given by

$$(f\hat{\otimes}g)_X\left([a\otimes b]_{(U,\phi)}\right) = [f_U(a)\otimes g_U(b)]_{(U,\phi)}$$

The lemma follows easily, since if f and g are morphisms of Green functors, then

$$f_{U\times V}(a\times c) = f_U(a)\times f_V(c) \qquad g_{U\times V}(b\times d) = g_V(b)\times g_V(d)$$

■

A special case of this lemma is the case when f is the identity morphism of A, and g is the unique (unitary) morphism of Green functors from b to B. since $A\hat{\otimes}b \simeq A$, what I get is a unitary morphism γ_A of Green functors from A to $A\hat{\otimes}B$, and it is easy to see that this morphism is given by

$$a \in A(X) \mapsto \gamma_{A,X}(a) = [a \otimes \varepsilon_{B,X}]_{(X,Id)} = \left[a \otimes B^*\begin{pmatrix} x \\ \bullet \end{pmatrix}(\varepsilon_B)\right]_{(X,Id)} \in A\hat{\otimes}B(X)$$

Similarly, there is a morphism γ_B from B to $A\hat{\otimes}B$ defined by

$$b \in B(X) \mapsto \gamma_{B,X}(b) = [\varepsilon_{A,X} \otimes b]_{(X,Id)} = \left[A^*\begin{pmatrix} x \\ \bullet \end{pmatrix}(\varepsilon_A) \otimes b\right]_{(X,Id)} \in A\hat{\otimes}B(X)$$

The notion of opposite functor of a Green functor leads to the following definition:

Definition: *Let A be a Green functor for the group G. If X and Y are G-sets, I will say that the element $\alpha \in A(X)$ commutes with $\beta \in A(Y)$ if $\alpha \times \beta = \alpha \times^{op} \beta$, i.e.*

$$\alpha \times \beta = A_*\begin{pmatrix} yx \\ xy \end{pmatrix}(\beta \times \alpha)$$

Similarly, I will say that a subset $P \subseteq A(X)$ commutes with a subset $Q \subseteq A(Y)$ if any element of P commutes with any element of Q. If M is a sub-Mackey functor of A, and N a sub-Mackey functor of B, I will say that M commutes with N if $M(X)$ commutes with $N(Y)$ for any G-sets X and Y.

It is clear that if α commutes with β, then β commutes with α, since taking the image of the above equality under $A_*\left(\begin{smallmatrix} xy \\ yx \end{smallmatrix}\right)$ exchanges the roles of X and Y and of α and β. I will also say that in a more symmetric way that α and β commute.

Lemma 6.3.3: The image of γ_A in $A\hat{\otimes}B$ commutes with the image of γ_B.

Proof: Let X and Y be G-sets. If $a \in A(X)$ and $b \in B(Y)$, then

$$\gamma_{A,X}(a) \times \gamma_{B,Y}(b) = [a \otimes \varepsilon_{B,X}]_{(X,Id)} \times [\varepsilon_{A,Y} \otimes b]_{(Y,Id)} = [(a \times \varepsilon_{A,Y}) \otimes (\varepsilon_{B,X} \times b)]_{(X\times Y,Id)}$$

Moreover

$$a \times \varepsilon_{A,Y} = a \times A^*\begin{pmatrix} y \\ \bullet \end{pmatrix}(\varepsilon_A) = A^*\begin{pmatrix} xy \\ x \end{pmatrix}(a)$$

Similarly, I have $\varepsilon_{B,X} \times b = B^*\left(\begin{smallmatrix} xy \\ y \end{smallmatrix}\right)(b)$. Then

$$\gamma_{A,X}(a) \times \gamma_{B,Y}(b) = \left[A^*\begin{pmatrix} xy \\ x \end{pmatrix}(a) \otimes B^*\begin{pmatrix} xy \\ y \end{pmatrix}(b)\right]_{(X\times Y,Id)}$$

On the other hand

$$\gamma_{B,Y}(b) \times \gamma_{A,X}(a) = [\varepsilon_{A,Y} \otimes b]_{(Y,Id)} \times [a \otimes \varepsilon_{B,X}]_{(X,Id)} = [(\varepsilon_{A,Y} \times a) \otimes (b \otimes \varepsilon_{B,X})]_{(Y\times X,Id)}$$

or

$$\gamma_{B,Y}(b) \times \gamma_{A,X}(a) = \left[A^*\begin{pmatrix} yx \\ x \end{pmatrix}(a) \otimes B^*\begin{pmatrix} yx \\ y \end{pmatrix}(b)\right]_{(Y\times X,Id)}$$

The image under $(A\hat{\otimes}B)_*\binom{yx}{xy}$ of this element is

$$\left[A^*\binom{yx}{x}(a)\otimes B^*\binom{yx}{y}(b)\right]_{(X\times Y,\binom{yx}{xy})}$$

Since moreover

$$A^*\binom{yx}{x}(a) = A^*\binom{yx}{xy}A^*\binom{xy}{x}(a)$$

it is also

$$\left[A^*\binom{xy}{x}(a)\otimes B_*\binom{yx}{xy}B^*\binom{yx}{y}(b)\right]_{(X\times Y,Id)}$$

I have also

$$B_*\binom{yx}{xy}B^*\binom{yx}{y}(b) = B^*\binom{xy}{yx}B^*\binom{yx}{y}(b) = B^*\binom{xy}{y}(b)$$

so finally

$$\gamma_{A,X}(a)\times\gamma_{B,Y}(b) = (A\hat{\otimes}B)_*\binom{yx}{xy}\big(\gamma_{B,Y}(b)\times\gamma_{A,X}(a)\big)$$

which proves the lemma. ■

This lemma leads naturally to the universal property of the tensor product of Green functors:

Proposition 6.3.4: Let A, B and C be Green functors for the group G. If f (resp. g) is a morphism of Green functors from A to C (resp. from B to C), and if the image of f commutes with the image of g, then there exists a unique morphism of Green functors h from $A\hat{\otimes}B$ to C such that $f = h\circ\gamma_A$ and $g = h\circ\gamma_B$.

Conversely, if h is a morphism of Green functors from $A\hat{\otimes}B$ to C, then $f = h\circ\gamma_A$ (resp. $g = h\circ\gamma_B$) is a morphism of Green functor from A to C (resp. from B to C), and the images of f and g commute.

Moreover h is unitary if and only if f and g are.

Proof: If (U,ϕ) is a G-set over X, if $a\in A(U)$ and $b\in B(U)$, then

$$[a\otimes b]_{(U,\phi)} = (A\hat{\otimes}B)_*\big([a\otimes b]_{(U,Id)}\big)$$

Moreover

$$\gamma_A(a)\times\gamma_B(b) = \left[A^*\binom{u_1u_2}{u_1}(a)\otimes B^*\binom{u_1u_2}{u_2}(b)\right]_{(U\times U,Id)}$$

The image of this element under $(A\hat{\otimes}B)^*\binom{u}{uu}$, which is the product $\gamma_A(a).\gamma_B(b)$ in $A\hat{\otimes}B(U)$, is obtained using the cartesian square

$$\begin{array}{ccc} U & \xrightarrow{\binom{u}{uu}} & U\times U \\ {\scriptstyle Id}\downarrow & & \downarrow{\scriptstyle Id} \\ U & \xrightarrow[\binom{u}{uu}]{} & U\times U \end{array}$$

So it is equal to

$$\gamma_A(a).\gamma_B(b) = \left[A^*\begin{pmatrix} u \\ uu \end{pmatrix} A^*\begin{pmatrix} u_1u_2 \\ u_1 \end{pmatrix}(a) \otimes B^*\begin{pmatrix} u \\ uu \end{pmatrix} B^*\begin{pmatrix} u_1u_2 \\ u_2 \end{pmatrix}(b)\right]_{(U,Id)} = [a\otimes b]_{(U,Id)}$$

Now if h is a morphism of $A\hat{\otimes}B$ to C, I have

$$h_X\Big([a\otimes b]_{(U,\phi)}\Big) = C_*(\phi)h_U\Big([a\otimes b]_{(U,Id)}\Big) = C_*(\phi)C^*\begin{pmatrix} u \\ uu \end{pmatrix} h_{U\times U}\Big(\gamma_A(a)\times\gamma_B(b)\Big)$$

Whence

$$h_X\Big([a\otimes b]_{(U,\phi)}\Big) = C_*(\phi)C^*\begin{pmatrix} u \\ uu \end{pmatrix}\Big(h_U\gamma_A(a)\times h_U\gamma_B(b)\Big)$$

and h is determined by $h\circ\gamma_A$ and $h\circ\gamma_B$.

Then if f is a morphism from A to C and g is a morphism from B to C, there is at most one morphism h from $A\hat{\otimes}B$ to C such that $f = h\circ\gamma_A$ and $g = h\circ\gamma_B$: the morphism h is given by

$$h_X\Big([a\otimes b]_{(U,\phi)}\Big) = C_*(\phi)C^*\begin{pmatrix} u \\ uu \end{pmatrix}\Big(f_U(a)\times g_U(b)\Big)$$

Now proposition 1.8.3 shows that a morphism h from $A\hat{\otimes}B$ to C is determined by the bilinear morphisms $\hat{h}_U$ from $A(U)\times B(U)$ to $C(U)$, defined for any G set U by

$$(a,b)\in A(U)\times B(U)\mapsto C^*\begin{pmatrix} u \\ uu \end{pmatrix} h_U\Big([a\otimes b]_{(U,Id)}\Big)$$

and h is a morphism of Mackey functors if and only if the morphisms $\hat{h}_U$ satisfy conditions $i)$, $ii)$ and $iii)$ of proposition 1.8.3. But here

$$\hat{h}_U(a,b) = f_U(a).g_U(b)$$

So let $k : U'\to U$ be a morphism of G-sets, and let $a\in A(Y)$, $b\in B(Y)$, $a'\in A(U')$ and $b'\in B(U')$. Condition $i)$ can be written as

$$\hat{h}_U\Big(A_*(k)(a'),b\Big) = C_*(k)\hat{f}_{U'}\Big(a',B^*(k)(b)\Big)$$

or

$$f_UA_*(k)(a').g_U(b) = C_*(k)\Big(f_{U'}(a').g_{U'}B^*(k)(b)\Big)$$

Since f is a morphism of Mackey functors, the left hand side is also equal to

$$C_*(k)f_{U'}(a').g_U(b)$$

But the product "." on C is the map associated by the proposition 1.8.3 to the bilinear morphism from C, C to C defined by the product of C. So relation $i)$ holds for this product, which gives

$$C_*(k)f_{U'}(a').g_U(b) = C_*(k)\Big(f_{U'}(a').B^*(k)g_U(b)\Big)$$

and this proves that $i)$ holds for $\hat{h}$. A similar argument proves relation $ii)$. Now relation $iii)$ can be written as

$$\hat{h}_{U'}\Big(A^*(k)(a),B^*(k)(b)\Big) = C^*(k)\hat{h}_U(a,b)$$

which gives here

$$f_{U'}A^*(k)(a).g_{U'}B^*(k)(b) = C^*(k)\big(f_U(a).g_U(b)\big)$$

The left hand side is also

$$C^*(k)f_U(a).C^*(k)g_U(b)$$

and since relation *iii*) holds for the product ".", I have

$$C^*(k)f_U(a).C^*(k)g_U(b) = C^*(k)\big(f_U(a).g_U(b)\big)$$

(in other words, the maps $C^*(k)$ are ring homomorphisms for the product ".", which is also a consequence of lemma 5.2.2). Thus the maps $\hat{f}$ satisfy *i*), *ii*), and *iii*), and it follows that h is a morphism of Mackey functors.

To prove that h is a morphism of Green functors, I must check that

$$h_{X\times Y}\Big([a\otimes b]_{(U,\phi)} \times [c\otimes d]_{(V,\psi)}\Big) = h_X\Big([a\otimes b]_{(U,\phi)}\Big) \times h_Y\Big([c\otimes d]_{(V,\psi)}\Big)$$

The left hand side is

$$h_{X\times Y}\Big([(a\times c)\otimes(b\times d)]_{(U\times V,\phi\times\psi)}\Big) = C_*(\phi\times\psi)\big(f_{U\times V}(a\times c).g_{U\times V}(b\times d)\big) \qquad (6.1)$$

and the right hand side is

$$C_*(\phi)\big(f_U(a).g_U(b)\big) \times C_*(\psi)\big(f_V(c).g_V(d)\big) = C_*(\phi\times\psi)\big(f_U(a).g_U(b) \times f_V(c).g_V(d)\big) \qquad (6.2)$$

Equality of (6.1) and (6.2) for all ϕ and ψ is equivalent to equality

$$f_{U\times V}(a\times c).g_{U\times V}(b\times d) = f_U(a).g_U(b) \times f_V(c).g_V(d)$$

The left hand side is

$$\big(f_U(a)\times f_V(c)\big).\big(g_U(b)\times g_V(d)\big) = C^*\begin{pmatrix} uv \\ uvuv \end{pmatrix}\big(f_U(a)\times f_V(c)\times g_U(b)\times g_V(d)\big) \qquad (6.3)$$

and the right hand side is

$$C^*\begin{pmatrix} u \\ uu \end{pmatrix}\big(f_U(a)\times g_U(b)\big) \times C^*\begin{pmatrix} v \\ vv \end{pmatrix}\big(f_V(c)\times g_V(d)\big) = \ldots$$

$$\ldots = C^*\begin{pmatrix} uv \\ uuvv \end{pmatrix}\big(f_U(a)\times g_U(b)\times f_V(c)\times g_V(d)\big) \qquad (6.4)$$

If the images of g and f commute, I have

$$g_U(b)\times f_V(c) = C_*\begin{pmatrix} vu \\ uv \end{pmatrix}\big(f_V(c)\times g_U(c)\big)$$

Then

$$f_U(a)\times g_U(b)\times f_V(c)\times g_V(d) = C_*\begin{pmatrix} u_1v_1u_2v_2 \\ u_1u_2v_1v_2 \end{pmatrix}\big(f_U(a)\times f_V(c)\times g_U(b)\times g_V(d)\big)$$

But since

$$C_*\begin{pmatrix} u_1v_1u_2v_2 \\ u_1u_2v_1v_2 \end{pmatrix} = C^*\begin{pmatrix} u_1u_2v_1v_2 \\ u_1v_1u_2v_2 \end{pmatrix}$$

the right hand side of (6.4) is equal to

$$C^*\begin{pmatrix} uv \\ uuvv \end{pmatrix} C^*\begin{pmatrix} u_1u_2v_1v_2 \\ u_1v_1u_2v_2 \end{pmatrix} \left(f_U(a)\times f_V(c)\times g_U(b)\times g_V(d)\right) = \ldots$$

$$\ldots = C^*\begin{pmatrix} uv \\ uvuv \end{pmatrix} \left(f_U(a)\times f_V(c)\times g_U(b)\times g_V(d)\right)$$

which is the right hand side of (6.3). Thus h is a morphism of Green functors.

Conversely, if h is a morphism of Green functors from $A\hat{\otimes}B$ to C, the images of $f = h\circ\gamma_A$ and $g = h\circ\gamma_B$ must commute: indeed, the images of γ_A and γ_B commute, and moreover, I have the following lemma

Lemma 6.3.5: Let X and Y be G-sets, and θ be a morphism of Green functors from D to C. If $\alpha \in D(X)$ commutes with $\beta \in D(Y)$, then $\theta_X(\alpha)$ commutes with $\theta_Y(\beta)$.

Proof: It suffices to write

$$\theta_Y(\beta)\times\theta_X(\alpha) = \theta_{Y\times X}(\beta\times\alpha) = \theta_{Y\times X}D_*\begin{pmatrix} yx \\ xy \end{pmatrix}(\alpha\times\beta) = \ldots$$

$$\ldots = C_*\begin{pmatrix} yx \\ xy \end{pmatrix}\theta_{X\times Y}(\alpha\times\beta) = C_*\begin{pmatrix} yx \\ xy \end{pmatrix}\left(\theta_X(\alpha)\times\theta_Y(\beta)\right)$$

The lemma follows. ∎

To complete the proof of the proposition, I must still observe that if f and g are unitary, then so is h: indeed, the unit of $A\hat{\otimes}B$ is $[\varepsilon_A\otimes\varepsilon_B]_{(\bullet,Id)}$, and its image under h is

$$C^*\begin{pmatrix} \bullet \\ \bullet\bullet \end{pmatrix}\left(f_\bullet(\varepsilon_A).g_\bullet(\varepsilon_B)\right) = f_\bullet(\varepsilon_A).g_\bullet(\varepsilon_B)$$

which is equal to the unit of C if f and g are unitary. Conversely, if h is unitary, then $f = h\circ\gamma_A$ and $g = h\circ\gamma_B$ are composition products of unitary morphisms, so they are unitary. ∎

Proposition 6.3.6: Let A and B be Green functors for the group G. Let M be an A-module and N be a B-module. Then $M\hat{\otimes}N$ has a natural structure of $A\hat{\otimes}B$-module, defined as follows: if X and Y are G-sets, if (U,ϕ) is a G-set over X, and (V,ψ) is a G-set over Y, if $a\in A(U)$, if $b\in B(U)$, if $m\in M(V)$ and $n\in N(V)$, then

$$[a\otimes b]_{(U,\phi)}\times[m\otimes n]_{(V,\psi)} = [(a\times m)\otimes(b\times n)]_{(U\times V,\phi\times\psi)}$$

Proof: To prove that this product is well defined, associative, and unitary, one just has to mimic the proof of proposition 6.3.1, replacing A by M and B by N in suitable places. ∎

6.4 Bimodules

The notion of right-module over a Green functor leads naturally to the notion of bimodule:

Definition: *Let A and B be Green functors for the group G. If M is an A-module, which is also a module-B, I will say that M is an A-module-B if for any G-sets X, Y, and Z, and any elements $a \in A(X)$, $m \in M(Y)$ and $b \in B(Z)$, I have*

$$(a \times m) \times b = a \times (m \times b) \quad in \quad M(X \times Y \times Z)$$

A morphism of A-modules-B from M to N is a morphism of Mackey functors from M to N, which is also a morphism of A-modules and a morphism of modules-B.

With those definitions, I can speak of the category of A-modules-B, that I will denote by A-**Mod**-B.

Proposition 6.4.1: The category A-Mod-B is equivalent to the category $A\hat{\otimes}B^{op}$-Mod.

Proof: To give M a structure of $A\hat{\otimes}B^{op}$-module is equivalent (see proposition 2.1.2) to give a unitary morphism of Green functors from $A\hat{\otimes}B^{op}$ to $\mathcal{H}(M, M)$. By proposition 6.3.4, this is equivalent to give unitary morphisms from A and B^{op} to $\mathcal{H}(M, M)$, the images of which commute. In particular, the module M is an A-module, and a B^{op}-module, i.e. a module-B.

If X is a G-set, and if $a \in A(X)$, then a defines an element

$$\lambda_a \in \mathrm{Hom}_{Mack(G)}(M, M_X)$$

by

$$m \in M(Y) \mapsto \lambda_a(m) = M_*\begin{pmatrix} xy \\ yx \end{pmatrix}(a \times m) \in M(Y \times X) = M_X(Y)$$

If Z is a G-set, and if $b \in B(Z)$, then b determines an element ρ_b of $\mathcal{H}(M, M)(Z) = \mathrm{Hom}_{Mack(G)}(M, M_Z)$, by

$$m \in M(Y) \mapsto \rho_b(m) = M_*\begin{pmatrix} zy \\ yz \end{pmatrix}(b \times^{op} m) = \ldots$$

$$\ldots = M_*\begin{pmatrix} zy \\ yz \end{pmatrix} M_*\begin{pmatrix} yz \\ zy \end{pmatrix}(m \times b) = m \times b \in M(Y \times Z) = M_Z(Y)$$

Now λ_a and ρ_b commute if and only if

$$\rho_b \times \lambda_a = \mathcal{H}(M, M)_*\begin{pmatrix} xz \\ zx \end{pmatrix}(\lambda_a \times \rho_b)$$

where the products $\times$ are in $C = \mathcal{H}(M, M)$. But if α is an element of $C(X)$, determined by morphisms

$$\alpha_Y : M(Y) \to M(YX) = M_X(Y)$$

and if β is an element of $C(Z)$, determined by morphisms

$$\beta_Y : M(Y) \to M(YZ) = M_Z(Y)$$

then the product $\alpha \times \beta$ is the element of $C(X \times Y)$ determined by the morphisms

$$(\alpha \times \beta)_Y = M_* \begin{pmatrix} yzx \\ yxz \end{pmatrix} \circ \alpha_{YZ} \circ \beta_Y$$

The product $\lambda_a \times \rho_b$ is then determined by the morphisms $(\lambda_a \times \rho_b)_Y = M_* \binom{yzx}{yxz} \circ \lambda_{a,YZ} \circ \rho_{b,Y}$:

$$m \in M(Y) \mapsto M_* \begin{pmatrix} yzx \\ yxz \end{pmatrix} M_* \begin{pmatrix} xyz \\ yzx \end{pmatrix} \Big(a \times (m \times b)\Big) = M_* \begin{pmatrix} xyz \\ yxz \end{pmatrix} \Big(a \times (m \times b)\Big)$$

On the other hand, the product $\rho_b \times \lambda_a$ is determined by

$$(\rho_b \times \lambda_a)_Y = M_* \begin{pmatrix} yxz \\ yzx \end{pmatrix} \circ \rho_{b,YX} \circ \lambda_{a,Y} : m \mapsto M_* \begin{pmatrix} yxz \\ yzx \end{pmatrix} \Big(M_* \begin{pmatrix} xy \\ yx \end{pmatrix} (a \times m) \times b\Big)$$

Now λ_a and ρ_b commute if and only if

$$(\rho_b \times \lambda_a)_Y(m) = M_* \begin{pmatrix} yxz \\ yzx \end{pmatrix} (\lambda_a \times \rho_b)_Y(m)$$

or equivalently

$$M_* \begin{pmatrix} yxz \\ yzx \end{pmatrix} \Big(M_* \begin{pmatrix} xy \\ yx \end{pmatrix} (a \times m) \times b\Big) = M_* \begin{pmatrix} yxz \\ yzx \end{pmatrix} \Big(M_* \begin{pmatrix} xyz \\ yxz \end{pmatrix} \Big(a \times (m \times b)\Big)$$

This is also

$$M_* \begin{pmatrix} yxz \\ yzx \end{pmatrix} M_* \begin{pmatrix} xyz \\ yxz \end{pmatrix} \Big((a \times m) \times b\Big) = M_* \begin{pmatrix} yxz \\ yzx \end{pmatrix} M_* \begin{pmatrix} xyz \\ yxz \end{pmatrix} \Big(a \times (m \times b)\Big)$$

thus, since $M_* \binom{xyz}{yzx}$ is bijective

$$a \times (m \times b) = (a \times m) \times b$$

and this proves the proposition. ∎

Proposition 6.4.2: Let A and B be Green functors for the group G. If M is an A-module, and N a B-module, then the product $\hat{\circ}$ and the morphisms from A to $\mathcal{H}(M,M)$ and from B to $\mathcal{H}(N,N)$ induce a natural structure of B-module-A on $\mathcal{H}(M,N)$.

Proof: By proposition 2.1.2, to say that M is an A-module (resp. that N is a B-module) is equivalent to give a unitary morphism of Green functors from A to $\mathcal{H}(M,M)$ (resp. from B to $\mathcal{H}(M,M)$). So it suffices to give $\mathcal{H}(M,N)$ a structure of $\mathcal{H}(N,N)$-module-$\mathcal{H}(M,M)$.

So let X, Y, and Z be G-sets. If $a \in \mathcal{H}(M,M)(X)$, if $f \in \mathcal{H}(M,N)(Y)$, and $b \in \mathcal{H}(N,N)(Z)$, I have seen that if U is a G-set, then $b \,\hat{\circ}\, f \,\hat{\circ}\, a$ is the element of $\mathcal{H}(M,P)(ZYX)$ defined on the G-set U by

$$(b \,\hat{\circ}\, f \,\hat{\circ}\, a)_U = N_* \begin{pmatrix} uxyz \\ uzyx \end{pmatrix} \circ b_{UXY} \circ f_{UX} \circ a_U$$

The product $\hat{\circ}$ turns $\mathcal{H}(M,N)$ into a bimodule, because it is bifunctorial and associative (proposition 6.1.2), and also unitary: if b is the unit of $\mathcal{H}(N,N)$, i.e. the identity morphism from N to $N_\bullet = N$, then it is clear that

$$(b\,\hat{\circ}\,f)_U = N_*\begin{pmatrix} ux\bullet \\ u\bullet x\end{pmatrix} \circ Id_{N(UX)} \circ f_U = f_U$$

Similarly, if a is the identity of M, then $m\,\hat{\circ}\,a = m$. This completes the proof of the proposition. ∎

6.5 Commutants

Definition: *Let A be a Green functor for the group G, and M an A-module-A. If U is a G-set, and $\alpha \in A(U)$, I set for any G-set X*

$$C_M(\alpha)(X) = \{m \in M(X) \mid \alpha \times m = M_*\begin{pmatrix} xu \\ ux\end{pmatrix}(m\times\alpha)\}$$

Similarly, if P is a subset of $A(U)$, I set

$$C_M(P)(X) = \{m \in M(X) \mid \alpha \times m = M_*\begin{pmatrix} xu \\ ux\end{pmatrix}(m\times\alpha) \quad \forall\alpha\in P\}$$

More generally, if L is a sub-Mackey functor of A, I set

$$C_M(L)(X) = \{m \in M(X) \mid \alpha \times m = M_*\begin{pmatrix} xu \\ ux\end{pmatrix}(m\times\alpha) \quad \forall U,\ \forall\alpha\in L(U)\}$$

Lemma 6.5.1: Let $<P>$ be the sub-Mackey functor of A generated by P (i.e. the intersection of the sub-Mackey functors L of A such that $L(U)\supseteq P$). Then for any X, I have $C_M(P)(X) = C_M(<P>)(X)$.

Proof: It is clear that $C_M(<P>)(X) \subseteq C_M(P)(X)$. Conversely, it is easy to see that for any G-set X, I have

$$<P>(Y) = \sum_{\substack{f:U\to Z\\ g:Z\to Y}} <A_*(g)A^*(f)(p)>_{p\in P} \subseteq A(Y)$$

so that any element α of $<P>(Y)$ can be written as

$$\alpha = \sum_i A_*(g_i)A^*(f_i)(p_i)$$

for suitable elements g_i, f_i and $p_i \in P$. But if $m \in C_M(P)(X)$, then

$$\alpha\times m = \sum_i A_*(g_i)A^*(f_i)(p_i)\times m = \sum_i M_*(g_i\times Id)M^*(f_i\times Id)(p_i\times m) = \ldots$$

$$\ldots = \sum_i M_*(g_i\times Id)M^*(f_i\times Id)M_*\begin{pmatrix} xu \\ ux\end{pmatrix}(m\times p_i)$$

And for $f: U\to Z$ and $g: Z\to Y$, I have

$$M_*(g\times Id)M^*(f\times Id)M_*\begin{pmatrix} xu \\ ux\end{pmatrix} = M_*(g\times Id)M^*(f\times Id)M^*\begin{pmatrix} ux \\ xu\end{pmatrix} = \ldots$$

$$\ldots = M_*(g\times Id)M^*\begin{pmatrix} zx \\ xz \end{pmatrix} M^*(Id\times f) = M_*(g\times Id)M_*\begin{pmatrix} xz \\ zx \end{pmatrix} M^*(Id\times f) = \ldots$$

$$\ldots = M_*\begin{pmatrix} xy \\ yx \end{pmatrix} M_*(Id\times g)M^*(Id\times f)$$

Thus

$$\alpha\times m = \sum_i M_*\begin{pmatrix} xy \\ yx \end{pmatrix} M_*(Id\times g_i)M^*(Id\times f_i)(m\times p_i) = \ldots$$

$$\ldots = M_*\begin{pmatrix} xy \\ yx \end{pmatrix}\Big(\sum_i m\times M_*(g_i)M^*(f_i)(p_i)\Big) = M_*\begin{pmatrix} xy \\ yx \end{pmatrix}(m\times\alpha)$$

which proves the lemma. ∎

Proposition 6.5.2: The previous definitions turn $C_M(L)$ into a sub-Mackey functor of M, called the commutant of L in M.

Proof: If $f : X \to Y$ is morphism of G-sets, and if $m \in C_M(X)$, then for $\alpha \in L(U)$, I have

$$\alpha\times M_*(f)(m) = M_*(Id\times f)(\alpha\times m) = M_*(Id\times f)M_*\begin{pmatrix} xu \\ ux \end{pmatrix}(m\times\alpha) = \ldots$$

$$\ldots = M_*\begin{pmatrix} yu \\ uy \end{pmatrix} M_*(f\times Id)(m\times\alpha) = M_*\begin{pmatrix} yu \\ uy \end{pmatrix}\big(M_*(f)(m)\times\alpha\big)$$

and this shows that $M_*(f)\big(C_M(L)(X)\big) \subseteq C_M(L)(Y)$. Similarly, if $m \in M(Y)$

$$\alpha\times M^*(f)(m) = M^*(Id\times f)(\alpha\times m) = M^*(Id\times f)M_*\begin{pmatrix} yu \\ uy \end{pmatrix}(m\times\alpha) = \ldots$$

$$\ldots = M^*(Id\times f)M^*\begin{pmatrix} uy \\ yu \end{pmatrix}(m\times\alpha) = M^*\begin{pmatrix} xu \\ ux \end{pmatrix} M^*(f\times Id)(m\times\alpha) = \ldots$$

$$\ldots = M_*\begin{pmatrix} ux \\ xu \end{pmatrix}\big(M^*(f)(m)\times\alpha\big)$$

which shows that $M^*(f)\big(C_M(L)(Y)\big) \subseteq C_M(L)(X)$, hence that $C_M(L)$ is a sub-Mackey functor of M. ∎

In the special case when M is the functor A, viewed as an A-module-A, there is a little more:

Proposition 6.5.3: Let L be a sub-Mackey functor of the Green functor A. Then $C_A(L)$ is a sub-Green functor of A.

Proof: I must check that if X, Y and U are G-sets, if $\alpha \in A(X)$ and $\beta \in A(Y)$ commute with $l \in L(U)$, then $\alpha\times\beta$ also commutes with l. But

$$A_*\begin{pmatrix} xyu \\ uxy \end{pmatrix}\big((\alpha\times\beta)\times l\big) = A_*\begin{pmatrix} xyu \\ uxy \end{pmatrix}\big(\alpha\times(\beta\times l)\big) = \ldots$$

$$\ldots = A_*\begin{pmatrix} xyu \\ uxy \end{pmatrix}\Big(\alpha \times A_*\begin{pmatrix} uy \\ yu \end{pmatrix}(l\times\beta)\Big) = A_*\begin{pmatrix} xyu \\ uxy \end{pmatrix}A_*\begin{pmatrix} xuy \\ xyu \end{pmatrix}(\alpha\times l\times\beta) = \ldots$$

$$\ldots = A_*\begin{pmatrix} xyu \\ uxy \end{pmatrix}A_*\begin{pmatrix} xuy \\ xyu \end{pmatrix}\Big(A_*\begin{pmatrix} ux \\ xu \end{pmatrix}(l\times\alpha)\times\beta\Big) = \ldots$$

$$\ldots = A_*\begin{pmatrix} xyu \\ uxy \end{pmatrix}A_*\begin{pmatrix} xuy \\ xyu \end{pmatrix}A_*\begin{pmatrix} uxy \\ xuy \end{pmatrix}(l\times\alpha\times\beta) = A_*\begin{pmatrix} uxy \\ uxy \end{pmatrix}(l\times\alpha\times\beta) = l\times(\alpha\times\beta)$$

which proves that $C_M(L)$ is closed under $\times$. Since is it clear that $\varepsilon \in C_M(L)(\bullet)$, the proposition follows. ∎

Notation: *If M and N are modules for the Green functor A, I denote by*

$$\mathcal{H}_A(M,N)$$

the commutant of A in the bimodule $\mathcal{H}(M,N)$.

By definition, if I set $\mathcal{H} = \mathcal{H}(M,N)$, then for any G-set X

$$\mathcal{H}_A(M,N)(X) = \{f \in \mathcal{H}(X) \mid \alpha \times f = \mathcal{H}_*\begin{pmatrix} xu \\ ux \end{pmatrix}(m \times \alpha) \quad \forall U,\ \forall \alpha \in A(U)\}$$

Let r_α (resp. l_α) be the element of $\mathcal{H}(M,M)(U)$ (resp. of $\mathcal{H}(N,N)(U)$) associated to α. Then $\alpha \times f$ is the morphism from M to N_{UX} defined for $m \in M(Z)$ by

$$(\alpha\times f)_Z(m) = (l_\alpha \hat{\circ} f)_Z(m) = N_*\begin{pmatrix} zxu \\ zux \end{pmatrix} \circ l_{\alpha,ZX} \circ f_Z(m) = \ldots$$

$$\ldots = N_*\begin{pmatrix} zxu \\ zux \end{pmatrix} \circ N_*\begin{pmatrix} uzx \\ zxu \end{pmatrix}\big(\alpha \times f_Z(m)\big) = N_*\begin{pmatrix} uzx \\ zux \end{pmatrix}\big(\alpha \times f_Z(m)\big)$$

Similarly, the morphism $f \times \alpha$ from M to N_{XU} is given by

$$(f\times\alpha)_Z(m) = N_*\begin{pmatrix} zux \\ zxu \end{pmatrix} \circ f_{ZU} \circ r_{\alpha,Z}(m) = N_*\begin{pmatrix} zux \\ zxu \end{pmatrix} \circ f_{ZU} \circ M_*\begin{pmatrix} uz \\ zu \end{pmatrix}(\alpha \times m)$$

Since f is a morphism of Mackey functors, I have

$$f_{ZU} \circ M_*\begin{pmatrix} uz \\ zu \end{pmatrix} = N_*\begin{pmatrix} uzx \\ zux \end{pmatrix} \circ f_{UZ}$$

and then

$$(f\times\alpha)_Z(m) = N_*\begin{pmatrix} zux \\ zxu \end{pmatrix} \circ N_*\begin{pmatrix} uzx \\ zux \end{pmatrix} \circ f_{UZ}(\alpha\times m) = N_*\begin{pmatrix} uzx \\ zxu \end{pmatrix} \circ f_{UZ}(\alpha \times m)$$

Thus f is in $\mathcal{H}_A(M,N)(X)$ if and only if for any U and any $\alpha \in A(U)$, I have

$$N_*\begin{pmatrix} uzx \\ zux \end{pmatrix}\big(\alpha\times f_Z(m)\big) = N_*\begin{pmatrix} zxu \\ zux \end{pmatrix}N_*\begin{pmatrix} uzx \\ zxu \end{pmatrix} f_{UZ}(\alpha\times m) = N_*\begin{pmatrix} uzx \\ zux \end{pmatrix} f_{UZ}(\alpha\times m)$$

which reduces to

$$\alpha \times f_Z(m) = f_{UZ}(\alpha \times m)$$

Thus $\mathcal{H}_A(M,N)(X)$ is just the set of A-module homomorphisms from M to N_X.

Proposition 6.5.4: Let A be a Green functor for the group G. If M and N are A-modules, and X is a G-set, then

$$\mathcal{H}_A(M,N)(X) = \mathrm{Hom}_{A\text{-}Mod}(M, N_X)$$

If moreover $M = N$, then the product $\hat{\circ}$ turns $\mathcal{H}_A(M,M)$ into a Green functor.

6.6 The functors $M \otimes_A N$

Let A and B be Green functors for the group G. If M is an A-module-B, and N an A-module, I have built $\mathcal{H}_A(M,N)$, which is a priori a Mackey functor. Actually, in that case, it is a B-module: if X and Y are G-sets, if $b \in B(X)$ and $f \in \mathrm{Hom}_A(M,N_Y)$, I denote by ρ_b the element of $\mathrm{Hom}_{Mack(G)}(M,M_Y)$ deduced from the action of B on the right on M. Then $f \hat{\circ} \rho_b \in \mathcal{H}(M,N)(YX)$. It is the morphism of Mackey functors from M to N_{YX} defined on the G-set U by

$$(f \hat{\circ} \rho_b)_U(m) = N_* \begin{pmatrix} uxy \\ uyx \end{pmatrix} \circ f_{UX} \circ \rho_{b,U}(m) = N_* \begin{pmatrix} uxy \\ uyx \end{pmatrix} \circ f_{UX}(m \times b)$$

It follows easily that if I set

$$(b \times f)_U(m) = N_* \begin{pmatrix} uyx \\ uxy \end{pmatrix} (f \hat{\circ} \rho_b)_U(m) = f_{UX}(m \times b)$$

I obtain a morphism from M to N_{XY}. Moreover, it is clear that this turns $\mathcal{H}(M,N)$ into a B-module. Finally, if f is a morphism of A-modules, so is $b \times f$, since if $a \in A(Z)$, then

$$(b \times f)_{ZU}(a \times m) = f_{ZUX}\big((a \times m) \times b\big) = f_{ZUX}\big(a \times (m \times b)\big) = \ldots$$

$$\ldots = a \times f_{UX}(m \times b) = a \times (b \times f)_U(m)$$

Now I have defined a B-module structure on $\mathcal{H}_A(M,N)$. The correspondence $N \mapsto \mathcal{H}_A(M,N)$ is moreover functorial in N: if θ is a morphism of A-modules from N to N', determined by morphisms θ_Y from $N(Y)$ to $N'(Y)$ for any G-set Y, then I define the morphism $\mathcal{H}_A(M,\theta)$ from $\mathcal{H}_A(M,N)$ to $\mathcal{H}_A(M,N')$ in the following way: if Y is a G-set, if f is an element of $\mathcal{H}_A(M,N)(Y) = \mathrm{Hom}_A(M,N_Y)$, and if U is a G-set, then

$$\big(\mathcal{H}_A(M,\theta)_Y(f)\big)_U = \theta_{UY} \circ f_U \in \mathrm{Hom}_R\big(M(U), N'(UY)\big)$$

It is then clear that $\mathcal{H}_A(M,\theta)_Y(f)$ is a morphism of Mackey functors from M to N'_Y, and since f and θ are morphisms of A-modules, it is actually a morphism of A-modules. So there is a map $\mathcal{H}_A(M,\theta)_Y$ from $\mathcal{H}_A(M,N)(Y)$ to $\mathcal{H}_A(M,N')(Y)$, which defines a morphism of Mackey functors from $\mathcal{H}_A(M,N)$ to $\mathcal{H}_A(M,N')$. In other words, the Mackey functor $\mathcal{H}_A(M,N)$ is a sub-functor of $\mathcal{H}(M,N)$, which is moreover invariant under $\mathcal{H}(M,\theta)$ if θ is a morphism of A-modules.

So the bimodule M defines a functor from A-**Mod** to B-**Mod**. A natural question is then to look for a left adjoint, as in proposition proposition 1.10.1. This question leads to the following definition:

Definition: *Let A and B be Green functors for the group G, and M be an A-module-B. If N is a B-module, and X is a G-set, I set*

$$(M\hat{\otimes}_B N)(X) = \Big(\bigoplus_{Y \xrightarrow{\phi} X} M(Y) \otimes_R N(Y)\Big)/\mathcal{J}$$

where $\mathcal{J}$ is the R-submodule generated by the elements

$$M_*(f)(m) \otimes n' - m \otimes N^*(f)(n') \text{ for } f : (Y, \phi) \to (Y', \phi'),\ m \in M(Y),\ n' \in N(Y')$$

$$M^*(f)(m') \otimes n' - m \otimes N_*(f)(n) \text{ for } f : (Y, \phi) \to (Y', \phi'),\ m' \in M(Y'),\ n \in N(Y)$$

$$m.b \otimes n - m \otimes b.n \text{ for } m \in M(Y),\ b \in B(Y),\ n \in N(Y)$$

Lemma 6.6.1: The projection $M\hat{\otimes}N(X) \to M\hat{\otimes}_B N(X)$ turns $M\hat{\otimes}_B N$ into a Mackey functor, and the morphism $A \to A\hat{\otimes}B$ turns it into an A-module, quotient of $M\hat{\otimes}N$. Moreover, the correspondence $N \mapsto M\hat{\otimes}_B N$ is a functor from B-Mod to A-Mod.

Proof: To say that the structure of Mackey functor of $M\hat{\otimes}N$ is compatible with the projection is equivalent to say that the elements

$$[m.b \otimes n]_{(Y,\phi)} - [m \otimes b.n]_{(Y,\phi)} \tag{E}$$

generate a sub-Mackey functor of $M\hat{\otimes}N$. But if $f : X \to X'$ is a morphism of G-sets, then

$$(M\hat{\otimes}N)_*(f)\Big([m.b \otimes n]_{(Y,\phi)} - [m \otimes b.n]_{(Y,\phi)}\Big) = [m.b \otimes n]_{(Y,f\phi)} - [m \otimes b.n]_{(Y,f\phi)}$$

Similarly, if now f is a morphism from X' to X, and if the square

$$\begin{array}{ccc} Y' & \xrightarrow{a} & Y \\ {\scriptstyle \phi'}\downarrow & & \downarrow{\scriptstyle \phi} \\ X' & \xrightarrow[f]{} & X \end{array}$$

is cartesian, then

$$(M\hat{\otimes}N)^*(f)\Big([m.b \otimes n]_{(Y,\phi)} - [m \otimes b.n]_{(Y,\phi)}\Big) = \ldots$$

$$\ldots = [M^*(a)(m.b) \otimes N^*(a)(n)]_{(Y',\phi')} - [M^*(a)(m) \otimes N^*(a)(b.n)]_{(Y',\phi')}$$

But the proof of lemma 5.2.2 shows more generally that

$$M^*(a)(m.b) = M^*(a)(m).B^*(a)(b) \qquad N^*(a)(b.n) = B^*(a)(b).N^*(a)(n)$$

which proves that $M\hat{\otimes}_B N$ is a quotient Mackey functor of $M\hat{\otimes}N$.

To prove that $M\hat{\otimes}_B N$ is a quotient A-module of $M\hat{\otimes}N$, I must check similarly that the elements (E) generate an A-submodule of $M\hat{\otimes}N$. But the product of the element $\alpha \in A(Z)$ by the element $[m \otimes n]_{(Y,\phi)}$ of $M\hat{\otimes}N(X)$ is given by

$$\alpha \times [m \otimes n]_{(Y,\phi)} = [\alpha \otimes \varepsilon_{B,Z}]_{(Z,Id)} \times [m \otimes n]_{(Y,\phi)} = [(\alpha \times m) \otimes (\varepsilon_{B,Z} \times n)]_{(Z\times Y, Id\times\phi)}$$

Since

$$\varepsilon_{B,Z} \times n = B^*\begin{pmatrix} z \\ \bullet \end{pmatrix}(\varepsilon_B) \times n = N^*\begin{pmatrix} zy \\ y \end{pmatrix}(n)$$

I have

$$\alpha \times [m \otimes n]_{(Y,\phi)} = [(\alpha \times m) \otimes N^* \begin{pmatrix} zy \\ y \end{pmatrix} (n)]_{(Z\times Y, Id\times\phi)}$$

In particular

$$\alpha \times [m \otimes b.n]_{(Y,\phi)} = [(\alpha \times m) \otimes N^* \begin{pmatrix} zy \\ y \end{pmatrix} (b.n)]_{(Z\times Y, Id\times\phi)}$$

Now the proof of lemma 5.2.2 shows that

$$N^* \begin{pmatrix} zy \\ y \end{pmatrix} (b.n) = B^* \begin{pmatrix} zy \\ y \end{pmatrix} (b).N^* \begin{pmatrix} zy \\ y \end{pmatrix} (n)$$

Moreover

$$(\alpha \times m).B_* \begin{pmatrix} zy \\ y \end{pmatrix} (b) = M^* \begin{pmatrix} zy \\ zyzy \end{pmatrix} \left(\alpha \times m \times B^* \begin{pmatrix} zy \\ y \end{pmatrix} (b)\right) = \ldots$$

$$\ldots = M^* \begin{pmatrix} zy \\ zyzy \end{pmatrix} M^* \begin{pmatrix} z_1y_1z_2y_2 \\ z_1y_1y_2 \end{pmatrix} (\alpha \times m \times b) = M^* \begin{pmatrix} zy \\ zyy \end{pmatrix} (\alpha \times m \times b) = \ldots$$

$$\ldots = \alpha \times M^* \begin{pmatrix} y \\ yy \end{pmatrix} (m \times b) = \alpha \times (m.b)$$

It follows that

$$\alpha \times \left([m.b \otimes n]_{(Y,\phi)} - [m \otimes b.n]_{(Y,\phi)}\right) = \ldots$$

$$\ldots = \left[\left(\alpha \times (m.b)\right) \otimes N^* \begin{pmatrix} zy \\ y \end{pmatrix} (n)\right]_{(ZY, Id\phi)} - \left[(\alpha \times m) \otimes N^* \begin{pmatrix} zy \\ y \end{pmatrix} (b.n)\right]_{(ZY, Id\phi)} = \ldots$$

$$= \left[\left((\alpha \times m).B_* \begin{pmatrix} zy \\ y \end{pmatrix} (b)\right) \otimes n\right]_{(ZY, Id\phi)} - \left[(\alpha \times m) \otimes B^* \begin{pmatrix} zy \\ y \end{pmatrix} (b).N^* \begin{pmatrix} zy \\ y \end{pmatrix} (n)\right]_{(ZY, Id\phi)}$$

which proves that the A-module structure of $M\hat{\otimes}_B N$ passes down to its quotient $M\hat{\otimes}_B N$.

It remains to observe that the construction $N \mapsto M\hat{\otimes}_B N$ is functorial in N: if $f : N \to N'$ is a morphism of B-modules, I can set for a G-set (Y, ϕ) over X

$$(M\hat{\otimes}_B f)\left([m \otimes n]_{(Y,\phi)}\right) = [m \otimes f_Y(n)]_{(Y,\phi)} \in M\hat{\otimes}_B N'(X)$$

The map $M\hat{\otimes}_B f$ is well defined, because

$$(M\hat{\otimes}_B f)\left([m.b\otimes n]_{(Y,\phi)} - [m\otimes b.n]_{(Y,\phi)}\right) = [m.b\otimes f_Y(n)]_{(Y,\phi)} - [m\otimes f_Y(b.n)]_{(Y,\phi)} = \ldots$$

$$\ldots = [m.b \otimes f_Y(n)]_{(Y,\phi)} - [m \otimes b.f_Y(n)]_{(Y,\phi)}$$

The lemma follows. ∎

Proposition 6.6.2: Let A and B Green functors for the group G. If M is an A-module-B, if N is a B-module and P an A-module, then there are isomorphisms of Mackey functors

$$\mathcal{H}_A(M\hat{\otimes}_B N, P) \simeq \mathcal{H}_B\big(N, \mathcal{H}_A(M,P)\big)$$

which are moreover natural in M,N and P.

In particular, the functor $N \mapsto M\hat{\otimes}_B N$ is left adjoint to the functor $P \mapsto \mathcal{H}_A(M,P)$.

Proof: I have the following diagram of Mackey functors:

$$\begin{array}{ccc}
\mathcal{H}_A(M\hat{\otimes}_B N, P) & & \mathcal{H}_B\big(N, \mathcal{H}_A(M,P)\big) \\
\downarrow & & \downarrow \\
\mathcal{H}(M\hat{\otimes}_B N, P) & & \mathcal{H}_B\big(N, \mathcal{H}(M,P)\big) \\
\downarrow & & \downarrow \\
\mathcal{H}(M\hat{\otimes} N, P) & \xrightarrow[\sigma]{\simeq} & \mathcal{H}\big(N, \mathcal{H}(M,P)\big)
\end{array}$$

where the bottom isomorphism σ comes from proposition 1.10.1. I need to show that this isomorphism maps the left column to the right one. Let X be a G-set, and f be a morphism of Mackey functors from $M\hat{\otimes}N$ to P_X. It follows from proposition 1.8.3 that f is determined by bilinear maps

$$\hat{f}_Y : M(Y) \times N(Y) \to P_X(Y) = P(YX)$$

More precisely, the image under f of the element $[m \otimes n]_{(U,\phi)}$ of $M\hat{\otimes}N(Y)$ is given by

$$f\big([m \otimes n]_{(Y,\phi)}\big) = P_{X,*}(\phi)\hat{f}_U(m,n)$$

To say that f passes down to the quotient $M\hat{\otimes}_B N(Y)$ is equivalent to say that

$$P_*\begin{pmatrix} ux \\ \phi(u)x \end{pmatrix} \hat{f}_U(m.b,n) = P_*\begin{pmatrix} ux \\ \phi(u)x \end{pmatrix} (\phi)\hat{f}_U(m,b.n)$$

for any G-set (U,ϕ) over Y and $b \in B(U)$. This is also equivalent to the special case $(U,\phi) = (Y, Id)$, i.e.

$$\hat{f}_U(m.b,n) = \hat{f}_U(m,b.n) \tag{6.5}$$

On the other hand, the element corresponding to f under σ in

$$\mathcal{H}\big(N, \mathcal{H}(M,P)\big)(X) = \mathrm{Hom}_{Mack(G)}\big(N, \mathcal{H}(M,P)_X\big) = \mathrm{Hom}_{Mack(G)}\big(N, \mathcal{H}(M,P_X)\big)$$

can also be defined using $\hat{f}$: if Y is a G-set, and if $n \in N(Y)$, then $\sigma(f)(n)$ is the morphism from M to $(P_X)_Y = P_{YX}$ defined for a G-set Z and $m \in M(Z)$ by

$$\sigma(f)(n)_Z(m) = \hat{f}_{ZY}\left(M^*\begin{pmatrix} zy \\ z \end{pmatrix}(m), N^*\begin{pmatrix} zy \\ y \end{pmatrix}(n)\right)$$

Now $\sigma(f)(n)$ is a morphism of B-modules if and only if for any G-set V and any $b \in B(V)$, I have

$$\sigma(f)(b \times n) = b \times \sigma(f)(n)$$

But

$$\sigma(f)(b \times n)_Z(m) = \hat{f}_{ZVY}\left(M^*\begin{pmatrix} zvy \\ z \end{pmatrix}(m), N^*\begin{pmatrix} zvy \\ y \end{pmatrix}(b \times n)\right) \tag{6.6}$$

whereas

$$\left(b \times \sigma(f)(n)\right)_Z(m) = \sigma(f)(n)_{ZV}(m \times b) = \hat{f}_{ZVY}\left(M^*\begin{pmatrix} zvy \\ zv \end{pmatrix}(m \times b), N^*\begin{pmatrix} zvy \\ y \end{pmatrix}(n)\right) \tag{6.7}$$

Moreover, the expression of the product $\times$ from the product ".", and lemma 5.2.2 show that

$$M^*\begin{pmatrix} zvy \\ zv \end{pmatrix}(m \times b) = M^*\begin{pmatrix} zvy \\ zv \end{pmatrix}\left(M^*\begin{pmatrix} zv \\ z \end{pmatrix}(m).B^*\begin{pmatrix} zv \\ v \end{pmatrix}(b)\right) = \ldots$$

$$\ldots = M^*\begin{pmatrix} zvy \\ z \end{pmatrix}(m).B^*\begin{pmatrix} zvy \\ v \end{pmatrix}(b)$$

Now equality of (6.6) and (6.7) becomes

$$\hat{f}_{ZVY}\left(M^*\begin{pmatrix} zvy \\ z \end{pmatrix}(m), B^*\begin{pmatrix} zvy \\ v \end{pmatrix}(b).N^*\begin{pmatrix} zvy \\ y \end{pmatrix}(n)\right) = \ldots$$

$$\ldots = \hat{f}_{ZVY}\left(M^*\begin{pmatrix} zvy \\ z \end{pmatrix}(m).B^*\begin{pmatrix} zvy \\ v \end{pmatrix}(b), N^*\begin{pmatrix} zvy \\ y \end{pmatrix}(n)\right) \tag{6.8}$$

It is now clear that (6.5) implies (6.8). Conversely, if I set

$$\tilde{f}_{Z,Y}(m,n) = \sigma(f)(n)_Z(m) \in P(ZYX)$$

I know that $\tilde{f}$ is a bilinear morphism from M, N to P_X, and that I can recover $\hat{f}$ by the formula

$$\hat{f}_Y(m,n) = P^*\begin{pmatrix} y \\ yy \end{pmatrix}\tilde{f}_{Y,Y}(m,n)$$

Equality of (6.6) and (6.7) can also be written

$$\tilde{f}_{ZV,Y}(m \times b, n) = \tilde{f}_{Z,VY}(m, b \times n)$$

Then if $Z = V = Y$, I have

$$\hat{f}_Y(m.b,n) = P^*\begin{pmatrix} y \\ yy \end{pmatrix}\tilde{f}_{Y,Y}(m.b,n) = P^*\begin{pmatrix} y \\ yy \end{pmatrix}\tilde{f}_{Y,Y}\left(M^*\begin{pmatrix} y \\ yy \end{pmatrix}(m \times b), n\right)$$

Since $\tilde{f}$ is bifunctorial, this is also

$$P^*\begin{pmatrix} y \\ yy \end{pmatrix}P^*\begin{pmatrix} y_1y_2 \\ y_1y_1y_2 \end{pmatrix}\tilde{f}_{Y^2,Y}(m \times b, n) = P^*\begin{pmatrix} y \\ yyy \end{pmatrix}\tilde{f}_{Y^2,Y}(m \times b, n)$$

By a similar computation

$$\hat{f}_Y(m, b.n) = P^*\begin{pmatrix} y \\ yy \end{pmatrix} \tilde{f}_{Y,Y}(m, b.n) = P^*\begin{pmatrix} y \\ yy \end{pmatrix} \tilde{f}_{Y,Y}\left(m, N^*\begin{pmatrix} y \\ yy \end{pmatrix}(b \times n)\right) = \ldots$$

$$\ldots = P^*\begin{pmatrix} y \\ yyy \end{pmatrix} \tilde{f}_{Y,Y^2}(m, b \times n)$$

which proves the equivalence of (6.5) and (6.8), and the isomorphism

$$\mathcal{H}(M\hat{\otimes}_B N, P) \simeq \mathcal{H}_B\big(N, \mathcal{H}(M, P)\big)$$

It remains to prove that this isomorphism maps $\mathcal{H}_A(M\hat{\otimes}_B N, P)$ into $\mathcal{H}_B\big(N, \mathcal{H}_A(M, P)\big)$. Let X be a G-set, and f be a morphism from $M\hat{\otimes}_B N$ to P_X, determined by

$$f_Y\big([m \otimes n]_{(U,\phi)}\big) = P_{X,*}(\phi)\hat{f}_U(m, n)$$

Then f is a morphism of A-modules if and only if for any G-set Z and any $a \in A(Z)$ I have

$$f_{ZY}\big(a \times [m \otimes n]_{(U,\phi)}\big) = a \times f_Y\big([m \otimes n]_{(U,\phi)}\big)$$

which can also be written

$$P_*\begin{pmatrix} zuyx \\ z\phi(u)yx \end{pmatrix} \hat{f}_{ZU}(a \times m, \varepsilon_{B,Z} \times n) = a \times P_*\begin{pmatrix} uyx \\ \phi(u)yx \end{pmatrix} \hat{f}_U(m, n)$$

The right hand side is equal to

$$P_*\begin{pmatrix} zuyx \\ z\phi(u)yx \end{pmatrix} \big(a \times \hat{f}_U(m, n)\big)$$

and f is a morphism of A-modules if and only if

$$\hat{f}_{ZU}(a \times m, \varepsilon_{B,Z} \times n) = a \times \hat{f}_U(m, n) \tag{6.9}$$

It corresponds to f under σ an element of $\mathrm{Hom}_B\big(N, \mathcal{H}(M, P_X)\big)$, which maps $n \in N(Y)$ on the morphism from M to $(P_X)_Y = P_{YX}$ defined by

$$m \in M(Z) \mapsto \hat{f}_{ZY}\left(M^*\begin{pmatrix} zy \\ z \end{pmatrix}(m), N^*\begin{pmatrix} zy \\ y \end{pmatrix}(n)\right)$$

This morphism is a morphism of A-modules if and only if for any G-set U and any $a \in A(U)$ I have

$$a \times \hat{f}_{ZY}\left(M^*\begin{pmatrix} zy \\ z \end{pmatrix}(m), N^*\begin{pmatrix} zy \\ y \end{pmatrix}(n)\right) = \ldots$$

$$\ldots = \hat{f}_{UZY}\left(M^*\begin{pmatrix} uzy \\ uz \end{pmatrix}(a \times m), N^*\begin{pmatrix} uzy \\ y \end{pmatrix}(n)\right) = \ldots$$

$$\ldots = \hat{f}_{UZY}\left(a \times M^*\begin{pmatrix} zy \\ z \end{pmatrix}(m), N^*\begin{pmatrix} uzy \\ y \end{pmatrix}(n)\right) \tag{6.10}$$

It is then clear that (6.9) implies (6.10), since

$$N^*\begin{pmatrix} uzy \\ y \end{pmatrix}(n) = \varepsilon_{B,U} \times N^*\begin{pmatrix} zy \\ y \end{pmatrix}(n)$$

Conversely, the image of the left hand side of equation (6.10) in the case $Y = Z$ under the map $P^*\binom{uy}{uyy}$ gives

$$P^*\begin{pmatrix} uy \\ uyy \end{pmatrix}\left[a \times \hat{f}_{Y^2}\left(M^*\begin{pmatrix} y_1y_2 \\ y_1 \end{pmatrix}(m), N^*\begin{pmatrix} y_1y_2 \\ y_2 \end{pmatrix}(n)\right)\right] = \dots$$

$$\dots = a \times P^*\begin{pmatrix} y \\ yy \end{pmatrix}\hat{f}_{Y^2}\left(M^*\begin{pmatrix} y_1y_2 \\ y_1 \end{pmatrix}(m), N^*\begin{pmatrix} y_1y_2 \\ y_2 \end{pmatrix}(n)\right)$$

which, thanks to condition *iii*) of proposition 1.8.3, can also be written

$$a \times \hat{f}_Y\left(M^*\begin{pmatrix} y \\ yy \end{pmatrix}M^*\begin{pmatrix} y_1y_2 \\ y_1 \end{pmatrix}(m), N^*\begin{pmatrix} y \\ yy \end{pmatrix}N^*\begin{pmatrix} y_1y_2 \\ y_2 \end{pmatrix}(n)\right) = a \times \hat{f}_Y(m,n)$$

The image of the right hand side is

$$P^*\begin{pmatrix} uy \\ uyy \end{pmatrix}\left[\hat{f}_{UY^2}\left(a \times M^*\begin{pmatrix} y_1y_2 \\ y_1 \end{pmatrix}(m), N^*\begin{pmatrix} uy_1y_2 \\ y_2 \end{pmatrix}(n)\right)\right] = \dots$$

$$\dots = \hat{f}_{UY}\left(M^*\begin{pmatrix} uy \\ uyy \end{pmatrix}\left(a \times M^*\begin{pmatrix} y_1y_2 \\ y_1 \end{pmatrix}(m)\right), N^*\begin{pmatrix} uy \\ uyy \end{pmatrix}N^*\begin{pmatrix} uy_1y_2 \\ y_2 \end{pmatrix}(n)\right) = \dots$$

$$\dots = \hat{f}_{UY}\left(a \times M^*\begin{pmatrix} y \\ yy \end{pmatrix}M^*\begin{pmatrix} y_1y_2 \\ y_1 \end{pmatrix}(m), N^*\begin{pmatrix} uy \\ y \end{pmatrix}(n)\right) = \dots$$

$$\dots = \hat{f}_{UY}(a \times m, \varepsilon_{B,Z} \times n)$$

and equation (6.10) is equivalent to equation (6.9), which proves the isomorphism

$$\mathcal{H}_A(M\hat{\otimes}_B N, P) \simeq \mathcal{H}_B\big(N, \mathcal{H}_A(M,P)\big)$$

Those isomorphisms are deduced from those of proposition 1.10.1. So they are natural in M, N and P. Now evaluation at • gives the claimed adjunction property, and completes the proof of the proposition. ∎

Chapter 7

A Morita theory

7.1 Construction of bimodules

Let A, B and C be Green functors for the group G. If M is an A-module-B and N is a B-module-C, then M is in particular an A-module, and N is a C^{op}-module, so $M\hat{\otimes}N$ is an $A\hat{\otimes}C^{op}$-module, that is an A-module-C. It is easy to check that this structure passes down to the quotient and turns $M\hat{\otimes}_B N$ into an A-module-C. Moreover

Proposition 7.1.1: Let A, B, C be Green functors for the group G.

1. **If M is an A-module-B, if N is a B-module-C, and if P is a C-module-D, then there are isomorphisms of A-modules-D**

$$(M\hat{\otimes}_B N)\hat{\otimes}_C P \simeq M\hat{\otimes}_B(N\hat{\otimes}_C P)$$

   which are moreover natural in M, N and P.

2. **If M is an A-module-B, then there are isomorphisms of A-modules-B**

$$A\hat{\otimes}_A M \simeq M \qquad M\hat{\otimes}_B B \simeq M$$

   which are natural in M.

Proof: The first assertion follows from proposition 5.3.2 and from the fact that the isomorphisms

$$(M\hat{\otimes}N)\hat{\otimes}P \simeq M\hat{\otimes}(N\hat{\otimes}P)$$

are compatible with taking quotient, and are natural in M, N, and P.

The second assertion follows easily by adjunction from the fact that

$$\mathrm{Hom}_A(A, M_X) = M_X(\bullet) = M(X)$$

which clearly implies $\mathcal{H}_A(A, M) \simeq M$. ■

Similarly, if P is an A-module-C, then $\mathcal{H}_A(M, P)$ is a B-module-C: if X, Y and Z are G-sets, if $b \in B(X)$, if $f \in \mathcal{H}_A(M, P_Y)$, and $c \in C(Z)$, I can define $b \times f \times c$ on the G-set U by

$$(b \times f \times c)_U(m) = f_{UX}(m \times b) \times c \in P(UXYZ) = P_{XYZ}(U)$$

7.2 Morita contexts

Those constructions of bimodules over Green functors lead to try to generalize the notion of a Morita context(see Curtis-Reiner [4] 3.53), in the following way:

Definition: *Let A and B be Green functors for the group G. A Morita context (M, N, Φ, Ψ) for A and B consists of an A-module-B M and a B-module-A N, and morphisms of bimodules $\Phi : M\hat{\otimes}_B N \to A$ and $\Psi : N\hat{\otimes}_A M \to B$, which are balanced in the sense that the bilinear morphisms $\tilde{\Phi}$ and $\tilde{\Psi}$ associated to them are such that for any G-sets X, Y and Z*

$$m \times \tilde{\Psi}_{Y,Z}(n, m') = \tilde{\Phi}_{X,Y}(m, n) \times m' \quad \forall m \in M(X),\ n \in N(Y),\ m' \in M(Z)$$

$$n \times \tilde{\Phi}_{Y,Z}(m, n') = \tilde{\Psi}_{X,Y}(n, m) \times n' \quad \forall n \in N(X),\ m \in M(Y),\ n' \in N(Z)$$

I will say that (M, N, Φ, Ψ) is a surjective Morita context if Φ and Ψ are surjective.

Lemma 7.2.1: Let (M, N, Φ, Ψ) be a Morita context for A and B. Let (U, ϕ) be a G-set over X, and (V, ψ) be a G-set over Y. Then if $m \in M(U)$, $n \in N(U)$, $p \in M(V)$ and $q \in N(V)$

$$[m \otimes n]_{(U,\phi)} \times \Phi_Y\Big([p \otimes q]_{(V,\psi)}\Big) = \Phi_X\Big([m \otimes n]_{(U,\phi)}\Big) \times [p \otimes q]_{(V,\psi)} \quad \textbf{in} \quad (M\hat{\otimes}_B N)(X \times Y)$$

Proof: The equation relating Φ and $\tilde{\Phi}$ is

$$\Phi_Y\Big([p \otimes q]_{(V,\psi)}\Big) = A_*(\psi) A^* \begin{pmatrix} v \\ vv \end{pmatrix} \tilde{\Phi}_{V,V}(p, q)$$

Setting

$$P = [m \otimes n]_{(U,\phi)} \times \Phi_Y\Big([p \otimes q]_{(V,\psi)}\Big)$$

I have

$$P = \left[(m \times \varepsilon_{B,Y}) \otimes \left(n \times A_*(\psi) A^* \begin{pmatrix} v \\ vv \end{pmatrix} \tilde{\Phi}_{V,V}(p, q)\right)\right]_{(U \times Y, \phi \times Id)}$$

On the other hand

$$n \times A_*(\psi) A^* \begin{pmatrix} v \\ vv \end{pmatrix} \tilde{\Phi}_{V,V}(p, q) = N_* \begin{pmatrix} uv \\ u\psi(v) \end{pmatrix} N^* \begin{pmatrix} uv \\ uvv \end{pmatrix} \Big(n \times \tilde{\Phi}_{V,V}(p, q)\Big) = \ldots$$

$$\ldots = N_* \begin{pmatrix} uv \\ u\psi(v) \end{pmatrix} N^* \begin{pmatrix} uv \\ uvv \end{pmatrix} \Big(\tilde{\Psi}_{U,V}(n, p) \times q\Big)$$

As $(\phi \times Id) \circ (Id \times \psi) = \phi \times \psi$, I have also

$$P = \left[M^* \begin{pmatrix} uv \\ u\psi(v) \end{pmatrix} (m \times \varepsilon_{B,Y}) \otimes N^* \begin{pmatrix} uv \\ uvv \end{pmatrix} \Big(\tilde{\Psi}_{U,V}(n, p) \times q\Big)\right]_{(U \times V, \phi \times \psi)}$$

Moreover

$$M^* \begin{pmatrix} uv \\ u\psi(v) \end{pmatrix} (m \times \varepsilon_{B,Y}) = M^* \begin{pmatrix} uv \\ u\psi(v) \end{pmatrix} M^* \begin{pmatrix} uy \\ u \end{pmatrix} (m) = M^* \begin{pmatrix} uv \\ u \end{pmatrix} (m)$$

whereas expressing the product $\times$ using the product ".", I have

$$N^*\begin{pmatrix} uv \\ uvv \end{pmatrix}\left(\tilde{\Psi}_{U,V}(n,p)\times q\right) = N^*\begin{pmatrix} uv \\ uvv \end{pmatrix}\left[B^*\begin{pmatrix} uv_1v_2 \\ uv_1 \end{pmatrix}\tilde{\Psi}_{U,V}(n,p).N^*\begin{pmatrix} uv_1v_2 \\ v_2 \end{pmatrix}(q)\right]$$

As the map $N^*\binom{uv}{uvv}$ is compatible with the product ".", it is also

$$B^*\begin{pmatrix} uv \\ uvv \end{pmatrix}B^*\begin{pmatrix} uv_1v_2 \\ uv_1 \end{pmatrix}\tilde{\Psi}_{U,V}(n,p).N^*\begin{pmatrix} uv \\ uvv \end{pmatrix}N^*\begin{pmatrix} uv_1v_2 \\ v_2 \end{pmatrix}(q) = \tilde{\Psi}_{U,V}(n,p).N^*\begin{pmatrix} uv \\ v \end{pmatrix}(q)$$

so

$$P = \left[M^*\begin{pmatrix} uv \\ u \end{pmatrix}(m)\otimes\tilde{\Psi}_{U,V}(n,p).N^*\begin{pmatrix} uv \\ v \end{pmatrix}(q)\right]_{(U\times V,\phi\times\psi)}$$

Similarly, if I set

$$Q = \Phi_X\left([m\otimes n]_{(U,\phi)}\right)\times[p\otimes q]_{(V,\psi)}$$

I have

$$\Phi_X\left([m\otimes n]_{(U,\phi)}\right) = A_*(\phi)A^*\begin{pmatrix} u \\ uu \end{pmatrix}\tilde{\Phi}_{U,U}(m,n)$$

so

$$Q = \left[\left(A_*(\phi)A^*\begin{pmatrix} u \\ uu \end{pmatrix}\tilde{\Phi}_{U,U}(m,n)\times p\right)\otimes(\varepsilon_{A,X}\times q)\right]_{(X\times V,Id\times\psi)}$$

Moreover

$$A_*(\phi)A^*\begin{pmatrix} u \\ uu \end{pmatrix}\tilde{\Phi}_{U,U}(m,n)\times p = A_*\begin{pmatrix} uv \\ \phi(u)v \end{pmatrix}A^*\begin{pmatrix} uv \\ uuv \end{pmatrix}\left(\tilde{\Phi}_{U,U}(m,n)\times p\right) = \ldots$$

$$\ldots = M_*\begin{pmatrix} uv \\ \phi(u)v \end{pmatrix}M^*\begin{pmatrix} uv \\ uuv \end{pmatrix}\left(m\times\tilde{\Psi}_{U,V}(n,p)\right)$$

As

$$M^*\begin{pmatrix} uv \\ uuv \end{pmatrix}\left(m\times\tilde{\Psi}_{U,V}(n,p)\right) = \ldots$$

$$\ldots = M^*\begin{pmatrix} uv \\ uuv \end{pmatrix}\left(M^*\begin{pmatrix} u_1u_2v \\ u_1 \end{pmatrix}(m).B^*\begin{pmatrix} u_1u_2v \\ u_2v \end{pmatrix}\tilde{\Psi}_{U,V}(n,p)\right) = \ldots$$

$$\ldots = M^*\begin{pmatrix} uv \\ u \end{pmatrix}(m).B^*\begin{pmatrix} uv \\ uuv \end{pmatrix}B^*\begin{pmatrix} u_1u_2v \\ u_2v \end{pmatrix}\tilde{\Psi}_{U,V}(n,p) = M^*\begin{pmatrix} uv \\ u \end{pmatrix}(m).\tilde{\Psi}_{U,V}(n,p)$$

I have also

$$A_*(\phi)A^*\begin{pmatrix} u \\ uu \end{pmatrix}\tilde{\Phi}_{U,U}(m,n)\times p = M_*\begin{pmatrix} uv \\ \phi(u)v \end{pmatrix}\left(M^*\begin{pmatrix} uv \\ u \end{pmatrix}(m).\tilde{\Psi}_{U,V}(n,p)\right)$$

Finally, I have

$$Q = \left[\left(M^*\begin{pmatrix} uv \\ u \end{pmatrix}(m).\tilde{\Psi}_{U,V}(n,p)\right)\otimes N^*\begin{pmatrix} uv \\ \phi(u)v \end{pmatrix}(\varepsilon_{A,X}\times q)\right]_{(U\times V,\phi\times\psi)}$$

and as $\varepsilon_{A,X}\times q = N^*\binom{xv}{v}(q)$, I have also

$$Q = \left[M^*\begin{pmatrix} uv \\ u \end{pmatrix}(m).\tilde{\Psi}_{U,V}(n,p)\otimes N^*\begin{pmatrix} uv \\ v \end{pmatrix}(q)\right]_{(U\times V,\phi\times\psi)}$$

so $P = Q$ in $M\hat{\otimes}_B N(X\times Y)$, and the lemma follows. ■

Proposition 7.2.2: Let A and B be Green functors for the group G, and let (M, N, Φ, Ψ) be a Morita context for A and B.

- **If $\Phi_\bullet$ is surjective, then Φ is an isomorphism.**
- **If $\Phi_\bullet$ and $\Psi_\bullet$ are surjective, then Φ and Ψ are isomorphisms. Moreover in that case**
  1. **The module M_Ω is a progenerator for A-Mod and for Mod-B, and the module N_Ω is a progenerator for B-Mod and for Mod-A.**
  2. **There are isomorphisms of bimodules**
  $$N \simeq \mathcal{H}_A(M, A) \simeq \mathcal{H}_{B^{op}}(M, B) \qquad M \simeq \mathcal{H}_B(N, B) \simeq \mathcal{H}_{A^{op}}(N, A)$$
  3. **There are isomorphisms of Green functors**
  $$A \simeq \mathcal{H}_{B^{op}}(M, M) \simeq \big(\mathcal{H}_B(N, N)\big)^{op} \qquad B \simeq \mathcal{H}_{A^{op}}(N, N) \simeq \big(\mathcal{H}_A(M, M)\big)^{op}$$
  4. **The functors $P \mapsto N \otimes_A P$ and $Q \mapsto M \hat{\otimes}_B Q$ are mutual inverse equivalences of categories between A-Mod and B-Mod.**

Proof: As Φ is a morphism of A-modules-A, and as A is generated as a bimodule by ε_A, if $\Phi_\bullet$ is surjective, then ε_A is in its image, and then Φ is surjective. To prove the first assertion, it suffices then to prove that Φ is injective.

Let X be a G-set. Let moreover $[m_i \otimes n_i]_{(U_i, \phi_i)}$, for $1 \leq i \leq m$ be elements of $M\hat{\otimes}_B N(X)$ such that
$$\Phi_X\left(\sum_{i=1}^{m}[m_i \otimes n_i]_{(U_i,\phi_i)}\right) = 0$$
By hypothesis, the map $\Phi_\bullet$ is surjective: let $[p_j \otimes q_j]_{(V_j,\psi_j)}$, for $1 \leq j \leq p$ be elements of $M\hat{\otimes}N(\bullet)$ such that
$$\Phi_\bullet\left(\sum_{j=1}^{p}[p_j \otimes q_j]_{(V_j,\psi_j)}\right) = \varepsilon_A$$
(the map ψ_j is then the unique map from V_j to $\bullet$). Setting
$$v = \sum_{i=1}^{m}[m_i \otimes n_i]_{(U_i,\phi_i)}$$
it follows from the previous lemma that
$$v = v \times \varepsilon_A = \sum_{i=1}^{m}\sum_{j=1}^{p}[m_i \otimes n_i]_{(U_i,\phi_i)} \times \Phi_\bullet\left([p_j \otimes q_j]_{(V_j,\psi_j)}\right) = \dots$$
$$\dots = \sum_{i=1}^{m}\sum_{j=1}^{p}\Phi_X\left([m_i \otimes n_i]_{(U_i,\phi_i)}\right) \times [p_j \otimes q_j]_{(V_j,\psi_j)} = \dots$$
$$\dots = \Phi_X\left(\sum_{i=1}^{m}[m_i \otimes n_i]_{(U_i,\phi_i)}\right) \times \sum_{j=1}^{p}[p_j \otimes q_j]_{(V_j,\psi_j)} = 0$$

which proves that Φ is injective, hence that it is an isomorphism.

Now if $\Phi_\bullet$ and $\Psi_\bullet$ are surjective, then Φ and Ψ are isomorphisms (indeed the 4-tuple (N, M, Ψ, Φ) is a Morita context for B and A, and Φ and Ψ play symmetric roles). Assertion 4) follows then from proposition 7.1.1, since

$$N\hat{\otimes}_A(M\hat{\otimes}_B P) \simeq (N\hat{\otimes}_A M)\hat{\otimes}_B P \simeq B\hat{\otimes}_B P \simeq P$$

Lemma 7.2.3: Let M and N be Mackey functors. If X and Y are G-sets, then there are isomorphisms of Mackey functors

$$(M\hat{\otimes}N)_{XY} \simeq M_X\hat{\otimes}N_Y$$

which are natural in M, N, X and Y.

Proof: This follows from the fact that

$$\mathcal{H}(M_X, N) \simeq \mathcal{H}(M, N_X)$$

naturally in M, N, and X. Then

$$\mathcal{H}(M_X\hat{\otimes}N_Y, P) \simeq \mathcal{H}\big(N_Y, \mathcal{H}(M_X, P)\big) \simeq \mathcal{H}\big(N, \mathcal{H}(M, P_X)_Y\big)$$

Moreover, for any G-set Z

$$\mathcal{H}(M, P_X)_Y(Z) = \mathcal{H}(M, P_X)(ZY) = \mathrm{Hom}_{Mack(G)}\big(M, (P_X)_{ZY}\big) = \dots$$

$$\dots = \mathrm{Hom}_{Mack(G)}(M, P_{ZYX}) = \mathrm{Hom}_{Mack(G)}\big(M, (P_{YX})_Z\big) = \mathcal{H}(M, P_{YX})(Z)$$

This gives the isomorphism $\mathcal{H}(M, P_X)_Y \simeq \mathcal{H}(M, P_{YX})$, and then

$$\mathcal{H}(M_X\hat{\otimes}N_Y, P) \simeq \mathcal{H}\big(N, \mathcal{H}(M, P_{YX})\big) \simeq \mathcal{H}(M\hat{\otimes}N, P_{YX}) \simeq \dots$$

$$\dots \simeq \mathcal{H}(M\hat{\otimes}N, P_{XY}) \simeq \mathcal{H}\big((M\hat{\otimes}N)_{XY}, P\big)$$

and the lemma follows. ∎

Remarks: 1) The isomorphisms of the lemma can be stated precisely. If Z is a G-set, then a G-set over ZXY is determined by a map $\left(\begin{smallmatrix} u \\ \gamma(u)\alpha(u)\beta(u)\end{smallmatrix}\right)$ from U to ZXY. I define then a map from $(M\hat{\otimes}N)_{XY}(Z) = M\hat{\otimes}N(ZXY)$ to $(M_X\hat{\otimes}N_Y)(Z)$ by

$$[m \otimes n]_{\left(U, \left(\begin{smallmatrix} u \\ \gamma(u)\alpha(u)\beta(u)\end{smallmatrix}\right)\right)} \mapsto \left[M_*\begin{pmatrix} u \\ u\alpha(u)\end{pmatrix}(m) \otimes N_*\begin{pmatrix} u \\ u\beta(u)\end{pmatrix}(n)\right]_{(U,\gamma)} \in (M_X\hat{\otimes}N_Y(Z)$$

which makes sense because $M_*\left(\begin{smallmatrix} u \\ u\alpha(u)\end{smallmatrix}\right)(m) \in M(UX) = M_X(U)$, and $N_*\left(\begin{smallmatrix} u \\ u\beta(u)\end{smallmatrix}\right)(n) \in N(UY) = N_Y(U)$.
The inverse isomorphism is defined by

$$[m \otimes n]_{(U,\phi)} \in (M_X\hat{\otimes}N_Y)(Z) \mapsto [M^*\begin{pmatrix} uxy \\ ux\end{pmatrix} \otimes N_*\begin{pmatrix} uxy \\ uy\end{pmatrix}(n)]_{\left(UXY, \left(\begin{smallmatrix} uxy \\ \phi(u)xy\end{smallmatrix}\right)\right)}$$

2) The case $X = \bullet$ gives in particular the isomorphism

$$(M\hat{\otimes}N)_X \simeq M\hat{\otimes}(N_X) \simeq (M_X)\hat{\otimes}N$$

and the naturality of this isomorphism shows that if M is an A-module-B, then for any B-module N

$$(M\hat{\otimes}_B N)_X \simeq M\hat{\otimes}_B(N_X) \simeq (M_X)\hat{\otimes}_B N$$

Proof of proposition 7.2.2 (part 2): Assertion 1) follows from the previous lemma, because an equivalence of categories preserves modules of finite type, and because M_Ω is the image of B_Ω under the functor $P \mapsto M\hat{\otimes}_B P$, since

$$M_\Omega \simeq (M\hat{\otimes}_B B)_\Omega \simeq M\hat{\otimes}_B(B_\Omega)$$

Concerning assertion 2), I observe first that evaluating at the set $\bullet$ the isomorphism

$$\mathcal{H}_B\Big(N, \mathcal{H}_A(M, A)\Big) \simeq \mathcal{H}_A(M\hat{\otimes}_B N, A)$$

gives

$$\mathrm{Hom}_B\Big(N, \mathcal{H}_A(M, A)\Big) \simeq \mathrm{Hom}_A(M\hat{\otimes}_B N, A)$$

Then the morphism Φ gives by adjunction a morphism Θ from N to $\mathcal{H}_A(M, A)$, very easy to describe: if $n \in N(X)$, then the image of n in $\mathrm{Hom}_A(M, A_X)$ is the morphism defined by

$$m \in M(Y) \mapsto \tilde{\Phi}_{Y,X}(m, n) \in A(YX) = A_X(Y)$$

The morphism Θ is injective: the element $n \in N(X)$ is in the kernel of Θ_X if and only if for any G-set Y and any element $m \in M(Y)$, I have

$$\tilde{\Phi}_{Y,X}(m, n) = 0$$

But as Ψ is surjective, there exists G-sets Y_i and elements $p_i \in N(Y_i)$ and $q_i \in M(Y_i)$, for $1 \leq i \leq n$, such that

$$\varepsilon_B = \sum_i B_*\begin{pmatrix} y_i \\ \bullet \end{pmatrix} B^*\begin{pmatrix} y_i \\ y_i y_i \end{pmatrix} \tilde{\Psi}_{Y_i,Y_i}(p_i, q_i)$$

Then

$$n = \varepsilon_B \times n = \sum_i N_*\begin{pmatrix} y_i x \\ x \end{pmatrix} N^*\begin{pmatrix} y_i x \\ y_i y_i x \end{pmatrix} \left[\tilde{\Psi}_{Y_i,Y_i}(p_i, q_i) \times n\right]$$

The expression inside hooks is also

$$p_i \times \tilde{\Phi}_{Y_i,X}(q_i, n)$$

So it is zero for all i, which proves that $n = 0$ and that Θ is injective.

To prove that Θ is also surjective, I must prove that if $f \in \mathrm{Hom}_A(M, A_X)$, then there exists an element $n \in N(X)$ such that for any G-set Y and any $m \in M(Y)$, I have

$$f_Y(m) = \tilde{\Phi}_{Y,X}(m, n)$$

If such an element exists, keeping the previous notations, I have

$$n = \varepsilon_B \times n = \sum_i N_*\begin{pmatrix} y_i x \\ x \end{pmatrix} N^*\begin{pmatrix} y_i x \\ y_i y_i x \end{pmatrix} \left[p_i \times \tilde{\Phi}_{Y_i,X}(q_i, n)\right] = \ldots$$

$$\ldots = \sum_i N_* \begin{pmatrix} y_i x \\ x \end{pmatrix} N^* \begin{pmatrix} y_i x \\ y_i y_i x \end{pmatrix} \left(p_i \times f_{Y_i}(q_i)\right)$$

Let then

$$n = \sum_i N_* \begin{pmatrix} y_i x \\ x \end{pmatrix} N^* \begin{pmatrix} y_i x \\ y_i y_i x \end{pmatrix} \left(p_i \times f_{Y_i}(q_i)\right)$$

Then for any Y and any $m \in M(Y)$, I have

$$\tilde{\Phi}_{Y,X}(m,n) = \sum_i \tilde{\Phi}_{Y,X}\left(m, N_* \begin{pmatrix} y_i x \\ x \end{pmatrix} N^* \begin{pmatrix} y_i x \\ y_i y_i x \end{pmatrix} \left(p_i \times f_{Y_i}(q_i)\right)\right)$$

As $\tilde{\Phi}$ is a bilinear morphism, it is also

$$\sum_i A_* \begin{pmatrix} yy_i x \\ yx \end{pmatrix} A^* \begin{pmatrix} yy_i x \\ yy_i y_i x \end{pmatrix} \tilde{\Phi}_{Y,Y_i^2X}\left(m, p_i \times f_{Y_i}(q_i)\right)$$

As Φ is a morphism of modules-A, I have

$$\tilde{\Phi}_{Y,Y_i^2X}\left(m, p_i \times f_{Y_i}(q_i)\right) = \tilde{\Phi}_{Y,Y_i}(m, p_i) \times f_{Y_i}(q_i)$$

and as f is a morphism of A-modules, it is also

$$f_{YY_i^2}\left(\tilde{\Phi}_{Y,Y_i}(m, p_i) \times q_i\right)$$

or

$$f_{YY_i^2}\left(m \times \tilde{\Psi}_{Y_i,Y_i}(p_i, q_i)\right)$$

Finally

$$\tilde{\Phi}_{Y,X}(m,n) = \sum_i A_* \begin{pmatrix} yy_i x \\ yx \end{pmatrix} A^* \begin{pmatrix} yy_i x \\ yy_i y_i x \end{pmatrix} f_{YY_i^2}\left(m \times \tilde{\Psi}_{Y_i,Y_i}(p_i, q_i)\right)$$

As f is a morphism of Mackey functors, it is also

$$f_Y \left[M_* \begin{pmatrix} yy_i \\ y \end{pmatrix} M^* \begin{pmatrix} yy_i \\ yy_i y_i \end{pmatrix} \left(m \times \tilde{\Psi}_{Y_i,Y_i}(p_i, q_i)\right) \right]$$

Expression inside hooks is equal to

$$m \times B_* \begin{pmatrix} y_i \\ \bullet \end{pmatrix} B^* \begin{pmatrix} y_i \\ y_i y_i \end{pmatrix} \tilde{\Psi}_{Y_i,Y_i}(p_i, q_i) = m \times \varepsilon_B = m$$

so I have

$$\tilde{\Phi}_{Y,X}(m,n) = f_Y(m)$$

which proves that Θ is an isomorphism. It is easy to see that it is an isomorphism of bimodules. The other isomorphisms of assertion 2) now follow, by switching the roles of A and B, or replacing them by their opposite.

Assertion 3) is proved by observing that

$$\mathcal{H}_A(A, A) \simeq A^{op}$$

Moreover, if I denote by F the functor $N\hat{\otimes}_A-$ from A-**Mod** to B-**Mod**, the isomorphisms

$$(N\hat{\otimes}_A P)_X \simeq N\hat{\otimes}_A(P_X)$$

show that $F(P_X) \simeq F(P)_X$. As F is an equivalence of categories, then

$$\mathrm{Hom}_B\big(F(P), F(Q)\big) = \mathrm{Hom}_A(P, Q)$$

for any P and Q. It follows that

$$\mathcal{H}_B\big(F(P), F(Q)\big)(X) = \mathrm{Hom}_B\big(F(P), F(Q)_X\big) = \mathrm{Hom}_B(F(P), F(Q_X)\big) = \ldots$$

$$\ldots = \mathrm{Hom}_A(P, Q_X) = \mathcal{H}_A(P, Q)(X)$$

and it is easy to see that those isomorphisms induce isomorphisms of Mackey functors

$$\mathcal{H}_B\big(F(P), F(Q)\big) \simeq \mathcal{H}_A(P, Q)$$

which are moreover compatible with the product $\hat{\circ}$. Thus for any A-module, the Green functors $\mathcal{H}_B\big(F(P), F(P)\big)$ and $\mathcal{H}_A(P, P)$ are isomorphic. Now for $P = A$, this gives

$$\mathcal{H}_B(N, N) \simeq A^{op}$$

and assertion 3) follows. This completes the proof of the proposition. ■

7.3 Converse

The previous proposition has a converse:

Proposition 7.3.1: Let A be a Green functor for the group G, and M be an A-module such that M_Ω is a progenerator of A-Mod. Let $N = \mathcal{H}_A(M, A)$ and $B = \big(\mathcal{H}_A(M, M)\big)^{op}$. Then there exists a surjective Morita context (M, N, Φ, Ψ) for A and B.

Proof: Let X and Y be G-sets. If $m \in M(X)$ and $f \in B(Y) = \mathrm{Hom}_A(M, M_Y)$, let

$$m \times f = f_X(m) \in M_Y(X) = M(XY)$$

It is easy to see that this definition turns M into an A-module-B. Then $N = \mathcal{H}_A(M, A)$ is a B-module-A. As moreover

$$\mathcal{H}_A(M\hat{\otimes}_B N, A) \simeq \mathcal{H}_B\big(N, \mathcal{H}_A(M, A)\big) = \mathcal{H}_B(N, N)$$

the identity map of N gives a morphism of A-modules from $M\hat{\otimes}_B N$ to A, which can be described as follows: if X and Y are G-sets, let $\tilde{\Phi}_{X,Y}$ be the bilinear map from $M(X) \times N(Y)$ to $A(XY)$ defined by

$$m \in M(X),\ \phi \in N(Y) = \mathrm{Hom}_A(M, A_Y) \mapsto \phi_X(m) \in A_Y(X) = A(XY)$$

It is easy to check that those maps define a bilinear morphism from M, N to A, associated to a morphism Φ from $M\hat{\otimes}_B N$ to A, which is a morphism of A-modules-A.

Conversely, there is a bilinear morphism $\tilde{\Psi}$ from N, M to B defined as follows: if X and Y are G-sets, if $f \in N(X) = \mathrm{Hom}_A(M, A_X)$ and $m \in M(Y)$, then $\tilde{\Psi}_{X,Y}(f,m)$ is the morphism from M to M_{XY} defined for a G-set Z and $m \in M(Z)$ by

$$\tilde{\Psi}_{X,Y}(f,m) : m_1 \in M(Z) \mapsto f_Z(m_1) \times m \in M(ZXY) = M_{XY}(Z)$$

The morphism from $N \hat{\otimes} M$ to B associated to it passes down to the quotient, and defines a morphism from $N \hat{\otimes}_A M$ to B, which is a morphism of B-modules-B.

Before proving the proposition, I will give an equivalent formulation of the hypothesis on M, independent of Ω:

Lemma 7.3.2: Let A be a Green functor for G, and M be an A-module. The following conditions are equivalent:

1. **The module M_Ω is a progenerator.**
2. **The module M is a finitely generated projective module, and there exists a G-set X such that A is a direct summand of M_X.**
3. **There exists G-sets X and Y such that A is a direct summand of M_X and M is a direct summand of A_Y.**

Proof: First I observe that if X and Y are G-sets such that X divides Y in $\mathcal{C}_A$, then M_X is a direct summand of M_Y: indeed, say that X divides Y in $\mathcal{C}_A$ is equivalent to say that A_X is a direct summand of A_Y. But M_X is isomorphic to $\mathcal{H}_A(A_X, M)$, which is a direct summand of $\mathcal{H}_A(A_Y, M) \simeq M_Y$.

So if M_Ω is a progenerator, as $\bullet$ divides Ω in $\mathcal{C}_A$, it follows that M is a direct summand of M_Ω, so M is a finitely generated projective module. Moreover, there exists a set I and a surjective morphism

$$p : \bigoplus_{i \in I} M_\Omega^{(i)} \to A$$

from a sum of copies of M_Ω to A. In particular, there is a finite subset $J \subseteq I$ and elements $m_j \in M_\Omega^{(j)}(\bullet)$, for $j \in J$, such that

$$\varepsilon_A = p_\bullet(\sum_{j \in J} m_j)$$

Then the image B of the restriction of p to $\oplus_{j \in J} M_\Omega^{(j)}$ is an A-submodule of A, and $B(\bullet)$ contains ε_A. This proves that there exists an integer n such that A is a quotient, hence a direct summand, of $nM_\Omega \simeq M_{n\Omega}$. Thus 1) implies 2).

If now hypothesis 2) holds, then as A_Ω is a progenerator, and as M is a finitely generated projective module, there exists an integer n such that M is a direct summand of $nA_\Omega \simeq A_{n\Omega}$. So 2) implies 3).

Finally if 3) holds, then M is a finitely generated projective module, because it is a direct summand of the projective module A_Y. Then M_Ω is a direct summand of $(A_Y)_\Omega \simeq A_{\Omega Y}$, and M_Ω is a finitely generated projective module. Moreover, the module A_Ω is a direct summand of $(M_X)_\Omega \simeq M_{\Omega X}$. As ΩX divides a multiple of Ω in $\mathcal{C}_A$, it follows that A_Ω is a direct summand of a direct sum of copies of M_Ω. As A_Ω is a progenerator, so is M_Ω, and this proves the lemma. ∎

Proof of proposition 7.3.1: I must prove that if M_Ω is a progenerator, then Φ and Ψ are surjective. It is equivalent to say that ε_A lies in the image of $\Phi_\bullet$, and ε_B in the image of $\Psi_\bullet$.

I will show that if there exists X such that A is a direct summand of M_X, then Φ is surjective, and that if there exists Y such that M is a direct summand of A_Y (which is equivalent to say that M is projective and of finite type), then Ψ is surjective.

So let X be such that A is a direct summand of M_X. Then there exists an element $\alpha \in Hom_A(A, M_X)$ and an element $\beta \in Hom_A(M_X, A)$ such that

$$\beta \circ \alpha = Id_A$$

But I have seen that $Hom_A(A, M_X) \simeq M(X)$. Let $m \in M(X)$ be the image of α under that isomorphism. Then α is defined on the G-set Z by

$$a \in A(Z) \mapsto a \times m \in M(ZX) = M_X(Z)$$

On the other hand

$$Hom_A(M_X, A) \simeq Hom_A(M, A_X)$$

Under this isomorphism, the element β maps to $\gamma \in Hom_A(M, A_X) = N(X)$, and I know that for any G-set Z and any $m' \in M_X(Z) = M(ZX)$, I have

$$\beta_Z(m') = A_* \begin{pmatrix} zx \\ z \end{pmatrix} A^* \begin{pmatrix} zx \\ zxx \end{pmatrix} \gamma_{ZX}(m')$$

Now say that $\beta \circ \alpha = Id_A$ is equivalent to say that for any Z and any $a \in A(Z)$

$$a = A_* \begin{pmatrix} zx \\ z \end{pmatrix} A^* \begin{pmatrix} zx \\ zxx \end{pmatrix} \gamma_{ZX}(a \times m)$$

which can be written as

$$a = a \times A_* \begin{pmatrix} x \\ \bullet \end{pmatrix} A^* \begin{pmatrix} x \\ xx \end{pmatrix} \gamma_X(m)$$

This is equivalent to

$$\varepsilon_A = A_* \begin{pmatrix} x \\ \bullet \end{pmatrix} A^* \begin{pmatrix} x \\ xx \end{pmatrix} \gamma_X(m)$$

Then let

$$\tau = [m \otimes \gamma]_{(X, \binom{x}{\bullet})} \in M \hat{\otimes}_B N(\bullet)$$

The image of τ under $\Phi_\bullet$ is precisely

$$\Phi_\bullet(\tau) = A_* \begin{pmatrix} x \\ \bullet \end{pmatrix} A^* \begin{pmatrix} x \\ xx \end{pmatrix} \tilde{\Phi}_{X,X}(m, \gamma) = A_* \begin{pmatrix} x \\ \bullet \end{pmatrix} A^* \begin{pmatrix} x \\ xx \end{pmatrix} \gamma_X(m) = \varepsilon_A$$

and this proves that $\Phi_\bullet$ is surjective.

Similarly, if M is a direct summand of M_Y, then there exists $\alpha \in Hom_A(A_Y, M)$ and $\beta \in Hom_A(M, A_Y)$ such that $\alpha \circ \beta = Id_M$. But

$$Hom_A(A_Y, M) \simeq M(Y)$$

and α is determined by an element $m \in M(Y)$, such that for $a \in A_Y(Z) = A(ZY)$

$$\alpha_Y(a) = M_*\begin{pmatrix} zy \\ z \end{pmatrix} M^*\begin{pmatrix} zy \\ zyy \end{pmatrix}(a \times m) = a \circ_Z m$$

Now say that $\alpha \circ \beta = Id_M$ means that for any G-set Z and any $m' \in M(Z)$, I have

$$m' = M_*\begin{pmatrix} zy \\ z \end{pmatrix} M^*\begin{pmatrix} zy \\ zyy \end{pmatrix}\Big(\beta_Z(m') \times m\Big)$$

Let

$$\tau' = [\beta \otimes m]_{(Y,\binom{y}{\bullet})} \in N\hat{\otimes}_A M(\bullet)$$

The image of τ' under Ψ is the element of $\mathcal{H}_A(M,M)(\bullet) = \mathrm{End}_A(M)$ defined by

$$\Psi_\bullet(\tau') = \mathcal{H}_A(M,M)_*\begin{pmatrix} y \\ \bullet \end{pmatrix} \mathcal{H}_A(M,M)^*\begin{pmatrix} y \\ yy \end{pmatrix} \tilde{\Psi}_{Y,Y}(\beta, m)$$

which can be evaluated at a G-set Z by

$$m' \in M(Z) \mapsto \Psi(\tau')_Z(m') = M_*\begin{pmatrix} zy \\ z \end{pmatrix} M^*\begin{pmatrix} zy \\ zyy \end{pmatrix}\Big(\beta_Z(m') \times m\Big)$$

It follows that $\Psi(\tau')(m') = m'$, so $\Psi(\tau')$ is the identity of M, that is the unit $\varepsilon_{\mathcal{H}_A(M,M)}$. This proves that Ψ is surjective, and completes the proof of the proposition. ■

7.4 A remark on bimodules

If A is a Green functor for G, theorem 4.3.1 states that evaluation at Ω is an equivalence of categories from A-**Mod** to $A(\Omega^2)$-**Mod**. If B is another Green functor, and M is an A-module-B, then $M(\Omega)$ is an $A(\Omega^2)$-module, and a $B^{op}(\Omega^2)$-module. As $B^{op}(\Omega^2) \simeq B(\Omega^2)^{op}$, the module $M(\Omega^2)$ is a module-$B(\Omega^2)$. However, the module $M(\Omega)$ is *not* in general an $A(\Omega^2)$-module-$B(\Omega^2)$: indeed, if $a \in A(\Omega^2)$, if $m \in A(\Omega)$, then

$$a \circ_\Omega m = A_*\begin{pmatrix} \omega_1\omega_2 \\ \omega_1 \end{pmatrix} A^*\begin{pmatrix} \omega_1\omega_2 \\ \omega_1\omega_2\omega_2 \end{pmatrix}(a \times m)$$

Then if $b \in B(\Omega^2)$

$$(a \circ_\Omega m) \circ_\Omega b = A_*\begin{pmatrix} \omega_1\omega_2 \\ \omega_2 \end{pmatrix} A^*\begin{pmatrix} \omega_1\omega_2 \\ \omega_1\omega_1\omega_2 \end{pmatrix}[(a \circ_\Omega m) \times b]$$

The product inside hooks is equal to

$$A_*\begin{pmatrix} \omega_1\omega_2\omega_3\omega_4 \\ \omega_1\omega_3\omega_4 \end{pmatrix} A^*\begin{pmatrix} \omega_1\omega_2\omega_3\omega_4 \\ \omega_1\omega_2\omega_2\omega_3\omega_4 \end{pmatrix}(a \times m \times b)$$

As the square

$$\begin{array}{ccc}
\Omega^3 & \xrightarrow{\begin{pmatrix} \omega_1\omega_2\omega_3 \\ \omega_1\omega_2\omega_1\omega_3 \end{pmatrix}} & \Omega^4 \\
\begin{pmatrix} \omega_1\omega_2\omega_3 \\ \omega_1\omega_3 \end{pmatrix}\Big\downarrow & & \Big\downarrow\begin{pmatrix} \omega_1\omega_2\omega_3\omega_4 \\ \omega_1\omega_3\omega_4 \end{pmatrix} \\
\Omega^2 & \xrightarrow[\begin{pmatrix} \omega_1\omega_2 \\ \omega_1\omega_1\omega_2 \end{pmatrix}]{} & \Omega^3
\end{array}$$

is cartesian, I have

$$A^*\begin{pmatrix}\omega_1\omega_2\\ \omega_1\omega_1\omega_2\end{pmatrix} A_*\begin{pmatrix}\omega_1\omega_2\omega_3\omega_4\\ \omega_1\omega_3\omega_4\end{pmatrix} = A_*\begin{pmatrix}\omega_1\omega_2\omega_3\\ \omega_1\omega_3\end{pmatrix} A^*\begin{pmatrix}\omega_1\omega_2\omega_3\\ \omega_1\omega_2\omega_1\omega_3\end{pmatrix}$$

and it follows that

$$(a \circ_\Omega m) \circ_\Omega b = \dots$$

$$\dots = A_*\begin{pmatrix}\omega_1\omega_2\\ \omega_2\end{pmatrix} A_*\begin{pmatrix}\omega_1\omega_2\omega_3\\ \omega_1\omega_3\end{pmatrix} A^*\begin{pmatrix}\omega_1\omega_2\omega_3\\ \omega_1\omega_2\omega_1\omega_3\end{pmatrix} A^*\begin{pmatrix}\omega_1\omega_2\omega_3\omega_4\\ \omega_1\omega_2\omega_2\omega_3\omega_4\end{pmatrix} (a\times m\times b) = \dots$$

$$\dots = A_*\begin{pmatrix}\omega_1\omega_2\omega_3\\ \omega_3\end{pmatrix} A^*\begin{pmatrix}\omega_1\omega_2\omega_3\\ \omega_1\omega_2\omega_2\omega_1\omega_3\end{pmatrix} (a\times m\times b) \qquad (E)$$

A similar computation gives

$$a \circ_\Omega (m \circ_\Omega b) = A_*\begin{pmatrix}\omega_1\omega_2\\ \omega_1\end{pmatrix} A^*\begin{pmatrix}\omega_1\omega_2\\ \omega_1\omega_2\omega_2\end{pmatrix} [a \times (m \circ_\Omega b)]$$

The product inside hooks is equal to

$$A_*\begin{pmatrix}\omega_1\omega_2\omega_3\omega_4\\ \omega_1\omega_2\omega_4\end{pmatrix} A^*\begin{pmatrix}\omega_1\omega_2\omega_3\omega_4\\ \omega_1\omega_2\omega_3\omega_3\omega_4\end{pmatrix} (a\times m\times b)$$

As the square

$$\begin{array}{ccc}
\Omega^3 & \xrightarrow{\begin{pmatrix}\omega_1\omega_2\omega_3\\ \omega_1\omega_2\omega_3\omega_2\end{pmatrix}} & \Omega^4\\
{\scriptstyle\begin{pmatrix}\omega_1\omega_2\omega_3\\ \omega_1\omega_2\end{pmatrix}}\Big\downarrow & & \Big\downarrow{\scriptstyle\begin{pmatrix}\omega_1\omega_2\omega_3\omega_4\\ \omega_1\omega_2\omega_4\end{pmatrix}}\\
\Omega^2 & \xrightarrow[\begin{pmatrix}\omega_1\omega_2\\ \omega_1\omega_2\omega_2\end{pmatrix}]{} & \Omega^3
\end{array}$$

is cartesian, I have

$$A^*\begin{pmatrix}\omega_1\omega_2\\ \omega_1\omega_2\omega_2\end{pmatrix} A_*\begin{pmatrix}\omega_1\omega_2\omega_3\omega_4\\ \omega_1\omega_2\omega_4\end{pmatrix} = A_*\begin{pmatrix}\omega_1\omega_2\omega_3\\ \omega_1\omega_2\end{pmatrix} A^*\begin{pmatrix}\omega_1\omega_2\omega_3\\ \omega_1\omega_2\omega_3\omega_2\end{pmatrix}$$

so

$$a \circ_\Omega (m \circ_\Omega b) = \dots$$

$$\dots = A_*\begin{pmatrix}\omega_1\omega_2\\ \omega_1\end{pmatrix} A_*\begin{pmatrix}\omega_1\omega_2\omega_3\\ \omega_1\omega_2\end{pmatrix} A^*\begin{pmatrix}\omega_1\omega_2\omega_3\\ \omega_1\omega_2\omega_3\omega_2\end{pmatrix} A^*\begin{pmatrix}\omega_1\omega_2\omega_3\omega_4\\ \omega_1\omega_2\omega_3\omega_3\omega_4\end{pmatrix} (a\times m\times b) = \dots$$

$$\dots = A_*\begin{pmatrix}\omega_1\omega_2\omega_3\\ \omega_1\end{pmatrix} A^*\begin{pmatrix}\omega_1\omega_2\omega_3\\ \omega_1\omega_2\omega_3\omega_3\omega_2\end{pmatrix} (a\times m\times b)$$

There is no obvious reason for which this expression should be equal to (E), so $M(\Omega)$ is not a bimodule in general. This is because $(A\hat{\otimes}B^{op})(\Omega^2)$ is the tensor product of $A(\Omega^2)$ and $B(\Omega^2)^{op}$ over the *non-commutative* algebra $b(\Omega^2)$ (i.e. Mackey algebra) as shown in the next proposition:

Proposition 7.4.1: Let A and B be Green functors for the group G. If Ω is a G-set, then the morphism from b to A (resp. from b to B) turns $A(\Omega^2)$ (resp. $B(\Omega^2)$) into a module-$b(\Omega^2)$ (resp. a $b(\Omega^2)$-module), and

$$(A\hat{\otimes}B)(\Omega^2) \simeq A(\Omega^2)\hat{\otimes}_{b(\Omega^2)}B(\Omega^2)$$

as R-modules.

Proof: Lemma 7.2.3 shows that

$$(A\hat{\otimes}B)_{\Omega^2} \simeq A_\Omega\hat{\otimes}B_\Omega$$

as Mackey functors. Evaluation at the trivial G-set gives

$$(A\hat{\otimes}B)_{\Omega^2}(\bullet) = (A\hat{\otimes}B)(\Omega^2) = (A_\Omega\hat{\otimes}B_\Omega)(\bullet)$$

and by definition of tensor product of Mackey functors

$$(A_\Omega\hat{\otimes}B_\Omega)(\bullet) = A_\Omega(\Omega) \otimes_{b(\Omega^2)} B_\Omega(\Omega) = A(\Omega^2) \otimes_{b(\Omega^2)} B(\Omega^2)$$

so proposition follows. ■

The algebra structure of $A(\Omega^2) \otimes_{b(\Omega^2)} B(\Omega^2)$ can be recovered using the isomorphism

$$\Phi : B(\Omega^2) \otimes_{b(\Omega^2)} A(\Omega^2) \simeq (B\hat{\otimes}A)(\Omega^2) \simeq (A\hat{\otimes}B)(\Omega^2) \simeq A(\Omega^2) \otimes_{b(\Omega^2)} B(\Omega^2)$$

If a and a' are in $A(\Omega^2)$, and b, b' in $B(\Omega^2)$, and if

$$\Phi(b \otimes a') = \sum_i a_i'' \otimes b_i''$$

then

$$(a \otimes b)(a' \otimes b') = \sum_i (aa_i'') \otimes (b_i'' b')$$

Chapter 8

Composition

8.1 Bisets

In [2], I have studied the following problem: if G and H are groups, what kind of functors F from G-**set** to H-**set** induce by composition a functor from $Mack(H)$ to $Mack(G)$? It seems natural to ask that the functor F transforms a disjoint union into a disjoint union, and a cartesian square into a cartesian square. The functors from G-**set** to H-**set** having those two properties can be completely classified up to isomorphism by the (isomorphism classes of) H-sets-G:

Definitions: *Let G and H be groups. An H-set-G is a set X with a left H-action and a right G-action, which commute, i.e. such that if $g \in G$, $h \in H$ and $x \in X$*

$$h.(x.g) = (h.x).g$$

If G, H, and K are groups, if X is an H-set-G, and Y a K-set-H, I denote by $Y \circ_H X$ the K-set-G defined by

$$Y \circ_H X = \{(y,x) \in Y \times X \mid \forall h \in H,\ yh = y \Rightarrow \exists g \in G,\ h.x = x.g\}/H$$

where the action of H is given by $(y,x).h = (y.h, h^{-1}x)$.

The action of K and G on $Y \circ_H X$ is given

$$k.(y,x).g = (k.y, x.g)$$

If X is an H-set-G, and if G and H are clear from context, I will also say that X is a set with a double action, or biset for short.

With those notations, if U is an H-set-G, and X is a G-set, or G-set-(1), then $U \circ_G X$ is an H-set-(1), that is an H-set. This construction gives a functor from G-**set** to H-**set**, that I denote by $U \circ_G -$. The precise statement proved in [2] is then

Theorem 8.1.1: Let G and H be finite groups.

- **If F is a functor from G-set to H-set which transforms disjoint unions into disjoint unions and cartesian squares into cartesian squares, then there exists an H-set-G U, unique up to isomorphism of H-sets-G, such that F is isomorphic to the functor $U \circ_G -$.**

- **Conversely, if U is an H-set-G, then the functor $U \circ_G -$ transforms disjoint unions into disjoint unions and cartesian squares into cartesian squares.**

I also proved that in these conditions, the set U induces a functor from $Mack(H)$ to $Mack(G)$, defined by composition, and denoted by

$$M \mapsto M \circ U$$

If M is a Mackey functor for H, the Mackey functor $M \circ U$ is defined over the G-set X by

$$(M \circ U)(X) = M(U \circ_G X)$$

If $f : X \to Y$ is a morphism of H-sets, then $U \circ_G f : U \circ_G X \to U \circ_G Y$ is defined by

$$(U \circ_G f)(u, x) = \big(u, f(x)\big)$$

and then

$$(M \circ U)_*(f) = M_*(U \circ_G f) \qquad (M \circ U)^*(f) = M^*(U \circ_G f)$$

Examples: 1) If H is a subgroup of G, and if U is the set G, viewed as an H-set-G by multiplication, then the functor $U \circ_G -$ is the restriction functor from G-**set** to H-**set**. If V is the set G viewed as a G-set-H, then $V \circ_H -$ is the induction functor from H-**set** to G-**set**. The functor $M \mapsto M \circ U$ is then the induction functor for Mackey functors, and the functor $N \mapsto N \circ V$ is the restriction functor for Mackey functors.
2) If N is a normal subgroup of the group G, if $H = G/N$, let U be the set H, viewed as an H-set-G, the group G acting by the projection $G \to G/N$, and let V be the same set viewed as a G-set-H. Then the functor $U \circ_G -$ is the "fixed points by N" functor. It is easy to identify the functor $M \mapsto M \circ U$ as the inflation functor from $Mack(H)$ to $Mack(G)$, defined by Thévenaz and Webb (see [14], [15]). The functor $V \circ_G -$ is the inflation functor from H-**set** to G-**set**. The functor $N \mapsto N \circ V$ is the "coinflation" functor for Mackey functors (that Thévenaz and Webb denote by $\beta^!$ in [15] Lemma 5.4, and I denote by $\rho^G_{G/N}$ in [2]).
3) If U is an H-set-G, then $U \circ_G \bullet \simeq U/G$.
4) If U is a G-set-G, and X a G-set, then

$$U \circ_G X = \{(u, x) \in U \times X \mid \forall g \in G,\ u.g = u \Rightarrow\ g.x = x\}/G$$

8.2 Composition and tensor product

Definition: *Let G and H be groups, and U be an H-set-G. If M and N are Mackey functors for H, and if X and Y are G-sets, I denote by $\tau^U_{X,Y}$ the map*

$$\tau^U_{X,Y} : M(U \circ_G X) \times N(U \circ_G Y) \to (M \hat{\otimes} N)\big(U \circ_G (X \times Y)\big)$$

defined by

$$\tau^U_{X,Y}(m, n) = \left[M^*\left(U \circ_G \binom{xy}{x}\right)(m) \otimes N^*\left(U \circ_G \binom{xy}{y}\right)(n)\right]_{(U \circ_G (X \times Y), Id)}$$

Lemma 8.2.1: The maps $\tau^U_{X,Y}$ form a bilinear morphism from $M\circ U$, $N\circ U$ to $(M\hat{\otimes}N)\circ U$.

Proof: The maps $\tau^U_{X,Y}$ being clearly bilinear, all I have to check is their bifunctoriality. So let $f: X\to X'$ and $g: Y\to Y'$ be morphisms of H-sets. If $m\in M(U\circ_G X)$ and $n\in N(U\circ_G X')$, setting $M'=M\circ U$ and $N'=N\circ U$, I have

$$\tau_{X',Y'}\Big(M'_*(f)(m), N'_*(g)(n)\Big) = \ldots$$

$$\ldots = \left[M'^*\begin{pmatrix} x'y' \\ x' \end{pmatrix} M'_*(f)(m)\otimes N'^*\begin{pmatrix} x'y' \\ y' \end{pmatrix} N'_*(g)(n)\right]_{(U\circ_G(X'\times Y'),Id)}$$

As the square

$$\begin{array}{ccc} X\times Y' & \xrightarrow{f\times Id} & X'\times Y' \\ {\scriptstyle\begin{pmatrix} xy' \\ x \end{pmatrix}}\Big\downarrow & & \Big\downarrow{\scriptstyle\begin{pmatrix} x'y' \\ x' \end{pmatrix}} \\ X & \xrightarrow[f]{} & X' \end{array}$$

is cartesian, its image under $U\circ_G -$ is also cartesian, and then

$$M^*\left(U\circ_G\begin{pmatrix} x'y' \\ x' \end{pmatrix}\right)M_*(U\circ_G f) = M_*\Big(U\circ_G(f\times Id)\Big)M^*\left(U\circ_G\begin{pmatrix} xy' \\ x \end{pmatrix}\right)$$

It follows that

$$\tau_{X',Y'}\Big(M'_*(f)(m), N'_*(g)(n)\Big) = \ldots$$

$$\ldots = \left[M_*\Big(U\circ_G(f\times Id)\Big)M'^*\begin{pmatrix} xy' \\ x \end{pmatrix}(m)\otimes N'^*\begin{pmatrix} x'y' \\ y' \end{pmatrix} N'_*(g)(n)\right]_{(U\circ_G(X'\times Y'),Id)} = \ldots$$

$$= \left[M'^*\begin{pmatrix} xy' \\ y' \end{pmatrix}(m)\otimes N^*\Big(U\circ_G(f\times Id)\Big)N'^*\begin{pmatrix} x'y' \\ y' \end{pmatrix} N'_*(g)(n)\right]_{(U\circ_G(X\times Y'),U\circ_G(f\times Id))} = \ldots$$

$$\ldots = \left[M'^*\begin{pmatrix} xy' \\ y' \end{pmatrix}(m)\otimes N'^*\begin{pmatrix} xy' \\ y' \end{pmatrix} N'_*(g)(n)\right]_{(U\circ_G(X\times Y'),U\circ_G(f\times Id))}$$

But for the same reason, I have

$$N^*\left(U\circ_G\begin{pmatrix} xy' \\ y' \end{pmatrix}\right)N_*(U\circ_G g) = N_*\Big(U\circ_G(Id\times g)\Big)N^*\left(U\circ_H\begin{pmatrix} xy \\ y \end{pmatrix}\right)$$

This gives

$$\tau_{X',Y'}\Big(M'_*(f)(m), N'_*(g)(n)\Big) = \ldots$$

$$\ldots = \left[M'^*\begin{pmatrix} xy' \\ y' \end{pmatrix}(m)\otimes N_*\Big(U\circ_G(Id\times g)\Big)N'^*\begin{pmatrix} xy \\ y \end{pmatrix}(n)\right]_{(U\circ_G(X\times Y'),U\circ_G(f\times Id))} = \ldots$$

$$= \left[M^*\Big(U\circ_G(Id\times g)\Big)M^*\left(U\circ_G\begin{pmatrix} xy' \\ y' \end{pmatrix}\right)(m)\otimes N'^*\begin{pmatrix} xy \\ y \end{pmatrix}(n)\right]_{(U\circ_G(X\times Y),U\circ_G(f\times g))} = \ldots$$

$$\dots = \left[M'^{*}\begin{pmatrix} xy \\ y \end{pmatrix}(m) \otimes N'^{*}\begin{pmatrix} xy \\ y \end{pmatrix}(n) \right]_{(U\circ_G(X\times Y), U\circ_G(f\times g))}$$

proving finally that

$$\tau_{X',Y'}\Big(M'_*(f)(m), N'_*(g)(n)\Big) = \Big((M\hat{\otimes}N)\circ U\Big)_*(f\times g)\Big(\tau^U_{X,Y}(m,n)\Big)$$

so τ is covariant.

Now if $m' \in M'(X')$ and $n' \in N'(Y')$, then

$$\tau^U_{X,Y}\Big(M'^*(f)(m'), N'^*(g)(n')\Big) = \dots$$

$$\dots = \left[M'^{*}\begin{pmatrix} xy \\ x \end{pmatrix} M'^*(f)(m') \otimes N'^{*}\begin{pmatrix} xy \\ y \end{pmatrix} N'^*(g)(n') \right]_{(U\circ(X\times Y), Id)}$$

On the other hand, as the square

$$\begin{array}{ccc} U\circ_G(X\times Y) & \xrightarrow{U\circ_G(f\times g)} & U\circ_G(X'\times Y') \\ {\scriptstyle Id}\downarrow & & \downarrow{\scriptstyle Id} \\ U\circ_G(X\times Y) & \xrightarrow[U\circ_G(f\times g)]{} & U\circ_G(X'\times Y') \end{array}$$

is trivially cartesian, I have

$$\Big((M\hat{\otimes}N)\circ U\Big)^*\left(\left[M'^{*}\begin{pmatrix} x'y' \\ x' \end{pmatrix}(m') \otimes N'^{*}\begin{pmatrix} x'y' \\ y' \end{pmatrix}(n') \right]_{(U\circ_G(X'\times Y'), Id)}\right) = \dots$$

$$\dots = \left[M'^*(f\times g)M'^{*}\begin{pmatrix} x'y' \\ x' \end{pmatrix}(m') \otimes N'^*(f\times g)N'^{*}\begin{pmatrix} x'y' \\ y' \end{pmatrix}(n') \right]_{(U\circ_G(X\times Y), Id)}$$

As moreover

$$f\circ\begin{pmatrix} xy \\ x \end{pmatrix} = \begin{pmatrix} xy \\ f(x) \end{pmatrix} = \begin{pmatrix} x'y' \\ x' \end{pmatrix}\circ(f\times g) \qquad g\circ\begin{pmatrix} xy \\ y \end{pmatrix} = \begin{pmatrix} xy \\ g(y) \end{pmatrix} = \begin{pmatrix} x'y' \\ y' \end{pmatrix}\circ(f\times g)$$

I have

$$\tau^U_{X,Y}\Big(M'^*(f)(m'), N^*(g)(n')\Big) = \Big((M\hat{\otimes}N)\circ U\Big)^*\tau_{X',Y'}(m',n')$$

and τ is also contravariant. The lemma follows. ■

8.3 Composition and Green functors

When A is a Green functor for the group H, the product on A gives a morphism from $A\hat{\otimes}A$ to A, hence a morphism from $(A\hat{\otimes}A)\circ U$ to $A\circ U$. Composing this morphism with the morphism

$$(A\circ U)\hat{\otimes}(A\circ U) \to (A\hat{\otimes}A)\circ U$$

deduced from τ gives a morphism

$$(A\circ U)\hat{\otimes}(A\circ U) \to A\circ U$$

I view it as a product on $A \circ U$, that I denote by $\times^U$. It is natural to ask if $A \circ U$ is a Green functor. To see this, I will first describe precisely the product: if X and Y are G-sets, if $a \in (A \circ U)(X)$ and $b \in (A \circ U)(Y)$, the product $a \times^U b$ is equal to

$$a \times^U b = A^*\Big(U \circ_H \begin{pmatrix} xy \\ x \end{pmatrix}\Big)(a).A^*\Big(U \circ_H \begin{pmatrix} xy \\ y \end{pmatrix}\Big)(b)$$

where the product "." on the right hand side is the product of A. Then

$$a \times^U b = A^*\begin{pmatrix} (u,x,y) \\ (u,x,y)(u,x,y) \end{pmatrix} \left[A^*\begin{pmatrix} (u,x,y) \\ (u,x) \end{pmatrix}(a) \times A^*\begin{pmatrix} (u,x,y) \\ (u,y) \end{pmatrix}(b)\right] = \ldots$$

$$\ldots = A^*\begin{pmatrix} (u,x,y) \\ (u,x,y)(u,x,y) \end{pmatrix} A^*\begin{pmatrix} (u_1,x_1,y_1)(u_2,x_2,y_2) \\ (u_1,x_1)(u_2,y_2) \end{pmatrix}(a \times b)$$

whence finally

$$a \times^U b = A^*\begin{pmatrix} (u,x,y) \\ (u,x)(u,y) \end{pmatrix}(a \times b)$$

This leads to the following definitions:

Definitions: *Let G and H be finite groups, and U be an H-set-G. If $X_1, \ldots X_n$ are G-sets, I denote by $\delta^U_{X_1,\ldots,X_n}$ the map from $U \circ_G (X_1 \ldots X_n)$ to $(U \circ_G X_1) \times \ldots \times (U \circ_G X_n)$, defined by*

$$\delta^U_{X_1,\ldots,X_n}\big(u,(x_1,\ldots,x_n)\big) = \big((u,x_1),\ldots,(u,x_n)\big)$$

If A is a Green functor for the group H, I set

$$\varepsilon_{A \circ U} = A_*(p_{U/G})(\varepsilon_A) \in (A \circ U)(\bullet) = A(U/G)$$

where $p_{U/G}$ is the unique map $U \circ \bullet \simeq U/G$ to $\bullet$.

With those notations, I have

$$a \times^U b = A^*(\delta^U_{X,Y})(a \times b)$$

where the product $\times$ on the right hand side is the product of A.

The following remark will be useful:

Lemma 8.3.1: The map $\delta^U_{X,Y}$ is injective.

Proof: Indeed, as the square

$$\begin{array}{ccc} X \times Y & \xrightarrow{\begin{pmatrix} xy \\ x \end{pmatrix}} & X \\ {\scriptstyle \begin{pmatrix} xy \\ y \end{pmatrix}}\big\downarrow & & \big\downarrow{\scriptstyle \begin{pmatrix} x \\ \bullet \end{pmatrix}} \\ Y & \xrightarrow[\begin{pmatrix} y \\ \bullet \end{pmatrix}]{} & \bullet \end{array}$$

is cartesian, so is its image under $U\circ_G -$

$$\begin{array}{ccc} U\circ_G(X\times Y) & \xrightarrow{U\circ_G\begin{pmatrix}xy\\x\end{pmatrix}} & U\circ_G X \\ {\scriptstyle U\circ_G\begin{pmatrix}xy\\y\end{pmatrix}}\Big\downarrow & & \Big\downarrow{\scriptstyle U\circ_G\begin{pmatrix}x\\\bullet\end{pmatrix}} \\ U\circ_G Y & \xrightarrow[U\circ_G\begin{pmatrix}y\\\bullet\end{pmatrix}]{} & U\circ_G\bullet \end{array}$$

which proves that $U\circ_G(X\times Y)$ maps into $(U\circ_G X)\times(U\circ_G Y)$, precisely by the map $\delta^U_{X,Y}$. So $\delta^U_{X,Y}$ is injective. ■

Proposition 8.3.2: Let G and H be finite groups, and U be an H-set-G. If A is a Green functor for the group H, then $A\circ U$ is a Green functor for G, for the product $\times^U$ and the unit $\varepsilon_{A\circ U}$.

Proof: The product $\times^U$ is bifunctorial by construction. So I must check that it is associative and unitary.

Let X, Y, and Z be G-sets. If $a\in(A\circ U)(X)$, if $b\in(A\circ U)(Y)$, and if $c\in(A\circ U)(Z)$, then

$$(a\times^U b)\times^U c = A^*(\delta^U_{X\times Y,Z})\big((a\times^U b)\times c\big) = A^*(\delta^U_{X\times Y,Z})\big(A^*(\delta^U_{X,Y})(a\times b)\times c\big) = \dots$$

$$\dots = A^*(\delta^U_{X\times Y,Z})A^*(\delta^U_{X,Y}\times Id_{U\circ_G Z})(a\times b\times c) = A^*(\delta^U_{X,Y,Z})(a\times b\times c)$$

because $(\delta^U_{X,Y}\times Id_{U\circ_G Z})\circ\delta^U_{X\times Y,Z} = \delta^U_{X,Y,Z}$, since for $(u,x,y,z)\in U\circ_G(X\times Y\times Z)$

$$(\delta^U_{X,Y}\times Id_{U\circ_G Z})\circ\delta^U_{X\times Y,Z}(u,x,y,z) = (\delta^U_{X,Y}\times Id_{U\circ_G Z})\big((u,x,y),(u,z)\big) = \dots$$

$$\dots = \big((u,x),(u,y),(u,z)\big) = \delta^U_{X,Y,Z}(u,x,y,z)$$

On the other hand

$$a\times^U(b\times^U c) = A^*(\delta^U_{X,Y\times Z})\big(a\times(b\times^U c)\big) = A^*(\delta^U_{X,Y\times Z})\big(a\times A^*(\delta^U_{Y,Z})(b\times c)\big) = \dots$$

$$\dots = A^*(\delta^U_{X,Y\times Z})A^*(Id_{U\circ_G X}\times\delta^U_{Y,Z})(a\times b\times c) = A^*(\delta^U_{X,Y,Z})(a\times b\times c)$$

since $\big(Id_{U\circ_G X}\times\delta^U_{Y,Z}\big)\circ\delta^U_{X,Y\times Z} = \delta^U_{X,Y,Z}$ by a similar computation. Finally

$$a\times^U(b\times^U c) = A^*(\delta^U_{X,Y,Z})(u,x,y,z) = (a\times^U b)\times^U c$$

and the product $\times^U$ is associative.

Moreover, if $a\in(A\circ U)(X)$, then

$$\varepsilon_{A\circ U}\times^U a = A^*(\delta^U_{\bullet,X})\big(A^*(p_{U/G})(\varepsilon)\times a\big) = A^*(\delta^U_{\bullet,X})A^*(p_{U/G}\times Id_X)(a)$$

But if $(u,x)\in U\circ_G X\simeq U\circ_G(\bullet\times X)$, then

$$(p_{U/G}\times Id_X)\circ\delta^U_{\bullet,X}(u,x) = p_{U/G}\times Id_X\big((u,\bullet),(u,x)\big) = \big(\bullet,(u,x)\big)$$

and with the usual identifications, this gives

$$\varepsilon_{A\circ U}\times a = a$$

A similar computation shows that $\varepsilon_{A\circ U}$ is a right unit for $A\circ U$, and the proposition follows. ■

8.4 Composition and associated categories

Let G and H be finite groups. If U is an H-set-G, and A is a Green functor for H, then $A \circ U$ is a Green functor for G. Each of these Green functors has an associated category, and the functor $U \circ_G -$ defines actually a functor from $\mathcal{C}_{A\circ U}$ to $\mathcal{C}_A$:

Proposition 8.4.1: Let G and H be finite groups, let U be an H-set-G, and A be a Green functor for H. Then the correspondence which maps the G-set X to the H-set $U[X] = U \circ_G X$, and the morphism $\alpha \in (A \circ U)(Y \times X)$ to $U[\alpha] = A_*(\delta^U_{Y,X})(\alpha) \in A\big((U \circ_G X) \times (U \circ_G Y)\big)$ is an R-additive functor from $\mathcal{C}_{A\circ U}$ to $\mathcal{C}$.

Proof: I must check that if X, Y, and Z are G-sets, if $\alpha \in A(YX)$ and $\beta \in A(ZY)$, then

$$U[\beta] \circ_{U\circ_G Y} U[\alpha] = U[\beta \circ^U_Y \alpha]$$

where $\circ^U_Y$ is the product $\circ$ for the category $\mathcal{C}_{A\circ U}$. This equality can also be written as

$$A_*(\delta^U_{Z,Y})(\beta) \circ_{U\circ_G Y} A_*(\delta^U_{Y,Z})(\alpha) = A_*(\delta^U_{Z,X})(\beta \circ^U_Y \alpha) \tag{8.1}$$

Let $X' = U[X]$, $Y' = U[Y]$, and $Z' = U[Z]$. The left hand side of (8.1) is equal to

$$A_*\begin{pmatrix} z'y'x' \\ z'x' \end{pmatrix} A^*\begin{pmatrix} z'y'x' \\ z'y'y'x' \end{pmatrix} A_*(\delta^U_{Z,Y} \times \delta^U_{Y,X})(\beta \times \alpha)$$

The square

$$\begin{array}{ccc}
U \circ_G (ZYX) & \xrightarrow{\begin{pmatrix} (u,z,y,x) \\ (u,z,y)(u,y,x) \end{pmatrix}} & \big(U \circ_G (ZY)\big) \times \big(U \circ_G (YX)\big) \\
\Big\downarrow \delta^U_{Z,Y,X} & & \Big\downarrow \delta_{Z,Y} \times \delta_{Y,X} \\
Z'Y'X' & \underset{\begin{pmatrix} z'y'x' \\ z'y'y'x' \end{pmatrix}}{\longrightarrow} & Z'Y'^2X'
\end{array} \tag{C}$$

is cartesian: indeed, if $(u_1, z, y_1) \in U \circ_G (ZY)$, $(u_2, y_2, x) \in U \circ_G (YX)$ and $(z', y', x') \in Z'Y'X'$ are such that

$$(z', y', y', x') = \big((u_1, z), (u_1, y_1), (u_2, y_2), (u_2, x)\big)$$

then there exists $s \in G$ such that

$$u_2 = u_1 s \qquad s y_2 = y_1$$

Thus

$$(u_2, y_2, x) = (u_1 s, y_2, x) = (u_1, s y_2, s x) = (u_1, y_1, s x)$$

The element (u_1, z, y_1, sx) is in $U \circ_G (ZYX)$: indeed, if $t \in G$ is such that $u_1 t = u_1$, then as $(u_1, z, y_1) \in Y \circ_G (ZY)$, I have $tz = z$ and $ty_1 = y_1$. As moreover $u_2 s^{-1} t s = u_1 t s = u_1 s = u_2$, and as $(u_2, y_2, s) \in U \circ_G (YX)$, I have also $s^{-1} t s x = x$, or $t.sx = sx$.

Moreover

$$\begin{pmatrix} (u,z,y,x) \\ (u,z,y)(u,y,x) \end{pmatrix} (u_1,z,y_1,sx) = \big((u_1,z,y_1),(u_1,y_1,sx)\big)$$

and $(u_1,y_1,sx) = (u_1s, s^{-1}y_1, x) = (u_2,y_2,x)$. Similarly $(u_1,sx) = (u_1s,x) = (u_2,x) = x'$, and then

$$\delta^U_{Z,Y,X}(u_1,z,y_1,sx) = \big((u_1,z),(u_1,y_1),(u_1,sx)\big) = (z',y',x')$$

The map $\delta^U_{Z,Y,X}$ is the product of two injective maps. So it is injective, and (C) is cartesian.

It follows that the left hand side of (8.1) is equal to

$$A_*\begin{pmatrix} z'y'x' \\ z'x' \end{pmatrix} A_*(\delta^U_{Z,Y,X})A^*\begin{pmatrix} (u,z,y,x) \\ (u,z,y)(u,y,x) \end{pmatrix}(\beta\times\alpha) = \ldots$$

$$\ldots = A_*\begin{pmatrix} (u,z,y,x) \\ (u,z)(u,x) \end{pmatrix} A^*\begin{pmatrix} (u,z,y,x) \\ (u,z,y)(u,y,x) \end{pmatrix}(\beta\times\alpha) \tag{8.2}$$

On the other hand

$$\beta\circ^U_Y\alpha = (A\circ U)_*\begin{pmatrix} zyx \\ zx \end{pmatrix}(A\circ U)^*\begin{pmatrix} zyx \\ zyyx \end{pmatrix}(\beta\times^U\alpha) = \ldots$$

$$\ldots = A_*\left(U\circ_G\begin{pmatrix} zyx \\ zx \end{pmatrix}\right) A^*\left(U\circ_G\begin{pmatrix} zyx \\ zyyx \end{pmatrix}\right) A^*(\delta^U_{ZY,YX})(\beta\times\alpha)$$

which gives

$$\beta\circ^U_Y\alpha = A_*\begin{pmatrix} (u,z,y,x) \\ (u,z,x) \end{pmatrix} A^*\begin{pmatrix} (u,z,y,x) \\ (u,z,y)(u,y,x) \end{pmatrix}(\beta\times\alpha)$$

The right hand side of (8.1) is then equal to

$$A_*(\delta^U_{Z,X})(\beta\circ^U_Y\alpha) = A_*\begin{pmatrix} (u,z,y,x) \\ (u,z)(u,x) \end{pmatrix} A^*\begin{pmatrix} (u,z,y,x) \\ (u,z,y)(u,y,x) \end{pmatrix}(\beta\times\alpha)$$

which is the right hand side of (8.2).

Moreover for any G-set X, I have

$$U[1_{(A\circ U)(X^2)}] = A_*(\delta^U_{X,X})(A\circ U)_*\begin{pmatrix} x \\ xx \end{pmatrix}(A\circ U)^*\begin{pmatrix} x \\ \bullet \end{pmatrix}(\varepsilon_{A\circ U}) = \ldots$$

$$\ldots = A_*(\delta^U_{X,X})A_*\begin{pmatrix} (u,x) \\ (u,x,x) \end{pmatrix} A^*\begin{pmatrix} (u,x) \\ (u,\bullet) \end{pmatrix} A^*\begin{pmatrix} (u,\bullet) \\ \bullet \end{pmatrix}(\varepsilon_A) = \ldots$$

$$\ldots = A_*\begin{pmatrix} (u,x) \\ (u,x)(u,x) \end{pmatrix} A^*\begin{pmatrix} (u,x) \\ \bullet \end{pmatrix}(\varepsilon_A) = 1_{A((U\circ_G X)^2)}$$

So I have defined a functor from $\mathcal{C}_{A\circ U}$ to $\mathcal{C}_A$. It is clear that this functor is R-additive, and this completes the proof of the proposition. ■

8.5 Composition and modules

Let G and H be finite groups, let U be an H-set-G, and A be a Green functor for H. The above functor from $\mathcal{C}_{A\circ U}$ to $\mathcal{C}_A$ induces by composition a functor between the associated categories of representations: if F is an R-additive functor on $\mathcal{C}_A$, then the functor $F\circ U$ is an R-additive functor on $\mathcal{C}_{A\circ U}$: if X is a G-set, then $(F\circ U)(X) = F(U\circ_G X)$. As the category of R-additive functors on $\mathcal{C}_A$ is equivalent to the category of A-modules, the functor F corresponds to an A-module N, and then the functor $F\circ U$ corresponds of course to the module $N\circ U$:

Proposition 8.5.1: Let G and H be finite groups, let U be an H-set-G, and A be a Green functor for H. If N is an A-module, then $N\circ U$ is an $A\circ U$-module, for the product $\times^U$ defined by

$$(\alpha, n) \in (A\circ U)(X)\times(N\circ U)(Y) \mapsto \alpha\times^U n = N^*(\delta^U_{X,Y})(\alpha\times n) \in (N\circ U)(X\times Y)$$

The correspondence $N\mapsto N\circ U$ is a functor from A-Mod to $A\circ U$-Mod.

Proof: With the notations of theorem 3.3.5, I have to find the $A\circ U$-module $N' = M_{F_N\circ U}$, and to prove that it coincides with $N\circ U$ as Mackey functor.

If X is a G-set, I have

$$N'(X) = (F_N\circ U)(X) = F_N(U\circ_G X) = N(U\circ_G X)$$

If $f: X\to Y$ is a morphism of G-sets, then by definition

$$N'_*(f) = (F_N\circ U)(f^U_*) = F_N(U[f^U_*])$$

where f^U_* is the element of $(A\circ U)(YX)$ associated to f by lemma 3.2.3, i.e.

$$f^U_* = (A\circ U)_*\begin{pmatrix} x \\ f(x)x\end{pmatrix}(A\circ U)^*\begin{pmatrix} x \\ \bullet\end{pmatrix}(\varepsilon_{A\circ U})$$

In other words

$$f^U_* = A_*\begin{pmatrix} (u,x) \\ (u,f(x),x)\end{pmatrix} A^*\begin{pmatrix} (u,x) \\ (u,\bullet)\end{pmatrix} A^*\begin{pmatrix} (u,\bullet) \\ \bullet\end{pmatrix}(\varepsilon_A) = \dots$$

$$\dots = A_*\begin{pmatrix} (u,x) \\ (u,f(x),x)\end{pmatrix} A^*\begin{pmatrix} (u,x) \\ \bullet\end{pmatrix}(\varepsilon_A)$$

then

$$U[f^U_*] = A_*(\delta^U_{Y,X})(f^U_*) = A_*\begin{pmatrix} (u,x) \\ (u,f(x))(u,x)\end{pmatrix} A^*\begin{pmatrix} (u,x) \\ \bullet\end{pmatrix}(\varepsilon) = (U\circ_G f)_*$$

so

$$N'_*(f) = F_N\big((U\circ_G f)_*\big) = N_*(U\circ_G f) = (N\circ U)_*(f)$$

A similar argument shows that if $f^{U,*}$ is the element of $(A\circ U)(XY)$ corresponding to f by lemma 3.2.3, then $f^{U,*} = (U\circ_G f)^*$, and then

$$N'^*(f) = F_N(f^{U,*}) = F_N\big((U\circ_G f)^*\big) = N^*(U\circ_G f) = (N\circ U)^*(f)$$

which proves that N' coincides as Mackey functor with $N \circ U$.

The structure of $A \circ U$-module of $N \circ U$ is then defined as follows: if X and Y are G-sets, if $\alpha \in (A \circ U)(X)$ and $n \in (N \circ U)(Y)$, then the product $\alpha \times^U n$ is the element of $(N \circ U)(X \times Y)$ given by

$$\alpha \times^U n = (F_N \circ U)\left[(A \circ U)_* \begin{pmatrix} xy \\ xyy \end{pmatrix} (A \circ U)^* \begin{pmatrix} xy \\ x \end{pmatrix} (\alpha)\right](n) = \dots$$

$$\dots = F_N\left(U\left[(A \circ U)_* \begin{pmatrix} xy \\ xyy \end{pmatrix} (A \circ U)^* \begin{pmatrix} xy \\ x \end{pmatrix} (\alpha)\right]\right)(n)$$

Let

$$\beta = U\left[(A \circ U)_* \begin{pmatrix} xy \\ xyy \end{pmatrix} (A \circ U)^* \begin{pmatrix} xy \\ x \end{pmatrix} (\alpha)\right] = \dots$$

$$\dots = A_*(\delta^U_{XY,Y})(A \circ U)_* \begin{pmatrix} xy \\ xyy \end{pmatrix} (A \circ U)^* \begin{pmatrix} xy \\ x \end{pmatrix} (\alpha) = \dots$$

$$\dots = A_* \begin{pmatrix} (u,x,y) \\ (u,x,y)(u,y) \end{pmatrix} A^* \begin{pmatrix} (u,x,y) \\ (u,x) \end{pmatrix} (\alpha)$$

then

$$\alpha \times^U n = F_N(\beta)(n) = \beta \circ_{U \circ_G Y} n = \dots$$

$$\dots = N_* \begin{pmatrix} (u_1,x,y_1)(u_2,y_2) \\ (u_1,x,y_1) \end{pmatrix} N^* \begin{pmatrix} (u_1,x,y_1)(u_2,y_2) \\ (u_1,x,y_1)(u_2,y_2)(u_2,y_2) \end{pmatrix} (\beta \times n)$$

Moreover

$$\beta \times n = N_* \begin{pmatrix} (u_1,x,y_1)(u_2,y_2) \\ (u_1,x,y_1)(u_1,y_1)(u_2,y_2) \end{pmatrix} N^* \begin{pmatrix} (u_1,x,y_1)(u_2,y_2) \\ (u_1,x)(u_2,y_2) \end{pmatrix} (\alpha \times n)$$

Let $f = \begin{pmatrix} (u_1,x,y_1)(u_2,y_2) \\ (u_1,x,y_1)(u_1,y_1)(u_2,y_2) \end{pmatrix}$, and $g = \begin{pmatrix} (u_1,x,y_1)(u_2,y_2) \\ (u_1,x,y_1)(u_2,y_2)(u_2,y_2) \end{pmatrix}$. As the square

$$\begin{array}{ccc} U \circ_G (XY) & \xrightarrow{\begin{pmatrix} (u,x,y) \\ (u,x,y)(u,y) \end{pmatrix}} & \big(U \circ_G (XY)\big) \times (U \circ_G Y) \\ {\scriptstyle \begin{pmatrix} (u,x,y) \\ (u,x,y)(u,y) \end{pmatrix}} \big\downarrow & & \big\downarrow {\scriptstyle f} \\ \big(U \circ_G (XY)\big) \times (U \circ_G Y) & \xrightarrow[g]{} & \big(U \circ_G (XY)\big) \times (U \circ_G Y)^2 \end{array}$$

is cartesian, I have

$$N^* \begin{pmatrix} (u_1,x,y_1)(u_2,y_2) \\ (u_1,x,y_1)(u_2,y_2)(u_2,y_2) \end{pmatrix} N_* \begin{pmatrix} (u_1,x,y_1)(u_2,y_2) \\ (u_1,x,y_1)(u_1,y_1)(u_2,y_2) \end{pmatrix} = \dots$$

$$\dots = N_* \begin{pmatrix} (u,x,y) \\ (u,x,y)(u,y) \end{pmatrix} N^* \begin{pmatrix} (u,x,y) \\ (u,x,y)(u,y) \end{pmatrix}$$

and finally $\alpha \times^U n$ is equal to the image of $\alpha \times n$ under the map

$$N_* \begin{pmatrix} (u_1,x,y_1)(u_2,y_2) \\ (u_1,x,y_1) \end{pmatrix} N_* \begin{pmatrix} (u,x,y) \\ (u,x,y)(u,y) \end{pmatrix} \circ \dots$$

$$\dots \circ N^*\begin{pmatrix} (u,x,y) \\ (u,x,y)(u,y) \end{pmatrix} N^*\begin{pmatrix} (u_1,x,y_1)(u_2,y_2) \\ (u_1,x)(u_2,y_2) \end{pmatrix} = \dots$$

$$\dots = N_*\begin{pmatrix} (u,x,y) \\ (u,x,y) \end{pmatrix} N^*\begin{pmatrix} (u,x,y) \\ (u,x)(u,y) \end{pmatrix}$$

which proves the claimed formula

$$\alpha \times^U n = N^*(\delta^U_{X,Y})(\alpha \times n)$$

8.6 Functoriality

If G and H are finite groups, and U is an H-set-G, then I have a functorial construction $X \mapsto U \circ_G X$ from G-**set** to H-**set**. This construction is not quite functorial in U: if $f : U \to V$ is a morphism of H-sets-G, there is in general no associated morphism $U \circ_G X \to V \circ_G X$: this is because if $(u,x) \in U \circ_G X$, i.e. if the right stabilizer of u is contained in the left stabilizer of x, generally, the right stabilizer of $f(u)$ is not.

I have studied this question in [2], and showed that it is natural to ask moreover that f is injective when restricted to each orbit of G (or equivalently to ask that the right stabilizer of $f(u)$ is equal to the right stabilizer of u, for any $u \in U$). Then, there is a morphism of functors $f \circ_G -$ from $U \circ_G -$ to $V \circ_G -$ defined on the G-set X by

$$(f \circ_G X)(u,x) = \big(f(u), x\big)$$

In those conditions (see [2] prop. 10 and 11), if M is a Mackey functor for H, then $M \circ U$ and $M \circ V$ are Mackey functors for G, and the morphism f induces two morphisms of Mackey functors (denoted by f_* and f^* in [2], but differently here to avoid confusion): a morphism M_f from $M \circ U$ to $M \circ V$, and a morphism M^f from $M \circ V$ to $M \circ U$. Those morphisms are defined for a G-set X by

$$M_{f,X} = M_*(f \circ_G X) : (M \circ U)(X) \to (M \circ V)(X)$$

$$M^f_X = M^*(f \circ_G X) : (M \circ V)(X) \to (M \circ U)(X)$$

With those notations:

Proposition 8.6.1: Let G and H be finite groups, and let A be a Green functor for H. Let moreover U and V be H-sets-G, and $f : U \to V$ be a morphism of H-sets-G, which is injective on each right orbit.

- **If A is a Green functor for H, then A^f is a unitary morphism of Green functors from $A \circ V$ to $A \circ U$.**
- **If M is an A-module, then restriction along A^f gives $M \circ V$ and $M \circ U$ structures of $A \circ V$-modules, and the morphisms M_f and M^f are morphisms of $A \circ V$-modules.**

Proof: For the first assertion, I must show that if X and Y are G-sets, if $a \in (A \circ V)(X)$ and $b \in (A \circ V)(Y)$, then

$$A^f_X(a) \times^U A^f_Y(b) = A^f_{X \times Y}(a \times^V b)$$

The left hand side is equal to

$$A^*(\delta^U_{X,Y})\Big(A^*(f\circ_G X)(a)\times A^*(f\circ_G Y)(b)\Big)=A^*(\delta^U_{X,Y})A^*\Big((f\circ_G X)\times(f\circ_G Y)\Big)(a\times b)$$

and the right hand side to

$$A^*\Big(f\circ_G(X\times Y)\Big)A^*(\delta^V_{X,Y})(a\times b)$$

Now equality follows from

$$\Big((f\circ_G X)\times(f\circ_G Y)\Big)\circ\delta^U_{X,Y}=\delta^V_{X,Y}\circ\Big(f\circ_G(X\times Y)\Big)$$

since for $(u,x,y)\in U\circ_G(X\times Y)$

$$\Big((f\circ_G X)\times(f\circ_G Y)\Big)\circ\delta^U_{X,Y}(u,x,y)=\Big((f\circ_G X)\times(f\circ_G Y)\Big)\Big((u,x),(u,y)\Big)=\dots$$

$$\dots=\Big((f(u),x),(f(u),y)\Big)=\delta^V_{X,Y}\Big(f(u),x,y\Big)=\delta^V_{X,Y}\circ\Big(f\circ_G(X\times Y)\Big)(u,x,y)$$

Moreover

$$A^f_\bullet(\varepsilon_{A\circ V})=A^*(f\circ_G\bullet)A^*(p_{V/G})(\varepsilon_A)=A^*(p_{U/G})(\varepsilon_A)=\varepsilon_{A\circ U}$$

since $p_{U/G}=p_{V/G}\circ(f\circ_G\bullet)$. So the morphism A^f is a unitary morphism of Green functors.

For the second assertion, I must show that if $a\in(A\circ V)(X)$ and $m\in(M\circ U)(Y)$, then

$$a\times^V M_{f,Y}(m)=M_{f,XY}(A^f(a)\times^U m)\tag{8.3}$$

and that if $m'\in(M\circ V)(Y)$, then

$$M^f(a\times^V m')=A^f(a)\times^U M^f(m')\tag{8.4}$$

But

$$a\times^V M_{f,Y}(m)=M^*(\delta^V_{X,Y})\Big(a\times M_*(f\circ_G Y)(m)\Big)=M^*(\delta^V_{X,Y})M_*\begin{pmatrix}(v,x)(u,y)\\(v,x)(f(u),y)\end{pmatrix}(a\times m)$$

The square

$$\begin{array}{ccc} U\circ_G(XY) & \xrightarrow{\begin{pmatrix}(u,x,y)\\(f(u),x)(u,y)\end{pmatrix}} & (V\circ_G X)\times(U\circ_G Y)\\ {\scriptstyle\begin{pmatrix}(u,x,y)\\(f(u),x,y)\end{pmatrix}}\downarrow & & \downarrow{\scriptstyle\begin{pmatrix}(v,x)(u,y)\\(v,x)(f(u),y)\end{pmatrix}}\\ V\circ_G(XY) & \xrightarrow[\begin{pmatrix}(v,x,y)\\(v,x)(v,y)\end{pmatrix}]{} & (V\circ_G X)\times(V\circ_G Y)\end{array}\qquad(C)$$

is cartesian: if $\Big((v,x),(u,y)\Big)\in(V\circ_G X)\times(U\circ_G Y)$ and $(v',x',y')\in V\circ_G(XY)$ are such that

$$(v',x')=(v,x)\qquad(v',y')=(f(u),y)$$

then there exists s and t in G such that

$$v' = vs \qquad sx' = x \qquad v' = f(u)t \qquad ty' = y$$

In those conditions, the element (u, tx', y) is in $U \circ_G (XY)$: indeed, if $r \in G$ is such that $ur = u$, then as $(u, y) \in U \circ_G Y$, I have $ry = y$. Moreover

$$v't^{-1}rt = f(u)rt = f(urt) = f(ut) = f(u)t = v'$$

and as $(v', x') \in V \circ_G X$, I have $t^{-1}rtx' = tx'$, or $r.tx' = tx'$. Moreover

$$\big(f(u), tx', y\big) = (v't^{-1}, tx', y) = (v', x', t^{-1}y) = (v', x', y')$$

and

$$\big((f(u), tx'), (u, y)\big) = \big((v't^{-1}, tx'), (u, y)\big) = \big((v', x'), (u, y)\big) = \ldots$$

$$\ldots = \big((vs^{-1}, sx), (u, y)\big) = \big((v, x), (u, y)\big)$$

Conversely, if $(u_1, x_1, y_1) \in U \circ_G (XY)$ is such that

$$\big(f(u_1), x_1, y_1\big) = (v', x', y') \qquad \big(f(u_1), x_1\big) = (v, x) \qquad (u_1, y_1) = (u, y)$$

then the last equality proves that replacing (u_1, x_1, y_1) by (u_1s^{-1}, sx_1, sy_1) for a suitable $s \in G$, I can suppose $u_1 = u$ and $y_1 = y$. Then $(f(u), x_1) = (v, x) = (f(u), tx')$. So there exists $r \in G$ such that

$$f(u)r = f(u) \qquad r^{-1}x_1 = tx'$$

As f is injective on the right orbits, the first equality shows that $ur = u$, and since $(u, x_1, y) \in U \circ_G (XY)$, I have $rx_1 = x_1 = tx'$, and

$$(u_1, x_1, y_1) = (u, tx', y)$$

which proves that (C) is cartesian.

It follows that

$$a \times^V M_{f,Y}(m) = M_*\begin{pmatrix} (u, x, y) \\ (f(u), x, y) \end{pmatrix} M^*\begin{pmatrix} (u, x, y) \\ (f(u), x)(u, y) \end{pmatrix} (a \times m)$$

But the right hand side of (8.3) is equal to

$$M_{f,XY}(A^f(a) \times^U m) = M_*\big(f \circ_G (XY)\big) M^*(\delta^U_{X,Y})\big(A^*(f \circ_G X)(a) \times m\big) = \ldots$$

$$\ldots = M_*\big(f \circ_G (XY)\big) M^*(\delta^U_{X,Y}) M^*\begin{pmatrix} (u_1, x), (u_2, y) \\ (f(u_1), x), (u_2, y) \end{pmatrix} (a \times m) = \ldots$$

$$\ldots = M_*\begin{pmatrix} (u, x, y) \\ (f(u), x, y) \end{pmatrix} M^*\begin{pmatrix} (u, x, y) \\ (f(u), x)(u, y) \end{pmatrix} (a \times m)$$

which proves equality (8.3).

Similarly, the left hand side of (8.4) is

$$M^f(a\times^V m')=M^*\big(f\circ_G(XY)\big)M^*(\delta^V_{X,Y})(a\times m')=M^*\begin{pmatrix}(u,x,y)\\ (f(u),x)(f(u),y)\end{pmatrix}(a\times m')$$

whereas the right hand side is

$$A^f(a)\times^U M^f(m')=M^*(\delta^U_{X,Y})\big(A^*(f\circ_G X)(a)\times^U M^*(f\circ_G Y)(m')\big)=\ldots$$

$$\ldots=M^*(\delta^U_{X,Y})M^*\begin{pmatrix}(u_1,x)(u_2,y)\\ (f(u_1),x)(f(u_2),y)\end{pmatrix}(a\times m)=M^*\begin{pmatrix}(u,x,y)\\ (f(u),x)(f(u),y)\end{pmatrix}(a\times m')$$

This proves equality (8.4), and the proposition. ■

8.7 Example: induction and restriction

Let G be a finite group, and H be a subgroup of G. Let U be the set G, viewed as a G-set-H, and V be the set G, viewed as an H-set-G. If X is an H-set, then $U\circ_G X$ identifies with $\mathrm{Ind}_H^G X$, functorially in X. It follows that if M is a Mackey functor for G, then $M\circ U$ is isomorphic to $\mathrm{Res}_H^G M$.

If Y is a G-set, then $V\circ_G Y$ identifies with $\mathrm{Res}_H^G Y$, functorially in Y, by the map $(g,y)\mapsto gy$. Thus if N is a Mackey functor for H, then $N\circ V$ identifies with $\mathrm{Ind}_H^G N$.

So if A is a Green functor for G, then $\mathrm{Res}_H^G A$ is a Green functor for H, and this is not surprising. The case of induction is less clear, but corresponds to what Thévenaz calls *coinduction* (see [13]):

Proposition 8.7.1: Let H be a subgroup of the group G, and B be a Green functor for H. then:

- **The functor $\mathrm{Ind}_H^G B$ is a Green functor for G. If K is a subgroup of G, then there is an isomorphism of rings (with unit)**

$$(\mathrm{Ind}_H^G B)(K)=\prod_{x\in H\backslash G/K} B(H\cap {}^xK)$$

- **The functor $B\mapsto \mathrm{Ind}_H^G B$, from the category $Green(H)$ of Green functors for H, to $Green(G)$, is right adjoint to the functor $A\mapsto \mathrm{Res}_H^G A$ from $Green(G)$ to $Green(H)$.**

Proof: The formula giving $(\mathrm{Ind}_H^G B)(K)$ is well known for Mackey functors. The point is that is is still true for Green functors.

By definition of induction

$$(\mathrm{Ind}_H^G B)(K)=B\big(\mathrm{Res}_H^G(G/K)\big)$$

But

$$\mathrm{Res}_H^G(G/K)\simeq \coprod_{x\in H\backslash G/K} H/(H\cap {}^xK)$$

the isomorphism (from right to left) mapping $h(H \cap {}^xK) \in H/(H \cap {}^xK)$ to $hxK \in \mathrm{Res}_H^G(G/K)$. Thus if a and a' are in $A(K) = A(G/K) = B\big(V \circ_G (G/K)\big)$, setting $\Gamma = G/K$, and denoting by "$.^V$" the product "." on A, I have

$$a.^V a' = A^*\begin{pmatrix}\gamma \\ \gamma\gamma\end{pmatrix}(a \times^V a') = A^*\begin{pmatrix}\gamma \\ \gamma\gamma\end{pmatrix} B^*(\delta^V_{\Gamma,\Gamma})(a \times a') = \ldots$$

$$\ldots = B^*\left(V \circ_G \begin{pmatrix}\gamma \\ \gamma\gamma\end{pmatrix}\right) B^*(\delta^V_{\Gamma,\Gamma})(a \times a') = B^*\begin{pmatrix}(v,\gamma) \\ (v,\gamma)(v,\gamma)\end{pmatrix}(a \times a') = a.a'$$

But I know that the ring $(B(V \circ_G \Gamma), .)$ is isomorphic to the direct product of rings $(B(\omega), .)$ for the various orbits ω of H on $V \circ_G \Gamma$. The first assertion follows.

Similarly, the second assertion is clear for Mackey functors. I must show that the adjunction passes down to Green functors. If A is a Green functor for G, if B is a Green functor for H, and Φ is a morphism from A to $\mathrm{Ind}_H^G B$, then for any subgroup K of G, I have a morphism

$$\Phi_K : A(K) \xrightarrow{\oplus \Phi_{K,x}} \bigoplus_{x \in H\backslash G/K} B(H \cap {}^xK)$$

Say that Φ is a morphism of Green functors means exactly by the previous remarks that $\Phi_{K,x}$ is a morphism of rings (with unit) from $(A(K), .^V)$ to $\big(B(H \cap {}^xK), .\big)$, for any x. But if L is a subgroup of H, the morphism Ψ_L from $A(L) = \mathrm{Res}_H^G(L)$ to $B(L)$ associated to Φ by adjunction is $\Phi_{L,1}$. So Ψ is a unitary morphism of Green functors.

Conversely, if Ψ is a unitary morphism of Green functors from $\mathrm{Res}_H^G A$ to B, then for any subgroup L of G, I have a morphism Ψ_L of rings (with unit) from $A(L)$ to $B(L)$. The morphism Φ associated to Ψ by adjunction is then defined on the subgroup K of G by

$$\Phi_{K,x}(a) = \Psi_{H \cap {}^xK}\big({}^x r^K_{H^x \cap K}(a)\big)$$

It is the product of three morphisms of rings (with unit). So it is a morphism of rings with unit, and Φ is a unitary morphism of Green functors. This proves the proposition. ∎

The adjunction property shows in particular that if A is a Green functor for G, then there is a unitary morphism η_A of Green functors, adjoint to the identity of $\mathrm{Res}_H^G A$

$$\eta_A : A \to \mathrm{Ind}_H^G \mathrm{Res}_H^G A$$

But

$$\mathrm{Ind}_H^G \mathrm{Res}_H^G A = (A \circ U) \circ V = A \circ (U \circ_H V)$$

It is easy to see that $U \circ_H V \simeq (G \times_H G)$, which is the quotient of $G \times G$ by the right action of H given by $(g_1, g_2)h = (g_1 h, h^{-1} g_2)$. On the other hand, if I denote by I the set G, viewed as a "identity " G-set-G, the functor A is equal to $A \circ I$, and the morphism η_A comes from the morphism

$$f : U \circ_H V \to I \qquad (g_1, g_2)H \mapsto g_1.g_2 \in I$$

This morphism is injective on the right orbits: indeed, if

$$f\big((g_1, g_2)H.g\big) = f\big((g_1, g_2 g)H\big) = g_1 g_2 g = f\big((g_1, g_2)H\big) = g_1 g_2$$

then

$$g_1.g_2.g = g_1.g_2$$

which proves that $g = 1$.

Now if M is an A-module, I know that $\mathrm{Res}_H^G M$ is a $\mathrm{Res}_H^G A$ module. Conversely, if N is a $\mathrm{Res}_H^G A$-module, then $\mathrm{Ind}_H^G N$ is a $\mathrm{Ind}_H^G \mathrm{Res}_H^G A$-module, so an A-module by the morphism η_A. The proposition 8.6.1 shows that those correspondences are functorial. Moreover

Proposition 8.7.2: Let A be a Green functor for the group G, and H be a subgroup of G. The functor $M \mapsto \mathrm{Res}_H^G M$ from A-Mod to $\mathrm{Res}_H^G A$-Mod is left and right adjoint to the functor $N \mapsto \mathrm{Ind}_H^G N$ from $\mathrm{Res}_H^G A$-Mod to A-Mod.

Proof: Here again, this property is well known for Mackey functors. All I have to do is to keep track of the adjunction procedure: if M is an A-module, if N is a $\mathrm{Res}_H^G A$-module, and f is a morphism of $\mathrm{Res}_H^G A$-modules from $\mathrm{Res}_H^G M$ to N, then $\mathrm{Ind}_H^G f$ is a morphism of $\mathrm{Ind}_H^G \mathrm{Res}_H^G A$-modules from $\mathrm{Ind}_H^G \mathrm{Res}_H^G M$ to $\mathrm{Ind}_H^G N$. Thus it is also a morphism of A modules from $\mathrm{Ind}_H^G \mathrm{Res}_H^G M$ to $\mathrm{Ind}_H^G N$. The adjoint of f is obtained by composing this morphism with the morphism from M to $\mathrm{Ind}_H^G \mathrm{Res}_H^G M$, which is a morphism of A-modules. So the adjunction at the level of Mackey functors maps morphisms of $\mathrm{Res}_H^G A$ modules to morphisms of A-modules.

Conversely, if g is a morphism of A-modules from M to $\mathrm{Ind}_H^G N$, then $\mathrm{Res}_H^G g$ is a morphism of $\mathrm{Res}_H^G A$-modules from $\mathrm{Res}_H^G M$ to $\mathrm{Res}_H^G \mathrm{Ind}_H^G N$. The adjoint of g is obtained by composing this morphism with the morphism ρ from $\mathrm{Res}_H^G \mathrm{Ind}_H^G N$ to N, which is a morphism of $\mathrm{Res}_H^G A$-modules: it follows indeed from the morphism of H-sets-H

$$\Theta : h \in H \mapsto (h,1) \in V \circ_G U$$

which is injective: indeed, if h_1 and h_2 in H are such that $(h_1,1) = (h_2,1)$ in $V \circ_G U$, then there exists $g \in G$ such that $h_2 = h_1 g$ and $1 = g.1$. So $g = 1$ and $h_1 = h_2$. If $B = \mathrm{Res}_H^G A$, the morphism ρ is then a morphism of $\mathrm{Res}_H^G \mathrm{Ind}_H^G B$-modules from $\mathrm{Res}_H^G \mathrm{Ind}_H^G N$ to N. As

$$\mathrm{Res}_H^G \mathrm{Ind}_H^G B = \mathrm{Res}_H^G \mathrm{Ind}_H^G \mathrm{Res}_H^G A$$

I have the morphism $\mathrm{Res}_H^G \eta_A$ from $\mathrm{Res}_H^G A$ to $\mathrm{Res}_H^G \mathrm{Ind}_H^G \mathrm{Res}_H^G A$, and the composite

$$\mathrm{Res}_H^G A \xrightarrow{\mathrm{Res}_H^G \eta_A} \mathrm{Res}_H^G \mathrm{Ind}_H^G \mathrm{Res}_H^G A \xrightarrow{B^\Theta} \mathrm{Res}_H^G A$$

is the identity of B, because it expresses the adjunction of Res_H^G and Ind_H^G at the level of Mackey functors.

So through $\mathrm{Res}_H^G \eta_A$ I recover the same $\mathrm{Res}_H^G A$-module structure on N, and this proves that ρ is a morphism of $\mathrm{Res}_H^G A$-modules. Now the adjunction at the level of Mackey functors maps morphisms of A-modules to morphisms of $\mathrm{Res}_H^G A$-modules, and this proves that the functor Res_H^G is left adjoint to the functor Ind_H^G. A similar argument shows that it is also right adjoint. This completes the proof of the proposition. ■

Chapter 9

Adjoint constructions

Notations: *If G and H are groups, and L is a subgroup of $H \times G$, I denote by $p_1(L)$ the projection from L to H, and $p_2(L)$ its projection onto G. I denote by $k_1(L)$ (resp. $k_2(L)$) the normal subgroup of $p_1(L)$ (resp. of $p_2(L)$) formed of elements $h \in H$ (resp. elements $g \in G$) such that $(h,1) \in L$ (resp. such that $(1,g) \in L$). The groups $p_1(L)/k_1(L)$ and $p_2(L)/k_2(L)$ are canonically isomorphic, and I denote by $q(L)$ this quotient.*

I denote by $(H \times G)/L$ the set of left classes of L in $H \times G$, viewed as an H-set-G by

$$h.(u,v)L.g = (hu, g^{-1}v)L$$

With those notations, I have shown ([2] Lemme 2) that if $U = (H \times G)/L$, and if X is a G-set, then

$$U \circ_G X \simeq \mathrm{Ind}_{p_1(L)}^{H} \mathrm{Inf}_{p_1(L)/k_1(L)}^{p_1(L)} \tau_L (\mathrm{Res}_{p_2(L)}^{G} X)^{k_2(L)}$$

denoting by τ_L the transport by the canonical isomorphism $p_1(L)/k_1(L) \simeq p_2(L)/k_2(L)$ of a $p_2(L)/k_2(L)$-set.

It follows (see [2] 4.1.2) that if M is a Mackey functor for H, then

$$M \circ U = \mathrm{Ind}_{p_2(L)}^{G} \mathrm{Inf}_{p_2(L)/k_2(L)}^{p_2(L)} \theta_L \rho_{p_1(L)/k_1(L)}^{p_1(L)} \mathrm{Res}_{p_1(L)}^{H} M$$

where θ_L denotes the transport by isomorphism of Mackey functors for $p_1(L)/k_1(L)$ to Mackey functors for $p_2(L)/k_2(L)$.

As any H-set-G U is a disjoint union of transitive H-sets-G, i.e. of the form $(H \times G)/L$, it follows that the functor $M \circ U$ is a direct sum of functors composed of restriction, inflation, coinflation, and induction.

I already know that the functors of restriction and inflation are mutual left and right adjoints. Thévenaz and Webb (see [14] 5.1, [15] 2.) have built left and right adjoints for inflation, and they also mention ([15] 5.4 and 5.6) that coinflation has a left adjoint (and a right one if the ground ring is a field). It follows that the functors $M \mapsto M \circ U$ always have a left adjoint. I will show that they always have a left and right adjoint, and describe their structure in terms of G-sets.

9.1 A left adjoint to the functor $Z \mapsto U \circ_H Z$

Notation: *Let U be a fixed G-set-H. I denote by $u \mapsto uH$ the projection from U to U/H. If (Y, f) is a G-set over U/H, I denote by $U.Y$ the pull-back product of U and*

Y over U/H, defined by

$$U.Y = \{(u,y) \in U \times Y \mid f(y) = uH\}$$

I view $U.Y$ as a G-set-H: if $g \in G$ and $h \in H$, then

$$g.(u,y).h = (guh, gy)$$

If $(u,y) \in U.Y$, I denote by $G(u,y)$ its orbit under G. I denote by $G\backslash U.Y$ the set of orbits of G on $U.Y$, viewed as an H-set by

$$h.G(u,y) = G(uh^{-1}, y)$$

This construction is functorial in Y: if $\alpha : (Y,f) \to (Y',f')$ is a morphism of G-sets over U/H, then the map

$$G\backslash U.\alpha : G\backslash U.Y \to G\backslash U.Y'$$

defined by

$$(G\backslash U.\alpha)\big(G(u,y)\big) = G\big(u, \alpha(y)\big)$$

is a morphism of H-sets.

Conversely, if Z is an H-set, and p_Z is the unique morphism from Z to $\bullet$, then $U \circ_H p_Z$ is a morphism from $U \circ_H Z$ to $U \circ_H \bullet = U/H$. Thus $Z \mapsto U \circ_H Z$ is a functor from H-**set** to G-**set**$\downarrow_{U/H}$. Actually:

Proposition 9.1.1: The functor $(Y,f) \mapsto G\backslash U.Y$ from G-set$\downarrow_{U/H}$ to H-set is left adjoint to the functor $Z \mapsto U \circ_H Z$.

Proof: Let (Y,f) be a G-set over U/H, and Z be an H-set. If ϕ is a morphism of G-sets over U/H from (Y,f) to $(U \circ_H Z, U \circ_H p_Z)$, and if $(u,y) \in U.Y$, then $f(y) = uH$. Moreover, as

$$(U \circ_H p_Z)\phi(y) = f(y) = uH$$

there exists $h \in H$ and $z \in Z$ such that $\phi(y) = (uh, z)$. If $h' \in H$ and $z' \in Z$ are such that $\phi(y) = (uh', z')$, then there exists $h" \in H$ such that

$$uh = uh'h" \qquad z' = h"z$$

In those conditions $uh = uh.h^{-1}h'h"$, and as $(uh, z) \in U \circ_H Z$, it follows that $h^{-1}h'h"z = z$, hence that $hz = h'h"z = h'z'$. In particular the element $hz \in Z$ is well defined by the condition $\phi(y) = (uh, z) = (u, hz)$.

I can then define $\theta(u,y) \in Z$ by the condition

$$\phi(y) = \big(u, \theta(u,y)\big)$$

Then as $g(u,y) = (gu, gy)$ for $g \in G$, if $z = \theta(u,y)$, I have

$$\phi(gy) = g\phi(y) = g(u,z) = (gu, z)$$

which proves that $\theta(gu, gy) = \theta(u,y)$. So θ is a map from $G\backslash U.Y$ to Z. As moreover $h.G(u,y) = (uh^{-1}, y)$, and as

$$\phi(y) = (u,z) = (uh^{-1}, hz)$$

I have $\theta\big(h.G(u,y)\big) = hz = h\theta(u,y)$. So θ is a morphism of H-sets from $G\backslash U.Y$ to Z.

Conversely, if θ is a morphism of H-sets from $G\backslash U.Y$ to Z, and if $y \in Y$, let $u \in U$ such that $f(y) = uH$. Then $(u,y) \in U.Y$. In those conditions, if $z = \theta\big(G(u,y)\big)$, the element (u,z) is in $U \circ_H Z$: indeed if $h \in H$ is such that $uh = u$, then

$$h^{-1}.G(u,y) = G(uh,y) = G(u,y)$$

thus $z = \theta\big(G(u,y)\big) = h\theta\big(G(u,y)\big) = hz$. The element (u,z) does not depend on the choice of u such that $f(y) = uH$: if I change u to uh, for $h \in H$, then

$$\theta\big(G(uh,y)\big) = h^{-1}\theta\big(G(u,y)\big) = h^{-1}z$$

and $(u,z) = (uh, h^{-1}z)$. I can then set $\phi(y) = (u,z)$. As $f(gy) = gf(y) = guH$, and as $\theta\big(G(gu,gy)\big) = \theta\big(G(u,y)\big) = z$, I have

$$\phi(gy) = (gu,z) = g(u,z) = g\phi(y)$$

Thus ϕ is a morphism of G-sets from Y to $U \circ_H Z$.

The correspondences $\phi \mapsto \theta$ and $\theta \mapsto \phi$ are inverse to each other: indeed, the equality

$$\phi(y) = \Big(u, \theta\big(G(u,y)\big)\Big)$$

defines θ from ϕ, and ϕ from θ. The proposition will then follow, if I know that the unit and co-unit of this adjunction are functorial:

Notation: *If (Y,f) is a G-set over U/H, I will denote by $\nu_{(Y,f)}$ (or ν_Y if f is clear from context) the unit of the previous adjunction: it is the morphism*

$$\nu_{(Y,f)} : Y \to U \circ_H (G\backslash U.Y)$$

defined by $\nu_{(Y,f)}(y) = \big(u, G(u,y)\big)$, if $f(y) = uH$.

If Z is an H-set, I will denote by η_Z the co-unit of this adjunction: it is the morphism

$$\eta_Z : G\backslash U.(U \circ_H Z) \to Z$$

defined by $\eta_Z\big(G(u',(u,z))\big) = h^{-1}z$ if $h \in H$ is such that $u' = uh$.

Then ν is functorial: if $\alpha : (Y,f) \to (Z,g)$ is a morphism of G-sets over U/H, and if $(u,y) \in U.Y$, then $\nu_{(Y,f)}(y) = \big(u, G(u,y)\big)$. Thus

$$\big(U \circ_H (G\backslash U.\alpha)\big)\nu_{(Y,f)}(y) = \big(u, G(u,\alpha(y))\big)$$

But $g\alpha(y) = f(y) = uH$, so $\nu_{(Z,g)}\alpha(y) = \big(u, G(u,\alpha(y))\big)$, which shows that

$$\nu_{(Z,g)}\alpha = \big(U \circ_H (G\backslash U.\alpha)\big)\nu_{(Y,f)}$$

Similarly η is functorial: if $f : Z \to T$ is a morphism of H-sets, and if $\big(u',(u,z)\big) \in U.(U \circ_H Z)$, then there exists $h \in H$ such that $u' = uh$. Then $\eta_Z\Big(G\big(u',(u,z)\big)\Big) = h^{-1}z$. As

$$G\backslash U.(U \circ_H f)\Big(G\big(u',(u,z)\big)\Big) = G\big(u',(u,f(z))\big)$$

I have

$$\eta_T\left[G\backslash U.(U\circ_H f)\Big(G\big(u',(u,z)\big)\Big)\right] = h^{-1}f(z) = f\left[\eta_Z\Big(G\big(u',(u,z)\big)\Big)\right]$$

so

$$\eta_T\Big(G\backslash U.(U\circ_H f)\Big) = f\eta_Z$$

9.2 The categories $\mathcal{D}_U(X)$

Definition: *If X is a G-set, let $\mathcal{D}_U(X)$ be the following category:*

- *The objects of $\mathcal{D}_U(X)$ are the G-sets over $X\times(U/H)$. If (Y,f) is an object of $\mathcal{D}_U(X)$, I denote by f_U the map from Y to U/H obtained by composing f with the projection from $X\times(U/H)$ on U/H, and f_X the composition product of f with the projection on X. Then (Y,f_U) is a G-set over U/H, to which I associate the pull-back product $U.Y$.*
- *A morphism α from (Y,f) to (Z,g) is a morphism of G-sets from Y to Z such that $g\circ\alpha=f$, and such that the morphism $U.\alpha$ from $U.Y$ to $U.Z$ associated to α is injective on each left orbit under G, i.e. such that*

$$(u,y)\in U.Y,\ g\in G,\ gu=u,\ \alpha(gy)=\alpha(y)\Rightarrow\ gy=y$$

Remark: In particular, if α is injective, then $U.\alpha$ is injective on each left orbit.

It is clear that $\mathcal{D}_U(X)$ is a category: the product of two morphisms which are injective on the left orbits is injective on the left orbits, and the identity morphism is injective on the left orbits.

If $\phi: X\to X'$ is a morphism of G-sets, let $\tilde{\phi}$ be the morphism $\phi\times Id_{U/H}$ from $X\times(U/H)$ to $X'\times(U/H)$. There is an obvious functor from $\mathcal{D}_U(X)$ to $\mathcal{D}_U(X')$, which maps (Y,f) to $(Y,\tilde{\phi}f)$, and the morphism $\alpha:(Y,f)\to(Z,g)$ to $\alpha:(Y,\tilde{\phi}f)\to(Z,\tilde{\phi}g)$. I will denote this functor by $\mathcal{D}_{U,*}(\phi)$. Similarly, there is a functor from $\mathcal{D}_U(X')$ to $\mathcal{D}_U(X)$, defined by inverse image along $\tilde{\phi}$: if (Y',f') is a G-set over $X'\times(U/H)$, let Y, f and a be such that the square

$$\begin{array}{ccc} Y & \xrightarrow{a} & Y' \\ {\scriptstyle f}\downarrow & & \downarrow{\scriptstyle f'} \\ X\times(U/H) & \xrightarrow[\tilde{\phi}]{} & X'\times(U/H) \end{array}$$

is cartesian. If $\alpha':(Y',f')\to(Z',g')$ is a morphism of G-sets over $X'\times(U/H)$, let Z, g, and b such that the square

$$\begin{array}{ccc} Z & \xrightarrow{b} & Z' \\ {\scriptstyle g}\downarrow & & \downarrow{\scriptstyle g'} \\ X\times(U/H) & \xrightarrow[\tilde{\phi}]{} & X'\times(U/H) \end{array}$$

is cartesian. Then as

$$g'\alpha' a = f'a = \tilde{\phi} f$$

there exists a unique morphism $\alpha : Y \to Z$ such that the diagram

$$\begin{array}{ccccc} Y & & \xrightarrow{\quad a \quad} & & Y' \\ & \searrow^{\alpha} & & & & \searrow^{\alpha'} \\ {\scriptstyle f}\downarrow & & Z & \xrightarrow{\quad b \quad} & {\scriptstyle f'}\downarrow & & Z' \\ & \swarrow_{g} & & & & \swarrow_{g'} \\ X \times (U/H) & & \xrightarrow{\quad \tilde{\phi} \quad} & & X' \times (U/H) \end{array}$$

is commutative. Moreover, in those conditions, the square

$$\begin{array}{ccc} Y & \xrightarrow{a} & Y' \\ {\scriptstyle \alpha}\downarrow & & \downarrow{\scriptstyle \alpha'} \\ Z & \xrightarrow[b]{} & Z' \end{array}$$

is cartesian. Indeed, if $z \in Z$ and $y' \in Y'$ are such that $b(z) = \alpha'(y')$, then

$$g'b(z) = \tilde{\phi}g(z) = g'\alpha'(y') = f'(y')$$

Then by definition of Y, there exists a unique $y \in Y$ such that

$$g(z) = f(y) \qquad y' = a(y)$$

Moreover

$$b\alpha(y) = \alpha' a(y) = \alpha'(y') = b(z) \qquad g\alpha(y) = f(y) = g(z)$$

Then z and $\alpha(y)$ have the same image under b and g, so $z = \alpha(y)$. And if another element $y_1 \in Y$ is such that $\alpha(y_1) = \alpha(y)$ and $a(y_1) = a(y)$, then

$$g\alpha(y_1) = g\alpha(y) = f(y_1) = f(y)$$

and unicity of y implies $y_1 = y$.

Moreover, the morphism $U.\alpha$ is injective on the left orbits:

Lemma 9.2.1: Let

$$\begin{array}{ccc} Y & \xrightarrow{a} & Y' \\ {\scriptstyle \alpha}\downarrow & & \downarrow{\scriptstyle \alpha'} \\ Z & \xrightarrow[b]{} & Z' \end{array}$$

be a cartesian square of G-sets over U/H. If $U.\alpha'$ is injective on each left orbit of G on $U.Y'$, then $U.\alpha$ is injective on each left orbit of G on $U.Y$

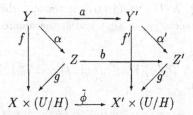

Proof: Indeed, if $(u,y) \in U.Y$ and (gu,gy) have the same image under $U.\alpha$, then $gu = u$ and $\alpha(gy) = \alpha(y)$. Taking the image under b of this relation gives

$$b\alpha(gy) = \alpha' a(gy) = b\alpha(y) = \alpha' a(y)$$

so $\alpha'\big(ga(y)\big) = \alpha'\big(a(y)\big)$. As $U.\alpha'$ is injective on the left orbits, it follows that $ga(y) = a(gy) = a(y)$. As moreover $\alpha(gy) = \alpha(y)$, and as the square Y, Y', Z, Z' is cartesian, I have $gy = y$. ∎

It follows that if I set $\mathcal{D}_U^*(\phi)(Y',f') = (Y,f)$ and $\mathcal{D}_U^*(\phi)(\alpha') = \alpha$, I get a functor from $\mathcal{D}_U(X')$ to $\mathcal{D}_U(X)$.

9.3 The functors $\mathcal{Q}_U(M)$

Definitions: *I will say that an H-set (Z,a) over $G\backslash U.Y$ is $\nu_{(Y,f)}$-disjoint (or ν-disjoint for short) if the square*

$$\begin{array}{ccc} \emptyset & \longrightarrow & Y \\ \downarrow & & \downarrow{\scriptstyle \nu_Y} \\ U \circ_H Z & \xrightarrow[U \circ_H a]{} & U \circ_H (G\backslash U.Y) \end{array}$$

is cartesian, or equivalently if

$$(U \circ_H a)(U \circ_H Z) \cap \nu_Y(Y) = \emptyset$$

If M is a Mackey functor for H, and (Y,f) is a G-set over U/H, I set

$$\mathcal{Q}_U(M)(Y,f) = M(G\backslash U.Y) / \sum_{(Z,a)} M_*(a)M(Z)$$

where the sum runs on the H-sets (Z,a) over $G\backslash U.Y$ which are ν-disjoint.

Remark: Say that (Z,a) is not ν-disjoint means that there exists $z \in Z$ and $y' \in Y$, such that if $u'H = f_U(y')$ and $a(z) = G(u,y)$, then there exists $v \in U$ such that $(v,z) \in U \circ_H Z$ and

$$\big(v, G(u,y)\big) = \big(u', G(u',y')\big)$$

This means that there exists $h \in H$ such that $v = u'h$ and

$$G(u,y) = h^{-1}G(u',y') = G(u'h,y')$$

So there exists $g \in G$ such that

$$u'h = gu \qquad y' = gy$$

Finally, there exists $g \in G$ and $h \in H$ such that

$$v = gu \qquad u' = guh^{-1} \qquad y' = gy$$

Now $(v,z) = (gu,z)$ belongs to $U \circ_H Z$ if and only if $(u,z) \in U \circ_H Z$ does.

Conversely, if $a(z) = G(u,y)$, and if $(u,z) \in U \circ_H Z$, then $(u,y) \in U.Y$, and

$$\nu_Y(y) = \big(u, G(u,y)\big) = (U \circ_H a)(u,z)$$

Thus (Z,a) is ν-disjoint if and only if

$$\forall z \in Z,\ a(z) = G(u,y) \Rightarrow (u,z) \notin U \circ_H Z$$

If X is a G-set, and Y is an object of $\mathcal{D}_U(X)$, I set

$$\mathcal{Q}_U(M)(Y,f) = \mathcal{Q}_U(M)(f_U)$$

Lemma 9.3.1: Let X be a G-set. The correspondence

$$(Y,f) \mapsto \mathcal{Q}_U(M)(Y,f)$$

$$\alpha : (Y,f) \to (Z,g) \quad \mapsto \quad M_*(G\backslash U.\alpha) : M(G\backslash U.Y) \to M(G\backslash U.Z)$$

induces a functor from $\mathcal{D}_U(X)$ to R-Mod.

Proof: I must prove that if α is a morphism in $\mathcal{D}_U(X)$ from (Y,f) to (Z,g), if (T,a) is a ν-disjoint H-set over $G\backslash U.Y$, and if $m \in M(T)$, then the image of $M_*(G\backslash U.\alpha)M_*(a)(m)$ in $\mathcal{Q}_U(M)(Z,g_U)$ is zero. This will be the case in particular if, denoting by β the morphism $G\backslash U.\alpha$, the H-set $(T,\beta a)$ is ν-disjoint over $G\backslash U.Z$. This is equivalent to say that the square

$$\begin{array}{ccc} \emptyset & \longrightarrow & Z \\ \big\downarrow & & \big\downarrow \nu_Z \\ U \circ_H T & \xrightarrow[U \circ_H (\beta a)]{} & U \circ_H (G\backslash U.Z) \end{array} \qquad (C)$$

is cartesian. But this square is composed of the two squares

$$\begin{array}{ccccc} \emptyset & \longrightarrow & Y & \xrightarrow{\alpha} & Z \\ \big\downarrow & & \big\downarrow \nu_Y & & \big\downarrow \nu_Z \\ U \circ_H T & \xrightarrow[U \circ_H a]{} & U \circ_H (G\backslash U.Y) & \xrightarrow[U \circ_H \beta]{} & U \circ_H (G\backslash U.Z) \end{array}$$

The left square is cartesian if (T,α) is ν-disjoint. Thus (C) is cartesian, since the right square is cartesian by the following lemma:

Lemma 9.3.2: Let $\alpha : (Y, f_U) \to (Z, g_U)$ be a morphism of G-sets over U/H, such that $U.\alpha$ is injective on each left orbit of G on $U.Y$. Then the square

$$\begin{array}{ccc} Y & \xrightarrow{\alpha} & Z \\ \nu_Y \big\downarrow & & \big\downarrow \nu_Z \\ U \circ_H (G\backslash U.Y) & \xrightarrow[U \circ_H (G\backslash U.\alpha)]{} & U \circ_H (G\backslash U.Z) \end{array}$$

is cartesian.

Proof: Let $\big(u_1, G(u_2,y)\big) \in U \circ_H (G\backslash U.Y)$ and $z \in Z$ such that

$$\Big(U \circ_H (G\backslash U.\alpha)\Big)\big(u_1, G(u_2,y)\big) = \nu_Z(z)$$

If $v \in U$ is such that $vH = g_U(z)$, then $\nu_Z(z) = \big(v, G(v,z)\big)$. The above equality means that there exists $h \in H$ such that

$$u_1 h = v \qquad h^{-1}G(u_2,\alpha(y)) = G(v,z) = G(u_2h,\alpha(y))$$

So there exists $x \in G$ such that

$$v = xu_2h \qquad z = x\alpha(y) = \alpha(xy)$$

The image of the second equality under g_U gives

$$g_U(z) = vH = g_U\alpha(xy) = f_U(xy)$$

Then

$$\nu_Y(xy) = \big(v, G(v,xy)\big) = \big(u_1h, G(x^{-1}v,y)\big) = \big(u_1, hG(x^{-1}v,y)\big) = \ldots$$
$$\ldots = \big(u_1, G(x^{-1}vh^{-1},y)\big) = \big(u_1, G(u_2,y)\big)$$

Conversely, let y_1 and y_2 be elements of Y having the same image under α and ν_Y. Then if $u \in U$ is such that $f_U(y_1) = uH$, I have also

$$f_U(y_2) = g_U\alpha(y_2) = g_U\alpha(y_1) = f_U(y_1) = uH$$

Then $\nu_Y(y_1) = \big(u, G(u,y_1)\big)$ and $\nu_Y(y_2) = \big(u, G(u,y_2)\big)$. These elements are equal if there exists $h \in H$ such that

$$u = uh \qquad h^{-1}G(u,y_1) = G(u,y_2) = G(uh,y_1) = G(u,y_1)$$

Then there exists $x \in G$ such that $u = xu$ and $y_2 = xy_1$. As moreover $\alpha(y_2) = \alpha(y_1)$, and as $U.\alpha$ is injective on the left orbits of G on $U.Y$, it follows that $y_2 = y_1$, which proves the lemma. ■

Let $\phi : X \to X'$ be a morphism of G-sets. If (Y,f) is an object of $\mathcal{D}_U(X)$, I denote by $(Y',f') = \mathcal{D}_{U,*}(\phi)(Y,f) = (Y,\tilde{\phi}f)$ its image in $\mathcal{D}_U(X')$. As $(\tilde{\phi}f)_U = f_U$, the sets $U.Y$ and $U.Y'$ are equal, and $M(G\backslash U.Y)$ identifies with $M(G\backslash U.Y')$. Similarly, the ν-disjoint H-sets over $G\backslash U.Y$ identify with ν-disjoint H-sets over $G\backslash U.Y'$. It follows that $\mathcal{Q}_U(M)(Y,f)$ is naturally isomorphic to $\mathcal{Q}_U(M)(Y',f')$, i.e.

$$\mathcal{Q}_U(M) \simeq \mathcal{Q}_U(M) \circ \mathcal{D}_{U,*}(\phi)$$

Conversely, if (Y',f') is an object of $\mathcal{D}_U(X')$, I set $(Y,f) = \mathcal{D}_U^*(Y',f')$. I have a cartesian square

$$\begin{array}{ccc} Y & \xrightarrow{\;a\;} & Y' \\ {\scriptstyle f}\downarrow & & \downarrow{\scriptstyle f'} \\ X \times (U/H) & \xrightarrow[\tilde{\phi}]{} & X' \times (U/H) \end{array}$$

If $m \in M(G\backslash U.Y')$, then $M^*(G\backslash U.a)(m) \in M(G\backslash U.Y)$. If (Z', α') is a ν-disjoint H-set over $G\backslash U.Y'$, I can fill with Z, β, α the cartesian square

$$\begin{array}{ccc} Z & \xrightarrow{\alpha} & G\backslash U.Y \\ {\scriptstyle\beta}\downarrow & & \downarrow{\scriptstyle G\backslash U.a} \\ Z' & \xrightarrow[\alpha']{} & G\backslash U.Y' \end{array}$$

Then (Z, α) is a ν-disjoint H-set over $G\backslash U.Y$: indeed, let T, γ, and δ be such that the square

$$\begin{array}{ccc} T & \xrightarrow{\gamma} & Y \\ {\scriptstyle\delta}\downarrow & & \downarrow{\scriptstyle \nu_Y} \\ U \circ_H Z & \xrightarrow[U \circ_H \alpha]{} & U \circ_H (G\backslash U.Y) \end{array}$$

is cartesian. I have then a commutative diagram

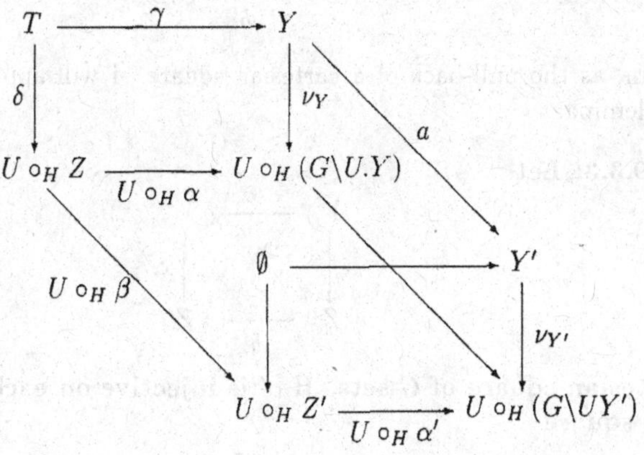

As the bottom square is cartesian, I can fill this diagram by a morphism from T to $\emptyset$, so T is empty, and (Z, α) is ν-disjoint.

As $M^*(G\backslash U.a)M_*(\alpha') = M_*(\alpha)M^*(\beta)$, the image of $M^*(G\backslash U.a)M_*(\alpha')$ in the quotient $Q_U(M)(Y, f)$ is zero. So I have built a morphism $T^{\phi}_{(Y',f')}$ from $Q_U(M)(Y', f')$ to $Q_U(M)(Y, f)$. This construction is moreover functorial in (Y', f'): if $\alpha' : (Y', f') \to (Z', g')$ is a morphism in $\mathcal{D}_U(X')$, then the square

$$\begin{array}{ccc} Q_U(M)(Y', f') & \xrightarrow{T^{\phi}_{(Y',f')}} & Q_U(M) \circ \mathcal{D}^*_U(\phi)(Y', f') \\ {\scriptstyle Q_U(M)(\alpha')}\downarrow & & \downarrow{\scriptstyle Q_U(M) \circ \mathcal{D}^*_U(\phi)(\alpha')} \\ Q_U(M)(Z', g') & \xrightarrow[T^{\phi}_{(Z',g')}]{} & Q_U(M) \circ \mathcal{D}^*_U(\phi)(Z', g') \end{array} \qquad (C)$$

is commutative: indeed, I have a diagram

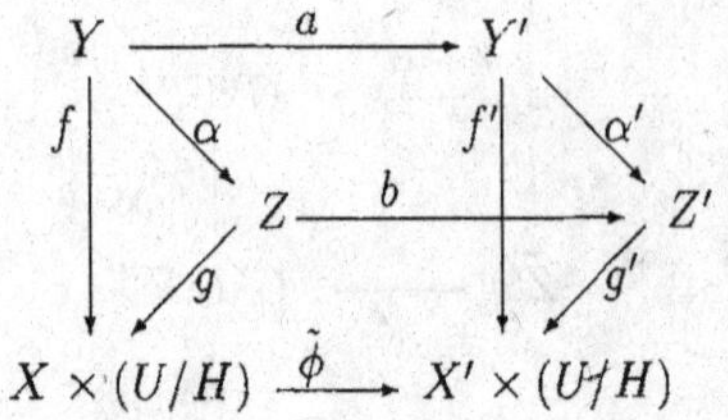

and to show that (C) is commutative, it suffices to show that

$$M_*(G\backslash U.\alpha)M^*(G\backslash U.a) = M^*(G\backslash U.b)M_*(G\backslash U.\alpha')$$

which will follow from the fact that the square

$$\begin{array}{ccc} U.Y & \xrightarrow{U.a} & U.Y' \\ {\scriptstyle U.\alpha}\downarrow & & \downarrow{\scriptstyle U.\alpha'} \\ U.Z & \xrightarrow[U.b]{} & U.Z' \end{array}$$

is cartesian, as the pull-back of a cartesian square. I will apply to this square the following lemma:

Lemma 9.3.3: Let

$$\begin{array}{ccc} Y & \xrightarrow{a} & Y' \\ {\scriptstyle \alpha}\downarrow & & \downarrow{\scriptstyle \alpha'} \\ Z & \xrightarrow[b]{} & Z' \end{array}$$

be a cartesian square of G-sets. If α' is injective on each orbit of G on Y', then the square

$$\begin{array}{ccc} G\backslash Y & \xrightarrow{G\backslash a} & G\backslash Y' \\ {\scriptstyle G\backslash\alpha}\downarrow & & \downarrow{\scriptstyle G\backslash\alpha'} \\ G\backslash Z & \xrightarrow[G\backslash b]{} & G\backslash Z' \end{array}$$

is cartesian.

Proof: Let $Gz \in G\backslash Z$ and $Gy' \in G\backslash Y'$ be such that

$$(G\backslash b)(Gz) = (G\backslash \alpha')(Gy') = Gb(z) = G\alpha'(y')$$

Then there exists an element $g \in G$ such that $gb(z) = b(gz) = \alpha'(y')$. So there exists a unique element $y \in Y$ such that

$$y' = a(y) \qquad gz = \alpha(y)$$

In those conditions

$$(G\backslash a)(Gy) = Ga(y) = Gy'$$

and moreover

$$(G\backslash\alpha)(Gy) = G\alpha(y) = Ggz = Gz$$

Now if Gy_1 and Gy_2 are elements of $G\backslash Y$ having the same image under $G\backslash a$ and $G\backslash\alpha$, I have

$$Ga(y_1) = Ga(y_2) \qquad G\alpha(y_1) = G\alpha(y_2)$$

So there exists elements g and g' of G such that

$$ga(y_1) = a(y_2) \qquad g'\alpha(y_1) = \alpha(y_2)$$

The image under b of this relation gives

$$b\big(g'\alpha(y_1)\big) = b\alpha(y_2) = \alpha'a(g'y_1) = \alpha'a(y_2) = \alpha'\big(g'a(y_1)\big)$$

Then as $a(y_1) = g^{-1}a(y_2)$, I have

$$\alpha'\big(g'g^{-1}a(y_2)\big) = \alpha'a(y_2)$$

As α' is injective on the left orbits, I have $g'g^{-1}a(y_2) = a(y_2)$, or

$$g^{-1}a(y_2) = a(g^{-1}y_2) = a(y_1) = g'^{-1}a(y_2)$$

As $\alpha(y_1) = \alpha(g'^{-1}y_2)$, it follows that $y_1 = g'^{-1}y_2$, and then $Gy_1 = Gy_2$, completing the proof of the lemma. ■

I have finally built a natural transformation T^ϕ from $\mathcal{Q}_U(M)$ to $\mathcal{Q}_U(M)\circ\mathcal{D}_U^*(\phi)$, so I have proved the following lemma:

Lemma 9.3.4: If $\phi : X \to X'$ is a morphism of G-sets, then ϕ induces an isomorphism of functors

$$T_\phi : \mathcal{Q}_U(M) \to \mathcal{Q}_U(M)\circ\mathcal{D}_{U,*}(\phi)$$

and a natural transformation

$$T^\phi : \mathcal{Q}_U(M) \to \mathcal{Q}_U(M)\circ\mathcal{D}_U^*(\phi)$$

9.4 The functors $\mathcal{L}_U(M)$

The previous lemma leads to the following definition:

Definition: *If X is a G-set, and M is a Mackey functor for H, I set*

$$\mathcal{L}_U(M)(X) = \varinjlim_{(Y,f)\in\mathcal{D}_U(X)} \mathcal{Q}_U(M)(Y,f)$$

If $\phi : X \to X'$ is a morphism of G-sets, then the isomorphism T_ϕ induces a morphism

$$\varinjlim_{(Y,f)\in\mathcal{D}_U(X)} \mathcal{Q}_U(M)(Y,f) \to \varinjlim_{(Y,f)\in\mathcal{D}_U(X)} \mathcal{Q}_U(M)\circ\mathcal{D}_{U,*}(\phi)(Y,f) \to \varinjlim_{(Y',f')\in\mathcal{D}_U(X')} \mathcal{Q}_U(M)(Y',f')$$

i.e. a morphism $\mathcal{L}_U(M)_*(\phi)$ from $\mathcal{L}_U(M)(X)$ to $\mathcal{L}_U(M)(X')$.

Conversely, the transformation T^ϕ induces a morphism

$$\varinjlim_{(Y',f')\in\mathcal{D}_U(X')} \mathcal{Q}_U(M)(Y',f') \to \varinjlim_{(Y',f')\in\mathcal{D}_U(X')} \mathcal{Q}_U(M)\circ\mathcal{D}_U^*(\phi)(Y',f') \to \varinjlim_{(Y,f)\in\mathcal{D}_U(X)} \mathcal{Q}_U(M)(Y,f)$$

i.e. a morphism $\mathcal{L}_U(M)^*(\phi)$ from $\mathcal{L}_U(M)(X')$ to $\mathcal{L}_U(M)(X)$

Proposition 9.4.1: The previous definitions turn $\mathcal{L}_U(M)$ into a Mackey functor.

Proof: It is clear that if $\phi : X \to X'$ and $\phi' : X' \to X"$ are morphisms of G-sets, then

$$\mathcal{L}_U(M)_*(\phi'\phi) = \mathcal{L}_U(M)_*(\phi')\mathcal{L}_U(M)_*(\phi) \qquad \mathcal{L}_U(M)^*(\phi'\phi) = \mathcal{L}_U(M)^*(\phi)\mathcal{L}_U(M)^*(\phi')$$

It is also clear that $\mathcal{L}_U(M)_*(Id)$ and $\mathcal{L}_U(M)^*(Id)$ are the identity morphisms.

It suffices then to check axioms (M1) and (M2) for $\mathcal{L}_U(M)$. I will start with (M2): let

$$\begin{array}{ccc} X & \xrightarrow{a} & Y \\ {\scriptstyle b}\downarrow & & \downarrow{\scriptstyle c} \\ Z & \xrightarrow[d]{} & T \end{array}$$

be a cartesian square of G-sets.

Notations: *Let (E,e) be a G-set over $Y \times (U/H)$, and $m \in M(G\backslash U.E)$. I denote by $m_{(E,e)}$ the image of m under the composite morphism*

$$M(G\backslash U.E) \to \mathcal{Q}_U(M)(E,e) \to \mathcal{L}_U(M)(Y)$$

*Furthermore, I denote by $X \mapsto \widetilde{X}$ the functor $X \mapsto X \times (U/H)$ from G-**set** to G-**set**.*

To compute $\mathcal{L}_U(M)^*(a)(m_{(E,e)})$, I fill the cartesian square

$$\begin{array}{ccc} F & \xrightarrow{\alpha} & E \\ {\scriptstyle f}\downarrow & & \downarrow{\scriptstyle e} \\ \widetilde{X} & \xrightarrow[\tilde{a}]{} & \widetilde{Y} \end{array}$$

The image of $m_{(E,e)}$ by $\mathcal{L}_U(M)^*(a)$ is then

$$\mathcal{L}_U(M)^*(a)(m_{(E,e)}) = M^*(G\backslash U.\alpha)(m)_{(F,f)}$$

The image of this element under $\mathcal{L}_U(M)_*(b)$ is

$$\mathcal{L}_U(M)_*(b)\mathcal{L}_U(M)^*(a)(m_{(E,e)}) = M^*(G\backslash U.\alpha)(m)_{(F,\tilde{b}f)}$$

On the other hand, the image of $m_{(E,e)}$ under $\mathcal{L}_U(M)_*(c)$ is

$$\mathcal{L}_U(M)_*(c)(m_{(E,e)}) = m_{(E,\tilde{c}e)}$$

As the square

$$\begin{array}{ccc} F & \xrightarrow{\alpha} & E \\ {\scriptstyle \tilde{b}f}\downarrow & & \downarrow{\scriptstyle \tilde{c}e} \\ \tilde{Z} & \xrightarrow[\tilde{d}]{} & \tilde{T} \end{array}$$

is cartesian, because it is composed of the previous one and of the cartesian square

$$\begin{array}{ccc} \tilde{X} & \xrightarrow{\tilde{a}} & \tilde{Y} \\ {\scriptstyle \tilde{b}}\downarrow & & \downarrow{\scriptstyle \tilde{c}} \\ \tilde{Z} & \xrightarrow[\tilde{d}]{} & \tilde{T} \end{array}$$

the image under $\mathcal{L}_U(M)^*(d)$ of $m_{(E,\tilde{c}e)}$ is then

$$\mathcal{L}_U(M)^*(d)\mathcal{L}_U(M)_*(c)(m_{(E,e)}) = M^*(G\backslash U.\alpha)(m)_{(F,\tilde{b}f)}$$

which proves that

$$\mathcal{L}_U(M)^*(d)\mathcal{L}_U(M)_*(c) = \mathcal{L}_U(M)_*(b)\mathcal{L}_U(M)^*(a)$$

So $\mathcal{L}_U(M)$ satisfies (M2).

To check axiom (M1), I consider two G-sets X and Y, and their disjoint union $Z = X \coprod Y$. As $\tilde{Z} = \widetilde{X} \coprod \tilde{Y}$, it is clear that a G-set (E, e) over $\tilde{Z}$ is the disjoint union of a G-set (E_1, e_1) over $\widetilde{X}$ and a G-set (E_2, e_2) over $\tilde{Y}$. Then $G\backslash U.E$ is the disjoint union of $G\backslash U.E_1$ and $G\backslash U.E_2$. Let i_1 and i_2 be the respective injections from E_1 and E_2 into E. If $m \in M(G\backslash U.E)$, let $m_1 = M^*(G\backslash U.i_1)(m)$ and $m_2 = M^*(G\backslash U.i_2)(m)$. Then

$$m = M_*(G\backslash U.i_1)(m_1) + M_*(G\backslash U.i_2)(m_2)$$

I can view this equality in $\mathcal{L}_U(M)(Z)$, as

$$m_{(E,e)} = M_*(G\backslash U.i_1)(m_1)_{(E,e)} + M_*(G\backslash U.i_2)(m_2)_{(E,e)}$$

But as i_1 is injective, it is a morphism in $\mathcal{D}_U(Z)$ from (E_1, ei_1) to (E, e), and in $\mathcal{L}_U(M)(Z)$, I have

$$M_*(G\backslash U.i_1)(m_1)_{(E,e)} = (m_1)_{(E_1,ei_1)}$$

Now if i_X is the injection from X to Z, I have

$$(m_1)_{(E_1,ei_1)} = \mathcal{L}_U(M)_*(i_X)\Big((m_1)_{(E_1,e_1)}\Big)$$

The same argument for i_2 shows then that, denoting by i_Y the injection from Y to Z, I have

$$m_{(E,e)} = \mathcal{L}_U(M)_*(i_X)\Big((m_1)_{(E_1,e_1)}\Big) + \mathcal{L}_U(M)_*(i_Y)\Big((m_2)_{(E_2,e_2)}\Big)$$

In particular, the map $\Big(\mathcal{L}_U(M)_*(i_X), \mathcal{L}_U(M)_*(i_Y)\Big)$ from $\mathcal{L}_U(M)(X) \oplus \mathcal{L}_U(M)(Y)$ to $\mathcal{L}_U(M)(Z)$ is surjective. The following lemma now shows that axiom (M1) holds for $\mathcal{L}_U(M)$, and this completes the proof of the proposition. ■

Lemma 9.4.2: Let L be a bifunctor on G-set, satisfying axiom (M2), and such that $L(\emptyset) = \{0\}$. If for any G-sets X and Y, the map

$$\big(L_*(i_X), L_*(i_Y)\big) : L(X) \oplus L(Y) \to L(X \coprod Y)$$

is surjective, or the map

$$L^*(i_X) \oplus L^*(i_Y) : L(X \coprod Y) \to L(X) \oplus L(Y)$$

injective, then L is a Mackey functor for G.

Proof: Let θ be the map $\big(L_*(i_X), L_*(i_Y)\big)$. As L satisfies (M2), as $L(\emptyset) = \{0\}$, and as the square

$$\begin{array}{ccc} \emptyset & \longrightarrow & X \\ \big\downarrow & & \big\downarrow{\scriptstyle i_X} \\ Y & \xrightarrow[i_Y]{} & X \coprod Y \end{array}$$

is cartesian, the products $L^*(i_X)L_*(i_Y)$ and $L^*(i_Y)L_*(i_X)$ are zero. As i_X is injective, the square

$$\begin{array}{ccc} X & \xrightarrow{Id} & X \\ {\scriptstyle Id}\big\downarrow & & \big\downarrow{\scriptstyle i_X} \\ X & \xrightarrow[i_X]{} & X \coprod Y \end{array}$$

is cartesian. It follows that $L^*(i_X)L_*(i_X)$ is the identity of $L(X)$, and that $L^*(i_Y)L_*(i_Y)$ is the identity of $L(Y)$. Now the map $\phi = L^*(i_X) \oplus L^*(i_Y)$ is such that $\phi\theta = Id_{L(X)\oplus L(Y)}$. Then $\theta\phi\theta = \theta$, and $\phi\theta\phi = \phi$. If θ is surjective (resp. if ϕ is injective), the first (resp. the second) of these equalities implies $\theta\phi = Id_{L(X \coprod Y)}$, so (M1) holds. This proves the lemma. ■

9.5 Left adjunction

Notation: *If $\theta : M \to M'$ is a morphism of Mackey functors for the group H, and if X is a G-set, I define a map $\mathcal{L}_U(\theta)_X$ from $\mathcal{L}_U(M)(X)$ to $\mathcal{L}_U(M')(X)$ by setting, for an object (Y, f) of $\mathcal{D}_U(X)$, and an element $m \in M(G\backslash U.Y)$*

$$\mathcal{L}_U(\theta)_X(m_{(Y,f)}) = \theta_{G\backslash U.Y}(m)_{(Y,f)}$$

Lemma 9.5.1: This definition turns $\mathcal{L}_U(\theta)$ into a morphism of Mackey functors from $\mathcal{L}_U(M)$ to $\mathcal{L}_U(M')$.

Proof: First, the map $\mathcal{L}_U(\theta)$ is well defined: if $a : Z \to G\backslash U.Y$ is ν-disjoint, then

$$\mathcal{L}_U(\theta)_X\big(M_*(G\backslash U.a)(m)_{(Y,f)}\big) = \theta_{G\backslash U.Y} M_*(a)(m)_{(Y,f)} = \big(M_*(a)\theta_Z(m)\big)_{(Y,f)} = 0$$

Moreover, if $\alpha : (Y, f) \to (Z, g)$ is a morphism in $\mathcal{D}_U(X)$, then

$$\mathcal{L}_U(\theta)_X M_*(\alpha)(m_{(Y,f)}) = \big(\theta_{G\backslash U.Z} M_*(G\backslash U.\alpha)(m)\big)_{(Z,g)} = \ldots$$

$$\dots = \Big(M_*(G\backslash U.\alpha)\theta_{G\backslash U.Y}(m)\Big)_{(Z,g)} = \theta_{G\backslash U.Y}(m)_{(Y,f)} = \mathcal{L}_U(\theta)_X(m_{(Y,f)})$$

Now let $\phi : X \to X'$ be a morphism of G-sets. Then

$$\mathcal{L}_U(M)_*(\phi)\mathcal{L}_U(\theta)_X(m_{(Y,f)}) = \mathcal{L}_U(M)_*(\phi)\Big(\theta_{G\backslash U.Y}(m)_{(Y,f)}\Big) = \dots$$

$$\dots = \theta_{G\backslash U.Y}(m)_{(Y,\tilde{\phi}f)} = \mathcal{L}_U(\theta)_{X'}(m_{(Y,\tilde{\phi}f)}) = \dots$$

$$\dots = \mathcal{L}_U(\theta)_{X'}\mathcal{L}_U(M)_*(\phi)(m_{(Y,f)})$$

If moreover (Y', f') is an object of $\mathcal{D}_U(X')$, and if $m' \in M(G\backslash U.Y')$, then let Y, f, and a filling the cartesian square

$$\begin{array}{ccc} Y & \xrightarrow{\;a\;} & Y' \\ {\scriptstyle f}\downarrow & & \downarrow{\scriptstyle f'} \\ X & \xrightarrow[\tilde{\phi}]{} & X' \end{array}$$

With those notations, I have

$$\mathcal{L}_U(M)^*(\phi)\mathcal{L}_U(\theta)_{X'}(m'_{(Y',f')}) = \mathcal{L}_U(M)^*(\phi)\Big(\theta_{G\backslash U.Y'}(m')_{(Y',f')}\Big) = \dots$$

$$\dots = \Big(M^*(G\backslash U.a)\theta_{G\backslash U.Y'}(m')\Big)_{(Y,f)} = \Big(\theta_{G\backslash U.Y}M^*(G\backslash U.a)(m')\Big)_{(Y,f)} = \dots$$

$$\dots = \mathcal{L}_U(\theta)_X\Big(M^*(G\backslash U.a)(m')_{(Y,f)}\Big) = \mathcal{L}_U(\theta)_X\mathcal{L}_U(M)^*(\phi)(m'_{(Y',f')})$$

which proves the lemma. ■

Theorem 9.5.2: Let G and H be finite groups, and U be a G-set-H. The correspondence

$$M \mapsto \mathcal{L}_U(M)$$

$$\theta \in \mathrm{Hom}_{Mack(H)}(M, M') \mapsto \mathcal{L}_U(\theta) \in \mathrm{Hom}_{Mack(G)}(\mathcal{L}_U(M), \mathcal{L}_U(M'))$$

is a functor from $Mack(H)$ to $Mack(G)$, which is left adjoint to the functor $N \mapsto N \circ U$.

Proof: It is clear that the correspondence $M \mapsto \mathcal{L}_U(M)$ is functorial in M. The main point is the adjunction property.

So let M be a Mackey functor for H, and N be a Mackey functor for G. If θ is a morphism of Mackey functors from M to $N \circ U$, then for any H-set Z, I have a morphism

$$\theta_Z : M(Z) \to (N \circ U)(Z) = N(U \circ_H Z)$$

In particular, if (Y, f) is a G-set over $X \times (U/H)$, I have a morphism

$$\theta_{G\backslash U.Y} : M(G\backslash U.Y) \to N\Big(U \circ_H (G\backslash U.Y)\Big)$$

Composing this morphism with $N^*(\nu_Y)$ gives

$$N^*(\nu_Y)\theta_{G\backslash U.Y} : M(G\backslash U.Y) \to N\Big(U \circ_H (G\backslash U.Y)\Big) \to N(Y)$$

If (Z,a) is a ν-disjoint H-set over $G\backslash U.Y$, I have the diagram

$$\begin{array}{ccccc} M(Z) & \xrightarrow{\theta_Z} & N(U\circ_H Z) & \longrightarrow & N(\emptyset) \\ {\scriptstyle M_*(a)}\downarrow & & \downarrow{\scriptstyle N_*(U\circ_H a)} & & \downarrow \\ M(G\backslash U.Y) & \xrightarrow[\theta_{G\backslash U.Y}]{} & N\big(U\circ_H (G\backslash U.Y)\big) & \xrightarrow[N^*(\nu_Y)]{} & N(Y) \end{array}$$

This diagram is commutative: the left square is because θ is a morphism of Mackey functors, and the right square is because (Z,a) is ν-disjoint. It follows a morphism from $\mathcal{Q}_U(M)(Y,f)$ to $N(Y)$. Now I can compose this morphism with the morphism $N_*(f_X): N(Y)\to N(X)$ deduced from f.

If $\alpha:(Y,f)\to(Z,g)$ is a morphism in $\mathcal{D}_U(X)$, I have the following diagram

$$\begin{array}{ccccccc} M(G\backslash U.Y) & \xrightarrow{\theta_{G\backslash U.Y}} & N\big(U\circ_H (G\backslash U.Y)\big) & \xrightarrow{N^*(\nu_Y)} & N(Y) & \xrightarrow{N_*(f_X)} & N(X) \\ {\scriptstyle M_*(G\backslash U.\alpha)}\downarrow & & \downarrow{\scriptstyle N_*\left(U\circ_H (G\backslash U.\alpha)\right)} & & \downarrow{\scriptstyle N_*(\alpha)} & & \downarrow{\scriptstyle Id} \\ M(G\backslash U.Z) & \xrightarrow[\theta_{G\backslash U.Z}]{} & N\big(U\circ_H (G\backslash U.Z)\big) & \xrightarrow[N^*(\nu_Z)]{} & N(Z) & \xrightarrow[N_*(g_X)]{} & N(X) \end{array}$$

The left square is commutative because θ is a morphism of Mackey functors. The middle square is commutative by lemma 9.3.2 because N is a Mackey functor. The right square is commutative because $g\alpha = f$.

It follows a morphism ψ_X from $\mathcal{L}_U(M)(X)$ to $N(X)$. Now if $\phi: X\to X'$ is a morphism of G-sets, if (Y,f) is a G-set over $X\times(U/H)$, and if $m\in M(G\backslash U.Y)$, then by definition of ψ_X

$$\psi_X(m_{(Y,f)}) = N_*(f_X)N^*(\nu_Y)\theta_{G\backslash U.Y}(m)$$

On the other hand

$$\mathcal{L}_U(M)_*(\phi)(m_{(Y,f)}) = m_{(Y,\tilde{\phi}f)}$$

so

$$\psi_{X'}\mathcal{L}_U(M)_*(\phi)(m_{(Y,f)}) = N_*(\phi f_X)N^*(\nu_Y)\theta_{G\backslash U.Y}(m) = N_*(\phi)\psi_X(m_{(Y,f)})$$

and then $\psi'_X\mathcal{L}_U(M)_*(\phi) = N_*(\phi)\psi_X$.

Conversely, if (Y',f') is a G-set over $X'\times(U/H)$, and if $m'\in M(G\backslash U.Y')$, then the image of $m'_{(Y',f')}$ under $\mathcal{L}_U(M)^*(\phi)$ is obtained by filling the cartesian square

$$\begin{array}{ccc} Y & \xrightarrow{a} & Y' \\ {\scriptstyle f}\downarrow & & \downarrow{\scriptstyle f'} \\ X & \xrightarrow[\tilde{\phi}]{} & X' \end{array}$$

More precisely, I have then

$$\mathcal{L}_U(M)^*(\phi)(m'_{(Y',f')}) = M^*(G\backslash U.a)(m')_{(Y,f)}$$

The image under ψ_X of this element is by definition

$$\psi_X \mathcal{L}_U(M)^*(\phi)(m'_{(Y',f')}) = N_*(f_X)N^*(\nu_Y)\theta_{G\backslash U.Y}M^*(G\backslash U.a)(m') \tag{9.1}$$

As θ is a morphism of Mackey functors, I have

$$\theta_{G\backslash U.Y}M^*(G\backslash U.a) = N^*\big(U \circ_H (G\backslash U.a)\big)\theta_{G\backslash U.Y'}$$

Moreover, as the square

$$\begin{array}{ccc}
Y & \xrightarrow{\quad a \quad} & Y' \\
{\scriptstyle \nu_Y}\big\downarrow & & \big\downarrow{\scriptstyle \nu_{Y'}} \\
U \circ_H (G\backslash U.Y) & \xrightarrow[U \circ_H (G\backslash U.a)]{} & U \circ_H (G\backslash U.Y')
\end{array}$$

is commutative, I have

$$N^*(\nu_Y)N^*\big(U \circ_H (G\backslash U.a)\big) = N^*(a)N^*(\nu_{Y'})$$

Furthermore, it is easy to see that the square

$$\begin{array}{ccc}
Y & \xrightarrow{\ a\ } & Y' \\
{\scriptstyle f_X}\big\downarrow & & \big\downarrow{\scriptstyle f'_{X'}} \\
X & \xrightarrow[\phi]{} & X'
\end{array}$$

is cartesian, and then $N_*(f_X)N^*(a) = N^*(\phi)N_*(f'_{X'})$. Finally, I can write equation (9.1) as

$$\psi_X \mathcal{L}_U(M)^*(\phi)(m'_{(Y',f')}) = N^*(\phi)N_*(f'_{X'})N^*(\nu_{Y'})\theta_{G\backslash U.Y'}(m')$$

On the other hand

$$\psi_{X'}(m_{(Y',f')}) = N_*(f'_{X'})N^*(\nu_{Y'})\theta_{G\backslash U.Y'}(m')$$

It follows that $\psi_X \mathcal{L}_U(M)^*(\phi) = N^*(\phi)\psi_{X'}$, and this shows that the morphisms ψ_X define a morphism ψ of Mackey functors from $\mathcal{L}_U(M)$ to N.

Conversely, let ψ be a morphism of Mackey functors from $\mathcal{L}_U(M)$ to N. Then for any G-set X, I have a morphism ψ_X from $\mathcal{L}_U(M)(X)$ to $N(X)$. In particular, if Z is an H-set, I have a morphism $\psi_{U \circ_H Z}$ from $\mathcal{L}_U(M)(U \circ_H Z)$ to $N(U \circ_H Z) = (N \circ U)(Z)$. But if $m \in M(Z)$, then $M^*(\eta_Z)(m) \in M\big(G\backslash U.(U \circ_H Z)\big)$. Moreover, setting $\pi_Z(u,z) = \big((u,z), uH\big)$, I define a morphism of G-sets π_Z from $U \circ_H Z$ to $(U \circ_H Z) \times (U/H)$, and then $U \circ_H Z$ is an object of $\mathcal{D}_U(U \circ_H Z)$. In particular, I can consider the element

$$\lambda_Z(m) = M^*(\eta_Z)(m)_{(U \circ_H Z, \pi_Z)} \in \mathcal{L}_U(M)(U \circ_H Z)$$

Finally, I have the composite morphism

$$\theta_Z : M(Z) \xrightarrow{\lambda_Z} \mathcal{L}_U(M)(U \circ_H Z) \xrightarrow{\psi_{U \circ_H Z}} N(U \circ_H Z)$$

If ϕ is a morphism of H-sets from Z to Z', and if $m' \in M(Z')$, then

$$\theta_Z M^*(\phi)(m') = \psi_{U\circ_H Z}\lambda_Z M^*(\phi)(m)$$

Moreover

$$\lambda_Z M^*(\phi)(m') = M^*(\eta_Z)M^*(\phi)(m)_{(U\circ_H Z,\pi_Z)}$$

It is clear moreover that the square

$$\begin{array}{ccc} G\backslash U.(U\circ_H Z) & \xrightarrow{\eta_Z} & Z \\ {\scriptstyle G\backslash U.(U\circ_H \phi)}\downarrow & & \downarrow{\scriptstyle \phi} \\ G\backslash U.(U\circ_H Z') & \xrightarrow[\eta_{Z'}]{} & Z' \end{array}$$

is commutative. So I have

$$\lambda_Z M^*(\phi)(m') = M^*\Big(G\backslash U.(U\circ_H \phi)\Big)M^*(\eta_{Z'})(m')_{(U\circ_H Z,\pi_Z)}$$

The square

$$\begin{array}{ccc} U\circ_H Z & \xrightarrow{U\circ_H \phi} & U\circ_H Z' \\ {\scriptstyle Id}\downarrow & & \downarrow{\scriptstyle Id} \\ U\circ_H Z & \xrightarrow[U\circ_H \phi]{} & U\circ_H Z' \end{array}$$

is trivially cartesian, so the square

$$\begin{array}{ccc} U\circ_H Z & \xrightarrow{U\circ_H \phi} & U\circ_H Z' \\ {\scriptstyle \pi_Z}\downarrow & & \downarrow{\scriptstyle \pi_{Z'}} \\ U\widetilde{\circ_H} Z & \xrightarrow[U\widetilde{\circ_H}\phi]{} & U\widetilde{\circ_H} Z' \end{array}$$

is also cartesian. It follows that

$$M^*\Big(G\backslash U.(U\circ_H\phi)\Big)M^*(\eta_{Z'})(m')_{(U\circ_H Z,\pi_Z)} = \mathcal{L}_U(M)^*(U\circ_H\phi)\Big(M^*(\eta_{Z'})(m')_{(U\circ_H Z',\pi_{Z'})}\Big) = .$$

$$\ldots = \mathcal{L}_U(M)^*(U\circ_H \phi)\lambda_{Z'}(m')$$

Finally, I have

$$\theta_Z M^*(\phi)(m') = \psi_{U\circ_H Z}\mathcal{L}_U(M)^*(U\circ_H \phi)\lambda_{Z'}(m')$$

and as ψ is a morphism of Mackey functors, it is also

$$\theta_Z M^*(\phi)(m') = (N\circ U)^*(\phi)\psi_{U\circ_H Z'}\lambda_{Z'}(m') = (N\circ U)^*(\phi)\theta_{Z'}(m')$$

Thus $\theta_Z M^*(\phi) = (N\circ U)^*(\phi)\theta_{Z'}$.

The previous proof shows that

$$\lambda_Z M^*(\phi) = \mathcal{L}_U(M)^*(U\circ_H \phi)\lambda_{Z'}$$

I will now prove that

$$\lambda_{Z'} M_*(\phi) = \mathcal{L}_U(M)_*(U \circ_H \phi)\lambda_Z$$

so that λ will be a morphism of Mackey functors from M to $\mathcal{L}_U(M) \circ U$. Then θ will be composed of λ and of the morphism $\psi \circ U$ from $\mathcal{L}_U(M) \circ U$ to $N \circ U$. It will also be a morphism of Mackey functors.

First I fill the cartesian square

$$
\begin{array}{ccc}
\Pi & \xrightarrow{\quad a \quad} & Z \\
{\scriptstyle b}\downarrow & & \downarrow{\scriptstyle \phi} \\
G\backslash U.(U \circ_H Z') & \xrightarrow[\eta_{Z'}]{} & Z'
\end{array}
$$

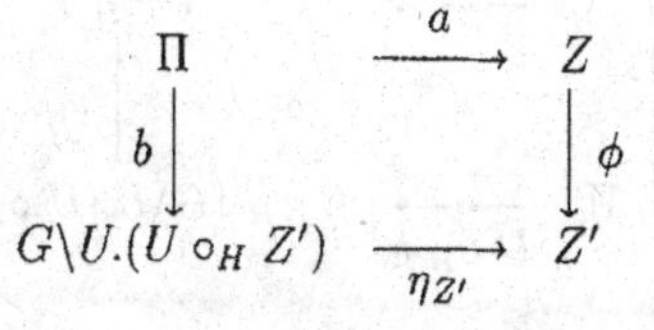

and I denote by i the morphism from $G\backslash U.(U \circ_H Z)$ to Π filling the commutative diagram

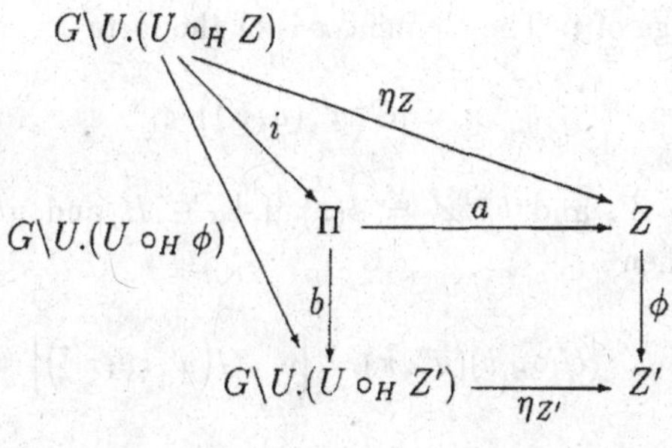

Lemma 9.5.3: In the previous diagram:

- **The morphism i is injective.**

- **If moreover $\Pi = \mathrm{Im}(i) \coprod \Pi'$, then Π' is a ν-disjoint H-set over the set $G\backslash U.(U \circ_H Z')$.**

Proof: Let $G\big(u',(u,z)\big)$ and $G\big(u'_1,(u_1,z_1)\big)$ be two elements having the same image under i. If $u' = uh$ and $u'_1 = u_1 h_1$, this means that

$$G\big(u',(u,\phi(z))\big) = G\big(u'_1,(u_1,\phi(z_1))\big) \qquad h^{-1}z = h_1^{-1}z_1$$

Then $z_1 = h_1 h^{-1}z$, and there exists $g \in G$ such that $\big(gu,\phi(z)\big) = \big(u_1,\phi(z_1)\big)$ and $gu' = u'_1$. So there exists $h' \in H$ such that $gu = u_1 h'$ and $\phi(z) = h'^{-1}\phi(z_1)$. In those conditions

$$G\big(u'_1,(u_1,z_1)\big) = G\big(gu',(u_1,h_1 h^{-1}z)\big) = \ldots$$

$$\ldots = G\big(u',(g^{-1}u_1,h_1 h^{-1}z)\big) = G\big(u',(uh'^{-1},h_1 h^{-1}z)\big)$$

But

$$u_1 h' h h_1^{-1} = guhh_1^{-1} = gu'h_1^{-1} = u'_1 h_1^{-1} = u_1$$

As $(u_1, z_1) \in U \circ_H Z$, it follows that $h'hh_1^{-1}z_1 = z_1$, i.e. $h'z = h_1h^{-1}z$. Then

$$G\big(u_1', (u_1, z_1)\big) = G\big(u', (uh'^{-1}, h'z)\big) = G\big(u', (u, z)\big)$$

and this proves that i is injective.

To prove the second assertion of the lemma, I must prove that the square

$$\begin{array}{ccc} \emptyset & \longrightarrow & U \circ_H Z' \\ \Big\downarrow & & \Big\downarrow \nu_{U \circ_H Z'} \\ U \circ_H \Pi' & \xrightarrow[U \circ_H b]{} & U \circ_H \big(G\backslash U.(U \circ_H Z')\big) \end{array}$$

is cartesian. This means that if $(u", \pi) \in U \circ_H \Pi$ and $(u_1, z_1') \in U \circ_H Z'$ are such that

$$(U \circ_H b)(u", \pi) = \nu_{U \circ_H Z'}(u_1, z_1') \tag{9.2}$$

then π is in the image of i. The element π is of the form

$$\pi = \Big[G\big(u', (u, z')\big), z\Big]$$

with $(u, z') \in U \circ_H Z'$, and $h_0^{-1}z' = \phi(z)$ if $h_0 \in H$ and $uh_0 = u'$. Now $b(\pi) = G\big(u', (u, z')\big)$, and then

$$(U \circ_H b)(u", \pi) = \Big[u", G\big(u', (u, z')\big)\Big]$$

On the other hand

$$\nu_{U \circ_H Z'}(u_1, z_1') = \Big[u_1, G\big(u_1, (u_1, z_1')\big)\Big]$$

Now equality (9.2) means that there exists $h \in H$ such that

$$u" = u_1h \qquad h^{-1}G\big(u_1, (u_1, z_1')\big) = G\big(u_1h, (u_1, z_1')\big) = G\big(u', (u, z')\big)$$

Then there exists $g \in G$ such that $(gu_1, z_1') = (u, z')$ and $gu_1h = u'$. Finally, there exists $g \in G$, and $h, h' \in H$ such that

$$u" = u_1h \qquad gu_1h = u' \qquad gu_1h' = u \qquad h'^{-1}z_1' = z'$$

In those conditions

$$\pi = \Big[G\big(u', (u, z')\big), z\Big] = \Big[G\big(gu_1h, (gu_1h', h_0\phi(z))\big), z\Big] = \ldots$$

$$\ldots = \Big[G\big(u_1h, (u_1h', h_0\phi(z))\big), z\Big] = \Big[G\big(u", (u"h^{-1}h', h_0\phi(z))\big), z\Big] = \ldots$$

$$\ldots = \Big[G\big(u", (u", h^{-1}h'h_0\phi(z))\big), z\Big]$$

Furthermore, as $(u", \pi) \in U \circ_H \Pi$, it follows that if $h_1 \in H$ is such that $u"h_1 = u"$, then $h_1\pi = \pi$, which implies in particular that $h_1z = z$. So $(u", z) \in U \circ_H Z$. Moreover

$$u"h^{-1}h'h_0 = u_1h'h_0 = g^{-1}uh_0 = g^{-1}u' = u_1h = u"$$

It follows that $h^{-1}h'h_0\phi(z) = \phi(h^{-1}h'h_0z) = \phi(z)$, and then

$$\pi = \Big[G\big(u", (u", \phi(z))\big), z\Big] = i(u", z)$$

which proves the lemma. ∎

To prove the equality

$$\lambda_{Z'}M_*(\phi) = \mathcal{L}_U(M)_*(U \circ_H \phi)\lambda_Z$$

I choose $m \in M(Z)$. Then

$$\mathcal{L}_U(M)_*(U \circ_H \phi)\lambda_Z(m) = \mathcal{L}_U(M)_*(U \circ_H \phi)\big(M^*(\eta_Z)(m)_{(U\circ_H Z, \pi_Z)}\big) = \dots$$

$$\dots = M^*(\eta_Z)(m)_{(U\circ_H Z, \widetilde{(U\circ_H \phi)}\pi_Z)}$$

Then $U \circ_H \phi$ is a morphism of G-sets over $\widetilde{U \circ_H Z'}$ from $(U \circ_H Z, \widetilde{(U \circ_H \phi)}\pi_Z)$ to $(U \circ_H Z', \pi_{Z'})$. Moreover $U.(U \circ_H \phi)$ is injective on the left orbits of G on $U.(U \circ_H Z)$: indeed, if $\big(u, (u', z)\big)$ has the same image than $g.\big(u, (u', z)\big) = \big(gu, (gu', z)\big)$, then $gu = u$ and there exists $h \in H$ such that $gu' = u'h$ and $h\phi(z) = \phi(z)$. But there exists h_0 such that $u = u'h_0$. As $gu = u$, I have also $gu' = u'$, and then $\big(gu, (gu', z)\big) = \big(u, (u', z)\big)$.

As $U \circ_H \phi$ is a morphism in $\mathcal{D}_U(U \circ_H Z')$, I have the following equality in $\mathcal{L}_U(M)(U \circ_H Z')$

$$M^*(\eta_Z)(m)_{(U\circ_H Z, \widetilde{(U\circ_H \phi)}\pi_Z)} = M_*(U \circ_H \phi)M^*(\eta_Z)(m)_{(U\circ_H Z', \pi_{Z'})}$$

Moreover

$$M_*(U \circ_H \phi)M^*(\eta_Z) = M_*(b)M_*(i)M^*(i)M^*(a) = \dots$$

$$\dots = M_*(b)M^*(a) + M_*(b)\big(1 - M_*(i)M^*(i)\big)M^*(a)$$

If j denotes the injection from Π' into Π, I have $1 - M_*(i)M^*(i) = M_*(j)M^*(j)$, and

$$M_*(b)\big(1 - M_*(i)M^*(i)\big)M^*(a) = M_*(bj)M^*(aj)$$

But (Π', bj) is a ν-disjoint G-set over $G\backslash U.(U \circ_H Z')$. The image of $M_*(bj)$ in $\mathcal{Q}_U(M)(U \circ_H Z')$ is then zero. It follows that

$$M_*(U \circ_H \phi)M^*(\eta_Z)(m)_{(U\circ_H Z', \pi_{Z'})} = M_*(b)M^*(a)(m)_{(U\circ_H Z', \pi_{Z'})} = \dots$$

$$\dots = M^*(\eta_{Z'})M_*(\phi)(m)_{(U\circ_H Z', \pi_{Z'})} = \lambda_{Z'}M_*(\phi)(m)$$

proving that

$$\lambda_{Z'}M_*(\phi) = \mathcal{L}_U(M)_*(U \circ_H \phi)\lambda_Z$$

so λ is a morphism of Mackey functors from M to $\mathcal{L}_U(M) \circ U$.

Now I have a correspondence $A : \theta \mapsto \psi$

$$\mathrm{Hom}_{Mack(H)}(M, N \circ U) \to \mathrm{Hom}_{Mack(G)}(\mathcal{L}_U(M), N)$$

and a correspondence B in the other direction. It is easy to see that those constructions are functorial in M and N. The theorem will now follow, if I prove that they are inverse to each other. It is equivalent to check the relations on unit and co-unit (see Mac-Lane [10] chp. IV): if θ is the identity of $N \circ U$, I must check that $B \circ A(\theta) = \theta$. Similarly, if ψ is the identity of $\mathcal{L}_U(M)$, I must check that $A \circ B(\psi) = \psi$.

But $A(\theta)$ is the morphism from $\mathcal{L}_U(N \circ U)$ to N defined on a G-set X by

$$m_{(Y,f)} \in \mathcal{L}_U(N \circ U)(X) \mapsto N_*(f_X)N^*(\nu_Y)\theta_{G\backslash U.Y}(m) \in N(X)$$

where (Y, f) is a G-set over $\widetilde{X}$, and m is an element of $M\big(G\backslash U.Y\big)$. In other words $A(\theta)_X$ is defined by

$$m_{(Y,f)} \in \mathcal{L}_U(N \circ U)(X) \mapsto N_*(f_X)N^*(\nu_Y)(m) \in N(X)$$

Then $B \circ A(\theta)$ is the endomorphism of $N \circ U$ defined on an H-set Z by

$$n \in (N \circ U)(Z) \mapsto A(\theta)_{U\circ_H Z}\lambda_Z(n)$$

Moreover I have here

$$\lambda_Z(n) = (N \circ U)^*(\eta_Z)(n)_{(U\circ_H Z,\pi_Z)} = N^*(U \circ_H \eta_Z)(n)_{(U\circ_H Z,\pi_Z)}$$

So

$$A(\theta)_{U\circ_H Z}\lambda_Z(n) = N_*\big((\pi_Z)_{U\circ_H Z}\big)N^*(\nu_{U\circ_H Z})N^*(U \circ_H \eta_Z)(n)$$

But $(\pi_Z)_{U\circ_H Z}$ is the identity of $U \circ_H Z$. And as ν and η are the unit and co-unit of the adjunction of $Z \mapsto U \circ_H Z$ and $Y \mapsto G\backslash U.Y$, I have

$$(U \circ_H \eta_Z)\nu_{U\circ_H Z} = Id_{U\circ_H Z}$$

So $B \circ A(\theta)$ is the identity of $N \circ U$.

Conversely, if ψ is the identity of $\mathcal{L}_U(M)$, then $B(\psi)$ is the morphism from M to $\mathcal{L}_U(M) \circ U$ defined on the H-set Z by

$$m \in M(Z) \mapsto \lambda_Z(m)$$

The endomorphism $A \circ B(\psi)$ of $\mathcal{L}_U(M)$ is then defined for a G set X, a set (Y, f) over $\widetilde{X}$, and an element $m \in M(G\backslash U.Y)$ by

$$m_{(Y,f)} \in \mathcal{L}_U(M)(X) \mapsto \mathcal{L}_U(M)_*(f_X)\mathcal{L}_U(M)^*(\nu_Y)\big(\lambda_{G\backslash U.\tilde{Y}}(m)\big)$$

But

$$\lambda_{G\backslash U.Y}(m) = M^*(\eta_{G\backslash U.Y})(m)_{(U\circ_H(G\backslash U.Y),\pi_{G\backslash U.Y})}$$

Let e be the map from Y to $\tilde{Y}$ defined by $e(y) = \big(y, f_U(y)\big)$. It is clear that the square

$$\begin{array}{ccc}
Y & \xrightarrow{\nu_Y} & U \circ_H (G\backslash U.Y) \\
{\scriptstyle e}\downarrow & & \downarrow{\scriptstyle \pi_{G\backslash U.Y}} \\
\tilde{Y} & \xrightarrow[\widetilde{\nu_Y}]{} & U \circ_H (G\backslash U.Y)
\end{array}$$

is cartesian, and then

$$\mathcal{L}_U(M)^*(\nu_Y)\lambda_{G\backslash U.Y}(m) = M^*(G\backslash U.\nu_Y)M^*(\eta_{G\backslash U.Y})(m)_{(Y,e)}$$

Thus

$$\mathcal{L}_U(M)_*(f_X)\mathcal{L}_U(M)^*(\nu_Y)\lambda_{G\backslash U.Y}(m) = M^*(G\backslash U.\nu_Y)M^*(\eta_{G\backslash U.Y})(m)_{(Y,\widetilde{f_X}e)}$$

But

$$\widetilde{f_X}e(y) = \widetilde{f_X}\big(y, f_U(y)\big) = \big(f_X(y), f_U(y)\big) = f(y)$$

so $\widetilde{f_X}e = f$. Moreover $\eta_{G\backslash U.Y}(G\backslash U.\nu_Y)$ is the identity map, because ν and η are the unit and co-unit of the adjunction of $Z \mapsto U \circ_H Z$ and $Y \mapsto G\backslash U.Y$. Finally

$$\mathcal{L}_U(M)_*(f_X)\mathcal{L}_U(M)^*(\nu_Y)\lambda_{G\backslash U.Y}(m) = m_{(Y,f)}$$

and $A \circ B(\psi)$ is the identity morphism of $\mathcal{L}_U(M)$. This completes the proof of theorem 9.5.2. ■

Remark: Let X be a G-set. The expression of the colimit over $\mathcal{D}_U(X)$ shows that if M is a Mackey functor for G, then

$$\mathcal{L}_U(M)(X) = \Big(\bigoplus_{Y \xrightarrow{f} X\times(U/H)} M(G\backslash U.Y)\Big)/\mathcal{J}$$

where $\mathcal{J}$ is the submodule generated by the submodule

$$M_*(a)\big(M(Z)\big)$$

whenever (Z,a) is a ν-disjoint H-set over $G\backslash U.Y$, and by the elements

$$m - M_*(G\backslash U.\alpha)(m)$$

whenever $m \in M(G\backslash U.Y)$ and $\alpha : (Y,f) \to (Y',f')$ is a morphism of G-sets over $X \times (U/H)$ which is injective on each G-orbit.

9.6 The functors $S_U(M)$

To build right adjoints, I need the following dual definition:

Definition: *If M is a Mackey functor for H, and (Y,f) is a G-set over Y/H, I set*

$$S_U(M)(Y,f) = \bigcap_{(Z,a)} \mathrm{Ker} M^*(a)$$

where the intersection runs over ν-disjoint H-sets (Z,a) over $G\backslash U.Y$.

Lemma 9.6.1: The correspondence

$$(Y,f) \mapsto S_U(M)(Y,f_U)$$

$$\alpha : (Y,f) \to (Z,g) \;\mapsto\; M^*(G\backslash U.\alpha) : M(G\backslash U.Z) \to M(G\backslash U.Y)$$

induces a contravariant functor from $\mathcal{D}_U(X)$ to R-Mod.

Proof: I have seen that if α is a morphism in $\mathcal{D}_U(X)$ from (Y,f) to (Z,g), and if (T,a) is a ν-disjoint H-set over $G\backslash U.Y$, then setting $\beta = G\backslash U.\alpha$, the H-set $(T,\beta a)$ is a ν-disjoint set over $G\backslash UZ$. Thus if $m \in \mathcal{S}_U(M)(Z,g)$, then $M^*(\beta a)(m) = 0 = M^*(a)M^*(\beta)(m)$, and this proves that $M^*(\beta)(m)$ is in $\mathcal{S}_U(M)(Y,f)$. ■

Lemma 9.6.2: If $\phi : X \to X'$ is a morphism of G-sets, then ϕ induces an isomorphism

$$S_\phi : \mathcal{S}_U(M) \circ \mathcal{D}_{U,*}(\phi) \to \mathcal{S}_U(M)$$

and a natural transformation

$$S^\phi : \mathcal{S}_U(M) \circ \mathcal{D}^*_U(\phi) \to \mathcal{S}_U(M)$$

Proof: The first assertion is clear: let (Y,f) be an object of $\mathcal{D}_U(X)$, and let $(Y',f') = \mathcal{D}_{U,*}(\phi)(Y,f) = (Y,\tilde{\phi}f)$. As $(\tilde{\phi}f)_U = f_U$, the sets $U.Y$ and $U.Y'$ coincide, and the ν-disjoint sets over $G\backslash U.Y$ and $G\backslash U.Y'$ are the same. Thus $\mathcal{S}_U(M)(Y,f) = \mathcal{S}_U(M)(Y',f')$.

For the second assertion, let (Y',f') be an object of $\mathcal{D}_U(X')$. I fill the cartesian square

$$\begin{array}{ccc} Y & \xrightarrow{a} & Y' \\ {\scriptstyle f}\downarrow & & \downarrow{\scriptstyle f'} \\ X & \xrightarrow[\tilde{\phi}]{} & X' \end{array}$$

Then if (Z',α') is a ν-disjoint H-set over $G\backslash UY'$, and if I fill the cartesian square

$$\begin{array}{ccc} Z & \xrightarrow{\alpha} & G\backslash U.Y \\ {\scriptstyle \beta}\downarrow & & \downarrow{\scriptstyle G\backslash U.a} \\ Z' & \xrightarrow[\alpha']{} & G\backslash U.Y' \end{array}$$

I have seen that (Z,α) is ν-disjoint over $G\backslash U.Y$. Then if $m \in \mathcal{S}_U(M)(Y,f)$, I have

$$M^*(\alpha')M_*(G\backslash U.a)(m) = M_*(\beta)M^*(\alpha)(m) = 0$$

and $M_*(G\backslash U.a)$ induces a morphism $S^\phi_{(Y',f')}$ from $\mathcal{S}_U(M)(Y,f) = \mathcal{S}_U(M)\circ\mathcal{D}^*_U(\phi)(Y',f')$ to $\mathcal{S}_U(M)(Y',f')$. It remains to see that this construction is functorial in (Y',f'): but if $\alpha' : (Y',f') \to (Z',g')$ is a morphism in $\mathcal{D}_U(X')$, then the diagram

$$\begin{array}{ccc} \mathcal{S}_U(M)\circ\mathcal{D}^*_U(\phi)(Y',f') & \xrightarrow{S^\phi_{(Y',f')}} & \mathcal{S}_U(M)(Y',f') \\ {\scriptstyle \mathcal{S}_U(M)\circ\mathcal{D}^*_U(\phi)(\alpha')}\downarrow & & \downarrow{\scriptstyle \mathcal{S}_U(M)(\alpha')} \\ \mathcal{S}_U(M)\circ\mathcal{D}^*_U(\phi)(Z',g') & \xrightarrow[S^\phi_{(Z',g')}]{} & \mathcal{S}_U(M)(Z',g') \end{array}$$

is commutative: indeed, this is equivalent to say that if $m \in \mathcal{S}_U(M)(Y,f)$, then

$$M^*(G\backslash U.\alpha')M_*(G\backslash U.a) = M_*(G\backslash U.b)M^*(G\backslash U.\alpha)$$

and this equality follows from lemma 9.3.3. This proves the second assertion. ■

9.7 The functors $\mathcal{R}_U(M)$

The previous section leads to the following definition, which is dual to the definition of $\mathcal{L}_U(M)$:

Definition: *If X is a G-set, and M a Mackey functor for H, I set*

$$\mathcal{R}_U(M)(X) = \lim_{\substack{\longleftarrow \\ (Y,f)\in\mathcal{D}_U(X)^{op}}} \mathcal{S}_U(M)(Y,f)$$

If $\phi : X \to X'$ is a morphism of G-sets, then the isomorphism S_ϕ induces a morphism

$$\lim_{\substack{\longleftarrow \\ (Y',f')\in\mathcal{D}_U(X')^{op}}} \mathcal{S}_U(M)(Y',f') \to \lim_{\substack{\longleftarrow \\ (Y,f)\in\mathcal{D}_U(X)^{op}}} \mathcal{S}_U(M)\circ\mathcal{D}_{U,*}(\phi)(Y,f) \to \lim_{\substack{\longleftarrow \\ (Y,f)\in\mathcal{D}_U(X)^{op}}} \mathcal{S}_U(M)(Y,f)$$

i.e. a morphism $\mathcal{R}_U(M)^*(\phi)$ from $\mathcal{R}_U(M)(X')$ to $\mathcal{R}_U(M)(X)$.

Conversely, the natural transformation S^ϕ induces a morphism

$$\lim_{\substack{\longleftarrow \\ (Y,f)\in\mathcal{D}_U(X)^{op}}} \mathcal{S}_U(M)(Y,f) \to \lim_{\substack{\longleftarrow \\ (Y',f')\in\mathcal{D}_U(X')^{op}}} \mathcal{S}_U(M)\circ\mathcal{D}_U^*(\phi)(Y',f') \to \lim_{\substack{\longleftarrow \\ (Y',f')\in\mathcal{D}_U(X')^{op}}} \mathcal{S}_U(M)(Y',f')$$

i.e. a morphism $\mathcal{R}_U(M)_*(\phi)$ from $\mathcal{R}_U(M)(X)$ to $\mathcal{R}_U(M)(X')$.

Proposition 9.7.1: The above definitions turn $\mathcal{R}_U(M)$ into a Mackey functor.

Proof: It is clear that $\mathcal{R}_U(M)$ is a bifunctor on G-**set**. It suffices then to check axioms (M1) and (M2) of Mackey functors. For (M2), let

$$\begin{array}{ccc} X & \xrightarrow{a} & Y \\ {\scriptstyle b}\downarrow & & \downarrow{\scriptstyle c} \\ Z & \xrightarrow[d]{} & T \end{array}$$

be a cartesian square of G-sets. The module $\mathcal{R}_U(M)(Z)$ identifies with the set of sequences $m_{(E,e)}$, indexed by the objects of $\mathcal{D}_U(Z)$, such that $m_{(E,e)} \in M(G\backslash U.E)$, and such that

$$M^*(G\backslash U.a)\big(m_{(E,e)}\big) = 0$$

whenever (T,a) is a ν-disjoint H-set over $G\backslash U.E$, and such that

$$M^*(G\backslash U.\alpha)(m_{(F,f)}) = m_{(E,e)}$$

if α is a morphism from (E,e) to (F,f) in the category $\mathcal{D}_U(X)$.

The image under $\mathcal{R}_U(M)^*(b)$ of the sequence $m_{(E,e)}$ is the sequence $m'_{(F,f)}$ indexed by the objects of $\mathcal{D}_U(X)$, and defined by

$$m'_{(F,f)} = m_{(F,\widetilde{bf})}$$

The image under $\mathcal{R}_U(M)_*(a)$ of this sequence is then the sequence $m''_{(F',f')}$ indexed by the objects of $\mathcal{D}_U(Y)$, and defined by filling the cartesian square

$$\begin{array}{ccc} F & \xrightarrow{\alpha} & F' \\ {\scriptstyle f}\downarrow & & \downarrow{\scriptstyle f'} \\ X & \xrightarrow[\tilde{a}]{} & Y \end{array}$$

and setting

$$m''_{(F',f')} = M_*(G\backslash U.\alpha)(m'_{(F,f)}) = M_*(G\backslash U.\alpha)(m_{(F,\tilde{b}f)})$$

On the other hand, the image under $\mathcal{R}_U(M)_*(d)$ of the sequence $m_{(E,e)}$ is the sequence $n'_{(E',e')}$ indexed by the objects of $\mathcal{D}_U(T)$, defined by filling the cartesian square

$$\begin{array}{ccc} E & \xrightarrow{\beta} & E' \\ {\scriptstyle e}\downarrow & & \downarrow{\scriptstyle e'} \\ Z & \xrightarrow[\tilde{d}]{} & T \end{array}$$

and setting

$$n'_{(E',e')} = M_*(G\backslash U.\beta)(m_{(E,e)})$$

The image under $\mathcal{R}_U(M)^*(c)$ of this sequence is the sequence $n''_{(F',f')}$ indexed by the objects of $\mathcal{D}_U(Y)$, defined by

$$n''_{(F',f')} = n'_{(F',\tilde{c}f')}$$

As the squares

$$\begin{array}{ccc} F & \xrightarrow{\alpha} & F' \\ {\scriptstyle \tilde{b}f}\downarrow & & \downarrow{\scriptstyle \tilde{c}f'} \\ Z & \xrightarrow[\tilde{d}]{} & T \end{array}$$

are cartesian, because they are composed of two cartesian squares, I have

$$n'_{(F',\tilde{c}f')} = M_*(G\backslash U.\alpha)(m_{(F,\tilde{b}f)})$$

so I have $\mathcal{R}_U(M)^*(c)\mathcal{R}_U(M)_*(d) = \mathcal{R}_U(M)_*(a)\mathcal{R}_U(M)^*(b)$, and (M2) holds.

To check (M1), I observe as before that an object (E,e) in $\mathcal{D}_U(X\coprod Y)$ is the disjoint union of an object (E_1,e_1) of $\mathcal{D}_U(X)$ and an object (E_2,e_2) of $\mathcal{D}_U(Y)$. Let i_1 and i_2 be the respective injections from E_1 and E_2 into E. If $m \in \mathcal{R}_U(X\coprod Y)$, and if (E,e) is an object of $\mathcal{D}_U(X\coprod Y)$, let $n = m_{(E,f)}$, and

$$n_1 = M^*(G\backslash U.i_1)(n) \qquad n_2 = M^*(G\backslash U.i_2)(n)$$

so that $n = M_*(G\backslash U.i_1)(n_1) + M_*(G\backslash U.i_2)(n_2)$.

As i_1 is injective, it is a morphism in $\mathcal{D}_U(X\coprod Y)$ from (E_1, fi_1) to (E, f). Thus $n_1 = m_{(E_1,fi_1)}$. But I have a cartesian square

$$\begin{array}{ccc} E_1 & \xrightarrow{i_1} & E \\ {\scriptstyle f_1}\downarrow & & \downarrow{\scriptstyle f} \\ X & \xrightarrow[\widetilde{i_X}]{} & X\coprod Y \end{array}$$

which proves that $fi_1 = \widetilde{i_X} f_1$, and that

$$\mathcal{R}_U(M)^*(i_X)(m)_{(E_1,f_1)} = m_{(E_1,\widetilde{i_X}f_1)} = m_{(E_1,fi_1)} = n_1$$

Similarly, for a suitable map f_2, I have

$$\mathcal{R}_U(M)^*(i_Y)(m)_{(E_2,f_2)} = n_2$$

Then if $\mathcal{R}_U(M)^*(i_X)(m) = 0$ and $\mathcal{R}_U(M)^*(i_Y)(m) = 0$, I have $n_1 = n_2 = 0$, and then $n = 0$ for any (E, e), so $m = 0$. Then the map

$$\mathcal{R}_U(M)^*(i_X) \oplus \mathcal{R}_U(M)^*(i_Y)$$

is injective, and lemma 9.4.2 shows that $\mathcal{R}_U(M)$ is a Mackey functor. ■

9.8 Right adjunction

Notation: *If $\theta : M \to M'$ is a morphism of Mackey functors for the group H, and if X is a G-set, I define a map $\mathcal{R}_U(\theta)_X$ from $\mathcal{R}_U(M)(X)$ to $\mathcal{R}_U(M')(X)$ by setting, for $m \in \mathcal{R}_U(M)(X)$, and for an object (Y, f) of $\mathcal{D}_U(X)$*

$$\mathcal{R}_U(\theta)_X(m)_{(Y,f)} = \theta_{G\backslash U.Y}(m_{(Y,f)})$$

Lemma 9.8.1: This definition turns $\mathcal{R}_U(\theta)$ into a morphism of Mackey functors from $\mathcal{R}_U(M)$ to $\mathcal{R}_U(M')$.

Proof: First, the map $\mathcal{R}_U(\theta)$ is well defined: if $a : Z \to G\backslash U.Y$ is ν-disjoint, then

$$M^*(a)\big(\mathcal{R}_U(\theta)_X(m)_{(Y,f)}\big) = M^*(a)\theta_{G\backslash U.Y}(m_{(Y,f)}) = \theta_{G\backslash U.Y}M^*(a)(m_{(Y,f)}) = 0$$

Moreover, if $\alpha : (Y, f) \to (Z, g)$ is a morphism in $\mathcal{D}_U(X)$, and if $m \in M(G\backslash U.Z)$, then

$$M^*(\alpha)\big(\mathcal{R}_U(\theta)_X(m)_{(Z,g)}\big) = M^*(\alpha)\theta_{G\backslash U.Z}(m_{(Z,g)}) = \dots$$

$$\dots = \theta_{G\backslash U.Y}M^*(\alpha)(m_{(Z,g)}) = \theta_{G\backslash U.Y}(m_{(Y,f)}) = \mathcal{R}_U(\theta)_X(m)_{(Y,f)}$$

Now let $\phi : X \to X'$ be a morphism of G-sets. Then, if $m' \in \mathcal{R}_U(M)(X')$,

$$\mathcal{R}_U(M)^*(\phi)\mathcal{R}_U(\theta)_{X'}(m')_{(Y,f)} = \mathcal{R}_U(\theta)_{X'}(m')_{(Y,\widetilde{\phi}f)} = \theta_{G\backslash U.Y}(m'_{(Y,\widetilde{\phi}f)}) = \dots$$

$$\dots = \theta_{G\backslash U.Y}\Big(\mathcal{R}_U(M)^*(\phi)(m')_{(Y,f)}\Big) = \Big(\mathcal{R}_U(\theta)_X\mathcal{R}_U(M)^*(\phi)(m')\Big)_{(Y,f)}$$

If moreover (Y', f') is an object of $\mathcal{D}_U(X')$, and if $m \in M(G\backslash U.Y')$, then let Y, f, and a fill the cartesian square

$$\begin{array}{ccc} Y & \xrightarrow{a} & Y' \\ {\scriptstyle f}\downarrow & & \downarrow{\scriptstyle f'} \\ X & \xrightarrow[\tilde{\phi}]{} & X' \end{array}$$

With those notations, I have

$$\mathcal{R}_U(M)_*(\phi)\mathcal{R}_U(\theta)_X(m)_{(Y',f')} = M_*(G\backslash U.a)\Big(\mathcal{R}_U(\theta)_X(m)_{(Y,f)}\Big) = \dots$$

$$\dots = M_*(G\backslash U.a)\theta_{G\backslash U.Y}(m_{(Y,f)}) = \theta_{G\backslash Y'}M_*(G\backslash U.a)(m_{(Y,f)}) = \dots$$

$$\dots = \theta_{G\backslash Y'}\Big(\mathcal{R}_U(M)(m)_{(Y',f')}\Big) = \mathcal{R}_U(\theta)_{X'}\mathcal{R}_U(M)(m)_{(Y',f')}$$

which proves the lemma. ∎

Theorem 9.8.2: Let G and H be finite groups, and U be a G-set-H. The functor $M \mapsto \mathcal{R}_U(M)$ is right adjoint to the functor $N \mapsto N \circ U$.

Proof: The proof is dual of the proof of theorem 9.5.2. Let θ be a morphism of Mackey functors from $M \circ U$ to N. Then for any H-set Z, I have a morphism

$$\theta_Z : (M \circ U)(Z) = M(U \circ_H Z) \to N(Z)$$

In particular, if (Y, f) is a G-set over $X \times (U/H)$, I have a morphism

$$\theta_{G\backslash U.Y} : M(U \circ_H G\backslash U.Y) \to N(G\backslash U.Y)$$

Composing this morphism with $M_*(\nu_Y)$, I have a morphism

$$\theta_{G\backslash U.Y}M_*(\nu_Y) : M(Y) \to M(U \circ_H G\backslash U.Y) \to N(G\backslash U.Y)$$

If (Z, a) is a ν-disjoint H-set over $G\backslash U.Y$, I have the diagram

$$\begin{array}{ccccc} M(Y) & \xrightarrow{M_*(\nu_Y)} & M(U \circ_H G\backslash U.Y) & \xrightarrow{\theta_{G\backslash U.Y}} & N(G\backslash U.Y) \\ \downarrow & & \downarrow{\scriptstyle M^*(U \circ_H a)} & & \downarrow{\scriptstyle N^*(a)} \\ M(\emptyset) & \longrightarrow & M(U \circ_H Z) & \xrightarrow[\theta_Z]{} & N(Z) \end{array}$$

This diagram is commutative: the left square is because (Z, a) is ν-disjoint, and the right square is because θ is a morphism of Mackey functors. So this gives a morphism from $M(Y)$ to $\mathcal{S}_U(N)(Y, f)$, that can be composed with the morphism $M^*(f_X)$ from $M(X)$ to $M(Y)$.

If $\alpha : (Y, f) \to (Z, g)$ is a morphism in $\mathcal{D}_U(X)$, I have the diagram

$$\begin{array}{ccccccc}
M(X) & \xrightarrow{M^*(g_X)} & M(Z) & \xrightarrow{M_*(\nu_Z)} & M(U \circ_H G\backslash U.Z) & \xrightarrow{\theta_{G\backslash U.Z}} & N(G\backslash U.Z) \\
{\scriptstyle Id}\downarrow & & \downarrow{\scriptstyle M^*(\alpha)} & & \downarrow{\scriptstyle M^*(U \circ_H G\backslash U.\alpha)} & & \downarrow{\scriptstyle N^*(G\backslash U.\alpha)} \\
M(X) & \xrightarrow[M^*(f_X)]{} & M(Y) & \xrightarrow[M_*(\nu_Y)]{} & M(U \circ_H G\backslash U.Y) & \xrightarrow[\theta_{G\backslash U.Y}]{} & N(G\backslash U.Y)
\end{array}$$

The left square is commutative because $g\alpha = f$. The middle square is commutative by lemma 9.3.2. The right square is commutative because θ is a morphism of Mackey functors.

So I have a morphism ψ_X from $M(X)$ to $\mathcal{R}_U(N)(X)$. If $\phi : X \to X'$ is a morphism of G-sets, and if $m \in M(X)$, then $\psi_X(m)$ is the sequence indexed by the objects (Y, f) of $\mathcal{D}_U(X)$ defined by

$$\psi_X(m)_{(Y,f)} = \theta_{G\backslash U.Y} M_*(\nu_Y) M^*(f_X)(m)$$

Then $\mathcal{R}_U(M)_*(\phi)\psi_X(m)$ is the sequence $m'_{(Y',f')}$ indexed by the objects of $\mathcal{D}_U(X')$, and defined by filling the cartesian square

$$\begin{array}{ccc}
Y & \xrightarrow{a} & Y' \\
{\scriptstyle f}\downarrow & & \downarrow{\scriptstyle f'} \\
X & \xrightarrow[\tilde{\phi}]{} & X'
\end{array}$$

More precisely

$$m'_{(Y',f')} = M_*(G\backslash U.a)\big(\psi_X(m)_{(Y,f)}\big) = N_*(G\backslash U.a)\theta_{G\backslash U.Y} M_*(\nu_Y) M^*(f_X)(m)$$

On the other hand, the element $\psi_{X'} M_*(\phi)(m)$ is the sequence $n_{(Y',f')}$ defined by

$$n_{(Y',f')} = \theta_{G\backslash U.Y'} M_*(\nu_{Y'}) M^*(f'_{X'}) M_*(\phi)$$

As the square

$$\begin{array}{ccc}
Y & \xrightarrow{a} & Y' \\
{\scriptstyle f_X}\downarrow & & \downarrow{\scriptstyle f'_{X'}} \\
X & \xrightarrow[\phi]{} & X'
\end{array}$$

is cartesian, I have

$$M^*(f'_{X'}) M_*(\phi) = M_*(a) M^*(f_X)$$

Moreover as $\nu_{Y'} \circ a = \big(U \circ_H (G\backslash U.a)\big) \circ \nu_Y$, I have

$$M_*(\nu_{Y'}) M_*(a) = M_*\big(U \circ_H (G\backslash U.a)\big) M_*(\nu_Y)$$

Finally as θ is a morphism of Mackey functors, I have

$$\theta_{G\backslash U.Y'} M_*\big(U \circ_H (G\backslash U.a)\big) = N_*(G\backslash U.a)\theta_{G\backslash U.Y}$$

and finally

$$n_{(Y',f')} = N_*(G\backslash U.a)\theta_{G\backslash U.Y} M_*(\nu_Y)M^*(f_X)(m) = m'_{(Y',f')}$$

This equality proves that

$$\mathcal{R}_U(M)_*(\phi)\psi_X = \psi_{X'}M_*(\phi)$$

Now if $m' \in M(X')$, then $\psi_{X'}(m')$ is the sequence indexed by the objects (Y', f') of $\mathcal{D}_U(X')$, defined by

$$\psi_{X'}(m')_{(Y',f')} = \theta_{G\backslash U.Y'} M_*(\nu_{Y'})M^*(f'_{X'})(m')$$

Its image under $\mathcal{R}_U(M)^*(\phi)$ is the sequence indexed by the objects (Y, f) of $\mathcal{D}_U(X)$, defined by

$$\Big(\mathcal{R}_U(M)^*(\phi)\psi_{X'}(m')\Big)_{(Y,f)} = \theta_{G\backslash U.Y} M_*(\nu_Y)M^*\Big((\tilde{\phi}f)_{X'}\Big)(m')$$

But $(\tilde{\phi}f)_{X'} = \phi f_X$, and then

$$\Big(\mathcal{R}_U(M)^*(\phi)\psi_{X'}(m')\Big)_{(Y,f)} = \theta_{G\backslash U.Y} M_*(\nu_Y)M^*(f_X)M^*(\phi)(m') = \psi_X\Big(M^*(\phi)(m')\Big)_{(Y,f)}$$

which proves that

$$\mathcal{R}_U(M)^*(\phi)\psi_{X'} = \psi_X M^*(\phi)$$

Thus ψ is a morphism of Mackey functors from M to $\mathcal{R}_U(M)$.

Conversely, if ψ is a morphism of Mackey functors from M to $\mathcal{R}_U(N)$, I have for any G-set X a morphism ψ_X from $M(X)$ to $\mathcal{R}_U(N)(X)$. In particular, if Z is an H-set, I have a morphism

$$\psi_{U\circ_H Z} : M(U \circ_H Z) \to \mathcal{R}_U(N)(U \circ_H Z)$$

An element of $\mathcal{R}_U(N)(U \circ_H Z)$ is a sequence $n_{(Y,f)}$ indexed by the objects of $\mathcal{D}_U(U \circ_H Z)$, with $n_{(Y,f)} \in N(G\backslash U.Y)$. I can then consider the element

$$n_{(U\circ_H Z,\pi_Z)} \in N\Big(G\backslash U.(U \circ_H Z)\Big)$$

and its image under $N_*(\eta_Z)$, which is an element of $N(Z)$. I get a morphism θ_Z from $M(U \circ_H Z) = (M \circ U)(Z)$ to $N(Z)$, defined by

$$\theta_Z(m) = N_*(\eta_Z)\Big(\psi_{U\circ_H Z}(m)_{(U\circ_H Z,\pi_Z)}\Big)$$

To prove that θ is a morphism of Mackey functors, it suffices to observe that θ is composed of the morphism $\psi \circ U$ from $M \circ U$ to $\mathcal{R}_U(M) \circ U$ deduced from ψ, and of the morphism Θ from $\mathcal{R}_U(N) \circ U$ to N, defined on the set Z by

$$n \in \Big(\mathcal{R}_U(N) \circ U\Big)(Z) = \mathcal{R}_U(N)(U \circ_H Z) \mapsto \Theta_Z(n) = N_*(\eta_Z)(n_{(U\circ_H Z,\pi_Z)})$$

It suffices then to prove that this is a morphism of Mackey functors. So let $\phi : Z \to Z'$ be a morphism of H-sets. The element

$$(\mathcal{R}_U(N) \circ U)_*(\phi)(n)_{(U\circ_H Z',\pi'_Z)}$$

is obtained by filling (trivially) the cartesian square

$$\begin{array}{ccc} U\circ_H Z & \xrightarrow{U\circ_H\phi} & U\circ_H Z' \\ \downarrow{\scriptstyle \pi_Z} & & \downarrow{\scriptstyle \pi_{Z'}} \\ \widetilde{U\circ_H Z} & \xrightarrow[\widetilde{U\circ_H\phi}]{} & \widetilde{U\circ_H Z'} \end{array}$$

Then:

$$(\mathcal{R}_U(N)\circ U)_*(\phi)(n)_{(U\circ_H Z',\pi'_Z)} = N_*\Big(G\backslash U.(U\circ_H\phi)\Big)\Big(n_{(U\circ_H Z,\pi_Z)}\Big)$$

so that

$$\Theta_{Z'}(\mathcal{R}_U(N)\circ U)_*(\phi)(n) = N_*(\eta_{Z'})N_*\Big(G\backslash U.(U\circ_H\phi)\Big)\Big(n_{(U\circ_H Z,\pi_Z)}\Big)$$

As $\eta_{Z'}\Big(G\backslash U.(U\circ_H\phi)\Big) = \phi\eta_Z$, this is also

$$\Theta_{Z'}\Big(\mathcal{R}_U(N)\circ U\Big)_*(\phi)(n) = N_*(\phi)N_*(\eta_Z)\Big(n_{(U\circ_H Z,\pi_Z)}\Big) = N_*(\phi)\Theta_Z(n)$$

thus $\Theta_{Z'}\Big(\mathcal{R}_U(N)\circ U\Big)_*(\phi) = N_*(\phi)\Theta_Z$.

Conversely, if $n'\in\mathcal{R}_U(N)(U\circ_H Z')$, then

$$N^*(\phi)\Theta_{Z'}(n') = N^*(\phi)N_*(\eta_{Z'})\Big(n'_{(U\circ_H Z',\pi'_Z)}\Big)$$

On the other hand $\Big(\mathcal{R}_U(N)\circ U\Big)^*(\phi)(n')$ is the sequence indexed by the objects (Y,f) of $\mathcal{D}_U(U\circ_H Z)$, defined by

$$\Big(\mathcal{R}_U(N)\circ U\Big)^*(\phi)(n')_{(Y,f)} = n'_{(Y,(\widetilde{U\circ_H\phi})f)}$$

It follows that

$$\Theta_Z\Big(\mathcal{R}_U(N)\circ U\Big)^*(\phi)(n') = N_*(\eta_Z)\Big(n'_{(U\circ_H Z,(\widetilde{U\circ_H\phi})\pi_Z)}\Big)$$

But I have already observed in the proof of theorem 9.5.2 that the morphism $U\circ_H\phi$ is a morphism in $\mathcal{D}_U(U\circ_H Z')$ from $\Big(U\circ_H Z,\Big((\widetilde{U\circ_H\phi})\pi_Z\Big)\Big)$ to $(U\circ_H Z',\pi_{Z'})$. Then in $\mathcal{R}_U(N)(U\circ_H Z')$, I have

$$n'_{(U\circ_H Z,(\widetilde{U\circ_H\phi})\pi_Z)} = N_*(U\circ_H\phi)\Big(n'_{(U\circ_H Z',\pi_{Z'})}\Big)$$

so that

$$\Theta_Z\Big(\mathcal{R}_U(N)\circ U\Big)^*(\phi)(n') = N_*(\eta_Z)N_*(U\circ_H\phi)\Big(n'_{(U\circ_H Z',\pi_{Z'})}\Big)$$

I have again the commutative diagram

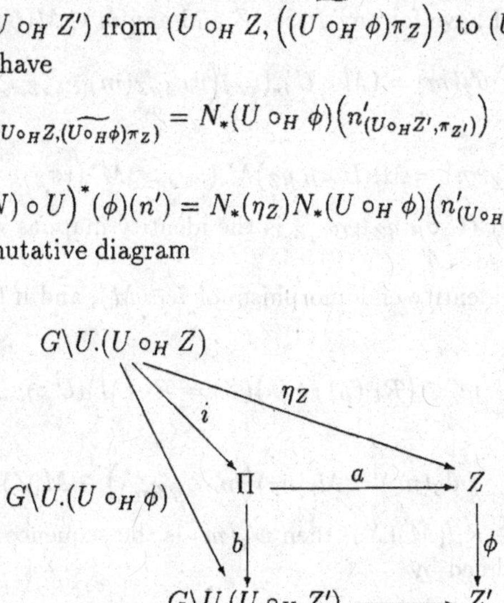

and lemma 9.5.3 shows that i is injective, and that if $\Pi = \mathrm{Im}(i) \coprod \Pi'$, denoting by j the injection from Π' into Π, the set (Π', bj) is ν-disjoint. I have then

$$N^*(\phi)N_*(\eta_{Z'})\Big(n'_{(U\circ_H Z',\pi'_Z)}\Big) = N_*(a)N^*(b)\Big(n'_{(U\circ_H Z',\pi'_Z)}\Big) = \dots$$

$$\dots = N_*(a)\Big(N_*(i)N^*(i)+N_*(j)N^*(j)\Big)N^*(b)\Big(n'_{(U\circ_H Z',\pi'_Z)}\Big) = \dots$$

$$\dots = N_*(ai)N^*(bi)\Big(n'_{(U\circ_H Z',\pi'_Z)}\Big) + N_*(aj)N^*(bj)\Big(n'_{(U\circ_H Z',\pi'_Z)}\Big)$$

As $n' \in \mathcal{R}_U(N)(U \circ_H Z')$, and as (Π', bj) is ν-disjoint, the second term is zero, and this gives

$$N^*(\phi)N_*(\eta_{Z'})\Big(n'_{(U\circ_H Z',\pi'_Z)}\Big) = N_*(ai)N^*(bi)\Big(n'_{(U\circ_H Z',\pi'_Z)}\Big) = \dots$$

$$\dots = N_*(\eta_Z)N^*\Big(G\backslash U.(U \circ_H \phi)\Big)\Big(n'_{(U\circ_H Z',\pi'_Z)}\Big)$$

Finally, I have proved that

$$\Theta_Z\Big(\mathcal{R}_U(N)\circ U\Big)^*(\phi) = N^*(\phi)\Theta_{Z'}$$

and Θ is a morphism of Mackey functors.

To complete the proof of the theorem, it remains to state that the correspondences $A : \theta \mapsto \psi$ and $B : \psi \mapsto \theta$, which are clearly functorial in M and N, are inverse to each other. It suffices to check that if θ is the identity, then so is $(B \circ A)(\theta)$, and that if ψ is the identity, so is $(A \circ B)(\psi)$.

Let θ be the identity endomorphism of $M \circ U$, and $\psi = A(\theta)$. If X is a G-set, the morphism ψ_X from $M(X)$ to $\mathcal{R}_U(M \circ U)(X)$ maps the element $m \in M(X)$ to the sequence $\psi_X(m)_{(Y,f)}$ defined by

$$\psi_X(m)_{(Y,f)} = M_*(\nu_Y)M^*(f_X)(m) \in M\big(U \circ_H (G\backslash U.Y)\big) = (M \circ U)(G\backslash U.Y)$$

Then if $\theta' = B(\psi)$, I have for any H-set Z and any $m \in M(U \circ_H Z)$

$$\theta'_Z(m) = (M \circ U)_*(\eta_Z)\Big(\psi_{U\circ_H Z}(m)_{(U\circ_H Z,\pi_Z)}\Big)$$

which gives

$$\theta'_Z(m) = M_*(U \circ_H \eta_Z)M_*(\nu_{U\circ_H Z})M^*\Big((\pi_Z)_X\Big)(m)$$

But I have seen that $(U \circ_H \eta_Z)\nu_{U\circ_H Z}$ is the identity map, as well as $(\pi_Z)_X$. Thus θ'_Z is the identity, and so is θ'.

Now if ψ is the identity endomorphism of $\mathcal{R}_U(M)$, and if $\theta = B(\psi)$, then for any H-set Z and any

$$m' \in \Big(\mathcal{R}_U(M) \circ U\Big)(Z) = \mathcal{R}_U(M)(U \circ_H Z)$$

I have

$$\theta_Z(m') = M_*(\eta_Z)\Big(m'_{(U\circ_H Z,\pi_Z)}\Big) \in M(Z)$$

If $\psi' = A(\theta)$, if $m \in \mathcal{R}_U(M)(X)$, then $\psi'_X(m)$ is the sequence indexed by the objects (Y, f) of $\mathcal{D}_U(X)$ defined by

$$\psi'_X(m)_{(Y,f)} = \theta_{G\backslash U.Y}\mathcal{R}_U(M)_*(\nu_Y)\mathcal{R}_U(M)^*(f_X)(m)$$

Setting $m' = \mathcal{R}_U(M)_*(\nu_Y)\mathcal{R}_U(M)^*(f_X)(m)$, I have then

$$\psi'_X(m)_{(Y,f)} = M_*(\eta_{G\backslash U.Y})\left[m'_{(U\circ_H(G\backslash U.Y),\pi_{G\backslash U.Y})}\right]$$

But $\mathcal{R}_U(M)^*(f_X)(m)$ is the sequence $n_{(E,e)}$ indexed by the objects of $\mathcal{D}_U(Y)$, defined by

$$n_{(E,e)} = m_{(E,\widetilde{f_X e})}$$

Moreover m' is the image of n under $\mathcal{R}_U(M)_*(\nu_Y)$. The component $m'_{(U\circ_H(G\backslash U.Y),\pi_{G\backslash U.Y})}$ is then obtained by observing that if e is the morphism from Y to $\tilde{Y}$ defined by $e(y) = \big((y, f_U(y)\big)$, then the square

$$\begin{array}{ccc} Y & \xrightarrow{\nu_Y} & U\circ_H(G\backslash U.Y) \\ {\scriptstyle e}\downarrow & & \downarrow{\scriptstyle \pi_{G\backslash U.Y}} \\ \tilde{Y} & \xrightarrow[\widetilde{\nu_Y}]{} & U\circ_H(G\backslash U.Y) \end{array}$$

is cartesian. Then

$$m'_{(U\circ_H(G\backslash U.Y),\pi_{G\backslash U.Y})} = M_*(G\backslash U.\nu_Y)\big(m_{(Y,\widetilde{f_X e})}\big)$$

As moreover $\widetilde{f_X}e = f$, I have then

$$\psi'_X(m)_{(Y,f)} = M_*(\eta_{G\backslash U.Y})M_*(G\backslash U.\nu_Y)(m_{(Y,f)})$$

As finally $\eta_{G\backslash U.Y}(G\backslash U.\nu_Y)$ is the identity map, I have $\psi'_X(m)_{(Y,f)} = m_{(Y,f)}$. Thus ψ' is the identity, which completes the proof of the theorem. ■

Remark: Let X be a G-set. The expression of the limit over $\mathcal{D}_U(X)$ shows that if M is a Mackey functor for G, then $\mathcal{R}_U(M)(X)$ is the set of sequences $m_{(Y,f)}$, indexed by G-sets (Y, f) over $X \times (U/H)$, such that $m_{(Y,f)} \in M(G\backslash U.Y)$, and

$$M^*(a)(m_{(Y,f)}) = 0$$

whenever (Z, a) is a ν-disjoint H-set over $G\backslash U.Y$, and moreover

$$m_{(Y,f)} = M^*(G\backslash U.\alpha)(m_{(Y',f')})$$

whenever $\alpha : (Y, f) \to (Y', f')$ is a morphism of G-sets over $X \times (U/H)$ which is injective on each G-orbit.

9.9 Examples

9.9.1 Induction and restriction

Let G be a group, and H be a subgroup of G. If U is the set G, viewed as a G-set-H, then the functor $N \mapsto N \circ U$ is the restriction functor for Mackey functors from G to H.

As $U/H = G/H$, an object (Y,f) over U/H is of the form $\mathrm{Ind}_H^G Z$, with $Z = f^{-1}(H)$ (see lemma 2.4.1). An object of $\mathcal{D}_U(X)$ is then an H-set Z, with a morphism from $\mathrm{Ind}_H^G Z$ to X, i.e. a morphism from Z to $\mathrm{Res}_H^G X$. Moreover, the group G acts freely on U, so on $U.Y$. Thus if α is a morphism of G-sets over $X \times (U/H)$, then $U.\alpha$ is injective on each left orbit. In other words, the category $\mathcal{D}_U(X)$ identifies with $H\text{-}\mathbf{set}\downarrow_{\mathrm{Res}_H^G X}$.

Moreover, the set $G\backslash U.Y$ identifies with Z, by the map

$$G(u,y) \in G\backslash U.Y \mapsto u^{-1}y \in Y$$

This map is indeed surjective, because if $z \in Z$, then z is the image of $G(1,z)$. Conversely, if $u^{-1}y = u'^{-1}y'$, then

$$G(u,y) = G(u, uu'^{-1}y') = G(1, u'^{-1}y') = G(u',y')$$

It follows that

$$U \circ_H (G\backslash U.Y) \simeq \mathrm{Ind}_H^G(G\backslash U.Y) \simeq \mathrm{Ind}_H^G Z \simeq Y$$

and that ν_Y is an isomorphism. Then if the H-set (T,a) over $G\backslash U.Y$ is ν-disjoint, the image of $U \circ_H a$ is disjoint of the image of ν_Y. As ν_Y is surjective, I have $U \circ_H T = \emptyset$. But $U \circ_H T = \mathrm{Ind}_H^G T$, and then $T = \emptyset$.

In particular, I see that

$$\mathcal{Q}_U(M)(Y,f) = M(G\backslash U.Y) = M(Z) = \mathcal{S}_U(M)(Y,f)$$

As $\mathrm{Res}_H^G X$ is a final object of $\mathcal{D}_U(X)$, I see that

$$\mathcal{L}_U(M)(X) = \varinjlim_{Z \in \mathcal{D}_U(X)} M(Z) = M(\mathrm{Res}_H^G X)$$

As $\mathrm{Res}_H^G X$ is an initial object of $\mathcal{D}_U(X)^{op}$, I have also

$$\mathcal{R}_U(M)(X) = M(\mathrm{Res}_H^G X)$$

and the following isomorphisms follow easily:

$$\mathcal{L}_U(M) \simeq \mathrm{Ind}_H^G(M) \simeq \mathcal{R}_U(M)$$

I recover that way the adjunction properties of induction and restriction.

Now switching the roles of H and G, I consider $V = G$ as an H-set-G. The functor $N \circ V$ is then the induction functor for Mackey functors from H to G. As $V/G = \bullet$, an H-set (Y,f) over V/G is just an H-set, and $V.Y = V \times Y$. The group H acts freely on V, so if α is a morphism of H-sets from Y to Z, then $V.\alpha$ is injective on each left orbit of H on $U.Y$ (because $h.u = u$ implies $h = 1$). The category $\mathcal{D}_V(X)$ identifies then with $H\text{-}\mathbf{set}\downarrow_X$, and has a final object X.

If Y is an H-set, then $H\backslash V.Y$ identifies with $\mathrm{Ind}_H^G Y$. Let (T,a) be a ν-disjoint G-set over $H\backslash V.Y$. If $t \in T$ and $a(t) = H(v,y)$, then as G acts freely on V, I have $(v,t) \in V \circ_G T$, which contradicts the hypothesis on (T,a). So $T = \emptyset$.

In those conditions, it is clear that

$$\mathcal{L}_V(M) \simeq \mathrm{Res}_H^G(M) \simeq \mathcal{R}_V(M)$$

and I recover once again the adjunction properties of induction and restriction.

9.9.2 Inflation

Let N be a normal subgroup of the group G, and H be the quotient G/N. If U is the set H, viewed as an H-set-G, then the functor $M \mapsto M \circ U$ is the inflation functor for Mackey functors: indeed, if X is a G-set, then $U \circ_G X = X^N$.

As $U/G = \bullet$, an object of $\mathcal{D}_U(X)$ is just an H-set over X. If (Y, f) is such a set, then $U.Y = U \times Y$, and $H\backslash U.Y$ identifies with $\mathrm{Inf}_H^G Y$. As H acts regularly on U, if $\alpha : (Y, f) \to (Z, g)$ is a morphism of sets over X, then $U.\alpha$ is injective on each orbit of H on $U.Y$. So X is a final object in $\mathcal{D}_U(X)$, and then for any X, I have

$$\mathcal{L}_U(M)(X) = \mathcal{Q}_U(M)(X) \qquad \mathcal{R}_U(M)(X) = \mathcal{S}_U(M)(X)$$

Furthermore as $U \circ_G (G\backslash U.Y) = (\mathrm{Inf}_H^G Y)^N \simeq Y$, the morphism ν_Y is an isomorphism for any Y. So an object (T, a) over $H\backslash U.Y = \mathrm{Inf}_H^G Y$ is ν-disjoint if and only if $U \circ_G T = T^N = \emptyset$. Finally:

Proposition 9.9.1: Let N be a normal subgroup of the group G, and $H = G/N$. If M is a Mackey functor for G, and X is an H-set, I set

$$M^N(X) = M(\mathrm{Inf}_H^G X) / \sum_{(T,a)} M_*(a)M(T) \qquad M_N(X) = \bigcap_{(T,a)} \mathrm{Ker} M^*(a)$$

where the sum and intersection run over the G-sets (T, a) over $\mathrm{Inf}_H^G X$ such that $T^N = \emptyset$.

If $\phi : X \to X'$ is a morphism of H-sets, then the maps $M_*(\mathrm{Inf}_H^G \phi)$ and $M^*(\mathrm{Inf}_H^G \phi)$ induce morphisms between $M^N(X)$ and $M^N(X')$, and between $M_N(X)$ and $M_N(X')$, which turn M^N and M_N into Mackey functors for H. The functor $M \mapsto M^N$ is left adjoint to the functor $L \mapsto \mathrm{Inf}_H^G L$, and the functor $M \mapsto M_N$ is right adjoint to it.

Remark: With the notations of Thévenaz and Webb (see [14], [15]), it is easy to identify M^N with M^+, and M_N with M^-: any set T such that $T^N = \emptyset$ is indeed isomorphic to a disjoint union of sets of the form G/K, for $N \not\subseteq K$, and it follows easily that if L/N is a subgroup of $H = G/N$

$$M^N(L/N) = M(L)/\sum_K t_K^L M(K) \qquad M_N(L/N) = \bigcap_K \mathrm{Ker}\, r_K^L$$

where the sum and intersection run on the subgroups K of L (which give morphisms from G/K to G/L) not containing N.

9.9.3 Coinflation

Let N be a normal subgroup of G, and $H = G/N$. Let V be the set H, viewed as a G-set-H. Then if Z is an H-set, the set $V \circ_H Z$ identifies with $\mathrm{Inf}_H^G Z$, and the functor $M \mapsto M \circ V$ is the functor that I have denoted by ρ_H^G (and denoted by $\beta^!$ by Thévenaz and Webb see[15].5).

Here again, the set V/H is trivial, so if X is a G-set, an object (Y, f) of $\mathcal{D}_V(X)$ is just a G-set over X. The set $V.Y$ is the product $V \times Y$. Let $\alpha : (Y, f) \to (Z, g)$ be a morphism of sets over X. Then $V.\alpha$ is injective on the left orbits of G on $V.Y$ if and only if the hypothesis

$$gv = v \qquad \alpha(gy) = \alpha(y)$$

imply $gy = y$. But the stabilizer in G of any point v of V is equal to N. Thus $V.\alpha$ is injective on the left orbits of G on $V.Y$ if and only if α is injective on the orbits of N on Y. A morphism in $\mathcal{D}_V(X)$ from (Y,f) to (Z,g) is then a morphism α of sets over X, which is moreover injective on each orbit of N.

Now the set $G\backslash V.Y$ identifies with $N\backslash Y$. Then $V \circ_H (G\backslash V.Y)$ identifies with $\mathrm{Inf}_H^G(N\backslash Y)$. The morphism ν_Y maps y to its orbit Ny by N. In particular, it is surjective. So an object (T,a) over $G\backslash V.Y$ is ν-disjoint if and only if $V \circ_H T = \mathrm{Inf}_H^G T = \emptyset$, i.e. if $T = \emptyset$. It follows that

$$\mathcal{Q}_V(M)(Y,f) = M(N\backslash Y) = \mathcal{S}_V(M)(Y,f)$$

Finally:

Proposition 9.9.2: Let N be a normal subgroup of G, and $H = G/N$. If V is the set H, viewed as a G-set-H, and if X is a G-set, then $\mathcal{D}_V(X)$ is isomorphic to the category which objects are the G-sets over X, and the morphisms are the morphisms of sets over X which are moreover injective on each orbit of N.

If M is a Mackey functor for H, then

$$\mathcal{L}_V(M)(X) = \varinjlim_{(Y,f)\in\mathcal{D}_V(X)} M(N\backslash Y) \qquad \mathcal{R}_V(M)(X) = \varprojlim_{(Y,f)\in\mathcal{D}_V(X)^{op}} M(N\backslash Y)$$

Notations: *Let K be a subgroup of G. I denote by $\omega_N(K)$ the set of subgroups of K, ordered by the following relation*

$$L \preceq L' \Leftrightarrow \begin{cases} L \subseteq L' \\ L\cap N = L'\cap N \end{cases}$$

If M is a Mackey functor for the group $H = G/N$, I denote by

$$\varinjlim_{L\in\omega_N(K)} M(LN/N)$$

the quotient of $\oplus_{L\subseteq K} M(LN/N)$ by the submodule generated by the elements of the form $t_{LN/N}^{L'N/N} m - m$, for $L \preceq L'$ and $m \in M(LN/N)$.

The group K acts on $\varinjlim_{L\in\omega_N(K)} M(LN/N)$, and I denote by

$$\iota_H^G(M)(K) = \Big(\varinjlim_{L\in\omega_N(K)} M(LN/N)\Big)_K$$

the biggest quotient on which K acts trivially. If L is a subgroup of K, and if m is an element of $M(LN/N)$, I denote by m_L^K the image of m in $\iota_H^G(M)(K)$.

If $K \subseteq K'$, I denote by $t_K^{K'}$ (resp. $r_K^{K'}$) the map from $\iota_H^G(M)(K)$ to $\iota_H^G(M)(K')$ (resp. from $\iota_H^G(M)(K')$ to $\iota_H^G(M)(K)$) defined by

$$t_K^{K'}(m_L^K) = m_L^{K'} \qquad r_K^{K'}(m'^{K'}_{L'}) = \sum_{x\in K\backslash K'/L'} \left({}^x r^{L'N/N}_{(K^x\cap L')N/N} m'\right)^K_{K\cap {}^xL'}$$

If $x \in G$, I set ${}^x(m_L^K) = ({}^x m)_{{}^xL}^{{}^xK}$.

Dually I denote by

$$\varprojlim_{L \in \omega_N(K)^{op}} M(LN/N)$$

the set of sequences m_L^K indexed by the subgroups of K, such that $m_L^K \in M(LN/N)$, and such that $r_{LN/N}^{L'N/N}(m_{L'}^K) = m_L^K$ whenever $L \preceq L'$. The group K acts on this set, and I denote by

$$\hat{\iota}_H^G(M)(K) = \Big(\varprojlim_{L \in \omega_N(K)^{op}} M(LN/N)\Big)^K$$

the set of its fixed points.

If $K \subseteq K'$, I denote by $t_K^{K'}$ (resp. $r_K^{K'}$) the map from $\hat{\iota}_H^G(M)(K)$ to $\hat{\iota}_H^G(M)(K')$ (resp. from $\hat{\iota}_H^G(M)(K')$ to $\hat{\iota}_H^G(M)(K)$) defined by

$$(t_K^{K'} m)_{L'}^{K'} = \sum_{x \in K\backslash K'/L'} t_{(L'\cap {}^xK)N/N}^{L'N/N}({}^x m_{L'^x \cap K}^K) \qquad (r_K^{K'} m')_L^{K'} = m'^K_L$$

Finally, if $x \in G$, and if m is an element of $\hat{\iota}_H^G(M)(K)$, I denote by ${}^x m$ the element of $\hat{\iota}_H^G(M)({}^xK)$ defined by $({}^x m)_L^{{}^xK} = {}^x(m_{L^x}^K)$.

Proposition 9.9.3: Those definitions turn $\iota_H^G(M)$ and $\hat{\iota}_H^G(M)$ into Mackey functors for G. The functor $M \mapsto \iota_H^G(M)$ is left adjoint to the functor $L \mapsto \rho(L)$, and the functor $M \mapsto \hat{\iota}_H^G(M)$ is right adjoint to it.

Proof: I have to identify $\mathcal{L}_V(M)(X)$ and $\mathcal{R}_V(M)(X)$ in the case $X = G/K$. If (Y, f) is a G-set over G/K, I can choose a system of representatives S of $G\backslash Y$ contained in $f^{-1}(K)$. If $s \in S$, then its stabilizer G_s is contained in K, and I have a natural map $\mu_s : gG_s \mapsto gs$ from G/G_s to Y. The union $\coprod_{s \in S} \mu_s$ is an isomorphism

$$\mu_Y : \coprod_{s \in S} G/G_s \to Y$$

If $m \in M(N\backslash Y)$, then $M^*(N\backslash\mu_s)(m) \in M\big(N\backslash(G/G_s)\big)$, and

$$N\backslash(G/G_s) \simeq (G/N)/(G_sN/N)$$

The map λ_Y defined by

$$\lambda_Y(m) = \sum_{s \in S} M^*(N\backslash\mu_s)(m)_{G_s}^K \in \iota_H^G(M)(K)$$

does not depend on the choice of the system S inside $f^{-1}(K)$: indeed, changing s in $g_s s$, so that $f(g_s s) = g_s f(s) \in K$ forces $g_s \in K$. Then $M^*(N\backslash\mu_s)(m)$ is replaced by its conjugate under g_s, which has the same image in $\iota_H^G(M)(K)$.

This map passes down to the quotient $\mathcal{L}_V(M)(G/K)$: if $\alpha : (Y, f) \to (Z, g)$ is a morphism of G-sets over G/K which is injective on each N-orbit, I choose a system of representatives T of $G\backslash Z$ contained in $g^{-1}(K)$. Then for any $s \in S$, there exists a unique $t_s \in T$ such that $\alpha(s) \in Gt_s$, i.e. $\alpha(s) = x_s t_s$, for $x_s \in G$. As $g\alpha = f$, I have $f(s) = x_s g(t_s)$, and then $x_s \in K$.

Moreover, I have $G_s \subseteq {}^{x_s}G_{t_s}$, and if $n \in {}^{x_s}G_{t_s} \cap N$, then

$$\alpha(ns) = n\alpha(s) = nx_s t_s = x_s . n^{x_s} t_s = x_s t_s = \alpha(s)$$

If α is injective on each N-orbit, I have $ns = s$, which proves that $N \cap {}^{x_s}G_{t_s} = N \cap G_s$, so

$$G_s \preceq {}^{x_s}G_{t_s}$$

Then let π_s be the map $gG_s \mapsto gx_sG_{t_s}$ from G/G_s to G/G_{t_s}, and π be the map

$$\coprod_{s \in S} \pi_s : \coprod_{s \in S} G/G_s \to \coprod_{t \in T} G/G_t$$

As

$$\alpha\mu_s(gG_s) = \alpha(gs) = gx_st_s = \mu_{t_s}(gx_sG_{t_s}) = \mu_Z\pi(gG_s)$$

The square

$$\begin{array}{ccc} \coprod_{s\in S} G/G_s & \xrightarrow{\mu_Y} & Y \\ {\scriptstyle \pi}\downarrow & & \downarrow{\scriptstyle \alpha} \\ \coprod_{t\in T} G/G_t & \xrightarrow[\mu_Z]{} & Z \end{array}$$

is commutative. As μ_Y and μ_Z are isomorphisms, it is cartesian. As moreover α is injective on N-orbits, by lemma 9.3.3 the square

$$\begin{array}{ccc} N\backslash \coprod_{s\in S} G/G_s & \xrightarrow{N\backslash\mu_Y} & N\backslash Y \\ {\scriptstyle N\backslash\pi}\downarrow & & \downarrow{\scriptstyle N\backslash\alpha} \\ N\backslash \coprod_{t\in T} G/G_t & \xrightarrow[N\backslash\mu_Z]{} & N\backslash Z \end{array}$$

is also cartesian. Then I have

$$M^*(N\backslash\mu_Z)M_*(N\backslash\alpha) = M_*(N\backslash\pi)M^*(N\backslash\mu_Y)$$

But if $m \in M(N\backslash Y)$, I have

$$M^*(N\backslash\mu_Y)(m) = \bigoplus_{s \in S} M^*(\mu_s)(m)$$

so

$$M_*(N\backslash\pi)M^*(N\backslash\mu_Y) = \bigoplus_{s \in S} \left[\left(t_{G_s}^{x_sG_{t_s}} M^*(N\backslash\mu_Y)(m)\right)^{x_s}\right]_{G_{t_s}}^K$$

As $x_s \in K$, this is also

$$\bigoplus_{s \in S} t_{G_s}^{x_sG_{t_s}} M^*(N\backslash\mu_Y)(m)_{x_sG_{t_s}}^K$$

and as $G_s \preceq {}^{x_s}G_{t_s}$, this is

$$\bigoplus_{s \in S} M^*(N\backslash\mu_Y)(m)_{G_s}^K$$

i.e. $\lambda_Y(m)$. Furthermore

$$M^*(N\backslash\mu_Z)M_*(N\backslash\alpha)(m) = \Big(\sum_{t \in T} M^*(\mu_t)\Big)M_*(N\backslash\alpha)(m) = \lambda_Z M_*(N\backslash\alpha)(m)$$

which proves that λ passes down to the quotient.

Conversely, to the element $m_L^K \in \iota_H^G(M)(K)$ I associate the image $\lambda'(m)$ of m in $\mathcal{L}_V(G/K)$ corresponding to the G-set $G/L \to G/K$ over G/K. This map is well defined, because if $L \preceq L'$ then the projection $G/L \to G/L'$ is injective on N-orbits: indeed, if $gL' = ngL'$ for $n \in N$, then $n^g \in L' \cap N = L \cap N$, so $gL = ngL$. Similarly, if $k \in K$, then the conjugation $G/L \to G/{}^kL$ is bijective, so it is injective on the N-orbits.

If $L \subseteq K$, I can take $\{L\}$ as a system of representatives of $G\backslash(G/L)$ for the computation of $\lambda_{G/L}$, and then it is clear that $\lambda\lambda'$ is the identity.

Conversely, the isomorphism $\mu_Y : \coprod_{s\in S} G/G_s \to Y$ is an isomorphism of sets over G/K, and it follows that $\lambda'\lambda$ is the identity. This states the isomorphism

$$\mathcal{L}_V(M)(G/K) \simeq \iota_H^G(M)(K)$$

The formulae giving the transfers and restrictions from $\iota_H^G(M)(K)$ to $\iota_H^G(M)(K')$ follow easily from isomorphisms λ and λ'. The isomorphism

$$\mathcal{R}_V(M)(G/K) \simeq \hat{\iota}_H^G(M)(K)$$

and the corresponding formulae can be proved by an argument dual to the previous one. ■

Remarks: 1) The set $\omega_N(K)$ is seldom connected: indeed if L and L' are in the same connected component, then $L \cap N = L' \cap N$. For a subgroup J of $K \cap N$, if I denote by $\omega_N(J, K)$ the set of subgroups L of K such that $L \cap N = J$, then $\omega_N(J, K)$ is the connected component of $\omega_N(K)$ containing J, and I have the isomorphism

$$\iota_H^G(M)(K) \simeq \bigoplus_{\substack{J \subseteq K\cap N \\ J \bmod. K}} \Big(\varinjlim_{L\in\omega_N(J,K)} M(LN/N)\Big)_{N_K(J)}$$

and a similar isomorphism for $\hat{\iota}(M)_H^G(K)$.

2) If $K \cap N = \{1\}$, then $\omega_N(K)$ has a biggest element K, and then

$$\iota_H^G(M)(K) = M(KN/N) = \hat{\iota}_H^G(M)(K)$$

It follows in particular that $\iota_H^G(M)(1) = M(N/N)$.

3) It is easy to see that for any Mackey functor L for H

$$\rho_H^G \mathrm{Inf}_H^G(L) \simeq L$$

This follows from the fact that, at the level of H-sets

$$(\mathrm{Inf}_H^G Z)^N \simeq Z$$

By adjunction, it follows that

$$\iota_H^G(L)^N \simeq L$$

That can be checked directly: let K be a subgroup of G containing N. If L is a subgroup of K not containing N, and if $m \in M(LN/N)$, the element m_L^K of $\iota_H^G(M)(K)$ is equal to $t_H^K(m_H^H)$. And if $H \supseteq N$, then $H \preceq K$, and then $m_H^K = t_{H/N}^{K/N}(m)_K^K$. So if π is the projection from $\iota_H^G(M)(K)$ onto its quotient $\iota_N^G(M)^N(K/N)$, the morphism

$$m \in M(K/N) \mapsto \pi(m_K^K) \in \iota_H^G(M)^N(K/N)$$

is surjective. Conversely, it is easy to see that the morphism

$$m_L^K \in \iota_H^G(M)(K) \mapsto \begin{cases} 0 \text{ if } N \not\subseteq L \\ t_{H/N}^{K/N}(m) \text{ otherwise} \end{cases}$$

passes down to the quotient $\iota_N^G(M)^N(K/N)$, and induces an isomorphism inverse of the previous one.
4) The associativity of the product $\circ_H$ and the adjunction show that if G, H, and K are groups, if U is a G-set-H, and V is an H-set-K, then

$$\mathcal{L}_U \circ \mathcal{L}_V \simeq \mathcal{L}_{U \circ_H V} \qquad \mathcal{R}_U \circ \mathcal{R}_V \simeq \mathcal{R}_{U \circ_H V}$$

5) The Burnside functor b for the group H is such that

$$\mathrm{Hom}_{Mack(H)}(b, N_Z) \simeq N(Z)$$

if N is a Mackey functor for H, and Z is an H-set. Then if U is a G-set-H, and M a Mackey functor for G

$$\mathrm{Hom}_{Mack(H)}\big(b, (M \circ U)_Z\big) \simeq M(U \circ_H Z) \simeq \mathrm{Hom}_{Mack(G)}(b, M_{U \circ_H Z}) \simeq \ldots$$

$$\ldots \simeq \mathrm{Hom}_{Mack(G)}(b_{U \circ_H Z}, M) \simeq \mathrm{Hom}_{Mack(H)}(b_Z, M \circ U) \simeq \ldots$$

$$\ldots \simeq \mathrm{Hom}_{Mack(G)}\big(\mathcal{L}_U(b_Z), M\big)$$

As this isomorphism is natural in M, it follows that

$$\mathcal{L}_U(b_Z) \simeq b_{U \circ_H Z}$$

where b in the left hand side is the Burnside functor for H, and in the right hand side the Burnside functor for G.

Chapter 10

Adjunction and Green functors

10.1 Frobenius morphisms

Let G and H be groups, and U be a G-set-H. If X and Y are H-sets, I have defined the maps

$$\delta^U_{X,Y} : U \circ_H (X \times Y) \to (U \circ_H X) \times (U \circ_H Y)$$

by

$$\delta^U_{X,Y}(u, x, y) = \big((u, x), (u, y)\big)$$

Lemma 10.1.1: If $f : X \to X'$ and $g : Y \to Y'$ are morphisms of H-sets, then the square

$$\begin{array}{ccc} U \circ_H (X \times Y) & \xrightarrow{\delta^U_{X,Y}} & (U \circ_H X) \times (U \circ_H Y) \\ {\scriptstyle U \circ_H (f \times g)} \downarrow & & \downarrow {\scriptstyle (U \circ_H f) \times (U \circ_H g)} \\ U \circ_H (X' \times Y') & \xrightarrow[\delta^U_{X',Y'}]{} & (U \circ_H X') \times (U \circ_H Y') \end{array} \qquad (C)$$

is cartesian.

Proof: Indeed, if

$$\delta_{X',Y'}\big(u, (x', y')\big) = \big((U \circ_H f) \times (U \circ_H g)\big)\big((u_1, x), (u_2, y)\big)$$

then in $U \circ_H X'$, I have $(u, x') = \big(u_1, f(x)\big)$, and in $U \circ_H Y'$ I have $(u, y') = \big(u_2, g(y)\big)$. Then there exists elements s and t of H such that

$$u.s = u_1 \qquad x' = s.f(x) \qquad u.t = u_2 \qquad y' = t.g(y)$$

In those conditions, the element $\big(u, (sx, ty)\big)$ is in $U \circ_H (X \times Y)$: indeed, if $r \in H$ is such that $u.r = r$, then $u_1 s^{-1} rs = u_1$. As $(u_1, x) \in U \circ_H X$, I have $s^{-1}rs.x = x$, so $r.sx = sx$. Similarly, since $u_2 t^{-1} rt = u_2$, and since $(u_2, t) \in U \circ_H Y$, I have $t^{-1}rt.y = y$, or $r.ty = ty$, which proves that $r.(sx, ty) = (sx, ty)$ if $u.r = u$.

Moreover

$$U \circ_H (f \times g)\big(u, (sx, ty)\big) = \big(u, (sf(x), tg(y))\big) = (u, x', y')$$

and

$$\delta^U_{X,Y}\big(u,(sx,ty)\big) = \big((u,sx),(u,ty)\big) = \big((us,x),(ut,y)\big) = \big((u_1,x),(u_2,y)\big)$$

The injectivity of $\delta^U_{X,Y}$ (see lemma 8.3.1) now shows that the square (C) is cartesian. The lemma follows. ■

Observing that if P is a Mackey functor for G, then

$$P\big(U\circ_H(X\times Y)\big) = (P\circ U)(X\times Y) = (P\circ U)_Y(X)$$

$$P(U\circ_H X\times U\circ_H Y) = P_{U\circ_H Y}(U\circ_H X) = (P_{U\circ_H Y}\circ U)(X)$$

lemma 10.1.1 shows that the maps

$$P_*(\delta^U_{X,Y}) : (P\circ U)_Y(X) \to (P_{U\circ_H Y}\circ U)(X)$$

and

$$P^*(\delta^U_{X,Y}) : (P_{U\circ_H Y}\circ U)(X) \to (P\circ U)_Y(X)$$

induce for any Y morphisms from $(P\circ U)_Y$ to $P_{U\circ_H Y}\circ U$. Those morphisms are moreover functorial in Y and P, in an obvious sense. Moreover, the injectivity of $\delta^U_{X,Y}$ shows that

$$P^*(\delta^U_{X,Y})P_*(\delta^U_{X,Y}) = Id$$

so $(P\circ U)_Y$ is a direct summand of $P_{U\circ_H Y}\circ U$.

Then if M is a Mackey functor for H, there are morphisms

$$\mathrm{Hom}_{Mack(H)}\big(M,(P\circ U)_Y\big) \rightleftarrows \mathrm{Hom}_{Mack(H)}(M, P_{U\circ_H Y}\circ U) \tag{10.1}$$

The left hand side is also

$$\mathrm{Hom}_{Mack(H)}(M_Y, P\circ U) \simeq \mathrm{Hom}_{Mack(G)}\big(\mathcal{L}_U(M_Y), P\big)$$

and the right hand side is

$$\mathrm{Hom}_{Mack(G)}\big(\mathcal{L}_U(M), P_{U\circ_H Y}\big) \simeq \mathrm{Hom}_{Mack(G)}\big(\mathcal{L}_U(M)_{U\circ_H Y}, P\big)$$

Thus the functor

$$\mathrm{Hom}_{Mack(G)}\big(\mathcal{L}_U(M_Y), -\big)$$

is a direct summand of the functor

$$\mathrm{Hom}_{Mack(G)}\big(\mathcal{L}_U(M)_{U\circ_H Y}, -\big)$$

Now Yoneda's lemma shows that $\mathcal{L}_U(M_Y)$ is a direct summand of $\mathcal{L}_U(M)_{U\circ_H Y}$.

The left hand side of (10.1) is also equal to $\mathcal{H}(M, P\circ U)(Y)$, and the right hand side is equal to

$$\mathrm{Hom}_{Mack(H)}(M, P_{U\circ_H Y}\circ U) \simeq \mathrm{Hom}_{Mack(G)}\big(\mathcal{L}_U(M), P_{U\circ_H Y}\big) = \ldots$$

$$\ldots = \mathcal{H}\big(\mathcal{L}_U(M), P\big)(U\circ_H Y) = \big[\mathcal{H}\big(\mathcal{L}_U(M), P\big)\circ U\big](Y)$$

The functoriality in Y shows that the above morphisms induce morphisms of Mackey functors

$$\Delta_{M,P} : \mathcal{H}(M, P \circ U) \to \mathcal{H}\big(\mathcal{L}_U(M), P\big) \circ U$$

and

$$\nabla_{M,P} : \mathcal{H}\big(\mathcal{L}_U(M), P\big) \circ U \to \mathcal{H}(M, P \circ U)$$

such that $\nabla_{M,P}\Delta_{M,P} = Id$. Those morphisms are moreover functorial in M and P.

If N is a Mackey functor for H, then by composition I have morphisms

$$\mathrm{Hom}_{Mack(H)}\big(N, \mathcal{H}(M, P \circ U)\big) \rightleftarrows \mathrm{Hom}_{Mack(H)}\left[N, \mathcal{H}\big(\mathcal{L}_U(M), P\big) \circ U\right]$$

By proposition 1.10.1, the left hand side is

$$\mathrm{Hom}_{Mack(H)}(M\hat{\otimes}N, P \circ U) \simeq \mathrm{Hom}_{Mack(G)}\big(\mathcal{L}_U(M\hat{\otimes}N), P\big)$$

and the right hand side is

$$\mathrm{Hom}_{Mack(G)}\big(\mathcal{L}_U(N), \mathcal{H}(\mathcal{L}_U(M), P)\big) \simeq \mathrm{Hom}_{Mack(G)}\big(\mathcal{L}_U(M)\hat{\otimes}\mathcal{L}_U(N), P\big)$$

Thus the functor

$$\mathrm{Hom}_{Mack(G)}\big(\mathcal{L}_U(M\hat{\otimes}N), -\big)$$

is a direct summand of the functor

$$\mathrm{Hom}_{Mack(G)}\big(\mathcal{L}_U(M)\hat{\otimes}\mathcal{L}_U(N), -\big)$$

Now Yoneda's lemma shows that the functor $\mathcal{L}_U(M\hat{\otimes}N)$ is a direct summand of the functor $\mathcal{L}_U(M)\hat{\otimes}\mathcal{L}_U(N)$.

Furthermore I have morphisms

$$\mathrm{Hom}_{Mack(H)}\big((P \circ U)_Y, M\big) \rightleftarrows \mathrm{Hom}_{Mack(H)}(P_{U\circ_H Y} \circ U, M) \qquad (10.2)$$

The left hand side is also

$$\mathrm{Hom}_{Mack(H)}(P \circ U, M_Y) \simeq \mathrm{Hom}_{Mack(G)}\big(P, \mathcal{R}_U(M_Y)\big)$$

and the right hand side is

$$\mathrm{Hom}_{Mack(G)}\big(P_{U\circ_H Y}, \mathcal{R}_U(M)\big) \simeq \mathrm{Hom}_{Mack(G)}\big(P, \mathcal{R}_U(M)_{U\circ_H Y}\big)$$

It follows that $\mathcal{R}_U(M_Y)$ is a direct summand of $\mathcal{R}_U(M)_{U\circ_H Y}$.

The left hand side of (10.2) can be written as

$$\mathrm{Hom}_{Mack(H)}(P \circ U, M_Y) = \mathcal{H}(P \circ U, M)(Y)$$

and the right hand side as

$$\mathrm{Hom}_{Mack(G)}\big(P_{U\circ_H Y}, \mathcal{R}_U(M)\big) \simeq \mathrm{Hom}_{Mack(G)}\big(P, \mathcal{R}_U(M)_{U\circ_H Y}\big) = \ldots$$

$$\ldots = \mathcal{H}\big(P, \mathcal{R}_U(M)\big)(U \circ_H Y) = \left[\mathcal{H}\big(P, \mathcal{R}_U(M)\big) \circ U\right](Y)$$

So $\mathcal{H}(P \circ U, M)$ is a direct summand of $\Big[\mathcal{H}\Big(P, \mathcal{R}_U(M)\Big)\Big] \circ U$. Then

$$\mathrm{Hom}_{Mack(H)}\Big(N, \mathcal{H}(P \circ U, M)\Big) \simeq \mathrm{Hom}_{Mack(H)}\Big((P \circ U)\hat{\otimes} N, M\Big)$$

is a direct summand of

$$\mathrm{Hom}_{Mack(H)}\Big(N, \mathcal{H}\Big(P, \mathcal{R}_U(M)\Big) \circ U\Big) \simeq \mathrm{Hom}_{Mack(G)}\Big(\mathcal{L}_U(N), \mathcal{H}\Big(P, \mathcal{R}_U(M)\Big)\Big) \simeq \ldots$$

$$\ldots \simeq \mathrm{Hom}_{Mack(G)}\Big(P\hat{\otimes}\mathcal{L}_U(M), \mathcal{R}_U(M)\Big) \simeq \mathrm{Hom}_{Mack(H)}\Big([P\hat{\otimes}\mathcal{L}_U(N)] \circ U, M\Big)$$

It follows that $(P \circ U)\hat{\otimes} N$ is a direct summand of $[P\hat{\otimes}\mathcal{L}_U(N)] \circ U$.

Then denoting by $M \mid N$ the relation "M is a direct summand of N", I have also

$$N\hat{\otimes}(P \circ U) \mid [\mathcal{L}_U(N)\hat{\otimes} P] \circ U$$

So for any M

$$\mathrm{Hom}_{Mack(H)}\Big(N\hat{\otimes}(P \circ U), M\Big) \mid \mathrm{Hom}_{Mack(H)}\Big([\mathcal{L}_U(N)\hat{\otimes} P] \circ U, M\Big)$$

which can be written as

$$\mathrm{Hom}_{Mack(H)}\Big(P \circ U, \mathcal{H}(N, M)\Big) \mid \mathrm{Hom}_{Mack(G)}\Big(\mathcal{L}_U(N)\hat{\otimes} P, \mathcal{R}_U(M)\Big)$$

or

$$\mathrm{Hom}_{Mack(G)}\Big[P, \mathcal{R}_U\Big(\mathcal{H}(N, M)\Big)\Big] \mid \mathrm{Hom}_{Mack(G)}\Big[P, \mathcal{H}\Big(\mathcal{L}_U(N), \mathcal{R}_U(M)\Big)\Big]$$

It follows that $\mathcal{R}_U\Big(\mathcal{H}(N, M)\Big)$ is a direct summand of $\mathcal{H}\Big(\mathcal{L}_U(N), \mathcal{R}_U(M)\Big)$.

In the case $U/H = \bullet$ (i.e. when H is transitive on U), as the square

$$\begin{array}{ccc}
U \circ_H (X \times Y) & \xrightarrow{U \circ_H \binom{xy}{x}} & U \circ_H X \\
{\scriptstyle U \circ_H \binom{xy}{y}} \downarrow & & \downarrow {\scriptstyle U \circ_H \binom{x}{\bullet}} \\
U \circ_H Y & \xrightarrow[U \circ_H \binom{y}{\bullet}]{} & U \circ_H \bullet \simeq \bullet
\end{array}$$

is cartesian, it follows that

$$U \circ_H (X \times Y) \simeq (U \circ_H X) \times (U \circ_H Y)$$

so $\delta^U_{X,Y}$ is an isomorphism. Then all the previous split monomorphisms are isomorphisms.

Finally, I have proved the

Proposition 10.1.2: Let G and H be groups, and U be a G-set-H. If P is a Mackey functor for G, if M and N are Mackey functors for H, and if Y is an H-set, then

$$(P \circ U)_Y \mid P_{U \circ_H Y} \circ U$$

$$\mathcal{L}_U(M_Y) \mid \mathcal{L}_U(M)_{U \circ_H Y} \qquad \mathcal{R}_U(M_Y) \mid \mathcal{R}_U(M)_{U \circ_H Y}$$

$$\mathcal{H}(M, P\circ U) \mid \mathcal{H}\big(\mathcal{L}_U(M), P\big)\circ U \qquad \mathcal{H}(P\circ U, M) \mid \mathcal{H}\big(P, \mathcal{R}_U(M)\big)\circ U$$

$$\mathcal{L}_U(M\hat{\otimes}N) \mid \mathcal{L}_U(M)\hat{\otimes}\mathcal{L}_U(N)$$

$$M\hat{\otimes}(P\circ U) \mid \big(\mathcal{L}_U(M)\hat{\otimes}P\big)\circ U$$

$$\mathcal{R}_U\big(\mathcal{H}(N,M)\big) \mid \mathcal{H}\big(\mathcal{L}_U(N), \mathcal{R}_U(M)\big)$$

Moreover, if $U/H = \bullet$, all these split monomorphisms are isomorphisms.

Example: In the case when H is a subgroup of G, if U is the set G, viewed as an H-set-G, I have $U/G = \bullet$ (the roles of H and G have to be switched in the proposition). The functor $-\circ U$ is the induction functor for Mackey functors, and the functors $\mathcal{L}_U$ and $\mathcal{R}_U$ are equal to the restriction functor. The previous isomorphisms give

$$(\mathrm{Ind}_H^G P)_Y \simeq \mathrm{Ind}_H^G(M_{\mathrm{Res}_H^G Y})$$

$$\mathrm{Res}_H^G(M_Y) \simeq (\mathrm{Res}_H^G M)_{\mathrm{Res}_H^G Y}$$

$$\mathcal{H}(M, \mathrm{Ind}_H^G P) \simeq \mathrm{Ind}_H^G \mathcal{H}(\mathrm{Res}_H^G M, P) \qquad \mathcal{H}(\mathrm{Ind}_H^G P, M) \simeq \mathrm{Ind}_H^G \mathcal{H}(P, \mathrm{Res}_H^G M)$$

$$\mathrm{Res}_H^G(M\hat{\otimes}N) \simeq \mathrm{Res}_H^G M\hat{\otimes}\mathrm{Res}_H^G N$$

$$M\hat{\otimes}\mathrm{Ind}_H^G P \simeq \mathrm{Ind}_H^G\big(\mathrm{Res}_H^G(M)\hat{\otimes}N\big)$$

$$\mathrm{Res}_H^G \mathcal{H}(N,M) \simeq \mathcal{H}(\mathrm{Res}_H^G N, \mathrm{Res}_H^G N)$$

The last but one of these relations explains the name of "Frobenius morphisms" for this section.

10.2 Left adjoints and tensor product

Let G and H be groups, and U be a G-set-H. If M and N are Mackey functors for H, proposition 10.1.2 shows that there are natural morphisms

$$\mathcal{L}_U(M)\hat{\otimes}\mathcal{L}_U(N) \to \mathcal{L}_U(M\hat{\otimes}N)$$

I will describe those morphisms explicitly. Let P be a Mackey functor for G. I have the following diagram

$$\begin{array}{ccc}
\mathrm{Hom}_{Mack(H)}\big(N, \mathcal{H}(M, P\circ U)\big) & \xrightarrow{\Phi} & \mathrm{Hom}_{Mack(H)}\big[N, \mathcal{H}\big(\mathcal{L}_U(M), P\big)\circ U\big] \\
\simeq \uparrow & & \downarrow \simeq \\
\mathrm{Hom}_{Mack(H)}(M\hat{\otimes}N, P\circ U) & & \mathrm{Hom}_{Mack(G)}\big[\mathcal{L}_U(N), \mathcal{H}\big(\mathcal{L}_U(M), P\big)\big] \\
\simeq \uparrow & & \downarrow \simeq \\
\mathrm{Hom}_{Mack(G)}\big(\mathcal{L}_U(M\hat{\otimes}N), P\big) & \xrightarrow[\Theta]{} & \mathrm{Hom}_{Mack(G)}\big(\mathcal{L}_U(M)\hat{\otimes}\mathcal{L}_U(N), P\big)
\end{array}$$

where $\Phi = \mathrm{Hom}_{Mack(H)}(N, \Delta_{M,P})$, and Θ is the map obtained by composition of Φ with the isomorphisms of the diagram. Then the morphism I am looking for is obtained by taking for P the functor $\mathcal{L}_U(M\hat{\otimes}N)$: it is the image under Θ of the identity map of $\mathcal{L}_U(M\hat{\otimes}N)$.

By adjunction, I have the morphism

$$\lambda_{M\hat{\otimes}N} : M\hat{\otimes}N \to \mathcal{L}_U(M\hat{\otimes}N)\circ U$$

that I already used in the proof of theorem theorem 9.5.2. It is defined for an H-set V by

$$t \in (M\hat{\otimes}N)(V) \mapsto (M\hat{\otimes}N)^*(\eta_V)(t)_{(U\circ_H V,\pi_V)}$$

The associated bilinear morphism is then defined for H-sets V and W by

$$m \in M(V),\ n \in N(W) \mapsto (M\hat{\otimes}N)^*(\eta_{V\times W})\Big(\omega_{V,W}(m,n)\Big)_{(U\circ_H(V\times W),\pi_{V\times W})}$$

where I set

$$\omega_{V,W}(m,n) = \left[M^*\binom{vw}{v}(m)\otimes N^*\binom{vw}{w}(n)\right]_{(V\times W,Id)} \in (M\hat{\otimes}N)(V\times W)$$

Now this gives through Φ a morphism from N to $\mathcal{H}\Big(\mathcal{L}_U(M),P\Big)\circ U$: it maps $n\in N(W)$ to the element

$$p \in \Big[\mathcal{H}\Big(\mathcal{L}_U(M),P\Big)\circ U\Big](W) = \mathrm{Hom}_{Mack(G)}\Big(\mathcal{L}_U(M),P_{U\circ_H W}\Big)$$

obtained by adjunction from the element

$$q \in \mathrm{Hom}_{Mack(H)}\Big(M,P_{U\circ_H W}\circ U\Big)$$

which maps the element $m\in M(V)$ to the element $r_{V,W}(m,n)$ of $\mathcal{L}_U(M\hat{\otimes}N)(U\circ_H V\times U\circ_H W)$ defined by

$$r_{V,W}(m,n) = P_*(\delta^U_{V,W})\Big[(M\hat{\otimes}N)^*(\eta_{V\times W})\Big(\omega_{V,W}(m,n)\Big)_{(U\circ_H(V\times W),\pi_{V\times W})}\Big]$$

Now the definition of $P_*(\delta^U_{V,W})$ for $P=\mathcal{L}_U(M\hat{\otimes}N)$ shows that

$$r_{V,W}(m,n) = (M\hat{\otimes}N)^*(\eta_{V\times W})\Big(\omega_{V,W}(m,n)\Big)_{(U\circ_H(V\times W),\widetilde{\delta^U_{V,W}\pi_{V\times W}})}$$

Then if X is a G-set, and (S,f) is an object of $\mathcal{D}_U(X)$, the morphism p maps the element $m\in M(G\backslash U.S)$ to

$$P_*(f_X\times Id_{U\circ_H W})P^*(\nu_{(S,f)}\times Id_{U\circ_H W})r_{G\backslash U.S,W}(m,n) \in P_{U\circ_H W}(X)$$

This morphism from N to $\mathcal{H}\Big(\mathcal{L}_U(M),P\Big)\circ U$ gives by adjunction a morphism from $\mathcal{L}_U(N)$ to $\mathcal{H}\Big(\mathcal{L}_U(M),P\Big)$, or a bilinear morphism from $\mathcal{L}_U(M)$, $\mathcal{L}_U(N)$ to P, defined as follows: if Y is a G-set, if (T,g) is an object of $\mathcal{D}_U(Y)$, and if $n\in N(G\backslash U.T)$, then the image of $m_{(S,f)}$, $n_{(T,g)}$ is the element

$$P_*(Id_X\times g_Y)P^*(Id_X\times\nu_{(T,g)})P_*(f_X\times Id_{U\circ_H(G\backslash U.T)})P^*(\nu_{(S,f)}\times Id_{U\circ_H(G\backslash U.T)})r_{G\backslash U.S,G\backslash U.T}(m,n)$$

of $P(X \times Y)$. As the square

$$\begin{array}{ccc}
S \times T & \xrightarrow{Id_S \times \nu_{(T,g)}} & S \times U \circ_H (G\backslash U.T) \\
{\scriptstyle f_X \times Id_T}\downarrow & & \downarrow{\scriptstyle f_X \times Id_{U \circ_H (G\backslash U.T)}} \\
X \times T & \xrightarrow[Id_X \times \nu_{(T,g)}]{} & X \times U \circ_H (G\backslash U.T)
\end{array}$$

is cartesian, it is also

$$P_*(Id_X \times g_Y)P_*(f_X \times Id_T)P^*(Id_S \times \nu_{(T,g)})P^*(\nu_{(S,f)} \times Id_{U \circ_H (G\backslash U.T)})r_{G\backslash U.S, G\backslash U.T}(m,n)$$

or

$$P_*(f_X \times g_Y)P^*(\nu_{(S,f)} \times \nu_{(T,g)})r_{G\backslash U.S, G\backslash U.T}(m,n)$$

To compute this expression, I must fill the cartesian square

$$\begin{array}{ccc}
F & \xrightarrow{a} & U \circ_H [(G\backslash U.S) \times (G\backslash U.T)] \\
{\scriptstyle b}\downarrow & & \downarrow{\scriptstyle d} \\
S \widetilde{\times} T & \xrightarrow[\nu_{(S,f)} \widetilde{\times} \nu_{(T,g)}]{} & U \circ_H [(G\backslash U.S) \times (G\backslash U.T)]
\end{array}$$

where I denote by d the map $(\widetilde{\delta^U_{G\backslash U.S, G\backslash U.T}})\pi_{G\backslash U.S, G\backslash U.T}$. Let $(s,t,u"H) \in S \widetilde{\times} T$, and u, $u' \in U$ such that $f_U(s) = uH$ and $g_U(t) = u'H$. Then

$$(\nu_{(S,f)} \widetilde{\times} \nu_{(T,g)})(s,t,u"H) = \Big((u, G(u,s)), (u', G(u',t)\Big)$$

Furthermore, let $\Big(u_1, G(u_2, s_0), G(u_3, t_0)\Big) \in U \circ_H [(G\backslash U.S) \times (G\backslash U.T)]$. Its image under d is equal to

$$\Big((u_1, G(u_2, s_0)), (u_1, G(u_3, t_0))\Big)$$

It is equal to $(\nu_{(S,f)} \widetilde{\times} \nu_{(T,g)})(s,t,u"H)$ if and only if

$$\Big(u, G(u,s)\Big) = \Big(u_1, G(u_2, s_0)\Big) \qquad \Big(u', G(u',t)\Big) = \Big(u_1, G(u_3, t_0)\Big) \qquad u"H = u_1 H$$

This is equivalent to say that there exists elements h and h' in H such that

$$u_1 = uh \qquad G(u_2, s_0) = G(uh, s) \qquad u_1 = u'h' \qquad G(u_3, t_0) = G(u'h, t) \qquad u"H = u_1 H$$

In those conditions, I have

$$uH = u_1 H = u'H = u"H$$

and then $f_U(s) = g_U(t)$. Thus (s,t) is in the pull-back of S and T over U/H.

Notations: *If (S,f) and (T,g) are G-sets over U/H, I denote by $S.T$ their pull-back over U/H. If $s \in S$ and $t \in T$ are such that $(s,t) \in S.T$, I denote by $s.t$ the couple (s,t), and $G.s.t$ its orbit by G. If X and Y are G-sets, if (S,f) is a G-set over*

$\widetilde{X} = X \times (U/H)$, and (T,g) is a G-set over $\tilde{Y} = Y \times (U/H)$, I denote by $f.g$ the map from $S.T$ to $X \times Y \times (U/H)$ defined by

$$(f.g)(s.t) = \big(f_X(s), g_Y(t), f_U(s)\big) = \big(f_X(s), g_Y(t), g_U(t)\big)$$

With these notations, I can define a map

$$\alpha : S.T \to U \circ_H [(G\backslash U.S) \times (G\backslash U.T)]$$

by setting

$$\alpha(s.t) = \big(u, G.u.s, G.u.t\big) \quad \text{if} \quad f_U(s) = g_U(t) = uH$$

Indeed, if $h \in H$ is such that $uh = u$, then

$$h(G.u.s, G.u.t) = (G.uh^{-1}.s, G.uh^{-1}.t) = (G.u.s, G.u.t)$$

Thus $\alpha(s.t) \in U \circ_H [(G\backslash U.S) \times (G\backslash U.T)]$. And if I replace u by uh, then

$$(uh, G.uh.s, G.uh.t) = (uh, h^{-1}G.u.s, h^{-1}G.u.t) = (u, G.u.s, G.u.t)$$

I have also a map β from $S.T$ to $S \widetilde{\times} T$, defined by

$$\beta(s.t) = \big(s, t, f_U(s)\big) = \big(s, t, g_U(t)\big)$$

It is clear that the square

$$\begin{array}{ccc}
S.T & \xrightarrow{\alpha} & U \circ_H [(G\backslash U.S) \times (G\backslash U.T)] \\
{\scriptstyle\beta}\downarrow & & \downarrow{\scriptstyle d} \\
S \widetilde{\times} T & \xrightarrow[\nu_{(S,f)} \widetilde{\times} \nu_{(T,g)}]{} & U \circ_H [(G\backslash U.S) \times (G\backslash U.T)]
\end{array}$$

is commutative. The previous argument shows that the associated map i from $S.T$ to F is surjective. As β is injective, and factors through i, it follows that i is injective. Thus F is isomorphic to $S.T$, and the above square is cartesian.

In those conditions, I have

$$P^*(\nu_{(S,f)} \times \nu_{(T,g)}) r_{G\backslash U.S, G\backslash U.T}(m,n) = \ldots$$

$$\ldots = (M\hat{\otimes}N)^*(G\backslash U.\alpha)(M\hat{\otimes}N)^*(\eta_{G\backslash U.S \times G\backslash U.T})\omega_{G\backslash U.S, G\backslash U.T}(m,n)$$

Furthermore, if $G.u.s.t \in G\backslash U.S.T$, then

$$\eta_{G\backslash U.S \times G\backslash U.T} \circ G\backslash U.\alpha(G.u.s.t) = \eta_{G\backslash U.S \times G\backslash U.T}\big(G.u.(u, G.u.s, G.u.t)\big) = (G.u.s, G.u.t)$$

Notation: *If (S,f) and (T,g) are G-sets over U/H, I denote by $\kappa^U_{S,T}$ the map*

$$\kappa^U_{S,T} : G\backslash U.(S.T) \to (G\backslash U.S) \times (G\backslash U.T)$$

defined by $\kappa^U_{S,T}(G.u.s.t) = (G.u.s, G.u.t)$.

So I have

$$P^*(\nu_{(S,f)} \times \nu_{(T,g)}) r_{G\backslash U.S, G\backslash U.T}(m,n) = (M\hat{\otimes}N)^*(\kappa^U_{S,T})\omega_{G\backslash U.S, G\backslash U.T}(m,n)$$

As

$$\omega_{G\backslash U.S,G\backslash U.T}(m,n) = \left[M^*\begin{pmatrix} G.u.s\ G.u.t \\ G.u.s \end{pmatrix}(m) \otimes N^*\begin{pmatrix} G.u.s\ G.u.t \\ G.u.t \end{pmatrix}(n)\right]_{(G\backslash U.S\times G\backslash U.T,Id)}$$

I have finally

$$(M\hat{\otimes}N)^*(\kappa^U_{S,T})\omega_{G\backslash U.S,G\backslash U.T}(m,n) = \ldots$$

$$\ldots = \left[M^*(\kappa^U_{S,T})M^*\begin{pmatrix} G.u.s\ G.u.t \\ G.u.s \end{pmatrix}(m) \otimes N^*(\kappa^U_{S,T})N^*\begin{pmatrix} G.u.s\ G.u.t \\ G.u.t \end{pmatrix}(n)\right]_{(G\backslash U.S.T,Id)}$$

It is also the element

$$\theta(m,n) = \left[M^*\begin{pmatrix} G.u.s.t \\ G.u.s \end{pmatrix}(m) \otimes N^*\begin{pmatrix} G.u.s.t \\ G.u.t \end{pmatrix}(n)\right]_{(G\backslash U.S.T,Id)}$$

Finally, I see that

$$P^*(\nu_{(S,f)} \times \nu_{(T,g)})r_{G\backslash U.S,G\backslash U.T}(m,n) = \theta(m,n)_{(S.T,\beta)}$$

and the image of $m_{(S,f)}$, $n_{(T,g)}$ is equal to

$$P_*(f_X \times g_Y)\big(\theta(m,n)_{(S.T,\beta)}\big) = \theta(m,n)_{(S.T,(\widetilde{f_X\times g_Y})\beta)}$$

Since

$$(\widetilde{f_X \times g_Y})\beta(s.t) = (\widetilde{f_X \times g_Y})\big(s,t,f_U(s)\big) = \big(f_X(s),g_Y(t),f_U(s)\big) = (f.g)(s.t)$$

I have proved the following lemma:

Lemma 10.2.1: Let X and Y be G-sets. Let (S,f) be an object of $\mathcal{D}_U(X)$ and (T,g) be an object of $\mathcal{D}_U(Y)$. Then the map which to $m \in M(G\backslash U.S)$ and $n \in N(G\backslash U.T)$ associates

$$\theta(m,n) = \left[M^*\begin{pmatrix} G.u.s.t \\ G.u.s \end{pmatrix}(m) \otimes N^*\begin{pmatrix} G.u.s.t \\ G.u.t \end{pmatrix}(n)\right]_{(G\backslash U.S.T,Id)} \in (M\hat{\otimes}N)(G\backslash U.S.T)$$

induces a bilinear morphism from $\mathcal{L}_U(M)$, $\mathcal{L}_U(N)$ to $\mathcal{L}_U(M\hat{\otimes}N)$, defined by

$$(m_{(S,f)}, n_{(T,g)}) \mapsto \theta(m,n)_{(S.T,f.g)}$$

10.3 The Green functors $\mathcal{L}_U(A)$

When A is a Green functor for H, and M is an A-module, I can compose the above bilinear morphism Θ

$$\mathcal{L}_U(A), \mathcal{L}_U(M) \to \mathcal{L}_U(A\hat{\otimes}M)$$

with the morphism from $\mathcal{L}_U(A\hat{\otimes}M)$ to $\mathcal{L}_U(M)$ induced by the product $A\hat{\otimes}M \to M$. I obtain a product that I denote by $^U\!\times$

$$\mathcal{L}_U(A)\hat{\otimes}\mathcal{L}_U(M) \to \mathcal{L}_U(M)$$

Lemma 10.3.1: Let X and Y be G-sets. If (Z,f) is a G-set over $\widetilde{X}$, if (T,g) is a G-set over $\tilde{Y}$, if $a \in A(G\backslash U.Z)$ and $m \in M(G\backslash U.T)$, then

$$a_{(Z,f)} \;{}^{U}\!\times m_{(T,g)} = M^*(\kappa^U_{Z,T})(a \times m)_{(Z.T,f.g)}$$

Proof: Indeed, by definition

$$\Theta(a_{(Z,f)}, m_{(T,g)}) = \theta(a,m)_{(Z.T,f.g)}$$

and

$$\theta(a,m) = \left[A^*\begin{pmatrix} G.u.z.t \\ G.u.z \end{pmatrix}(a) \otimes M^*\begin{pmatrix} G.u.z.t \\ G.u.t \end{pmatrix}(m)\right]_{(G\backslash U.Z.T,Id)} \in (A\hat{\otimes}M)(G\backslash U.Z.T)$$

By composition with the product, I obtain the element

$$M_*(Id)\left[A^*\begin{pmatrix} G.u.z.t \\ G.u.z \end{pmatrix}(a).M^*\begin{pmatrix} G.u.z.t \\ G.u.t \end{pmatrix}(m)\right]$$

But

$$A^*\begin{pmatrix} G.u.z.t \\ G.u.z \end{pmatrix}(a).M^*\begin{pmatrix} G.u.z.t \\ G.u.t \end{pmatrix}(m) = \ldots$$

$$\ldots = M^*\begin{pmatrix} G.u.z.t \\ G.u.z.t\; G.u.z.t \end{pmatrix}\left(A^*\begin{pmatrix} G.u.z.t \\ G.u.z \end{pmatrix}(a) \times M^*\begin{pmatrix} G.u.z.t \\ G.u.t \end{pmatrix}(m)\right) = \ldots$$

$$\ldots = M^*\begin{pmatrix} G.u.z.t \\ G.u.z.t\; G.u.z.t \end{pmatrix} M^*\begin{pmatrix} G.u_1.z_1.t_1\; G.u_2.z_2.t_2 \\ G.u_1.z_1\; G.u_2.t_2 \end{pmatrix}(a \times m) = \ldots$$

$$\ldots = M^*\begin{pmatrix} G.u.z.t \\ G.u.z\; G.u.t \end{pmatrix}(a \times m) = M^*(\kappa^U_{Z,T})(a \times m)$$

and the lemma follows. ∎

Notation: *Let $p_{G\backslash U}$ be the unique morphism from $G\backslash U$ to $\bullet$. Then $(U/H, Id)$ is a G-set over $\bullet \times (U/H)$, that is an object of $\mathcal{D}_U(\bullet)$. Moreover $U.(U/H) \simeq U$. I denote by*

$$\varepsilon_{\mathcal{L}_U(A)} = A^*(p_{G\backslash U})(\varepsilon_A)_{(U/H,Id)}$$

the image of the element $A^*(p_{G\backslash U})(\varepsilon_A)$ of $A(G\backslash U)$ in $\mathcal{L}_U(\bullet)$.

Proposition 10.3.2: Let G and H be groups, and U be a G-set-H.

- **If A is a Green functor for H, then $\mathcal{L}_U(A)$ is a Green functor for G for the product ${}^{U}\!\times$, with unit $\varepsilon_{\mathcal{L}_U(A)}$. The correspondence $A \mapsto \mathcal{L}_U(A)$ is a functor from $Green(H)$ to $Green(G)$.**

- **If M is an A-module, then $\mathcal{L}_U(M)$ is an $\mathcal{L}_U(A)$-module, and the correspondence $M \mapsto \mathcal{L}_U(M)$ is a functor from A-Mod to $\mathcal{L}_U(A)$-Mod.**

Proof: The product ${}^U\!\times$ is bifunctorial by construction. I must check that it is associative and unitary.

To check associativity, I consider G-sets X, X' and $X"$. Let (Y,f), (Y',f') and $(Y",f")$ be G-sets over $X\times(U/H)$, $X'\times(U/H)$ and $X"\times(U/H)$ respectively. Let $a\in A(G\backslash U.Y)$, $a'\in A(G\backslash U.Y')$, and $m"\in M(G\backslash U.Y")$. Then

$$a'\ {}^U\!\times m" = M^*(\kappa^U_{Y',Y"})(a'\times m")$$

So

$$a\ {}^U\!\times (a'\ {}^U\!\times m") = M^*(\kappa^U_{Y,Y'.Y"})\Big(a\times M^*(\kappa^U_{Y',Y"})(a'\times m")\Big) = \dots$$

$$\dots = M^*(\kappa^U_{Y,Y'.Y"})M^*(Id\times\kappa^U_{Y',Y"})(a\times a'\times m")$$

On the other hand

$$a\ {}^U\!\times a' = A^*(\kappa^U_{Y,Y'})(a\times a')$$

which gives

$$(a\ {}^U\!\times a')\ {}^U\!\times m" = M^*(\kappa^U_{Y.Y',Y"})\Big(A^*(\kappa^U_{Y,Y'})(a\times a')\times m"\Big) = \dots$$

$$\dots = M^*(\kappa^U_{Y.Y',Y"})M^*(\kappa^U_{Y,Y'}\times Id)(a\times a'\times m")$$

But after identification of $(Y.Y').Y"$ with $Y.(Y'Y")$, I have

$$(\kappa^U_{Y,Y'}\times Id)\kappa^U_{Y.Y',Y"} = (Id\times\kappa^U_{Y',Y"})\kappa^U_{Y,Y'.Y"}$$

which proves that $a\ {}^U\!\times (a'\ {}^U\!\times m") = (a\ {}^U\!\times a')\ {}^U\!\times m"$.

Moreover, if (Y,f) is an object of $\mathcal{D}_U(X)$, and if $m\in M(G\backslash U.Y)$, then

$$\varepsilon_{\mathcal{L}_U(A)}\ {}^U\!\times m = M^*(\kappa^U_{U/H,Y})\Big(A^*(p_{G\backslash U})(\varepsilon)\times m\Big) = M^*(\kappa^U_{U/H,Y})M^*(p)(m)$$

where p is the projection from $(G\backslash U)\times(G\backslash U.Y)$ onto $G\backslash U.Y$. But identifying $(U/H).Y$ with Y, and $G\backslash U.\big((U/H).Y\big)$ with $G\backslash U.Y$, and $U.(U/H)$ with U, the map $\kappa^U_{U/H,Y}$ is the map

$$G\backslash U.Y \to (G\backslash U)\times(G\backslash U.Y)$$

which maps $G(u,y)$ to $\big(Gu, G(u,y)\big)$. Then $p\circ\kappa^U_{U/H,Y}$ is the identity, and it follows that

$$\varepsilon_{\mathcal{L}_U(A)}\ {}^U\!\times m = m$$

The previous arguments, in the case $M=A$, show that $\mathcal{L}_U(A)$ is a Green functor (it remains to check by a similar argument that $\varepsilon_{\mathcal{L}_U(A)}$ is also a right unit). This proves the first part of the first assertion. The case of an arbitrary A-module M proves the first part of the second.

If $\theta: A\to B$ is a unitary morphism of Green functors from A to B, then θ induces a morphism of Mackey functors $\mathcal{L}_U(\theta)$ from $\mathcal{L}_U(A)$ to $\mathcal{L}_U(B)$. If X and Y are G-sets, if (Z,f) is an object of $\mathcal{D}_U(X)$ and (T,g) an object of $\mathcal{L}_U(Y)$, if $a\in A(G\backslash U.Z)$ and $a'\in A(G\backslash U.T)$, then

$$\mathcal{L}_U(\theta)_{X\times Y}(a_{(Z,f)}\ {}^U\!\times a'_{(T,g)}) = \theta_{G\backslash U.Z.T}\Big(A^*(\kappa^U_{Z,T})(a\times b)_{(Z.T,f.g)}\Big)$$

As θ is a morphism of Mackey functors, this is also

$$B^*(\kappa^U_{Z,T})\theta_{G\backslash U.Z\times G\backslash U.T}(a\times a')_{(Z.T,f.g)}$$

and as θ is a morphism of Green functors, it is

$$B^*(\kappa^U_{Z,T})\Big(\theta_{G\backslash U.Z}(a)\times\theta_{G\backslash U.T}(a')\Big)_{(Z.T,f.g)}$$

so that

$$\mathcal{L}_U(\theta)_{X\times Y}(a_{(Z,f)}\ {}^U\!\times\ a'_{(T,g)})=\mathcal{L}_U(\theta)_X(a_{(Z,f)})\ {}^U\!\times\ \mathcal{L}_U(\theta)_Y(a'_{(T,g)})$$

Moreover

$$\mathcal{L}_U(\theta)(\varepsilon_{\mathcal{L}_U(A)})=\mathcal{L}_U(\theta)_\bullet\Big(A^*(p_{G\backslash U})(\varepsilon_A)_{(U/H,Id)}\Big)=\theta_{G\backslash U}A^*(p_{G\backslash U})(\varepsilon_A)_{(U/H,Id)}=\ldots$$

$$\ldots=B^*(p_{G\backslash U})\theta_\bullet(\varepsilon_A)_{(U/H,Id)}=B^*(p_{G\backslash U})(\varepsilon_B)_{(U/H,Id)}=\varepsilon_{\mathcal{L}_U(B)}$$

Thus $\mathcal{L}_U(\theta)$ is a unitary morphism of Green functors, if θ is. This proves the first assertion.

And if $\theta : M\to N$ is a morphism of A-modules, then θ induces in particular a morphism of Mackey functors $\mathcal{L}_U(\theta)$ from $\mathcal{L}_U(M)$ to $\mathcal{L}_U(N)$, defined for an object (Y,f) of $\mathcal{D}_U(X)$ and a element m of $M(G\backslash U.Y)$ by

$$\mathcal{L}_U(\theta)_X(m_{(Y,f)})=\theta_{G\backslash U.Y}(m)_{(Y,f)}$$

Then if (Y',f') is an object of $\mathcal{D}_U(X')$, and if $a'\in A(G\backslash U.Y')$, I have

$$a'_{(Y',f')}\ {}^U\!\times\ \mathcal{L}_U(\theta)_X(m_{(Y,f)})=N^*(\kappa^U_{Y',Y})\Big(a'\times\theta_{G\backslash U.Y}(m)\Big)_{(Y'.Y,f'.f)}=\ldots$$

$$\ldots=N^*(\kappa^U_{Y',Y})\theta_{(G\backslash U.Y')\times(G\times U.Y)}(a'\times m)_{(Y'.Y,f'.f)}=\ldots$$

$$\ldots=\theta_{G\backslash U.(Y'.Y)}M^*(\kappa^U_{Y',Y})(a'\times m)_{(Y'.Y,f'.f)}=\theta_{G\backslash U.(Y'.Y)}(a'\ {}^U\!\times\ m)_{(Y'.Y,f'.f)}=\ldots$$

$$\ldots=\mathcal{L}_U(\theta)_{X'\times X}(a'_{(Y',f')}\ {}^U\!\times\ m_{(Y,f)})$$

which proves that the correspondence $M\mapsto\mathcal{L}_U(M)$ is functorial in M. ■

10.4 $\mathcal{L}_U(A)$-modules and adjunction

Let G and H be groups, and U be a G-set-H. Let moreover A be a Green functor for the group H. In the proof of theorem 9.5.2, I have considered the morphism deduced by adjunction from the identity morphism of $\mathcal{L}_U(A)$. It is the morphism

$$\lambda_A : A\to\mathcal{L}_U(A)\circ U$$

defined as follows: if Z is an H-set, then $(U\circ_H Z,\pi_Z)$ is an object of $\mathcal{D}_U(U\circ_H Z)$. If $a\in A(Z)$, then

$$A^*(\eta_Z)(a)\in A\Big(G\backslash U.(U\circ_H Z)\Big)$$

and I set

$$\lambda_{A,Z}(a)=A^*(\eta_Z)(a)_{(U\circ_H Z,\pi_Z)}$$

Similarly, if M is an A-module, I denote by λ_M the morphism of Mackey functors $M\to\mathcal{L}_U(M)\circ U$. The module $\mathcal{L}_U(M)$ is an $\mathcal{L}_U(A)$-module, so the module $\mathcal{L}_U(M)\circ U$ is an $\mathcal{L}_U(A)\circ U$-module. In those conditions:

Lemma 10.4.1: Let Z and Z' be H-sets. If $a \in A(Z)$ and $m \in M(Z')$, then

$$\lambda_{M,Z\times Z'}(a \times m) = \lambda_{A,Z}(a) \times^U \lambda_{M,Z'}(m)$$

where the product in the right hand side is the product of $\mathcal{L}_U(M) \circ U$ as $\mathcal{L}_U(A) \circ U$-module.

Proof: With the notations of the lemma, I have

$$\lambda_{M,Z\times Z'}(a \times m) = M^*(\eta_{Z\times Z'})(a \times m)_{(U\circ_H(Z\times Z'),\pi_{Z\times Z'})}$$

On the other hand, I have $\lambda_{A,Z}(a) \in \Big(\mathcal{L}_U(A) \circ U\Big)(Z)$, and $\lambda_{M,Z'}(m) \in \Big(\mathcal{L}_U(M) \circ U\Big)(Z')$. Their product for the functor $\mathcal{L}_U(M) \circ U$ is equal to

$$\lambda_{A,Z}(a) \times^U \lambda_{M,Z'}(m) = \mathcal{L}_U(M)^*(\delta^U_{Z,Z'})\Big(\lambda_{A,Z}(a)\ {}^U\!\!\times \lambda_{M,Z'}(m)\Big)$$

where the product ${}^U\!\!\times$ in the right hand side is computed inside $\mathcal{L}_U(M)$, i.e.

$$\lambda_{A,Z}(a)\ {}^U\!\!\times \lambda_{M,Z'}(m) = A^*(\eta_Z)(a)_{(U\circ_H Z,\pi_Z)}\ {}^U\!\!\times M^*(\eta_{Z'})(m)_{(U\circ_H Z',\pi_{Z'})} = \ldots$$
$$\ldots = M^*(\kappa^U_{U\circ_H Z, U\circ_H Z'})\Big(A^*(\eta_Z)(a) \times M^*(\eta_{Z'})(m)\Big)_{((U\circ_H Z).(U\circ_H Z'),\pi_Z.\pi_{Z'})} = \ldots$$
$$\ldots = M^*(\kappa^U_{U\circ_H Z, U\circ_H Z'})M^*(\eta_Z \times \eta_{Z'})(a \times m)_{((U\circ_H Z).(U\circ_H Z'),\pi_Z.\pi_{Z'})}$$

Finally

$$\lambda_{A,Z}(a) \times^U \lambda_{M,Z'}(m) = \ldots$$
$$\ldots = \mathcal{L}_U(M)^*(\delta^U_{Z,Z'})\Big(M^*(\kappa^U_{U\circ_H Z, U\circ_H Z'})M^*(\eta_Z \times \eta_{Z'})(a \times m)_{((U\circ_H Z).(U\circ_H Z'),\pi_Z.\pi_{Z'})}\Big) \tag{10.3}$$

To compute this expression, I note that the square

$$\begin{array}{ccc}
U\circ_H(Z\times Z') & \xrightarrow{U\circ_H\binom{zz'}{z}} & U\circ_H Z \\
{\scriptstyle U\circ_H\binom{zz'}{z'}}\Big\downarrow & & \Big\downarrow{\scriptstyle U\circ_H\binom{z}{\bullet}} \\
U\circ_H Z' & \xrightarrow[U\circ_H\binom{z'}{\bullet}]{} & U\circ_H \bullet \simeq U/H
\end{array}$$

is cartesian, as the image under $U\circ_H -$ of a cartesian square. Thus $\delta^U_{Z,Z'}$ induces an isomorphism

$$\theta^U_{Z,Z'} : U\circ_H(Z\times Z') \simeq (U\circ_H Z).(U\circ_H Z')$$

Then the square

$$\begin{array}{ccc}
U\circ_H(Z\times Z') & \xrightarrow{\theta^U_{Z,Z'}} & (U\circ_H Z).(U\circ_H Z') \\
{\scriptstyle \pi_{Z\times Z'}}\Big\downarrow & & \Big\downarrow{\scriptstyle \pi_Z.\pi_{Z'}} \\
U\circ_H(\widetilde{Z\times Z'}) & \xrightarrow[\widetilde{\delta^U_{Z,Z'}}]{} & \widetilde{U\circ_H Z}\times\widetilde{U\circ_H Z'}
\end{array}$$

is cartesian, and expression (10.3) can be written

$$\lambda_{A,Z}(a) \times^U \lambda_{M,Z'}(m) = \ldots$$

$$\ldots = M^*(G\backslash U.\theta^U_{Z,Z'})M^*(\kappa^U_{U\circ_H Z, U\circ_H Z'})M^*(\eta_Z \times \eta_{Z'})(a \times m)_{(Z\times Z', \pi_{Z\times Z'})}$$

Let $G\big(u', (u, (z, z'))\big)$ be an element of $G\backslash U.\big(U \circ_H (Z \times Z')\big)$. Then there exists $h \in H$ such that $u' = uh$. Moreover

$$\big(G\backslash U.\theta^U_{Z,Z'}\big)\Big[G\big(u', (u, (z, z'))\big)\Big] = G\big(u', ((u, z), (u, z'))\big)$$

Hence

$$\kappa^U_{U\circ_H Z, U\circ_H Z'}\big(G\backslash U.\theta^U_{Z,Z'}\big)\Big[G\big(u', (u, (z, z'))\big)\Big] = \big(G(u', (u, z)), G(u', (u, z))\big)$$

and finally

$$\big(\eta_Z \times \eta_{Z'}\big)\kappa^U_{U\circ_H Z, U\circ_H Z'}\big(G\backslash U.\theta^U_{Z,Z'}\big)\Big[G\big(u', (u, (z, z'))\big)\Big] = \ldots$$

$$\ldots = (h^{-1}z, h^{-1}z') = h^{-1}(z, z') = \eta_{Z\times Z'}\Big[G\big(u', (u, (z, z'))\big)\Big]$$

Now it follows that

$$\lambda_{A,Z}(a) \times^U \lambda_{M,Z'}(m) = M^*(\eta_{Z\times Z'})(a \times m)_{(Z\times Z', \pi_{Z\times Z'})} = \lambda_{M,Z\times Z'}(a \times m)$$

which proves the lemma. ∎

It follows in particular that the morphism λ_A is a morphism of Green functors from A to $\mathcal{L}_U(A) \circ A$, non unitary in general: indeed, by definition

$$\lambda_{A,\bullet}(\varepsilon_A) = A^*(\eta_\bullet)(\varepsilon_A)_{(U\circ_H \bullet, \pi_\bullet)}$$

Identifying $U \circ_H \bullet$ with U/H, the map $\pi_\bullet$ is the diagonal injection from U/H into $U/H \times U/H$, and $\eta_\bullet$ is the (unique) morphism from $G\backslash U.(U/H) \simeq G\backslash U$ to $\bullet$, that I denote by $p_{G\backslash U}$. Thus

$$\lambda_{A,\bullet}(\varepsilon_A) = A^*(p_{G\backslash U})_{(U/H, \pi_\bullet)}$$

On the other hand, the unit of $\mathcal{L}_U(A) \circ U$ is by definition

$$\mathcal{L}_U(A)^*(p_{U/H})(\varepsilon_{\mathcal{L}_U(A)} = \mathcal{L}_U(A)^*(p_{U/H})\big(A^*(p_{G\backslash U})(\varepsilon_A)_{(U/H, Id)}\big)$$

I compute this element with the cartesian square

$$\begin{array}{ccc} U/H \times U/H & \xrightarrow{\begin{pmatrix} u_1H\ u_2H \\ u_2H \end{pmatrix}} & U/H \\ {\scriptstyle Id}\Big\downarrow & & \Big\downarrow{\scriptstyle Id} \\ \widetilde{U/H} & \xrightarrow[\widetilde{p_{U/H}}]{} & \tilde{\bullet} \end{array}$$

So

$$\varepsilon_{\mathcal{L}_U(A)\circ U} = A^*(G\backslash U.\begin{pmatrix} u_1H\ u_2H \\ u_2H \end{pmatrix})A^*(p_{G\backslash U})(\varepsilon_A)_{(U/H\times U/H, Id)}$$

or

$$\varepsilon_{\mathcal{L}_U(A)\circ U} = A^*(p_{G\backslash U.(U/H)^2})(\varepsilon_A)_{(U/H\times U/H,Id)}$$

which is not equal to $\lambda_{A,\bullet}(\varepsilon_A)$ in general.

Notation: *If N is an $\mathcal{L}_U(A)$-module, the module $N\circ U$ is an $\mathcal{L}_U(A)\circ U$ module, which gives an A-module by restriction along λ_A, that I will denote by $N\circ U_{|A}$, defined on the H-set Z by*

$$(N\circ U_{|A})(Z) = \lambda_{A,\bullet}(\varepsilon_A)\times^U (N\circ U)(Z)\subseteq (N\circ U)(Z)$$

where the product in the right hand side is the product of $N\circ U$ as an $\mathcal{L}_U(A)\circ U$-module.

Proposition 10.4.2: Let G and H be groups, and U be a G-set-H. Let A be a Green functor for H. Then the correspondence

$$N\mapsto N\circ U_{|A}$$

is a functor from $\mathcal{L}_U(A)$-Mod to A-Mod, which is right adjoint to the functor $M\mapsto \mathcal{L}_U(M)$.

Proof: It is clear that the correspondence $N\mapsto N\circ U_{|A}$ is a functor, because it is composed of the functor $N\mapsto N\circ U$ and of the functor of restriction to A along λ_A.

I must then prove the adjunction property. Let ψ be a morphism of $\mathcal{L}_U(A)$-modules from $\mathcal{L}_U(M)$ to N. Then $\psi\circ U$ is a morphism of $\mathcal{L}_U(A)\circ U$-modules from $\mathcal{L}_U(M)\circ U$ to $N\circ U$. Restricting along λ, I obtain a morphism $\psi\circ U_{|A}$ of A-modules from $\mathcal{L}_U(M)\circ U_{|A}$ to $N\circ U_{|A}$. But lemma 10.4.1 shows that the morphism from M to $\mathcal{L}_U(M)\circ U_{|A}$ is a morphism of A-modules. Composing with $\psi\circ U_{|A}$, I obtain a morphism θ of A-modules from M to $N\circ U_{|A}$.

The square

$$\begin{array}{ccc}
\mathrm{Hom}_{\mathcal{L}_U(A)}(\mathcal{L}_U(M),N) & \xrightarrow{\hookrightarrow} & \mathrm{Hom}_{Mack(G)}(\mathcal{L}_U(M),N)\\
\downarrow & & \downarrow\simeq\\
\mathrm{Hom}_A(M,N\circ U|A) & \xrightarrow[\hookrightarrow]{} & \mathrm{Hom}_{Mack(H)}(M,N\circ U)
\end{array}$$

is commutative: indeed, any homomorphism θ from M to $N\circ U$ which is compatible with the product of A is a morphism of A-modules from M to $N\circ U_{|A}$, since if Z is an H-set, and if $m\in M(Z)$, then

$$\theta_Z(m) = \theta_Z(\varepsilon_A\times m) = \lambda_{A,\bullet}(\varepsilon_A)\times^U \theta_Z(m)\in (N\circ U_{|A})(Z)$$

It follows that the correspondence $\psi\mapsto\theta$ defined above is injective.

Conversely, if θ is a morphism of A-modules from M to $N\circ U_{|A}$, then I can compose θ with the inclusion $N\circ U_{|A}\hookrightarrow N\circ U$, and obtain by duality a morphism of Mackey functors ψ from $\mathcal{L}_U(M)$ to N. If I know that ψ is a morphism of $\mathcal{L}_U(A)$-modules, the proposition will follow.

Let X be a G-set. If (Y,f) is an object of $\mathcal{D}_U(X)$, and if $m\in M(G\backslash U.Y)$, then by construction of ψ, I have

$$\psi_X(m_{(Y,f)}) = N_*(f_X)N^*(\nu_{(Y,f)})\theta_{G\backslash U.Y}(m) \tag{10.4}$$

Lemma 10.4.3: Let $d_Y : Y \to Y \times (U/H)$ be the map $\binom{y}{y\, f_U(y)}$. Then if $a \in A(G\backslash U.Y)$ and $m \in M(G\backslash U.Y)$, and if $\theta \in \mathrm{Hom}_A(M, N \circ U_{|A})$, I have

$$N^*(\nu_{(Y,f)})\theta_{G\backslash U.Y}(a.m) = a_{Y,d_Y}.N^*(\nu_{(Y,f)})\theta_{G\backslash U.Y}(m)$$

where the product in the right hand side is the product "." for the $\mathcal{L}_U(A)$-module N.

Proof: I denote by Z the set $G\backslash U.Y$. If $\theta \in \mathrm{Hom}_A(M, N \circ U_{|A})$, then

$$\theta_Z(a.m) = a.^U\theta_Z(m)$$

where the product "$.^U$" is the product of the A-module $N \circ U$. By definition, I have

$$a.^U\theta_Z(m) = (N \circ U)^*\binom{z}{zz}\Big(a \times^U \theta_Z(m)\Big) = \ldots$$

$$\ldots = N^*\Big(U \circ_H \binom{z}{zz}\Big)\, N^*(\delta^U_{Z,Z})\Big(\lambda_{A,Z}(a) \times \theta_Z(m)\Big)$$

where the product in the right hand side is the product of the $\mathcal{L}_U(A)$-module N. As

$$\lambda_{A,Z}(a) = A^*(\eta_Z)(a)_{(U\circ_H Z,\pi_Z)}$$

I have finally

$$N^*(\nu_{(Y,f)})\theta_Z(a.m) = N^*(\nu_{(Y,f)})N^*\Big(U \circ_H \binom{z}{zz}\Big)N^*(\delta^U_{Z,Z})\Big(A^*(\eta_Z)(a)_{(U\circ_H Z,\pi_Z)} \times \theta_Z(m)\Big)$$

As moreover, if $y \in Y$ and $f_U(y) = uH$, I have

$$\delta^U_{Z,Z}\Big(U \circ_H \binom{z}{zz}\Big)\,\nu_{(Y,f)}(y) = \delta^U_{Z,Z}\Big(U \circ_H \binom{z}{zz}\Big)\Big(u, G(u,y)\Big) = \ldots$$

$$\ldots = \delta^U_{Z,Z}\Big(u, G(u,y), G(u,y)\Big) = \Big((u,G(u,y)),(u,G(u,y))\Big) = \Big(\nu_Y(y), \nu_Y(y)\Big)$$

this gives

$$N^*(\nu_{(Y,f)})\theta_Z(a.m) = N^*\begin{pmatrix} y \\ \nu_Y(y)\ \nu_U(y)\end{pmatrix}\Big(A^*(\eta_Z)(a)_{(U\circ_H Z,\pi_Z)} \times \theta_Z(m)\Big) = \ldots$$

$$\ldots = N^*\binom{y}{yy}N^*(\nu_Y \times \nu_Y)\Big(A^*(\eta_Z)(a)_{(U\circ_H Z,\pi_Z)} \times \theta_Z(m)\Big)$$

and as N is an $\mathcal{L}_U(A)$-module, it is also

$$N^*(\nu_{(Y,f)})\theta_Z(a.m) = N^*\binom{y}{yy}\Big[\mathcal{L}_U(A)^*(\nu_Y)\Big(A^*(\eta_Z)(a)_{(U\circ_H Z,\pi_Z)}\Big) \times N^*(\nu_Y)\theta_Z(m)\Big] = \ldots$$

$$\ldots = \mathcal{L}_U(A)^*(\nu_Y)\Big(A^*(\eta_Z)(a)_{(U\circ_H Z,\pi_Z)}\Big).N^*(\nu_Y)\theta_Z(m)$$

Now the lemma follows from the next lemma:

Lemma 10.4.4: Let $Z = G\backslash U.Y$. Then

$$\mathcal{L}_U(A)^*(\nu_Y)\Big(A^*(\eta_Z)(a)_{(U\circ_H Z,\pi_Z)}\Big) = a_{(Y,d_Y)}$$

Proof: Indeed the square

$$\begin{array}{ccc} Y & \xrightarrow{\nu_Y} & U \circ_H Z \\ d_Y \big\downarrow & & \big\downarrow \pi_Z \\ \tilde{Y} & \xrightarrow[\widetilde{\nu_Y}]{} & U \circ_H Z \end{array}$$

is cartesian. Then

$$\mathcal{L}_U(A)^*(\nu_Y)\Big(A^*(\eta_Z)(a)_{(U\circ_H Z,\pi_Z)}\Big) = \Big(A^*(G\backslash U.\nu_Y)A^*(\eta_Z)(a)\Big)_{(Y,d_Y)}$$

and the lemma follows, since $\eta_Z \circ (G\backslash U.\nu_Y)$ is the identity. ∎

Let then X' be a G-set. If (Y',f') is an object of $\mathcal{D}_U(X')$, and if $a \in A(G\backslash U.Y')$, then the product of equality (10.4) by $a_{(Y',f')}$ gives

$$a_{(Y',f')} \times \psi_X(m_{(Y,f)}) = a_{(Y',f')} \times N_*(f_X)N^*(\nu_{(Y,f)})\theta_{G\backslash U.Y}(m) \qquad (10.5)$$

Moreover, setting $d_{Y'} = \begin{pmatrix} y' \\ y' \, f'_{Y'}(y') \end{pmatrix}$, I have

$$a_{(Y',f')} = \mathcal{L}_U(A)_*(f'_{X'})(a_{(Y',d_{Y'})})$$

As N is an $\mathcal{L}_U(A)$-module, equality (10.5) becomes

$$a_{(Y',f')} \times \psi_X(m_{(Y,f)}) = N_*(f'_{X'} \times f_X)\Big(a_{(Y',d_{Y'})} \times N^*(\nu_{(Y,f)})\theta_{G\backslash U.Y}(m)\Big)$$

Let e be the unit of the ring $\big(A(G\backslash U.Y),.\big)$, equal to

$$e = A^*(p_{G\backslash U.Y})(\varepsilon_A)$$

denoting by $p_{G\backslash U.Y}$ the unique morphism from $G\backslash U.Y$ to $\bullet$. Then by lemma 10.4.3

$$N^*(\nu_{(Y,f)})\theta_{G\backslash U.Y}(m) = N^*(\nu_{(Y,f)})\theta_{G\backslash U.Y}(e.m) = e_{(Y,d_Y)}.N^*(\nu_{(Y,f)})\theta_{G\backslash U.Y}(m)$$

and then

$$a_{(Y',f')} \times \psi_X(m_{(Y,f)}) = N_*(f'_{X'} \times f_X)\Big[a_{(Y',d_{Y'})} \times e_{(Y,d_Y)}.N^*(\nu_{(Y,f)})\theta_{G\backslash U.Y}(m)\Big] \quad (10.6)$$

Lemma 10.4.5: Let A be a Green functor for G. If X and Y are G-sets, if $a \in A(X)$ and $b, c \in A(Y)$, then

$$a \times (b.c) = (a \times b).A^*\begin{pmatrix} xy \\ y \end{pmatrix}(c)$$

Proof: It suffices to compute

$$a \times (b.c) = a \times A^*\begin{pmatrix} y \\ yy \end{pmatrix}(b \times c) = A^*\begin{pmatrix} xy \\ xyy \end{pmatrix}(a \times b \times c) = \ldots$$

$$\ldots = A^*\begin{pmatrix} xy \\ xyy \end{pmatrix}\left(A^*\begin{pmatrix} xy_1y_2 \\ xy_1 \end{pmatrix}(a \times b).A^*\begin{pmatrix} xy_1y_2 \\ y_2 \end{pmatrix}(c)\right) = \ldots$$

$$\dots = A^*\begin{pmatrix} xy \\ xy \end{pmatrix}(a\times b).A^*\begin{pmatrix} xy \\ y \end{pmatrix}(c) = (a\times b).A^*\begin{pmatrix} xy \\ y \end{pmatrix}(c)$$

Thus in equality (10.6), the expression inside hooks is equal to

$$a_{(Y',d_{Y'})}\times e_{(Y,d_Y)}.N^*(\nu_{(Y,f)})\theta_{G\backslash U.Y}(m) = (a_{(Y',d_{Y'})}\times e_{(Y,d_Y)}).N^*\begin{pmatrix} y'y \\ y \end{pmatrix}N^*(\nu_{(Y,f)})\theta_{G\backslash U.Y}(m)$$

I set

$$T = N^*\begin{pmatrix} y'y \\ y \end{pmatrix}N^*(\nu_{(Y,f)})\theta_{G\backslash U.Y}(m)$$

Now equality (10.6) becomes

$$a_{(Y',f')}\times\psi_X(m_{(Y,f)}) = N_*(f'_{X'}\times f_X)\Big((a_{(Y',d_{Y'})}\times e_{(Y,d_Y)}).T\Big)$$

By definition, I have

$$a_{(Y',d_{Y'})}\times e_{(Y,d_Y)} = A^*(\kappa^U_{Y',Y})(a\times e)_{(Y'.Y,d_{Y'}.d_Y)}$$

I set $P = A^*(\kappa^U_{Y',Y})(a\times e)$. I denote by i the injection from $Y'.Y$ into $Y'\times Y$, and J the complement of $Y'.Y$ in $Y'\times Y$. If j is the injection from J to $Y'\times Y$, and $d_{Y'.Y}$ is the map from $Y'.Y$ to $(Y'.Y)\times(U/H)$ defined by

$$d_{Y'.Y}(y',y) = \big(y',y,f_U(y)\big) = \big(y',y,f'_U(y')\big)$$

then

$$a_{(Y',d_{Y'})}\times e_{(Y,d_Y)} = \mathcal{L}_U(A)_*(i)\Big(P_{(Y'.Y,d_{Y'.Y})}\Big)$$

As

$$N^*(j)\Big((a_{(Y',d_{Y'})}\times e_{(Y,d_Y)}).T\Big) = \mathcal{L}_U(A)^*(j)(a_{(Y',d_{Y'})}\times e_{(Y,d_Y)}).N^*(j)(T)$$

and as

$$\mathcal{L}_U(A)^*(j)(a_{(Y',d_{Y'})}\times e_{(Y,d_Y)}) = \mathcal{L}_U(A)^*(j)\mathcal{L}_U(A)_*(i)\Big(P_{(Y'.Y,d_{Y'.Y})}\Big) = 0$$

since the images of i and j are disjoint, I have

$$N^*(j)\Big((a_{(Y',d_{Y'})}\times e_{(Y,d_Y)}).T\Big) = 0$$

But the identity of $N(Y'\times Y)$ is equal to $N_*(i)N^*(i)+N_*(j)N^*(j)$. It follows that

$$a_{(Y',f')}\times\psi_X(m_{(Y,f)}) = N_*(f'_{X'}\times f_X)\Big(N_*(i)N^*(i)+N_*(j)N^*(j)\Big)\Big((a_{(Y',d_{Y'})}\times e_{(Y,d_Y)}).T\Big) = \dots$$

$$\dots = N_*(f'_{X'}\times f_X)N_*(i)N^*(i)\Big((a_{(Y',d_{Y'})}\times e_{(Y,d_Y)}).T\Big)$$

As moreover by lemma 10.4.5

$$(a_{(Y',d_{Y'})}\times e_{(Y,d_Y)}).T = a_{(Y',d_{Y'})}\times e_{(Y,d_Y)}.N^*(\nu_{(Y,f)})\theta_{G\backslash U.Y}(m) = \dots$$

$$\dots = a_{(Y',d_{Y'})}\times N^*(\nu_{(Y,f)})\theta_{G\backslash U.Y}(m)$$

I get finally

$$a_{(Y',f')}\times\psi_X(m_{(Y,f)}) = N_*(f'_{X'}\times f_X)N_*(i)N^*(i)\Big(a_{(Y',d_{Y'})}\times N^*(\nu_{(Y,f)})\theta_{G\backslash U.Y}(m)\Big) \quad (10.7)$$

On the other hand

$$a_{(Y',f')}\times m_{(Y,f)} = M^*(\kappa^U_{Y',Y})(a\times m)_{(Y'.Y,f'.f)}$$

Thus

$$\psi_{X'\times X}(a_{(Y',f')}\times m_{(Y,f)}) = N_*\Big((f'.f)_{X'\times X}\Big)N^*(\nu_{Y'.Y})\theta_{G\backslash U.(Y'.Y)}M^*(\kappa^U_{Y',Y})(a\times m)$$

As θ is a morphism of Mackey functors, I have

$$\theta_{G\backslash U.(Y'.Y)}M^*(\kappa^U_{Y',Y}) = (N\circ U)^*(\kappa^U_{Y',Y})\theta_{(G\backslash U.Y')\times(G\backslash U.Y)}(a\times m)$$

As θ is a morphism of A-modules, I have

$$\theta_{(G\backslash U.Y')\times(G\backslash U.Y)}(a\times m) = a\times^U\theta_{G\backslash U.Y}(m)$$

where the product $\times^U$ is the product of the A-module $N\circ U$. Then

$$a\times^U\theta_{G\backslash U.Y}(m) = N^*(\delta^U_{G\backslash U.Y',G\backslash U.Y})\Big(\lambda_{A,G\backslash U.Y'}(a)\times\theta_{G\backslash U.Y}(m)\Big)$$

It follows that

$$\psi_{X'\times X}(a_{(Y',f')}\times m_{(Y,f)}) = \dots$$

$$= N_*\Big((f'.f)_{X'\times X}\Big)N^*(\nu_{Y'.Y})(N\circ U)^*(\kappa^U_{Y',Y})N^*(\delta^U_{G\backslash U.Y',G\backslash U.Y})\Big(\lambda_{A,G\backslash U.Y'}(a)\times\theta_{G\backslash U.Y}(m)\Big)$$

But the composite map

$$\delta^U_{G\backslash U.Y',G\backslash U.Y}(U\circ_H\kappa^U_{Y',Y})\nu_{Y'.Y}$$

has the following effect on $(y',y)\in Y'.Y$, if $f_U(y)=f'_U(y')=uH$

$$(y',y)\mapsto\Big(u,G(u,y',y)\Big)\mapsto\Big(u,G(u,y'),G(u,y)\Big)\mapsto\Big((u,G(u,y')),(u,G(u,y))\Big)=\dots$$

$$\dots=\Big(\nu_{Y'}(y'),\nu_Y(y)\Big)$$

So it is the restriction to $Y'.Y$ of $\nu_{Y'}\times\nu_Y$, or $(\nu_{Y'}\times\nu_Y)\circ i$. Then

$$\psi_{X'\times X}(a_{(Y',f')}\times m_{(Y,f)}) = \dots$$

$$\dots = N_*\Big((f'.f)_{X'\times X}\Big)N^*(i)N^*(\nu_{Y'}\times\nu_Y)\Big(\lambda_{A,G\backslash U.Y'}(a)\times\theta_{G\backslash U.Y}(m)\Big) = \dots$$

$$\dots = N_*\Big((f'.f)_{X'\times X}\Big)N^*(i)\Big[\mathcal{L}_U(A)^*(\nu_{Y'})\lambda_{A,G\backslash U.Y'}(a)\times N^*(\nu_Y)\theta_{G\backslash U.Y}(m)\Big]$$

It remains to observe that setting $Z'=G\backslash U.Y'$, lemma 10.4.4 shows that

$$\mathcal{L}_U(A)^*(\nu_{Y'})\lambda_{A,Z'}(a) = \mathcal{L}_U(A)^*(\nu_{Y'})\Big(A^*(\eta_{Z'})(a)_{(U\circ_H Z',\pi_{Z'})}\Big) = a_{(Y',d_{Y'})}$$

Finally

$$\psi_{X'\times X}(a_{(Y',f')}\times m_{(Y,f)}) = N_*\Big((f'.f)_{X'\times X}\Big)N^*(i)\Big(a_{(Y',d_{Y'})}\times N^*(\nu_Y)\theta_{G\backslash U.Y}(m)\Big)$$

As moreover $(f'.f)_{X'\times X}=(f'_{X'}\times f_X)\circ i$, this expression is equal to the right hand side of (10.7), and then

$$\psi_{X'\times X}(a_{(Y',f')}\times m_{(Y,f)}) = a_{(Y',f')}\times\psi_X\Big(m_{(Y,f)}\Big)$$

which proves that ψ is a morphism of $\mathcal{L}_U(A)$-modules, and completes the proof of the proposition. ■

10.5 Right adjoints and tensor product

Let G and H be groups, and U be an H-set G. If M is a Mackey functor for H, and P is a Mackey functor for G, I have built morphisms

$$\left(\mathcal{L}_U(M)\hat{\otimes}P\right)\circ U \rightleftarrows M\hat{\otimes}(P\circ U)$$

In the case $P=\mathcal{R}_U(N)$, for a Mackey functor N for the group H, I have in particular a morphism

$$\left(\mathcal{L}_U(M)\hat{\otimes}\mathcal{R}_U(N)\right)\circ U \to M\hat{\otimes}\left(\mathcal{R}_U(N)\circ U\right)$$

Composing this morphism with the morphism

$$M\hat{\otimes}\left(\mathcal{R}_U(N)\circ U\right)\to M\hat{\otimes}N$$

deduced from the co-unit $\mathcal{R}_U(N)\circ U\to N$, I obtain a morphism

$$\left(\mathcal{L}_U(M)\hat{\otimes}\mathcal{R}_U(N)\right)\circ U \to M\hat{\otimes}N$$

By adjunction, I have a morphism

$$\mathcal{L}_U(M)\hat{\otimes}\mathcal{R}_U(N)\to \mathcal{R}_U(M\hat{\otimes}N)$$

A tedious computation shows that this morphism can be described as follows:

Notation: *Let G and H be groups, and U be an H-set G. If X is a G-set, if (Z,f) and (T,g) are objects of $\mathcal{D}_U(X)$, I denote by $Z\natural T$ their pullback over $\widetilde{X}$*

$$Z\natural T=\{(z,t)\in Z\times T\mid f(z)=g(t)\}$$

I denote by $z\natural t$ the couple (z,t) of $Z\natural T$. I denote by $f\natural g$ the map from $Z\natural T$ to $\widetilde{X}$ defined by

$$(f\natural g)(z\natural t)=f(z)=g(t)$$

Definition: *If M and N are Mackey functors for H, if X is a G-set, if (Z,f) is an object of $\mathcal{D}_U(X)$, if $m\in M(G\backslash U.Z)$, and if $n\in\mathcal{R}_U(N)(X)$, I define a sequence $\theta_X(a_{(Y,f)},n)_{(T,g)}$ indexed by the objects (T,g) of $\mathcal{D}_U(X)$, such that $\theta_X(a_{(Y,f)},n)_{(T,g)}\in(M\hat{\otimes}N)(G\backslash U.T)$, by setting*

$$\theta_X(a_{(Y,f)},n)_{(T,g)}=\left[M^*\begin{pmatrix}G.u.(z\natural t)\\ G.u.z\end{pmatrix}(m)\otimes n_{(Z\natural T,f\natural g)}\right]_{(G\backslash U.(Z\natural T),\left(\substack{G.u.(z\natural t)\\ G.u.t}\right))}$$

Lemma 10.5.1: The previous equality defines a bilinear map

$$\theta_X:\mathcal{L}_U(M)(X)\times\mathcal{R}_U(N)(X)\to\mathcal{R}_U(N)(X)$$

Proof: First I must check that θ_X is well defined. So let (S,α) be a ν-disjoint H-set over $G\backslash U.Z$, and let $m\in M(S)$. Then

$$\theta_X\left(M_*(\alpha)(m)_{(Z,f)},n\right)=\left[M^*\begin{pmatrix}G.u.(z\natural t)\\ G.u.z\end{pmatrix}M_*(\alpha)(m)\otimes n_{(Z\natural T,f\natural g)}\right]_{(G\backslash U.(Z\natural T),\left(\substack{G.u.(z\natural t)\\ G.u.t}\right))}$$

Let S', α' and β filling the cartesian square

$$\begin{array}{ccc} S' & \xrightarrow{\alpha'} & G\backslash U.(Z\natural T) \\ \beta\downarrow & & \downarrow\begin{pmatrix} G.u.(z\natural t) \\ G.u.z \end{pmatrix} \\ S & \xrightarrow[\alpha]{} & G\backslash U.Z \end{array}$$

Then $M^*\begin{pmatrix} G.u.(z\natural t) \\ G.u.z \end{pmatrix} M_*(\alpha) = M_*(\alpha')M^*(\beta)$, and

$$\theta_X\Big(M_*(\alpha)(m)_{(Z,f)}, n\Big) = \Big[M_*(\alpha')M^*(\beta)(m) \otimes n_{(Z\natural T, f\natural g)}\Big]_{(G\backslash U.(Z\natural T), \left(\begin{smallmatrix} G.u.(z\natural t) \\ G.u.t \end{smallmatrix}\right))} = \ldots$$

$$\ldots = \Big[M^*(\beta)(m) \otimes N^*(\alpha')(n_{(Z\natural T, f\natural g)})\Big]_{(S', \left(\begin{smallmatrix} G.u.(z\natural t) \\ G.u.t \end{smallmatrix}\right)\circ\alpha')}$$

But (S', α') is a ν-disjoint H-set over $G\backslash U.(Z\natural T)$: indeed, if $\alpha'(s') = G.u.(z\natural t)$ and if $(u, s') \in U \circ_H S'$, then

$$\begin{pmatrix} G.u.(z\natural t) \\ G.u.z \end{pmatrix}(s') = \alpha\beta(s') = G.u.z$$

and $\big(u, \beta(s')\big) \in U \circ_H S$, which is impossible if (S, α) is ν-disjoint. Then as $n \in \mathcal{S}_U(N)\big((G\backslash U.(Z\natural T)\big)$, I have

$$N^*(\alpha')(n_{(Z\natural T, f\natural g)}) = 0$$

so $\theta_X\Big(M_*(\alpha)(m)_{(Z,f)}, n\Big) = 0$.

Now if $\alpha : (Z, f) \to (Z', f')$ is a morphism in $\mathcal{D}_U(X)$, then

$$\theta_X\Big(M_*(G\backslash U.\alpha)(m)_{(Z',f')}, n\Big)_{(T,g)} = \ldots$$

$$\ldots = \left[M^*\begin{pmatrix} G.u.(z'\natural t) \\ G.u.z' \end{pmatrix} M_*(G\backslash U.\alpha)(m) \otimes n_{(Z'\natural T, f'\natural g)}\right]_{(G\backslash U.(Z'\natural T), \left(\begin{smallmatrix} G.u.(z'\natural t) \\ G.u.t \end{smallmatrix}\right))}$$

It is clear that the squares

$$\begin{array}{ccc} Z\natural T & \xrightarrow{\alpha\natural T} & Z'\natural T \\ \begin{pmatrix} z\natural t \\ z \end{pmatrix}\downarrow & & \downarrow\begin{pmatrix} z'\natural t \\ z' \end{pmatrix} \\ Z & \xrightarrow[\alpha]{} & Z' \end{array} \qquad \begin{array}{ccc} U.(Z\natural T) & \xrightarrow{U.(\alpha\natural T)} & U.(Z'\natural T) \\ \begin{pmatrix} u.(z\natural t) \\ u.z \end{pmatrix}\downarrow & & \downarrow\begin{pmatrix} u.(z'\natural t) \\ u.z' \end{pmatrix} \\ U.Z & \xrightarrow[U.\alpha]{} & U.Z' \end{array}$$

are cartesian, and as $U.\alpha$ is injective on the orbits of G, lemma 9.3.3 shows that the square

$$\begin{array}{ccc} G\backslash U.(Z\natural T) & \xrightarrow{G\backslash U.(\alpha\natural T)} & G\backslash U.(Z'\natural T) \\ \begin{pmatrix} G.u.(z\natural t) \\ G.u.z \end{pmatrix}\downarrow & & \downarrow\begin{pmatrix} G.u.(z'\natural t) \\ G.u.z' \end{pmatrix} \\ G\backslash U.Z & \xrightarrow[G\backslash U.\alpha]{} & G\backslash U.Z' \end{array}$$

is cartesian. Then

$$\theta_X\Big(M_*(G\backslash U.\alpha)(m)_{(Z',f')},n\Big)_{(T,g)} = \ldots$$

$$\ldots = \left[M_*\begin{pmatrix} G.u.(z\natural t) \\ G.u.\alpha(z)\end{pmatrix} M^*\begin{pmatrix} G.u.(z\natural t) \\ G.u.z\end{pmatrix}(m)\otimes n_{(Z'\natural T,f'\natural g)}\right]_{(G\backslash U.(Z'\natural T),\left(\begin{smallmatrix} G.u.(z'\natural t) \\ G.u.t\end{smallmatrix}\right))} = \ldots$$

$$\ldots = \left[M^*\begin{pmatrix} G.u.(z\natural t) \\ G.u.z\end{pmatrix}(m)\otimes N^*\begin{pmatrix} G.u.(z\natural t) \\ G.u.\alpha(z)\end{pmatrix}(n_{(Z'\natural T,f'\natural g)})\right]_{(G\backslash U.(Z\natural T),\left(\begin{smallmatrix} G.u.(z\natural t) \\ G.u.t\end{smallmatrix}\right))}$$

But the morphism

$$\alpha\natural T = \begin{pmatrix} G.u.(z\natural t) \\ G.u.\alpha(z)\end{pmatrix} : (Z\natural T, f\natural g) \to (Z'\natural T, f'\natural g)$$

is a morphism in $\mathcal{D}_U(X)$: indeed, the morphism $U.(\alpha\natural T)$ is injective on the orbits of G on $U.(Z\natural T)$: if $x \in G$, and if

$$U.(\alpha\natural T)\big(xu.(xz\natural xt)\big) = U.(\alpha\natural T)\big(u.(z\natural t)\big)$$

then $xu = u$, $xt = t$ and $\alpha(xz) = \alpha(z)$. Thus $xz = z$, since $U.\alpha$ is injective on the orbits of G.

It follows that

$$N^*\begin{pmatrix} G.u.(z\natural t) \\ G.u.\alpha(z)\end{pmatrix}(n_{(Z'\natural T,f'\natural g)}) = n_{(Z\natural T,f\natural g)}$$

and then

$$\theta_X\Big(M_*(G\backslash U.\alpha)(m)_{(Z',f')},n\Big)_{(T,g)} = \ldots$$

$$\ldots = \left[M^*\begin{pmatrix} G.u.(z\natural t) \\ G.u.z\end{pmatrix}(m)\otimes n_{(Z\natural T,f\natural g)}\right]_{(G\backslash U.(Z\natural T),\left(\begin{smallmatrix} G.u.(z\natural t) \\ G.u.t\end{smallmatrix}\right))} = \theta_X(m_{(Z,f)},n)$$

Finally, the map θ_X is well defined.

Now if (S,α) is a ν-disjoint H-set over $G\backslash U.T$, and if $m \in M(S)$, then I fill the cartesian square

$$\begin{array}{ccc} S' & \xrightarrow{\alpha'} & G\backslash U.(Z\natural T) \\ {\scriptstyle\beta}\big\downarrow & & \big\downarrow{\scriptstyle\begin{pmatrix} G.u.(z\natural t) \\ G.u.t\end{pmatrix}} \\ S & \xrightarrow[\alpha]{} & G\backslash U.T \end{array}$$

so that

$$(M\hat{\otimes}N)^*(\alpha)\Big(\theta_X(m_{(Z,f)},n)_{(T,g)}\Big) = \ldots$$

$$\ldots = \left[M^*(\alpha')M^*\begin{pmatrix} G.u.(z\natural t) \\ G.u.z\end{pmatrix}(m)\otimes N^*(\alpha')(n_{(Z\natural T,f\natural g)})\right]_{(S',\beta)}$$

But (S',α') is as before a ν-disjoint H-set over $G\backslash U.(Z\natural T)$, and then $N^*(\alpha')(n_{(Z\natural T,f\natural g)}) = 0$, so

$$(M\hat{\otimes}N)^*(\alpha)\Big(\theta_X(m_{(Z,f)},n)_{(T,g)}\Big) = 0$$

Finally if $\alpha : (T', g') \to (T, g)$ is a morphism in $\mathcal{D}_U(X)$, then as the square

$$\begin{array}{ccc} G\backslash U.(Z\natural T') & \xrightarrow{G\backslash U.(Z\natural\alpha)} & G\backslash U.(Z\natural T) \\ \begin{pmatrix} G.u.(z\natural t') \\ G.u.t' \end{pmatrix}\Big\downarrow & & \Big\downarrow\begin{pmatrix} G.u.(z\natural t) \\ G.u.t \end{pmatrix} \\ G\backslash U.T' & \xrightarrow[G\backslash U.\alpha]{} & G\backslash U.T \end{array}$$

is cartesian, setting $\psi = G\backslash U.(Z\natural\alpha)$, I have

$$(M\hat{\otimes}N)^*(G\backslash U.\alpha)\big(\theta_X(m_{(Z,f)}, n)_{(T,g)}\big) = \dots$$

$$\dots = \left[M^*(\psi)M^*\begin{pmatrix} G.u.(z\natural t) \\ G.u.z \end{pmatrix}(m) \otimes N^*(\psi)(n_{(Z\natural T, f\natural g)})\right]_{(G\backslash U.(Z\natural T'), \left(\begin{smallmatrix} G.u.(z\natural t') \\ G.u.t \end{smallmatrix}\right))}$$

Now the morphism $U.(Z\natural\alpha)$ is injective on the orbits of G, thus

$$N^*(\psi)(n_{(Z\natural T, f\natural g)}) = n_{(Z\natural T', f\natural g')}$$

and moreover

$$M^*(\psi)M^*\begin{pmatrix} G.u.(z\natural t) \\ G.u.z \end{pmatrix} = M^*\begin{pmatrix} G.u.(z\natural t) \\ G.u.t \end{pmatrix}$$

Finally

$$(M\hat{\otimes}N)^*(G\backslash U.\alpha)\big(\theta_X(m_{(Z,f)}, n)_{(T,g)}\big) = \theta_X(m_{(Z,f)}, n)_{(T',g')}$$

which proves that the image of θ_X is contained in $\mathcal{R}_U(M\hat{\otimes}N)(X)$, and the lemma follows. ■

Lemma 10.5.2: The maps θ_X define a bilinear morphism from $\mathcal{L}_U(M)$, $\mathcal{R}_U(N)$ to $\mathcal{R}_U(M\hat{\otimes}N)$.

Proof: I must check the three conditions of the proposition 1.8.3. Let then $\phi : X \to X'$ be a morphism of G-sets. If (Z', f') is an object of $\mathcal{D}_U(X')$, if $m \in M(G\backslash U.Z')$, and if $n \in \mathcal{R}_U(M)(X')$, then to compute $\mathcal{L}_U(M)^*(\phi)(m_{(Z',f')})$, I fill the cartesian square

$$\begin{array}{ccc} Z & \xrightarrow{a} & Z' \\ f\Big\downarrow & & \Big\downarrow f' \\ X & \xrightarrow[\tilde{\phi}]{} & X' \end{array}$$

and then

$$\mathcal{L}_U(M)^*(\phi)(m_{(Z',f')}) = M^*(G\backslash U.a)(m)_{(Z,f)}$$

In those conditions

$$\theta_X\big(\mathcal{L}_U(M)^*(\phi)(m_{(Z',f')}), \mathcal{R}_U(N)^*(\phi)(n)\big)_{(T,g)} = \dots$$

$$= \left[M^*\begin{pmatrix} G.u.(z\natural t) \\ G.u.z \end{pmatrix} M^*(G\backslash U.a)(m) \otimes \mathcal{R}_U(M)^*(\phi)(n)_{(Z\natural T, f\natural g)}\right]_{(G\backslash U.(Z\natural T), \left(\begin{smallmatrix} G.u.(z\natural t) \\ G.u.t \end{smallmatrix}\right))} = \dots$$

$$\dots = \left[M^* \begin{pmatrix} G.u.(z\natural t) \\ G.u.a(z) \end{pmatrix} (m) \otimes n_{(Z\natural T, \tilde{\phi}(f\natural g))} \right]_{(G\backslash U.(Z\natural T), \left(\begin{smallmatrix} G.u.(z\natural t) \\ G.u.t \end{smallmatrix}\right))}$$

On the other hand

$$\Big(\mathcal{R}_U(M\hat{\otimes}N)^*(\phi)\theta_{X'}(m_{(Z',f')}, n) \Big)_{(T,g)} = \theta_{X'}(m_{(Z',f')}, n)_{(T,\tilde{\phi}g)} = \dots$$

$$\dots = \left[M^* \begin{pmatrix} G.u.(z'\natural t) \\ G.u.z' \end{pmatrix} (m) \otimes n_{(Z'\natural T, f'\natural(\tilde{\phi}g))} \right]_{(G\backslash U.(Z'\natural T), \left(\begin{smallmatrix} G.u.(z'\natural t) \\ G.u.t \end{smallmatrix}\right))} \tag{10.8}$$

But the diagram

$$\begin{array}{ccccc}
Z\natural T & \xrightarrow{\begin{pmatrix} z\natural t \\ z \end{pmatrix}} & Z & \xrightarrow{a} & Z' \\
{\scriptstyle \left(\begin{smallmatrix} z\natural t \\ t \end{smallmatrix}\right)} \big\downarrow & & f \big\downarrow & & \big\downarrow f' \\
T & \xrightarrow[g]{} & X & \xrightarrow[\tilde{\phi}]{} & X'
\end{array}$$

is composed of two cartesian squares. Thus the square

$$\begin{array}{ccc}
Z\natural T & \xrightarrow{a\begin{pmatrix} z\natural t \\ z \end{pmatrix}} & Z' \\
\begin{pmatrix} z\natural t \\ t \end{pmatrix} \big\downarrow & & \big\downarrow f' \\
T & \xrightarrow[\tilde{\phi}g]{} & X'
\end{array}$$

is cartesian. It follows that in equality (10.8), I have $Z\natural T = Z'\natural T$, and with this identification, the map $\left(\begin{smallmatrix} G.u.(z'\natural t) \\ G.u.z' \end{smallmatrix}\right)$ becomes the map $\left(\begin{smallmatrix} G.u.(z\natural t) \\ G.u.a(z) \end{smallmatrix}\right)$. Those remarks prove that

$$\theta_X\Big(\mathcal{L}_U(M)^*(\phi)(m_{(Z',f')}), \mathcal{R}_U(N)^*(\phi)(n)\Big) = \mathcal{R}_U(M\hat{\otimes}N)^*(\phi)\theta_{X'}(m_{(Z',f')}, n)$$

which is condition *iii*) of proposition 1.8.3.

Now if $m \in M(G\backslash U.Z)$, then

$$\mathcal{L}_U(M)_*(\phi)(m_{(Z,f)}) = m_{(Z,\tilde{\phi}f)}$$

so that for an object (T,g) of $\mathcal{D}_U(X')$, and $n \in \mathcal{R}_U(N)(X')$, I have

$$\theta_{X'}\Big(\mathcal{L}_U(M)_*(\phi)(m_{(Z,f)}), n\Big)_{(T,g)} = \dots$$

$$\dots = \left[M^* \begin{pmatrix} G.u.(z\natural t) \\ G.u.z \end{pmatrix} (m) \otimes n_{(Z\natural T, (\tilde{\phi}f)\natural g)} \right]_{(G\backslash U.(Z\natural T), \left(\begin{smallmatrix} G.u.(z\natural t) \\ G.u.t \end{smallmatrix}\right))}$$

where the pull-back product $Z\natural T$ fills the cartesian square

$$\begin{array}{ccc} Z\natural T & \xrightarrow{\begin{pmatrix} z\natural t \\ t \end{pmatrix}} & T \\ {\scriptstyle\begin{pmatrix} z\natural t \\ z \end{pmatrix}}\Big\downarrow & & \Big\downarrow{\scriptstyle g} \\ Z & \xrightarrow[\widetilde{\phi}f]{} & X' \end{array} \qquad (C)$$

I must compare this element with

$$E = \Big(\mathcal{R}_U(M\hat{\otimes}N)_*(\phi)\theta_X\big(m_{(Z,f)}, \mathcal{R}_U(N)^*(\phi)(n)\big)\Big)_{(T,g)}$$

To compute this element, I fill the cartesian square

$$\begin{array}{ccc} Z' & \xrightarrow{\alpha} & T \\ {\scriptstyle\beta}\Big\downarrow & & \Big\downarrow{\scriptstyle g} \\ X & \xrightarrow[\widetilde{\phi}]{} & X' \end{array}$$

so that

$$E = (M\hat{\otimes}N)_*(G\backslash U.\alpha)\Big(\theta_X\big(m_{(Z,f)}, \mathcal{R}_U(N)^*(\phi)(n)\big)_{(Z',\beta)}\Big)$$

Moreover

$$\theta_X\big(m_{(Z,f)}, \mathcal{R}_U(N)^*(\phi)(n)\big)_{(Z',\beta)} = \ldots$$

$$\ldots = \left[M^*\begin{pmatrix} G.u.(z\natural z') \\ G.u.z \end{pmatrix}(m) \otimes \mathcal{R}_U(N)^*(\phi)(n)_{(Z\natural Z', f\natural\beta)}\right]_{\left(G\backslash U.(Z\natural Z'), \begin{pmatrix} G.u.(z\natural z') \\ G.u.z' \end{pmatrix}\right)}$$

On the other hand

$$\mathcal{R}_U(N)^*(\phi)(n)_{(Z\natural Z', f\natural\beta)} = n_{(Z\natural Z', \widetilde{\phi}(f\natural\beta))}$$

This gives

$$E = (M\hat{\otimes}N)_*(G\backslash U.\alpha)\left[M^*\begin{pmatrix} G.u.(z\natural z') \\ G.u.z \end{pmatrix}(m) \otimes n_{(Z\natural Z', \widetilde{\phi}(f\natural\beta))}\right]_{\left(G\backslash U.(Z\natural Z'), \begin{pmatrix} G.u.(z\natural z') \\ G.u.z' \end{pmatrix}\right)} = \ldots$$

$$\ldots = \left[M^*\begin{pmatrix} G.u.(z\natural z') \\ G.u.z \end{pmatrix}(m) \otimes n_{(Z\natural Z', \widetilde{\phi}(f\natural\beta))}\right]_{\left(G\backslash U.(Z\natural Z'), \begin{pmatrix} G.u.(z\natural z') \\ G.u.\alpha(z') \end{pmatrix}\right)}$$

Then the following diagram is composed of two cartesian squares

$$\begin{array}{ccccc} Z\natural Z' & \xrightarrow{\begin{pmatrix} z\natural z' \\ z' \end{pmatrix}} & Z' & \xrightarrow{\alpha} & T \\ {\scriptstyle\begin{pmatrix} z\natural z' \\ z \end{pmatrix}}\Big\downarrow & & {\scriptstyle\beta}\Big\downarrow & & \Big\downarrow{\scriptstyle g} \\ Z & \xrightarrow[f]{} & X & \xrightarrow[\widetilde{\phi}]{} & X' \end{array}$$

It follows that that in (C), I can identify $Z\natural T$ with $Z\natural Z'$. With this identification, I have

$$(\tilde{\phi}f)\natural g = \tilde{\phi}(f\natural\beta) \qquad \begin{pmatrix} G.u.(z\natural t) \\ G.u.t \end{pmatrix} = \begin{pmatrix} G.u.(z\natural z') \\ G.u.\alpha(z') \end{pmatrix}$$

so that I have

$$\theta_{X'}\Big(\mathcal{L}_U(M)_*(\phi)(m_{(Z,f)}, n\Big) = \mathcal{R}_U(M\hat{\otimes}N)_*(\phi)\theta_X\Big(m_{(Z,f)}, \mathcal{R}_U(N)^*(\phi)(n)\Big)$$

which is condition $i)$ of proposition 1.8.3.

Finally for condition $ii)$ I must compare

$$E' = \theta_{X'}\Big(m_{(Z,f)}, \mathcal{R}_U(N)_*(\phi)(n)\Big)$$

and

$$E" = \mathcal{R}_U(M\hat{\otimes}N)_*(\phi)\theta_X\Big(\mathcal{L}_U(M)^*(\phi)(m_{(Z,f)}), n\Big)$$

when (Z,f) is an object of $\mathcal{D}_U(X')$ and $m \in M(G\backslash U.Z)$, and $n \in \mathcal{R}_U(N)(X)$. For an object (T,g) of $\mathcal{D}_U(X')$, I have

$$E'_{(T,g)} = \left[M^*\begin{pmatrix} G.u.(z\natural t) \\ G.u.z \end{pmatrix}(m) \otimes \mathcal{R}_U(N)_*(\phi)(n)_{(Z\natural T, f\natural g)}\right]_{(G\backslash U.(Z\natural T, \left(\begin{smallmatrix} G.u.(z\natural t) \\ G.u.t \end{smallmatrix}\right)))}$$

To compute this expression, I fill the cartesian square

$$\begin{array}{ccc} S & \xrightarrow{\alpha} & Z\natural T \\ {\scriptstyle\beta}\downarrow & & \downarrow{\scriptstyle f\natural g} \\ X & \xrightarrow[\tilde{\phi}]{} & X' \end{array} \qquad (C_1)$$

so that

$$\mathcal{R}_U(N)_*(\phi)(n)_{(Z\natural T, f\natural g)} = N_*(G\backslash U.\alpha)(n_{(S,\beta)})$$

Thus

$$E'_{(T,g)} = \left[M^*\begin{pmatrix} G.u.(z\natural t) \\ G.u.z \end{pmatrix}(m) \otimes N_*(G\backslash U.\alpha)(n_{(S,\beta)})\right]_{(G\backslash U.(Z\natural T), \left(\begin{smallmatrix} G.u.(z\natural t) \\ G.u.t \end{smallmatrix}\right))}$$

which is also

$$E'_{(T,g)} = \left[M^*(G\backslash U.\alpha)M^*\begin{pmatrix} G.u.(z\natural t) \\ G.u.z \end{pmatrix}(m) \otimes n_{(S,\beta)}\right]_{(G\backslash U.S, \left(\begin{smallmatrix} G.u.(z\natural t) \\ G.u.t \end{smallmatrix}\right)\circ(G\backslash U.\alpha))}$$

On the other hand, to compute $\mathcal{L}_U(M)^*(\phi)(m_{(Z,f)})$, I fill the cartesian square

$$\begin{array}{ccc} S' & \xrightarrow{a} & Z \\ {\scriptstyle b}\downarrow & & \downarrow{\scriptstyle f} \\ X & \xrightarrow[\tilde{\phi}]{} & X' \end{array} \qquad (C_2)$$

so that

$$\mathcal{L}_U(M)^*(\phi)(m_{(Z,f)}) = M^*(G\backslash U.a)(m)_{(S',b)}$$

Let

$$F = \theta_X\Big(\mathcal{L}_U(M)^*(\phi)(m_{(Z,f)}), n\Big)$$

Then

$$E''_{(T,g)} = \Big(\mathcal{R}_U(M\hat{\otimes}N)_*(\phi)(F)\Big)_{(T,g)}$$

To compute this expression, I fill the cartesian square

$$\begin{array}{ccc} S'' & \xrightarrow{\;a'\;} & T \\ {\scriptstyle b'}\downarrow & & \downarrow{\scriptstyle g} \\ \widetilde{X} & \xrightarrow[\;\tilde{\phi}\;]{} & \widetilde{X}' \end{array} \qquad (C_3)$$

Then

$$E''_{(T,g)} = (M\hat{\otimes}N)_*(G\backslash U.a')(F_{(S'',b')})$$

Moreover

$$F_{(S'',b')} = \left[M^*\begin{pmatrix} G.u.(s'\natural s'') \\ G.u.s' \end{pmatrix} M^*(G\backslash U.a)(m) \otimes n_{(S'\natural S'', b\natural b')}\right]_{\left(G\backslash U.(S'\natural S''), \begin{pmatrix} G.u.(s'\natural s'') \\ G.u.s'' \end{pmatrix}\right)}$$

Hence

$$E''_{(T,g)} = \left[M^*\begin{pmatrix} G.u.(s'\natural s'') \\ G.u.s' \end{pmatrix} M^*(G\backslash U.a)(m) \otimes n_{(S'\natural S'', b\natural b')}\right]_{\left(G\backslash U.(S'\natural S''), \begin{pmatrix} G.u.(s'\natural s'') \\ G.u.a'(s'') \end{pmatrix}\right)}$$

In those conditions, the set S identifies with the pull-back $S'\natural S''$ of S' and S'' over $\widetilde{X}$: indeed, if $s'\natural s'' \in S'\natural S''$, then $b(s') = b'(s'')$, so

$$\tilde{\phi}b(s') = fa(s') = ga'(s'') = \tilde{\phi}b'(s'')$$

and then I have the element $a(s')\natural a'(s'')$ of $Z\natural T$. Moreover

$$(f\natural g)\Big(a(s')\natural a'(s'')\Big) = fa(s') = \tilde{\phi}b(s')$$

Then as (C_1) is cartesian, there exists a unique $s \in S$ such that

$$b(s') = \beta(s) \qquad a(s')\natural a'(s'') = \alpha(s)$$

So I have a map $s'\natural s'' \mapsto s$ from $S'\natural S''$ to S. Conversely, if $s \in S$, and if $\alpha(s) = z\natural t$, then

$$(f\natural g)(z\natural t) = f(z) = (f\natural g)\alpha(s) = \tilde{\phi}\beta(s)$$

As (C_2) is cartesian, there exists a unique $s' \in S'$ such that

$$b(s') = \beta(s) \qquad a(s') = z$$

Similarly, as

$$(f\natural g)(z\natural t) = g(t) = (f\natural g)\alpha(s) = \tilde{\phi}\beta(s)$$

and as (C_3) is cartesian, there exists a unique $s" \in S"$ such that

$$b'(s") = \beta(s) \qquad a'(s") = t$$

Moreover $b(s') = b'(s")$, so I have the element $s'\natural s"$ of $S'\natural S"$. This correspondence $s \mapsto s'\natural s"$ is clearly inverse of the previous one. As they are morphisms of G-sets, I have $S \simeq S'\natural S"$.

With this identification, I have

$$\alpha(s'\natural s") = a(s')\natural a'(s") \qquad \beta(s'\natural s") = b(s') = b'(s")$$

and I can write $E'_{(T,g)}$ as

$$E'_{(T,g)} = \left[M^*(G\backslash U.\alpha)M^*\begin{pmatrix} G.u.(z\natural t) \\ G.u.z \end{pmatrix}(m) \otimes n_{(S'\natural S",b\natural b')}\right]_{(G\backslash U.S,\begin{pmatrix} G.u.(s'\natural s") \\ G.u.a'(s") \end{pmatrix})}$$

Moreover

$$\begin{pmatrix} G.u.(z\natural t) \\ G.u.z \end{pmatrix}(G\backslash U.\alpha)\big(G.u.(s'\natural s")\big) = \begin{pmatrix} G.u.(z\natural t) \\ G.u.z \end{pmatrix}\big(G.u.(a(s')\natural a'(s"))\big) = \ldots$$

$$\ldots = G.u.a(s') = (G\backslash U.a)\begin{pmatrix} G.u.(s'\natural s") \\ G.u.s' \end{pmatrix}$$

Then

$$M^*(G\backslash U.\alpha)M^*\begin{pmatrix} G.u.(z\natural t) \\ G.u.z \end{pmatrix}(m) = M^*\begin{pmatrix} G.u.(s'\natural s") \\ G.u.s' \end{pmatrix}M^*(G\backslash U.a)(m)$$

whence $E'_{(T,g)} = E"_{(T,g)}$, so $E' = E"$, and condition *ii)* of proposition 1.8.3 holds. This proves lemma 10.5.2. ■

10.6 $\mathcal{R}_U(M)$ as $\mathcal{L}_U(A)$-module

Let G and H be groups, and U be a G-set-H. If A is a Green functor for H, then $\mathcal{L}_U(A)$ is a Green functor for G. If M is an A-module, then $\mathcal{R}_U(M)$ is a Mackey functor for G. I have moreover a bilinear morphism

$$\mathcal{L}_U(A) \times \mathcal{R}_U(M) \to \mathcal{R}_U(A\hat{\otimes}M)$$

that I can compose with the morphism

$$\mathcal{R}_U(A\hat{\otimes}M) \to \mathcal{R}_U(M)$$

deduced of the product $A\hat{\otimes}M \to M$. Now I obtain a product

$$\mathcal{L}_U(A)\hat{\otimes}\mathcal{R}_U(M) \to \mathcal{R}_U(M)$$

Lemma 10.6.1: Let X be a G-set, let (Z,f) be an object of $\mathcal{D}_U(X)$, and $a \in A(G\backslash U.Z)$. If $m \in \mathcal{R}_U(M)(X)$, the above product is such that for any object (T,g) of $\mathcal{D}_U(X)$

$$(a_{(Z,f)}.m)_{(T,g)} = M^*\begin{pmatrix} G.u.(z\natural t) \\ G.u.t \end{pmatrix}\left(A^*\begin{pmatrix} G.u.(z\natural t) \\ G.u.z \end{pmatrix}(a).m_{(Z\natural T,f\natural g)}\right)$$

where the product in the right hand side is the product "." for the A-module M.

Proof: This follows from the definition, and from the formula

$$\theta_X(a_{(Z,f)},m)_{(T,g)} = \left[A^*\begin{pmatrix} G.u.(z\natural t) \\ G.u.z \end{pmatrix}(a)\otimes n_{(Z\natural T,f\natural g)}\right]_{(G\backslash U.(Z\natural T),\left(\begin{smallmatrix} G.u.(z\natural t) \\ G.u.t \end{smallmatrix}\right))}$$

■

Lemma 10.6.2: Let (Z,f) and (Z',f') be objects of $\mathcal{D}_U(X)$. If $a \in A(G\backslash U.Z)$ and $a' \in A(G\backslash U.Z')$, then in $\mathcal{L}_U(A)(X)$, I have

$$a_{(Z,f)}.a'_{(Z',f')} = A^*\begin{pmatrix} G.u.(z\natural z') \\ G.u.z \end{pmatrix}(a).A^*\begin{pmatrix} G.u.(z\natural z') \\ G.u.z' \end{pmatrix}(a')_{(Z\natural Z',f\natural f')}$$

Proof: Indeed, by definition

$$a_{(Z,f)}.a'_{(Z',f')} = \mathcal{L}_U(A)^*\begin{pmatrix} x \\ xx \end{pmatrix}(a_{(Z,f)}\times a'_{(Z',f')}) = \ldots$$

$$\ldots = \mathcal{L}_U(A)^*\begin{pmatrix} x \\ xx \end{pmatrix}\left(A^*(\kappa^U_{Z,Z'})(a\times a')_{(Z.Z',f.f')}\right)$$

To compute this expression, I fill the cartesian square

$$\begin{array}{ccc} P & \xrightarrow{\ i\ } & Z.Z' \\ {\scriptstyle j}\downarrow & & \downarrow{\scriptstyle f.f'} \\ \widetilde{X} & \xrightarrow[\begin{pmatrix} x \\ xx \end{pmatrix}]{} & X\times X \end{array}$$

It is then clear that P is isomorphic to the pull-back of Z and Z' over $\widetilde{X}$, with

$$i(z\natural z') = z.z' \qquad j(z\natural z') = f(z) = f'(z')$$

In other words, the map j identifies with $f\natural f'$. Then

$$a_{(Z,f)}.a'_{(Z',f')} = A^*\begin{pmatrix} G.u.(z\natural z') \\ G.u.z.z' \end{pmatrix}A^*\begin{pmatrix} G.u.z.z' \\ G.u.z\ G.u.z' \end{pmatrix}(a\times a')_{(Z\natural Z',f\natural f')} = \ldots$$

$$\ldots = A^*\begin{pmatrix} G.u.(z\natural z') \\ G.u.z\ G.u.z' \end{pmatrix}(a\times a')_{(Z\natural Z',f\natural f')}$$

Moreover

$$a \times a' = A^*\begin{pmatrix} G.u_1.z\ G.u_2.z' \\ G.u_1.z \end{pmatrix}(a).A^*\begin{pmatrix} G.u_1.z\ G.u_2.z' \\ G.u_2.z' \end{pmatrix}(a')$$

and then denoting by ψ the map $\begin{pmatrix} G.u.(z\natural z') \\ G.u.z\ G.u.z' \end{pmatrix}$, I have

$$a_{(Z,f)}.a'_{(Z',f')} = \dots$$

$$\dots = A^*(\psi)A^*\begin{pmatrix} G.u_1.z\ G.u_2.z' \\ G.u_1.z \end{pmatrix}(a).A^*(\psi)A^*\begin{pmatrix} G.u_1.z\ G.u_2.z' \\ G.u_2.z' \end{pmatrix}(a') = \dots$$

$$\dots = A^*\begin{pmatrix} G.u.(z\natural z') \\ G.u.z \end{pmatrix}(a).A^*\begin{pmatrix} G.u.(z\natural z') \\ G.u.z' \end{pmatrix}(a')_{(Z\natural Z', f\natural f')}$$

which proves the lemma. ■

Proposition 10.6.3: The above product

$$\mathcal{L}_U(A)\hat{\otimes}\mathcal{R}_U(M) \to \mathcal{R}_U(M)$$

turns $\mathcal{R}_U(M)$ into an $\mathcal{L}_U(A)$-module.

Proof: The product "." satisfies by construction the three conditions of proposition 1.8.3. Then the associated product "×" is bifunctorial. It suffices then to check that it is associative and unitary, which is equivalent to check that the product "." itself is associative and unitary.

Let (Z, f) and (Z', f') be objects of $\mathcal{D}_U(X)$, and let $a \in A(G\backslash U.Z)$ and $a' \in A(G\backslash U.Z')$. If $m \in \mathcal{R}_U(M)(X)$, then for any object (T, g) of $\mathcal{D}_U(X)$, I must compute

$$P = \Big(a_{(Z,f)}.(a'_{(Z',f')}.m)\Big)_{(T,g)} = \dots$$

$$\dots = M_*\begin{pmatrix} G.u.(z\natural t) \\ G.u.t \end{pmatrix}\left[A^*\begin{pmatrix} G.u.(z\natural t) \\ G.u.z \end{pmatrix}(a).\Big(a'_{(Z',f')}.m\Big)_{(Z\natural T, f\natural g)}\right]$$

Moreover, identifying $Z'\natural(Z\natural T)$ and $(Z'\natural Z)\natural T$ with $Z'\natural Z\natural T$

$$(a'_{(Z',f')}.m)_{(Z\natural T, f\natural g)} = M_*\begin{pmatrix} G.u.(z'\natural z\natural t) \\ G.u.(z\natural t) \end{pmatrix}\left(A^*\begin{pmatrix} G.u.(z'\natural z\natural t) \\ G.u.z' \end{pmatrix}(a').m_{(Z'\natural Z\natural T, f'\natural f\natural g)}\right)$$

I set

$$E = A^*\begin{pmatrix} G.u.(z'\natural z\natural t) \\ G.u.z' \end{pmatrix}(a').m_{(Z'\natural Z\natural T, f'\natural f\natural g)}$$

and

$$F = A^*\begin{pmatrix} G.u.(z\natural t) \\ G.u.z \end{pmatrix}(a)$$

Then

$$A^*\begin{pmatrix} G.u.(z\natural t) \\ G.u.z \end{pmatrix}(a).(a'_{(Z',f')}.m)\Big)_{(Z\natural T, f\natural g)} = F.M_*\begin{pmatrix} G.u.(z'\natural z\natural t) \\ G.u.(z\natural t) \end{pmatrix}(E) = \dots$$

$$\dots = M_* \begin{pmatrix} G.u.(z'\natural z\natural t) \\ G.u.(z\natural t) \end{pmatrix} \left[A^* \begin{pmatrix} G.u.(z'\natural z\natural t) \\ G.u.(z\natural t) \end{pmatrix} (F).E \right]$$

As

$$A^* \begin{pmatrix} G.u.(z'\natural z\natural t) \\ G.u.(z\natural t) \end{pmatrix} (F) = A^* \begin{pmatrix} G.u.(z'\natural z\natural t) \\ G.u.z \end{pmatrix} (a)$$

I have

$$A^* \begin{pmatrix} G.u.(z'\natural z\natural t) \\ G.u.(z\natural t) \end{pmatrix} (F).E = A^* \begin{pmatrix} G.u.(z'\natural z\natural t) \\ G.u.z \end{pmatrix} (a).A^* \begin{pmatrix} G.u.(z'\natural z\natural t) \\ G.u.z' \end{pmatrix} (a').m_{(Z'\natural Z\natural T, f'\natural f\natural g)}$$

Moreover

$$A^* \begin{pmatrix} G.u.(z'\natural z\natural t) \\ G.u.z \end{pmatrix} (a).A^* \begin{pmatrix} G.u.(z'\natural z\natural t) \\ G.u.z' \end{pmatrix} (a') = \dots$$

$$= A^* \begin{pmatrix} G.u.(z'\natural z\natural t) \\ G.u.(z\natural z') \end{pmatrix} A^* \begin{pmatrix} G.u.(z\natural z') \\ G.u.z \end{pmatrix} (a).A^* \begin{pmatrix} G.u.(z'\natural z\natural t) \\ G.u.(z\natural z') \end{pmatrix} A^* \begin{pmatrix} G.u.(z\natural z') \\ G.u.z' \end{pmatrix} (a) = \dots$$

$$\dots = A^* \begin{pmatrix} G.u.(z'\natural z\natural t) \\ G.u.(z\natural z') \end{pmatrix} \left[A^* \begin{pmatrix} G.u.(z\natural z') \\ G.u.z \end{pmatrix} (a).A^* \begin{pmatrix} G.u.(z\natural z') \\ G.u.z' \end{pmatrix} (a') \right]$$

Setting

$$F' = A^* \begin{pmatrix} G.u.(z'\natural z\natural t) \\ G.u.(z\natural z') \end{pmatrix} \left(A^* \begin{pmatrix} G.u.(z\natural z') \\ G.u.z \end{pmatrix} (a).A^* \begin{pmatrix} G.u.(z\natural z') \\ G.u.z' \end{pmatrix} (a') \right) .m_{(Z'\natural Z\natural T, f'\natural f\natural g)}$$

this gives

$$P = M_* \begin{pmatrix} G.u.(z\natural t) \\ G.u.t \end{pmatrix} M_* \begin{pmatrix} G.u.(z'\natural z\natural t) \\ G.u.(z\natural t) \end{pmatrix} (F')$$

or

$$P = M_* \begin{pmatrix} G.u.(z'\natural z\natural t) \\ G.u.t \end{pmatrix} (F')$$

Setting

$$E' = A^* \begin{pmatrix} G.u.(z'\natural z\natural t) \\ G.u.(z\natural z') \end{pmatrix} \left(A^* \begin{pmatrix} G.u.(z\natural z') \\ G.u.z \end{pmatrix} (a).A^* \begin{pmatrix} G.u.(z\natural z') \\ G.u.z' \end{pmatrix} (a') \right)$$

I have

$$P = M_* \begin{pmatrix} G.u.(z'\natural z\natural t) \\ G.u.t \end{pmatrix} \left[E'.m_{(Z'\natural Z\natural T, f'\natural f\natural g)} \right] = \dots$$

$$\dots = M_* \begin{pmatrix} G.u.(z\natural z'\natural t) \\ G.u.t \end{pmatrix} M_* \begin{pmatrix} G.u.(z'\natural z\natural t) \\ G.u.(z\natural z'\natural t) \end{pmatrix} \left[E'.m_{(Z'\natural Z\natural T, f'\natural f\natural g)} \right]$$

Moreover

$$M_* \begin{pmatrix} G.u.(z'\natural z\natural t) \\ G.u.(z\natural z'\natural t) \end{pmatrix} \left[E'.m_{(Z'\natural Z\natural T, f'\natural f\natural g)} \right] = M^* \begin{pmatrix} G.u.(z\natural z'\natural t) \\ G.u.(z'\natural z\natural t) \end{pmatrix} \left[E'.m_{(Z'\natural Z\natural T, f'\natural f\natural g)} \right] = \dots$$

$$\dots = A^* \begin{pmatrix} G.u.(z\natural z'\natural t) \\ G.u.(z'\natural z\natural t) \end{pmatrix} (E').M^* \begin{pmatrix} G.u.(z\natural z'\natural t) \\ G.u.(z'\natural z\natural t) \end{pmatrix} (m_{(Z'\natural Z\natural T, f'\natural f\natural g)})$$

Finally

$$A^*\begin{pmatrix}G.u.(z\natural z'\natural t)\\ G.u.(z'\natural z\natural t)\end{pmatrix}(E') = \ldots$$

$$\ldots = A^*\begin{pmatrix}G.u.(z\natural z'\natural t)\\ G.u.(z\natural z')\end{pmatrix}\left(A^*\begin{pmatrix}G.u.(z\natural z')\\ G.u.z\end{pmatrix}(a).A^*\begin{pmatrix}G.u.(z\natural z')\\ G.u.z'\end{pmatrix}(a')\right)$$

As $m \in \mathcal{R}_U(M)(X)$, and as the map $\begin{pmatrix}z\natural z'\natural t\\ z'\natural z\natural t\end{pmatrix}$ is bijective, hence a morphism in $\mathcal{D}_U(X)$, I have also

$$M^*\begin{pmatrix}G.u.(z\natural z'\natural t)\\ G.u.(z'\natural z\natural t)\end{pmatrix}(m_{(Z'\natural Z\natural T,f'\natural f\natural g)}) = m_{(Z\natural Z'\natural T,f\natural f'\natural g)}$$

Finally, setting

$$a" = A^*\begin{pmatrix}G.u.(z\natural z')\\ G.u.z\end{pmatrix}(a).A^*\begin{pmatrix}G.u.(z\natural z')\\ G.u.z'\end{pmatrix}(a')$$

this gives

$$P = M_*\begin{pmatrix}G.u.(z\natural z'\natural t)\\ G.u.t\end{pmatrix}\left[A^*\begin{pmatrix}G.u.(z\natural z'\natural t)\\ G.u.(z\natural z')\end{pmatrix}(a").m_{(Z\natural Z'\natural T,f\natural f'\natural g)}\right]$$

and since

$$a_{(Z,f)}.a'_{(Z',g')} = a"_{(Z\natural Z',f\natural f')}$$

I have

$$P = \Big((a_{(Z,f)}.a'_{(Z',g')}).n\Big)_{(T,g)}$$

which proves associativity of the product.

Moreover the unit e_X of $\big(\mathcal{L}_U(A)(X),.\big)$ is equal to

$$e_X = \mathcal{L}_U(A)^*\begin{pmatrix}x\\ \bullet\end{pmatrix}(\varepsilon_{\mathcal{L}_U(A)}) = \mathcal{L}_U(A)^*\begin{pmatrix}x\\ \bullet\end{pmatrix}\Big(A^*(p_{G\backslash U})(\varepsilon_A)_{(U/H,Id)}\Big)$$

As the square

$$\begin{array}{ccc}
\widetilde{X} & \xrightarrow{\widetilde{\begin{pmatrix}x\\ \bullet\end{pmatrix}}} & U/H \\
{\scriptstyle Id}\downarrow & & \downarrow{\scriptstyle Id} \\
\widetilde{X} & \xrightarrow[\widetilde{\begin{pmatrix}x\\ \bullet\end{pmatrix}}]{} & \tilde{\bullet}
\end{array}$$

is trivially cartesian, I have also

$$e_X = A^*(G\backslash U.\widetilde{\begin{pmatrix}x\\ \bullet\end{pmatrix}})A^*(p_{G\backslash U})(\varepsilon_A)_{(\widetilde{X},Id)} = A^*\begin{pmatrix}G.u.\tilde{x}\\ \bullet\end{pmatrix}(\varepsilon_A)_{(\widetilde{X},Id)}$$

Then if $m \in \mathcal{R}_U(M)(X)$, I have, for an object (T,g) of $\mathcal{D}_U(X)$

$$(e_X.m)_{(T,g)} = M^*\begin{pmatrix}G.u.(\tilde{x}\natural t)\\ G.u.t\end{pmatrix}\left(A^*\begin{pmatrix}G.u.(\tilde{x}\natural t)\\ G.u.\tilde{x}\end{pmatrix}A^*\begin{pmatrix}G.u.\tilde{x}\\ \bullet\end{pmatrix}(\varepsilon_A).m_{(\widetilde{X}\natural T,Id\natural g)}\right)$$

As the square

$$\begin{array}{ccc} T & \xrightarrow{Id} & T \\ {\scriptstyle g}\downarrow & & \downarrow{\scriptstyle g} \\ X & \xrightarrow[Id]{} & X \end{array}$$

is trivially cartesian, I have $\widetilde{X}\natural T = T$, and then

$$(e_X.m)_{(T,g)} = A^*\begin{pmatrix} G.u.(\tilde{x}\natural t) \\ \bullet \end{pmatrix}(\varepsilon_A).m_{(T,g)} = A^*\begin{pmatrix} G.u.t \\ \bullet \end{pmatrix}(\varepsilon_A).m_{(T,g)} = m_{(T,g)}$$

since $A^*\begin{pmatrix} G.u.t \\ \bullet \end{pmatrix}(\varepsilon_A)$ is the unit of the ring $\big(A(G\backslash U.T),.\big)$. So the product is unitary, and this completes the proof of the proposition. ∎

10.7 $\mathcal{L}_U(A)$-modules and right adjoints

Let G and H be groups, and U be a G-set-H. If A is a Green functor for H, I have defined $\mathcal{L}_U(A)$, which is a Green functor for the group G, and the functor $N \mapsto N \circ U_{|A}$ from $\mathcal{L}_U(A)$-**Mod** to A-**Mod**: if N is an $\mathcal{L}_U(A)$-module and Z is an H-set, then

$$(N \circ U_{|A})(Z) = \lambda_{A,\bullet}(\varepsilon_A) \times^U (N \circ U)(Z)$$

where the product in the right hand side is the product of the $\mathcal{L}_U(A) \circ U$-module $N \circ U$.

Lemma 10.7.1: Let $n \in (N \circ U)(Z)$. Then denoting by $p_{G\backslash U.(U \circ_H Z)}$ the unique morphism from $G\backslash U.(U \circ_H Z)$ to $\bullet$, I have

$$\lambda_{A,\bullet}(\varepsilon_A) \times^U n = A^*\big(p_{G\backslash U.(U\circ_H Z)}\big)(\varepsilon_A)_{(U\circ_H Z,\pi_Z)}.n$$

where the product in the right hand side is the product "." for the $\mathcal{L}_U(A)$-module N.

Proof: Indeed by definition

$$\lambda_{A,\bullet}(\varepsilon_A) \times^U n = N^*(\delta^U_{\bullet,Z})\big(\lambda_{A,\bullet}(\varepsilon_A) \times n\big)$$

where the product in the right hand side is is the product of the $\mathcal{L}_U(A)$-module N. Let $X = U \circ_H Z$. Then $\lambda_{A,\bullet}(\varepsilon_A) \in \mathcal{L}_U(A)(U/H)$, and $n \in N(X)$. Thus

$$\lambda_{A,\bullet}(\varepsilon_A) \times n = \mathcal{L}_U(A)^*\begin{pmatrix} uH\ x \\ uH \end{pmatrix}\big(\lambda_{A,\bullet}(\varepsilon_A)\big).N^*\begin{pmatrix} uH\ x \\ x \end{pmatrix}(n)$$

So

$$\lambda_{A,\bullet}(\varepsilon_A) \times^U n = \mathcal{L}_U(A)^*(\delta^U_{\bullet,Z})\mathcal{L}_U(A)^*\begin{pmatrix} uH\ x \\ uH \end{pmatrix}\big(\lambda_{A,\bullet}(\varepsilon_A)\big).N^*(\delta^U_{\bullet,Z})N^*\begin{pmatrix} uH\ x \\ x \end{pmatrix}(n)$$

But if $x = (u,z) \in X$, then

$$\begin{pmatrix} uH\ x \\ x \end{pmatrix}\delta^U_{\bullet,Z}(x) = \begin{pmatrix} uH\ x \\ x \end{pmatrix}(uH,x) = x$$

and then

$$N^*(\delta^U_{\bullet,Z})N^*\begin{pmatrix} uH\,x \\ x \end{pmatrix}(n) = n$$

Moreover

$$\begin{pmatrix} uH\,x \\ uH \end{pmatrix}\delta^U_{\bullet,Z}(x) = uH = \left(U \circ_H \begin{pmatrix} z \\ \bullet \end{pmatrix}\right)(x)$$

Hence

$$\mathcal{L}_U(A)^*(\delta^U_{\bullet,Z})\mathcal{L}_U(A)^*\begin{pmatrix} uH\,x \\ uH \end{pmatrix}\left(\lambda_{A,\bullet}(\varepsilon_A)\right) = \mathcal{L}_U(A)^*\left(U \circ_H \begin{pmatrix} z \\ \bullet \end{pmatrix}\right)\left(\lambda_{A,\bullet}(\varepsilon_A)\right) = \dots$$

$$\dots = \mathcal{L}_U(A)^*\left(U \circ_H \begin{pmatrix} z \\ \bullet \end{pmatrix}\right)\left(A^*(p_{G\backslash U})(\varepsilon_A)_{(U/H,\pi_\bullet)}\right)$$

To compute this expression, I use the cartesian square

$$\begin{array}{ccc} U \circ_H Z & \xrightarrow{\;U \circ_H \binom{Z}{\bullet}\;} & U/H \\ {\scriptstyle \pi_Z}\downarrow & & \downarrow{\scriptstyle \pi_\bullet} \\ U \circ_H Z & \xrightarrow[\;U \circ_H \binom{Z}{\bullet}\;]{} & U/H \end{array}$$

and then

$$\mathcal{L}_U(A)^*(\delta^U_{\bullet,Z})\mathcal{L}_U(A)^*\begin{pmatrix} uH\,x \\ uH \end{pmatrix}\left(\lambda_{A,\bullet}(\varepsilon_A)\right) = \dots$$

$$\dots = A^*\left[G\backslash U.\left(U \circ_H \begin{pmatrix} z \\ \bullet \end{pmatrix}\right)\right]A^*(p_{G\backslash U})(\varepsilon_A)_{(U\circ_H Z,\pi_Z)} = \dots$$

$$\dots = A^*\left(p_{G\backslash U.(U\circ_H Z)}\right)(\varepsilon_A)_{(U\circ_H Z,\pi_Z)}$$

which proves the lemma. ■

Proposition 10.7.2: Let G and H be groups, and U be a G-set-H. If A is a Green functor for H, then the correspondence $M \mapsto \mathcal{R}_U(M)$ is a functor from A-Mod to $\mathcal{L}_U(A)$-Mod, which is right adjoint to the functor $N \mapsto N \circ U_{|A}$.

Proof: I already know that the correspondence $M \mapsto \mathcal{R}_U(M)$ is a functor between the categories of Mackey functors. Moreover, if $\phi : M \to M'$ is a morphism of A-modules, then the morphism $\mathcal{R}_U(\phi)$ is given for a G-set X, an element $m \in \mathcal{R}_U(M)(X)$, and an object (T,g) of $\mathcal{D}_U(X)$ by

$$\mathcal{R}_U(\phi)(m)_{(T,g)} = \phi_{G\backslash U.T}(m_{(T,g)})$$

Then if (Z,f) is an object of $\mathcal{D}_U(X)$ and if $m \in M(G\backslash U.Z)$, I have

$$\mathcal{R}_U(\phi)(a_{(Z,f)}.m)_{(T,g)} = \phi_{G\backslash U.T}\left((a_{(Z,f)}.m)_{(T,g)}\right) = \dots$$

$$\dots = \phi_{G\backslash U.T}M^*\begin{pmatrix} G.u.(z\natural t) \\ G.u.t \end{pmatrix}\left[A^*\begin{pmatrix} G.u.(z\natural t) \\ G.u.z \end{pmatrix}(a).m_{(Z\natural T,f\natural g)}\right]$$

As ϕ is a morphism of Mackey functors, it is also

$$\mathcal{R}_U(\phi)(a_{(Z,f)}.m)_{(T,g)} = M^*\begin{pmatrix} G.u.(z\natural t) \\ G.u.t \end{pmatrix} \phi_{G\backslash U.(Z\natural T)}\left[A^*\begin{pmatrix} G.u.(z\natural t) \\ G.u.z \end{pmatrix}(a).m_{(Z\natural T, f\natural g)}\right]$$

Finally as ϕ is a morphism of A-modules, I have

$$\mathcal{R}_U(\phi)(a_{(Z,f)}.m)_{(T,g)} = \dots$$

$$\dots = M^*\begin{pmatrix} G.u.(z\natural t) \\ G.u.t \end{pmatrix}\left[A^*\begin{pmatrix} G.u.(z\natural t) \\ G.u.z \end{pmatrix}(a).\phi_{G\backslash U.(Z\natural T)}(m_{(Z\natural T, f\natural g)})\right] = \dots$$

$$\dots = \Big(a_{(Z,f)}.\mathcal{R}_U(\phi)(m)\Big)_{(T,g)}$$

which proves that $\mathcal{R}_U(\phi)$ is a morphism of $\mathcal{L}_U(A)$-modules.

Let θ be a morphism of A-modules from $N \circ U_{|A}$ to M. As $N \circ U_{|A}$ is a direct summand of $N \circ U$ as Mackey functor, I have a morphism θ' of Mackey functors from $N \circ U$ to M, defined on the H-set Z by

$$\theta'_Z(m) = \theta\Big(\lambda_{A,\bullet}(\varepsilon_A) \times^U m\Big)$$

By adjunction, it correspond to θ' a morphism ψ of Mackey functors from N to $\mathcal{R}_U(M)$, defined as follows: if X is a G-set and (T,g) an object of $\mathcal{D}_U(X)$, and if $n \in N(X)$, then

$$\psi(n)_{(T,g)} = \theta'_{G\backslash U.T} N_*(\nu_T) N^*(g_X)(n) \in M(G\backslash U.T)$$

Thus

$$\psi(n)_{(T,g)} = \theta_{G\backslash U.T}\Big(\lambda_{A,\bullet}(\varepsilon_A) \times^U N_*(\nu_T)N^*(g_X)(n)\Big)$$

The previous lemma shows that setting $T' = G\backslash U.T$

$$\lambda_{A,\bullet}(\varepsilon_A) \times^U N_*(\nu_T)N^*(g_X)(n) = A^*(p_{G\backslash U.(U\circ_H T')})(\varepsilon_A)_{(U\circ_H T', \pi_{T'})}.N_*(\nu_T)N^*(g_X)(n)$$

But as N is an $\mathcal{L}_U(A)$-module

$$A^*(p_{G\backslash U.(U\circ_H T')})(\varepsilon_A)_{(U\circ_H T', \pi_{T'})}.N_*(\nu_T)N^*(g_X)(n) = \dots$$

$$\dots = N_*(\nu_T)\Big[\mathcal{L}_U(A)^*(\nu_T)\Big(A^*(p_{G\backslash U.(U\circ_H T')})(\varepsilon_A)_{(U\circ_H T', \pi_{T'})}\Big).N^*(g_X)(n)\Big]$$

Let d_T be the map from T to $\tilde{T}$ defined by $d_T(t) = \big(t, g_U(t)\big)$. The square

$$\begin{array}{ccc} T & \xrightarrow{\nu_T} & U \circ_H T' \\ {\scriptstyle d_T}\downarrow & & \downarrow{\scriptstyle \pi_{T'}} \\ \tilde{T} & \xrightarrow[\tilde{\nu_T}]{} & U \circ_H T' \end{array}$$

is cartesian. So

$$\mathcal{L}_U(A)^*(\nu_T)\Big(A^*(p_{G\backslash U.(U\circ_H T')})(\varepsilon_A)_{(U\circ_H T', \pi_{T'})}\Big) = \dots$$

$$\ldots = A^*(G\backslash U.\nu_T)A^*(p_{G\backslash U.(U\circ_H T')})(\varepsilon_A)_{(T,d_T)} = \ldots$$
$$\ldots = A^*(p_{G\backslash U.T})(\varepsilon_A)_{(T,d_T)}$$

This gives finally

$$\psi(n)_{(T,g)} = \theta_{G\backslash U.T}N_*(\nu_T)\Big(A^*(p_{G\backslash U.T})(\varepsilon_A)_{(T,d_T)}.N^*(g_X)(n)\Big)$$

I must show that ψ is a morphism of $\mathcal{L}_U(A)$-modules from N to $\mathcal{R}_U(M)$. So let (Z,f) be an object of $\mathcal{D}_U(X)$, and $a \in A(G\backslash U.Z)$. Then by definition of the product on $\mathcal{R}_U(M)$, I have

$$\Big(a_{(Z,f)}.n\Big)_{(T,g)} = M_*\begin{pmatrix} G.u.(z\natural t) \\ G.u.t \end{pmatrix}\left[A^*\begin{pmatrix} G.u.(z\natural t) \\ G.u.z \end{pmatrix}(a).\psi(n)_{(Z\natural T, f\natural g)}\right]$$

Moreover

$$\psi(n)_{(Z\natural T,f\natural g)} = \theta_{G\backslash U.(Z\natural T)}N_*(\nu_{Z\natural T})\Big(A^*(p_{G\backslash U.(Z\natural T)})(\varepsilon_A)_{(Z\natural T,d_{Z\natural T})}.N^*((f\natural g)_X)(n)\Big)$$

I set

$$E = N_*(\nu_{Z\natural T})\Big(A^*(p_{G\backslash U.(Z\natural T)})(\varepsilon_A)_{(Z\natural T,d_{Z\natural T})}.N^*\big((f\natural g)_X\big)(n)\Big)$$

so that

$$\Big(a_{(Z,f)}.n\Big)_{(T,g)} = M_*\begin{pmatrix} G.u.(z\natural t) \\ G.u.t \end{pmatrix}\left[A^*\begin{pmatrix} G.u.(z\natural t) \\ G.u.z \end{pmatrix}(a).\theta_{G\backslash U.(Z\natural T)}(E)\right]$$

As θ is a morphism of A-modules, I have

$$A^*\begin{pmatrix} G.u.(z\natural t) \\ G.u.z \end{pmatrix}(a).\theta_{G\backslash U.(Z\natural T)}(E) = \theta_{G\backslash U.(Z\natural T)}\left(A^*\begin{pmatrix} G.u.(z\natural t) \\ G.u.z \end{pmatrix}(a).^U E\right)$$

where the product $.^U$ is the product of the A-module $N\circ U_{|A}$. Setting $Z' = G\backslash U.(Z\natural T)$, and

$$a' = \lambda_{A,Z'}A^*\begin{pmatrix} G.u.(z\natural t) \\ G.u.z \end{pmatrix}(a)$$

I have

$$A^*\begin{pmatrix} G.u.(z\natural t) \\ G.u.z \end{pmatrix}(a).^U E = a'.^U E$$

Then by definition

$$a'.^U E = (N\circ U)^*\begin{pmatrix} z' \\ z'z' \end{pmatrix}(a' \times^U E) = (N\circ U)^*\begin{pmatrix} z' \\ z'z' \end{pmatrix}N^*(\delta^U_{Z',Z'})(a'\times E)$$

where the product in the right hand side is the product of the $\mathcal{L}_U(A)$-module N. But if $(u,z') \in U \circ_H Z'$, then

$$\delta^U_{Z',Z'}\left(U\circ_H\begin{pmatrix} z' \\ z'z' \end{pmatrix}\right)(u,z') = \big((u,z'),(u,z')\big)$$

It follows that

$$a'.^U E = N^*\begin{pmatrix} (u,z') \\ (u,z')(u,z') \end{pmatrix}(a'\times E) = a'.E$$

where the product in the right hand side is the product "." of the $\mathcal{L}_U(A)$-module N.

Now setting

$$E' = A^*(p_{G\backslash U.(Z\natural T)})(\varepsilon_A)_{(Z\natural T, d_{Z\natural T})}.N^*\Big((f\natural g)_X\Big)(n)$$

I have $E = N_*(\nu_{Z\natural T})(E')$. But as N is an $\mathcal{L}_U(A)$-module

$$a'.E = N_*(\nu_{Z\natural T})\Big(\mathcal{L}_U(A)^*(\nu_{Z\natural T})(a').E'\Big)$$

Moreover

$$a' = A^*(\eta_{Z'})A^*\begin{pmatrix} G.u.(z\natural t) \\ G.u.z \end{pmatrix}(a)_{(U\circ_H Z', \pi_{Z'})}$$

To compute $\mathcal{L}_U(A)^*(\nu_{Z\natural T})(a')$, I use the cartesian square

$$\begin{array}{ccc} Z\natural T & \xrightarrow{\nu_{Z\natural T}} & U\circ_H Z' \\ {\scriptstyle d_{Z\natural T}}\downarrow & & \downarrow{\scriptstyle \pi_{Z'}} \\ \widetilde{Z\natural T} & \xrightarrow[\widetilde{\nu_{Z\natural T}}]{} & U\circ_H Z' \end{array}$$

which shows that

$$\mathcal{L}_U(A)^*(\nu_{Z\natural T})(a') = A^*(G\backslash U.\nu_{Z\natural T})A^*(\eta_{Z'})A^*\begin{pmatrix} G.u.(z\natural t) \\ G.u.z \end{pmatrix}(a)_{(Z\natural T, d_{Z\natural T})}$$

As $A^*(G\backslash U.\nu_{Z\natural T})A^*(\eta_{Z'})$ is the identity (by an adjunction argument), it follows that

$$\mathcal{L}_U(A)^*(\nu_{Z\natural T})(a') = A^*\begin{pmatrix} G.u.(z\natural t) \\ G.u.z \end{pmatrix}(a)_{(Z\natural T, d_{Z\natural T})}$$

Then

$$\mathcal{L}_U(A)^*(\nu_{Z\natural T})(a').E' = \ldots$$

$$\ldots = A^*\begin{pmatrix} G.u.(z\natural t) \\ G.u.z \end{pmatrix}(a)_{(Z\natural T, d_{Z\natural T})}.A^*(p_{G\backslash U.(Z\natural T)})(\varepsilon_A)_{(Z\natural T, d_{Z\natural T})}.N^*\Big((f\natural g)_X\Big)(n)$$

Moreover, the square

$$\begin{array}{ccc} Z\natural T & \xrightarrow{Id} & Z\natural T \\ {\scriptstyle Id}\downarrow & & \downarrow{\scriptstyle d_{Z\natural T}} \\ Z\natural T & \xrightarrow[d_{Z\natural T}]{} & \widetilde{Z\natural T} \end{array}$$

is cartesian, because $d_{Z\natural T}$ is injective. Then by lemma 10.6.2

$$A^*\begin{pmatrix} G.u.(z\natural t) \\ G.u.z \end{pmatrix}(a)_{(Z\natural T, d_{Z\natural T})}.A^*(p_{G\backslash U.(Z\natural T)})(\varepsilon_A)_{(Z\natural T, d_{Z\natural T})} = A^*\begin{pmatrix} G.u.(z\natural t) \\ G.u.z \end{pmatrix}(a)_{(Z\natural T, d_{Z\natural T})}$$

so

$$\mathcal{L}_U(A)^*(\nu_{Z\natural T})(a').E' = A^*\begin{pmatrix} G.u.(z\natural t) \\ G.u.z \end{pmatrix}(a)_{(Z\natural T, d_{Z\natural T})}.N^*\Big((f\natural g)_X\Big)(n)$$

As moreover the square

$$\begin{array}{ccc} Z\natural T & \xrightarrow{\begin{pmatrix} z\natural t \\ z \end{pmatrix}} & Z \\ {\scriptstyle d_{Z\natural T}}\downarrow & & \downarrow{\scriptstyle d_Z} \\ \widetilde{Z\natural T} & \xrightarrow[\widetilde{\begin{pmatrix} z\natural t \\ z \end{pmatrix}}]{} & \tilde{Z} \end{array}$$

is cartesian, I have

$$A^*\begin{pmatrix} G.u.(z\natural t) \\ G.u.z \end{pmatrix}(a)_{(Z\natural T, d_{Z\natural T})} = \mathcal{L}_U(A)^*\begin{pmatrix} z\natural t \\ z \end{pmatrix}(a_{(Z,d_Z)})$$

Then

$$\mathcal{L}_U(A)^*(\nu_{Z\natural T})(a').E' = \mathcal{L}_U(A)^*\begin{pmatrix} z\natural t \\ z \end{pmatrix}(a_{(Z,d_Z)}).N^*\big((f\natural g)_X\big)(n)$$

Moreover $(f\natural g)_X = f_X \circ \binom{z\natural t}{z}$, so

$$\mathcal{L}_U(A)^*(\nu_{Z\natural T})(a').E' = \mathcal{L}_U(A)^*\begin{pmatrix} z\natural t \\ z \end{pmatrix}(a_{(Z,d_Z)}).N^*\begin{pmatrix} z\natural t \\ z \end{pmatrix}N^*(f_X)(n) = \ldots$$

$$\ldots = N^*\begin{pmatrix} z\natural t \\ z \end{pmatrix}\Big[a_{(Z,d_Z)}.N^*(f_X)(n)\Big]$$

Finally, I have

$$\Big(a_{(Z,f)}.n\Big)_{(T,g)} = M_*\begin{pmatrix} G.u.(z\natural t) \\ G.u.t \end{pmatrix}\theta_{G\backslash U.(Z\natural T)}N_*(\nu_{Z\natural T})N^*\begin{pmatrix} z\natural t \\ z \end{pmatrix}\Big[a_{(Z,d_Z)}.N^*(f_X)(n)\Big]$$

As θ is a morphism of Mackey functors, I have also

$$M_*\begin{pmatrix} G.u.(z\natural t) \\ G.u.t \end{pmatrix}\theta_{G\backslash U.(Z\natural T)} = \theta_{G\backslash U.T}N_*\left(U\circ_H\begin{pmatrix} G.u.(z\natural t) \\ G.u.t \end{pmatrix}\right)$$

Moreover

$$N_*\left(U\circ_H\begin{pmatrix} G.u.(z\natural t) \\ G.u.t \end{pmatrix}\right)N_*(\nu_{Z\natural T}) = N_*(\nu_T)N_*\begin{pmatrix} z\natural t \\ t \end{pmatrix}$$

This gives finally

$$\Big(a_{(Z,f)}.n\Big)_{(T,g)} = \theta_{G\backslash U.T}N_*(\nu_T)N_*\begin{pmatrix} z\natural t \\ t \end{pmatrix}N^*\begin{pmatrix} z\natural t \\ z \end{pmatrix}\Big[a_{(Z,d_Z)}.N^*(f_X)(n)\Big]$$

The square

$$\begin{array}{ccc} Z\natural T & \xrightarrow{\begin{pmatrix} z\natural t \\ z \end{pmatrix}} & Z \\ {\scriptstyle \begin{pmatrix} z\natural t \\ t \end{pmatrix}}\downarrow & & \downarrow{\scriptstyle f} \\ T & \xrightarrow[g]{} & X \end{array}$$

being cartesian by definition, I have

$$N_*\begin{pmatrix} z\natural t \\ t \end{pmatrix} N^*\begin{pmatrix} z\natural t \\ z \end{pmatrix} = N^*(g)N_*(f)$$

and then

$$\left(a_{(Z,f)}.n\right)_{(T,g)} = \theta_{G\backslash U.T}N_*(\nu_T)N^*(g)N_*(f)\left[a_{(Z,d_Z)}.N^*(f_X)(n)\right]$$

Let p be the projection from $\widetilde{X}$ onto X. Then $f_X = f \circ p$, and

$$N_*(f)\left[a_{(Z,d_Z)}.N^*(f_X)(n)\right] = N_*(f)\left[a_{(Z,d_Z)}.N^*(f)N^*(p)(n)\right] = \dots$$
$$\dots = \mathcal{L}_U(A)_*(f)(a_{(Z,d_Z)}).N^*(p)(n)$$

Moreover

$$N^*(g)\left[\mathcal{L}_U(A)_*(a_{(Z,d_Z)}).N^*(p)(n)\right] = \dots$$
$$\dots = \mathcal{L}_U(A)^*(g)\mathcal{L}_U(A)_*(f)(a_{(Z,d_Z)}).N^*(g)N^*(p)(n) = \dots$$
$$\dots = \mathcal{L}_U(A)^*(g)\mathcal{L}_U(A)_*(f)(a_{(Z,d_Z)}).N^*(g_X)(n)$$

Finally $\mathcal{L}_U(A)_*(f)(a_{(Z,d_Z)}) = a_{\tilde{f}d_Z}$, and the square

$$\begin{array}{ccc} Z\natural T & \xrightarrow{\begin{pmatrix} z\natural t \\ z \end{pmatrix}} & Z \\ {\scriptstyle \begin{pmatrix} z\natural t \\ d_T(t) \end{pmatrix}}\Big\downarrow & & \Big\downarrow{\scriptstyle \tilde{f}d_Z} \\ \tilde{T} & \xrightarrow[\tilde{g}]{} & \widetilde{X} \end{array}$$

is cartesian. Then

$$\mathcal{L}_U(A)^*(g)(a_{\tilde{f}d_Z}) = A^*\begin{pmatrix} G.u.(z\natural t) \\ G.u.z \end{pmatrix}(a)_{\left(Z\natural T,\begin{pmatrix} z\natural t \\ d_T(t) \end{pmatrix}\right)}$$

whence

$$\left(a_{(Z,f)}.n\right)_{(T,g)} = \theta_{G\backslash U.T}N_*(\nu_T)\left[A^*\begin{pmatrix} G.u.(z\natural t) \\ G.u.z \end{pmatrix}(a)_{\left(Z\natural T,\begin{pmatrix} z\natural t \\ d_T(t) \end{pmatrix}\right)}.N^*(g_X)(n)\right]$$

I must compare this element with $\psi(a_{(Z,f)}.n)_{(T,g)}$. By definition of ψ, I have

$$\psi(a_{(Z,f)}.n)_{(T,g)} = \theta_{G\backslash U.T}N_*(\nu_T)\left(A^*(p_{G\backslash U.T})(\varepsilon_A)_{(T,d_T)}.N^*(g_X)(a_{(Z,f)}.n)\right)$$

But

$$N^*(g_X)(a_{(Z,f)}.n) = \mathcal{L}_U(A)^*(g_X)(a_{(Z,f)}).N^*(g_X)(n)$$

Let P, α and β fill the cartesian square

$$\begin{array}{ccc} P & \xrightarrow{\alpha} & Z \\ {\scriptstyle\beta}\downarrow & & \downarrow{\scriptstyle f} \\ T & \xrightarrow[\widetilde{g_X}]{} & X \end{array}$$

Then

$$\mathcal{L}_U(A)^*(g_X)(a_{(Z,f)}) = A^*(G\backslash U.\alpha)(a)_{(P,\beta)}$$

and I can write

$$\psi(a_{(Z,f)}.n)_{(T,g)} = \theta_{G\backslash U.T} N_*(\nu_T)\left[A^*(p_{G\backslash U.T})(\varepsilon_A)_{(T,d_T)}.A^*(G\backslash U.\alpha)(a)_{(P,\beta)}.N^*(g_X)(n)\right]$$

To compute this product by lemma 10.6.2, I fill the cartesian square

$$\begin{array}{ccc} T\natural P & \xrightarrow{\alpha'} & P \\ {\scriptstyle\beta'}\downarrow & & \downarrow{\scriptstyle\beta} \\ T & \xrightarrow[d_T]{} & \tilde{T} \end{array}$$

so that

$$A^*(p_{G\backslash U.T})(\varepsilon_A)_{(T,d_T)}.A^*(G\backslash U.\alpha)(a)_{(P,\beta)} = \ldots$$

$$\ldots = A^*\begin{pmatrix} G.u.(t\natural p) \\ G.u.t \end{pmatrix}(\varepsilon_A).A^*\begin{pmatrix} G.u.(t\natural p) \\ p \end{pmatrix} A^*(G\backslash U.\alpha)(a)_{(T\natural P,d_T\natural\beta)}$$

Then the square

$$\begin{array}{ccc} T\natural P & \xrightarrow{\alpha\alpha'} & Z \\ {\scriptstyle\beta'}\downarrow & & \downarrow{\scriptstyle f} \\ T & \xrightarrow[\widetilde{g_X} d_T]{} & X \end{array}$$

is composed of the two previous cartesian squares. It is then cartesian, which proves that $T\natural P$ identifies with $Z\natural T$. With this identification, I have $\beta'(z\natural t) = t$, and $\alpha'\alpha(z\natural t) = z$. Then

$$\psi(a_{(Z,f)}.n)_{(T,g)} = \theta_{G\backslash U.T} N_*(\nu_T)\left[A^*\begin{pmatrix} G.u.(z\natural t) \\ G.u.t \end{pmatrix}(a)_{\left(Z\natural T,\left(\substack{z\natural t \\ d_T(t)}\right)\right)}.N^*(g_X)(n)\right]$$

which proves finally that

$$\psi(a_{(Z,f)}.n) = a_{(Z,f)}.\psi(n)$$

and ψ is a morphism of $\mathcal{L}_U(A)$-modules.

Conversely, being given a morphism of $\mathcal{L}_U(A)$-modules ψ from N to $\mathcal{R}_U(M)$, I have a morphism of $\mathcal{L}_U(A)\circ U$-modules $\psi\circ U$ from $N\circ U$ to $\mathcal{R}_U(M)\circ U$. Composing this morphism with the morphism of Mackey functors from $\mathcal{R}_U(M)\circ U$ to M, I obtain a morphism of Mackey functors from $N\circ U$ to M. If I know that this morphism is compatible with the action of A, then it will be entirely determined by its restriction to $N\circ U_{|A}$: indeed, the image of an element is then equal to the image of its product by $\lambda_{A,\bullet}(\varepsilon_A)$. In other words, the proposition will be proved if I know that the co-unit morphism

$$\Phi : \mathcal{R}_U(M)\circ U \to M$$

is compatible with the action of A. But Φ is defined for an H-set Z and an element $m\in(\mathcal{R}_U(M)\circ U)(Z)=\mathcal{R}_U(M)(U\circ_H Z)$ by

$$\Phi_Z(m)=M_*(\eta_Z)(m_{(U\circ_H Z,\pi_Z)})\in M(Z)$$

Let $a\in A(Z)$. Then a acts on $(\mathcal{R}_U(M)\circ U)(Z)$ through the morphism $\lambda : A\to \mathcal{L}_U(A)\circ U$. In other words

$$a.m=\lambda_{A,Z}(a).^U m$$

I already observed that $\lambda_{A,Z}(a).^U m=\lambda_{A,Z}(a).m$, this product being the product of $\mathcal{R}_U(M)$ as an $\mathcal{L}_U(A)$-module. Then

$$\Phi_Z(a.m)=M_*(\eta_Z)\Big(\big(\lambda_{A,Z}(a).m\big)_{(U\circ_H Z,\pi_Z)}\Big)$$

But $\lambda_{A,Z}(a)=A^*(\eta_Z)(a)_{(U\circ_H Z,\pi_Z)}$, so setting $T=U\circ_H Z$, I have

$$\big(\lambda_{A,Z}(a).m\big)_{(U\circ_H Z,\pi_Z)}=\Big(A^*(\eta_Z)(a)_{(T,\pi_Z)}.m\Big)_{(T,\pi_Z)}$$

To compute this product, I must build $T\natural T$: but the square

$$\begin{array}{ccc} T & \xrightarrow{Id} & T\\ {\scriptstyle Id}\downarrow & & \downarrow{\scriptstyle \pi_Z}\\ T & \xrightarrow[\pi_Z]{} & \tilde{T}\end{array}$$

is cartesian since π_Z is injective. Thus $T\natural T=T$, and then

$$\Big(A^*(\eta_Z)(a)_{(T,\pi_Z)}.m\Big)_{(T,\pi_Z)}=M_*(Id)\Big[A^*(Id)A^*(\eta_Z)(a).m_{(T\natural T,\pi_Z\natural\pi_Z)}\Big]=\ldots$$

$$\ldots=A^*(\eta_Z)(a).m_{(T,\pi_Z)}$$

In those conditions

$$\Phi_Z(a.m)=M_*(\eta_Z)\Big(A^*(\eta_Z)(a).m_{(T,\pi_Z)}\Big)=a.M_*(\eta_Z)(m_{(T,\pi_Z)})=a.\Phi_Z(m)$$

which proves that Φ is compatible with the action of A, and completes the proof of the proposition. ∎

10.8 Examples and applications

10.8.1 Induction and restriction

Let G be a group, and H be a subgroup of G. Let U be the set G, viewed as a G-set-H, and V be the set G, viewed as an H-set-G. Then I know that if A is a Green functor for G, the functors $A \circ U$ and $\mathcal{L}_V(A)$ are both equal to the restriction of A to H. It is not clear a priori that the products defined on $A \circ U$ and $\mathcal{L}_V(A)$ are the same.

But they are: the isomorphism

$$\mathcal{L}_V(A)(X) \simeq \mathrm{Res}_H^G A(X)$$

comes from the fact that the category $\mathcal{D}_V(X)$ identifies with H-**set**$\downarrow_X$, which has a final object (X, Id). Moreover, the set $H\backslash V.X$ identifies with $\mathrm{Ind}_H^G X$. If a and a' are elements of $A(\mathrm{Ind}_H^G X)$, then

$$a_{(X,Id)}.a'_{(X,Id)} = A^*\begin{pmatrix} H.v.(x_1 \natural x_2) \\ H.v.x_1 \end{pmatrix}(a).A^*\begin{pmatrix} H.v.(x_1 \natural x_2) \\ H.v.x_2 \end{pmatrix}(a')_{(X\natural X, Id\natural Id)}$$

But obviously $X \natural X = X$, so

$$a_{(X,Id)}.a'_{(X,Id)} = a.a'_{(X,Id)}$$

and the product of $\mathcal{L}_V(A)$ is the product of $\mathrm{Res}_H^G A$.

A similar argument shows that if B is a Green functor for H, then the products defined on

$$\mathcal{L}_U(B) \simeq \mathrm{Ind}_H^G B \simeq B \circ V$$

are the same.

10.8.2 The case $U/H = \bullet$

Proposition 10.8.1: Let G and H be groups, and U be a G-set-H. The following conditions are equivalent:

1. **The group H is transitive on U, i.e. $U/H = \bullet$.**
2. **For any Green functor A for H, the morphism**

$$\lambda_A : A \to \mathcal{L}_U(A) \circ U$$

   is unitary.

3. **The functor $A \mapsto \mathcal{L}_U(A)$ from $Green(H)$ to $Green(G)$ is left adjoint to the functor $B \mapsto B \circ U$.**

Proof: I recall the formulae

$$\lambda_{A,\bullet}(\varepsilon_A) = A^*(p_{G\backslash U})(\varepsilon_A)_{(U/H,\pi_\bullet)} \qquad \varepsilon_{\mathcal{L}_U(A)\circ U} = A^*(p_{G\backslash U.(U/H)^2})(\varepsilon_A)_{((U/H)^2,Id)}$$

Moreover, as $\pi_\bullet$ is the diagonal injection from U/H to $(U/H)^2$, it is a morphism from $(U/H,\pi_\bullet)$ to $\big((U/H)^2, Id\big)$ in the category $\mathcal{D}_U(U/H)$. So in $\mathcal{L}_U(A)(U/H)$, I have also

$$\lambda_{A,\bullet}(\varepsilon_A) = A_*(\pi_\bullet)A^*(p_{G\backslash U})(\varepsilon_A)_{((U/H)^2,Id)}$$

If 1) holds, then $U/H \simeq (U/H)^2 \simeq \bullet$, and in this case

$$\lambda_{A,\bullet}(\varepsilon_A) = \varepsilon_{\mathcal{L}_U(A)\circ U}$$

so 2) holds. Conversely if 2) holds, then for any Green functor A for H and any $\mathcal{L}_U(A)$-module N, I have

$$N \circ U_{|A} = N \circ U$$

Then

$$\mathrm{Hom}_A(A, N\circ U_{|A}) \simeq (N\circ U)(\bullet) = N(U/H) \simeq \mathrm{Hom}_{\mathcal{L}_U(A)}(\mathcal{L}_U(A), N) \simeq N(\bullet)$$

Taking for A the Burnside functor b, and for N the module $\mathcal{L}_U(b) \simeq b_{U/H}$, this gives

$$b(U/H) \simeq b\big((U/H)^2\big)$$

But $(U/H)^2$ is the disjoint union of its diagonal, isomorphic to U/H, and of its complement C in $(U/H)^2$. Then the above isomorphism shows that $b(C) = 0$. But the only set X such that $b(X) = 0$ is the empty set. Then $U/H \simeq (U/H)^2$, and $U/H \simeq \bullet$, so 1) holds.

Now if 3) holds, then as λ_A is the unit of the adjunction, it is a unitary morphism of Green functors. So 2) holds.

And if 2) holds, then to prove 3), I must prove that the unitary morphisms of Green functors from $\mathcal{L}_U(A)$ to B are sent by adjunction to unitary morphisms of Green functors from A to $B\circ U$, and conversely. But if ϕ is a unitary morphism of Green functors from $\mathcal{L}_U(A)$ to B, then $\phi\circ U$ is a unitary morphism of Green functors from $\mathcal{L}_U(A)\circ U$ to $B\circ U$. Composing this morphism with λ_A, which is a unitary morphism of Green functors by hypothesis, I obtain the morphism from A to $B\circ U$ which is associated to ϕ, hence this is a unitary morphism of Green functors.

To prove the inverse correspondence, it suffices similarly to prove that the co-unit of the adjunction

$$\Theta : \mathcal{L}_U(B\circ U) \to B$$

is a unitary morphism of Green functors. But if X is a G-set, as $U/H = \bullet$, the objects of $\mathcal{D}_U(X)$ are just the G-sets over X. Let (Z,f) be such a set, and $b \in (B\circ U)(G\backslash U.Z)$ (furthermore the set $U.Z$ is equal to $U\times Z$ in this case). Then

$$\Theta_X(b_{(Z,f)}) = B_*(f)B^*(\nu_Z)(b)$$

The unit of $\mathcal{L}_U(B\circ U)$ is by definition

$$\varepsilon_{\mathcal{L}_U(B\circ U)} = (B\circ U)^*(p_{G\backslash U})(\varepsilon_{B\circ U})_{(U/H,Id)} = B^*(U\circ_H p_{G\backslash U})B^*(p_{U/H})(\varepsilon_B)_{(\bullet,Id)} = \cdots$$
$$\cdots = B^*(p_{U\circ_H(G\backslash U)})(\varepsilon_B)_{(\bullet,Id)}$$

So

$$\Theta_\bullet(\varepsilon_{\mathcal{L}_U(B\circ U)}) = B_*(Id)B^*(\nu_\bullet)B^*(p_{U\circ_H(G\backslash U)})(\varepsilon_B) = B^*(p_\bullet)(\varepsilon_B) = \varepsilon_B$$

Thus Θ is unitary. It remains to see that it is a morphism of Green functors. So let Y be a G-set, let (T, g) be a set over Y, and let $b' \in B(G\backslash U.T)$. Then

$$b_{(Z,f)} \times b'_{(T,g)} = B^*(\kappa^U_{Z,T})(b \times b')_{(Z.T,f.g)}$$

But $Z.T = Z \times T$ and $f.g = f \times g$. Then

$$\Theta_{X\times Y}(b_{(Z,f)} \times b'_{(T,g)}) = B_*(f \times g)B^*(\nu_{Z\times T})B^*(\kappa^U_{Z,T})(b \times b')$$

Let moreover u be any element of H. Then

$$\nu_{Z\times T}(z,t) = \big(u, G(u,z,t)\big)$$

so

$$\kappa^U_{Z,T}\nu_{Z\times T}(z,t) = \Big[\big(u, G(u,z)\big), \big(u, G(u,t)\big)\Big] = \big((\nu_Z(z), \nu_T(t)\big)$$

Finally

$$\Theta_{X\times Y}(b_{(Z,f)} \times b'_{(T,g)}) = B_*(f \times g)B^*(\nu_Z \times \nu_T)(b \times b') = \ldots$$

$$\ldots = B_*(f)B^*(\nu_Z)(b) \times B_*(g)B^*(\nu_T)(b') = \Theta_X(b_{(Z,f)}) \times \Theta_Y(b_{(T,g)})$$

Thus Θ is a unitary morphism of Green functors. The proposition follows. ■

10.8.3 Adjunction and Morita contexts

Let G and H be groups, and U be a G-set-H. Let A be a Green functor for H, and M be a module-A. If N is a Mackey functor for H, then $\mathcal{H}(M, N)$ has a natural structure of A-module (see proposition 6.4.2), defined as follows: if X and Y are H-sets, if $a \in A(X)$ and

$$\phi \in \mathcal{H}(M, N)(Y) = \mathrm{Hom}_{Mack(H)}(M, N_Y)$$

then $a \times \phi$ is the morphism from M to $N_{X\times Y}$ defined for an H-set Z and $m \in M(Z)$ by

$$(a \times \phi)_Z(m) = \phi_{Z\times X}(m \times a) \in N(Z \times X \times Y) = N_{X\times Y}(Z)$$

The Mackey functor $\mathcal{L}_U(M)$ has also a structure of module-$\mathcal{L}_U(A)$. Then if P is a Mackey functor for G, the functor $\mathcal{H}\big(\mathcal{L}_U(M), P\big)$ is an $\mathcal{L}_U(A)$-module. So the module $\mathcal{H}\big(\mathcal{L}_U(M), P\big) \circ U$ is an $\mathcal{L}_U(A) \circ U$-module. By restriction along the morphism $\lambda_A : A \to \mathcal{L}_U(A) \circ U$, I obtain an A-module $\mathcal{H}\big(\mathcal{L}_U(M), P) \circ U_{|A}$.

On the other hand, the functor $\mathcal{H}(M, P \circ U)$ is also an A-module. I have built morphisms of Mackey functors from $\mathcal{H}(M, P \circ U)$ to $\mathcal{H}\big(\mathcal{L}_U(M), P) \circ U$. It is natural to ask if these morphisms are compatible with the product of A.

Proposition 10.8.2: Let G and H be groups, and U be a G-set-H. Let A be a Green functor for H, and M be a module-A. If P is a Mackey functor, then the morphisms $\mathcal{H}(M, P \circ U) \to \mathcal{H}\big(\mathcal{L}_U(M), P) \circ U$ induce isomorphisms of A-modules

$$\mathcal{H}(M, P \circ U) \xrightarrow{\cong} \mathcal{H}\big(\mathcal{L}_U(M), P\big) \circ U_{|A}$$

Proof: Let Θ be the morphism from $\mathcal{H}(M, P\circ U)$ to $\mathcal{H}\big(\mathcal{L}_U(M), P\big)\circ U$. If I prove that Θ is compatible with the product of A, then as the module in the left hand side is unitary, the image of Θ will be contained in $\mathcal{H}\big(\mathcal{L}_U(M), P\big)\circ U_{|A}$.

So let X and Y be H-sets. Let $a\in A(X)$ and

$$\phi\in\mathcal{H}(M, P\circ U)(Y) = \mathrm{Hom}_{Mack(H)}\big(M, (P\circ U)_Y\big)$$

Then let

$$\psi = \Theta(\phi)\in\Big[\mathcal{H}\big(\mathcal{L}_U(M), P\big)\circ U\Big](Y) = \mathrm{Hom}_{Mack(G)}\big(\mathcal{L}_U(M), P_{U\circ_H Y}\big)$$

If S is a G-set, and (T,g) is an object of $\mathcal{D}_U(S)$, and if $m\in M(G\backslash U.T)$, then the image $\Theta(\phi)_S(m_{(T,g)})$ is given by the following diagram

$$M(G\backslash U.T)\xrightarrow{\phi_{G\backslash U.T}} P\big(U\circ[(G\backslash U.T)\times Y]\big)\xrightarrow{P_*(\delta^U_{G\backslash U.T,Y})} P\big(U\circ_H(G\backslash U.T)\times U\circ_H T\big)\to\ldots$$

$$\ldots\xrightarrow{\nu_{(T,g)}\times Id_{U\circ_H Y}} P\big(T\times(U\circ_H Y)\big)\xrightarrow{g_S\times Id_{U\circ_H Y}} P\big((X\times(U\circ_H Y)\big)$$

In other words, denoting by γ the functor from H-**set** to G-**set**$\downarrow_{U/H}$ defined by $\gamma(X) = U\circ_H X$, and ω its left adjoint, defined by $\omega(T) = G\backslash U.T$, I have

$$\psi_S(m_{(T,g)}) = P_*(g_S\times Id_{\gamma(Y)})P^*(\nu_{(T,g)}\times Id_{\gamma(Y)})P_*(\delta^U_{\omega(T),Y})\phi_{\omega(T)}(m)$$

To simplify this expression, I fill the cartesian square

$$\begin{array}{ccc} F & \xrightarrow{\alpha} & \gamma\big(\omega(T)\times Y\big)\\ \beta\big\downarrow & & \big\downarrow\delta^U_{\omega(T),Y}\\ T\times\gamma(Y) & \xrightarrow[\nu_{(T,g)}\times Id_{\gamma(Y)}]{} & \gamma\omega(T)\times\gamma(Y)\end{array}$$

Let $(t,u,y)\in T\times\gamma(Y)$. If $u'\in U$ is such that $g_U(t) = u'H$, then

$$(\nu_{(T,g)}\times Id_{\gamma(Y)})(t,u,y) = \big((u', G.u'.t), (u,y)\big)$$

Let $(u_1, G.u_2.t_1, y_1)\in\gamma\big(\omega(T)\times Y\big)$. Then

$$\delta^U_{\omega(T),Y}(u_1, G.u_2.t_1, y_1) = \big((u_1, G.u_2.t_1), (u_1, y_1)\big)$$

Those images are equal if and only if

$$(u', G.u'.t) = (u_1, G.u_2.t_1)\qquad (u,y) = (u_1,y_1)$$

This is equivalent to say that there exists $h, h'\in H$ such that

$$u_1 = u'h\qquad G.u_2.t_1 = G.u'h.t\qquad uh' = u_1\qquad h'y_1 = y$$

It follows in particular that $u'H = u_1H = uH$. Moreover $uh' = u'h$, thus

$$(u_1, G.u_2.t_1, y_1) = (uh', G.u'h.t, y_1) = (uh', G.uh'.t, y_1) = (u, G.u.t, h'y_1) = (u, G.u.t, y)$$

Then the element (t, u, y) is in $T.\gamma(Y)$. If $\alpha_{T,Y}$ is the map from $T.\gamma(Y)$ to $\gamma\big(\omega(T)\times Y\big)$ defined by

$$\alpha_{T,Y}(t, u, y) = (u, G.u.t, y)$$

I have $\alpha_{T,Y}(t, u, y) = (u_1, G.u_2.t_1, y_1)$. If moreover $\beta_{T,Y}$ is the inclusion from $T.\gamma(Y)$ into $T \times \gamma(Y)$, then the square

$$\begin{array}{ccc} T.\gamma(Y) & \xrightarrow{\alpha_{T,Y}} & \gamma\big(\omega(T) \times Y\big) \\ \beta_{T,Y}\Big\downarrow & & \Big\downarrow \delta^U_{\omega(T),Y} \\ T \times \gamma(Y) & \xrightarrow[\nu_{(T,g)} \times Id_{\gamma(Y)}]{} & \gamma\omega(T) \times \gamma(Y) \end{array}$$

is commutative. The above argument shows that the associated morphism s from $T.\gamma(Y)$ to F is surjective. As $\beta_{T,Y}$ factors through s, the morphism s is also injective. Thus F is isomorphic to $T.\gamma(Y)$.

In those conditions, I can write

$$\psi_S(m_{(T,g)}) = P_*(g_S \times Id_{\gamma(Y)})P_*(\beta_{T,Y})P^*(\alpha_{T,Y})\phi_{\omega(T)}(m)$$

As the composite map

$$(g_S \times Id_{\gamma(Y)}) \circ \beta_{T,Y}$$

is nothing but $(g.\pi_Y)_{S\times\gamma(Y)}$, I have finally

$$\psi_S(m_{(T,g)}) = P_*\big((g.\pi_Y)_{S\times\gamma(Y)}\big)P^*(\alpha_{T,Y})\phi_{\omega(T)}(m)$$

On the other hand, if Z is an H-set, and if $m' \in M(Z)$, then

$$(a \times \phi)_Z(m') = \phi_{Z\times X}(m' \times a)$$

It follows that

$$\Theta(a \times \phi)_S(m_{(T,g)}) = P_*\big((g.\pi_{X\times Y})_{S\times\gamma(X\times Y)}\big)P^*(\alpha_{T,X\times Y})\phi_{\omega(T)\times X}(m \times a) \tag{10.9}$$

The image of a under λ_A is the element

$$b = \lambda_{A,X}(a) = A^*(\eta_X)(a)_{(\gamma(X),\pi_X)} \in \big(\mathcal{L}_U(A) \circ U\big)(X) = \mathcal{L}_U\big(\gamma(X)\big)$$

Then, by definition of the product on $\mathcal{H}\big(\mathcal{L}_U(M), P\big) \circ U$, I have

$$a \times \psi = \mathcal{H}\big(\mathcal{L}_U(M), P\big)^*(\delta^U_{X,Y})(b \times \psi)$$

where the product in the right hand side is the product of $\mathcal{H}\big(\mathcal{L}_U(M), P\big)$ as $\mathcal{L}_U(A)$-module. Then

$$(a \times \psi)_S(m_{(T,g)}) = P^*(Id_S \times \delta^U_{X,Y})\psi_{S\times\gamma(X)}(m_{(T,g)} \times b) \tag{10.10}$$

Moreover

$$m_{(T,g)} \times b = M^*(\kappa^U_{T,\gamma(X)})\big(m \times A^*(\eta_X)(a)\big)_{(T.\gamma(X),g.\pi_X)} = \ldots$$

$$\ldots = M^*(\kappa^U_{T,\gamma(X)})M^*(Id_{\omega(T)}\times\eta_X)(m\times a)_{(T.\gamma(X),g.\pi_X)} = M^*(r)(m\times a)_{(T.\gamma(X),g.\pi_X)}$$

where I denote by r the composite map

$$r = (Id_{\omega(T)} \times \eta_X) \circ \kappa^U_{T,\gamma(X)}$$

Finally,

$$(a \times \psi)_S(m_{(T,g)}) = \ldots$$

$$\ldots = P^*(Id_S \times \delta^U_{X,Y})P_*\Big((g.\pi_X.\pi_Y)_{S\times\gamma(X)\times\gamma(Y)}\Big)P^*(\alpha_{T.\gamma(X),Y})\phi_{\omega(T\gamma(X))}M^*(r)(m\times a)$$

As ϕ is a morphism of Mackey functors, I have

$$\phi_{\omega(T.\gamma(X))}M^*(r) = (P\circ U)^*(r\times Id_Y)\phi_{\omega(T)\times X}$$

so

$$(a \times \psi)_S(m_{(T,g)}) = \ldots$$

$$= P^*(Id_S\times\delta^U_{X,Y})P_*\Big((g.\pi_X.\pi_Y)_{S\times\gamma(X)\times\gamma(Y)}\Big)P^*(\alpha_{T.\gamma(X),Y})(P\circ U)^*(r\times Id_Y)\phi_{\omega(T)\times X}(m\times a)$$

I must compare this expression with (10.9). I have already observed that

$$\gamma(X).\gamma(Y) \simeq \gamma(X\times Y)$$

Indeed, the map $\rho : \gamma(X\times Y) \to \gamma(X).\gamma(Y)$ defined by

$$\rho(u,x,y) = (u,x).(u,y)$$

is an isomorphism. Now it follows from the injectivity of $\delta^U_{X,Y}$ that the square

$$\begin{array}{ccc}
T.\gamma(X\times Y) & \xrightarrow{\ \rho\ } & T.\gamma(X).\gamma(Y) \\
{\scriptstyle (g.\pi_{X\times Y})_{S\times\gamma(X\times Y)}}\Big\downarrow & & \Big\downarrow{\scriptstyle (g.\pi_X.\pi_Y)_{S\times\gamma(X)\times\gamma(Y)}} \\
S\times\gamma(X\times Y) & \xrightarrow[Id_S\times\delta^U_{X,Y}]{} & S\times\gamma(X)\times\gamma(Y)
\end{array}$$

is cartesian. Thus

$$P^*(Id_S\times\delta^U_{X,Y})P_*\Big((g.\pi_X.\pi_Y)_{S\times\gamma(X)\times\gamma(Y)}\Big) = P_*\Big((g.\pi_{X\times Y})_{S\times\gamma(X\times Y)}\Big)P^*(\rho)$$

and then

$$(a \times \psi)_S(m_{(T,g)}) = \ldots$$

$$\ldots = P_*\Big((g.\pi_{X\times Y})_{S\times\gamma(X\times Y)}\Big)P^*(\rho)P^*(\alpha_{T.\gamma(X),Y})(P\circ U)^*(r\times Id_Y)\phi_{\omega(T)\times X}(m\times a)$$

Let $t.(u,x,y) \in T.\gamma(X\times Y)$. Then

$$\rho\big(t.(u,x,y)\big) = t.(u,x).(u,y)$$

Thus

$$\alpha_{T.\gamma(X),Y}\rho\big(t.(u,x,y)\big) = \big(u, G.u.t.(u,x), y\big)$$

Then

$$[U \circ_H (\kappa^U_{T,\gamma(X)} \times Id_Y)]\alpha_{T.\gamma(X),Y}\rho\Big(t.(u,x,y)\Big) = \Big(u, G.u.t, G.u.(u,x), y\Big)$$

As $r = (Id_{\omega(T)} \times \eta_X) \circ \kappa^U_{T,\gamma(X)}$, and as $\eta_X\Big(G.u.(u,x)\Big) = x$, this gives

$$[U \circ_H (r \times Id_Y)]\alpha_{T.\gamma(X),Y}\rho\Big(t.(u,x,y)\Big) = \Big(u, G.u.t, x, y\Big) = \alpha_{T,X\times Y}\Big(t.(u,x,y)\Big)$$

Finally

$$(a \times \psi)_S(m_{(T,g)}) = P_*\Big((g.\pi_{X\times Y})_{S\times\gamma(X\times Y)}\Big)P^*(\alpha_{T,X\times Y})\phi_{\omega(T)\times X}(m \times a)$$

The comparison with (10.9) shows that

$$a \times \psi = \Theta(a \times \phi) = a \times \Theta(\phi)$$

hence Θ is compatible with the product of A. In particular

$$\Theta\Big(\mathcal{H}(M, P \circ U)\Big) \subseteq \mathcal{H}\Big(\mathcal{L}_U(M), P\Big) \circ U_{|A}$$

Conversely, let

$$\psi \in \mathcal{H}\Big(\mathcal{L}_U(M), P\Big) \circ U(Y)$$

I can apply formula (10.10) in the case $X = \bullet$, and $a = \varepsilon_A$: this gives

$$b = A^*(\eta_\bullet)(\varepsilon_A)_{(\gamma(\bullet),\pi_\bullet)}$$

so

$$(\varepsilon_A \times \psi)_S(m_{(T,g)}) = P^*(Id_S \times \delta_{\bullet,Y})\psi_{S\times\gamma(\bullet)}(m_{(T,g)} \times b)$$

Moreover

$$m_{(T,g)} \times b = M^*(r)(m \times a)_{(T.\gamma(\bullet),g.\pi_\bullet)}$$

Here $m \times a = m$, and $T.\gamma(\bullet) = T.(U/H) \simeq T$. The map $g.\pi_\bullet$ identifies then with the map

$$t \in T \mapsto (g(t), g_U(t)) \in S \times (U/H)^2$$

Now the map r is

$$r = (Id_{\omega(T)} \times \eta_\bullet) \circ \kappa^U_{T,\gamma(\bullet)}$$

Then identifying $\omega\Big(T.\gamma(\bullet)\Big)$ with $\omega(T)$, the map r becomes the identity of $\omega(T)$, and then

$$(\varepsilon_A \times \psi)_S(m_{(T,g)}) = P^*(Id_S \times \delta_{\bullet,Y})\psi_{S\times\gamma(\bullet)}(m_{(T,g.\pi_\bullet)})$$

Moreover

$$m_{(T,g.\pi_\bullet)} = \mathcal{L}_U(M)_*(f)(m_{(T,d_T)})$$

where d_T is the map from T to $T \times (U/H)$ defined by $d_T(t) = \Big(t, g_U(t)\Big)$. As ψ is a morphism of Mackey functors, I have

$$\psi_{S\times\gamma(\bullet)}\mathcal{L}_U(M)_*(f) = P_*(f \times Id_{\gamma(Y)})\psi_T$$

and then

$$(\varepsilon_A \times \psi)_S(m_{(T,g)}) = P^*(Id_S \times \delta_{\bullet,Y})P_*(f \times Id_{\gamma(Y)})\psi_T(m_{T,d_T})$$

It is easy to see that the square

$$\begin{array}{ccc} T.\gamma(Y) & \xrightarrow{\beta_{T,Y}} & T\times\gamma(Y) \\ {\scriptstyle (g.\pi_Y)_{S\times\gamma(Y)}}\downarrow & & \downarrow{\scriptstyle f\times Id_{\gamma(Y)}} \\ T\times\gamma(Y) & \xrightarrow[Id_T\times\delta^U_{\bullet,Y}]{} & T\times(U/H)\times\gamma(Y) \end{array}$$

is cartesian. Finally, this gives

$$(\varepsilon_A\times\psi)_S(m_{(T,g)}) = P_*\Big((g.\pi_Y)_{S\times\gamma(Y)}\Big)P^*(\beta_{T,Y})\psi_T(m_{T,d_T})$$

The question is now to know if there exists $\phi\in\mathcal{H}(M,P\circ U)(Y)$ such that

$$(\varepsilon_A\times\psi)_S(m_{(T,g)}) = P_*\Big((g.\pi_Y)_{S\times\gamma(Y)}\Big)P^*(\alpha_{T,Y})\phi_{\omega(T)}(m)$$

This relation must hold for any T and g. This forces the equality

$$P^*(\beta_{T,Y})\psi_T(m_{T,d_T}) = P^*(\alpha_{T,Y})\phi_{\omega(T)}(m) \qquad (10.11)$$

In the case $T=\gamma(X)$ for an H-set X, and $m=M^*(\eta_X)(m')$, this relation gives

$$P^*(\beta_{\gamma(X),Y})\psi_{\gamma(X)}(M^*(\eta_X)(m')_{(\gamma(X),\pi_X)}) = P^*(\alpha_{\gamma(X),Y})\phi_{\omega\gamma(X)}M^*(\eta_X)(m')$$

As ϕ is a morphism of Mackey functors, I have

$$\phi_{\omega\gamma(X)}M^*(\eta_X) = P^*\Big(U\circ_H(\eta_X\times Id_Y)\Big)\phi_X$$

Thus

$$P^*(\beta_{\gamma(X),Y})\psi_{\gamma(X)}(M^*(\eta_X)(m')_{(\gamma(X),\pi_X)}) = P^*(\alpha_{\gamma(X),Y})P^*(U\circ_H(\eta_X\times Id_Y))\phi_X(m')$$

The map $\alpha_{\gamma(X),Y}$ is a map from $\gamma(X).\gamma(Y)$ to $\gamma\Big(\omega(\gamma(X))\times Y\Big)$. I can compose the previous equality with $P^*(\rho)$, where ρ is thc isomorphism from $\gamma(X\times Y)$ to $\gamma(X).\gamma(Y)$. Then, as

$$\Big(U\circ_H(\eta_X\times Id_Y)\Big)\circ\alpha_{\gamma(X),Y}\circ\rho(u,x,y) = \Big(U\circ_H(\eta_X\times Id_Y)\Big)\circ\alpha_{\gamma(X),Y}\Big((u,x)(u,y)\Big) = \ldots$$
$$\ldots = \Big(U\circ_H(\eta_X\times Id_Y)\Big)\Big(u,G.u.(u,x),y\Big) = (u,x,y)$$

this gives

$$\phi_X(m') = P^*(\rho)P^*(\beta_{\gamma(X),Y})\psi_{\gamma(X)}(M^*(\eta_X)(m')_{(\gamma(X),\pi_X)})$$

Moreover, the map $\beta_{\gamma(X),Y}\circ\rho$ is the inclusion from $\gamma(X\times Y)$ into $\gamma(X)\times\gamma(Y)$, that is $\delta^U_{X,Y}$. Thus for any X and any $m\in M(X)$, I have

$$\phi_X(m) = P^*(\delta^U_{X,Y})\psi_{\gamma(X)}(M^*(\eta_X)(m)_{(\gamma(X),\pi_X)})$$

which proves at least unicity of ϕ (I know that Θ is a split monomorphism). I observe that this equality can be written as

$$\phi_X(m) = P^*(\delta^U_{X,Y})\psi_{U\circ_H X}\lambda_{M,X}(m)$$

In this form, It is easy to check that this equality defines a morphism of Mackey functors from M to $P \circ U$. Moreover, relation (10.11) holds: indeed, with this definition of ϕ, I have

$$P^*(\alpha_{T,Y})\phi_{\omega(T)}(m) = P^*(\alpha_{T,Y})P^*(\delta^U_{\omega(T),Y})\psi_{\gamma\omega(T)}(M^*(\eta_{\omega(T)})(m)_{(\gamma\omega(T),\pi_{\omega(T)})})$$

Moreover, if $t.(u,y) \in T.\gamma(Y)$, then

$$\delta^U_{\omega(T),Y} \circ \alpha_{T,Y}\Big(t.(u,y)\Big) = \delta^U_{\omega(T),Y}(u, G.u.t, y) = \Big((u, G.u.t), (u,y)\Big)$$

Thus $\delta^U_{\omega(T),Y} \circ \alpha_{T,Y}$ is the restriction to $T.\gamma(Y)$ of $\nu_{(T,g)} \times Id_{\gamma(Y)}$, that is

$$\delta^U_{\omega(T),Y} \circ \alpha_{T,Y} = (\nu_{(T,g)} \times Id_{\gamma(Y)}) \circ \beta_{T,Y}$$

In those conditions,

$$P^*(\alpha_{T,Y})\phi_{\omega(T)}(m) = P^*(\beta_{T,Y})P^*(\nu_{(T,g)} \times Id_{\gamma(Y)})\psi_{\gamma\omega(T)}(M^*(\eta_{\omega(T)})(m)_{(\gamma\omega(T),\pi_{\omega(T)})})$$

As ψ is by hypothesis a morphism of Mackey functors from $\mathcal{L}_U(M)$ to $P_{U \circ Y}$, I have

$$P^*(\nu_{(T,g)} \times Id_{\gamma(Y)})\psi_{\gamma\omega(T)} = \psi_T \mathcal{L}_U(M)^*(\nu_T)$$

So

$$P^*(\alpha_{T,Y})\phi_{\omega(T)}(m) = P^*(\beta_{T,Y})\psi_T\mathcal{L}_U(M)^*(\nu_T)(M^*(\eta_{\omega(T)})(m)_{(\gamma\omega(T),\pi_{\omega(T)})})$$

As the square

$$\begin{array}{ccc} T & \xrightarrow{\nu_T} & \gamma\omega(T) \\ {\scriptstyle d_T}\downarrow & & \downarrow{\scriptstyle \pi_{\omega(T)}} \\ \tilde{T} & \xrightarrow[\widetilde{\nu_T}]{} & \gamma\omega(T) \end{array}$$

is cartesian, it follows that

$$\mathcal{L}_U(M)^*(\nu_T)(M^*(\eta_{\omega(T)})(m)_{(\gamma\omega(T),\pi_{\omega(T)})}) = M^*(G\backslash\nu_T)M^*(\eta_{\omega(T)})(m)_{(T,d_T)} = m_{(T,d_T)}$$

Then

$$P^*(\alpha_{T,Y})\phi_{\omega(T)}(m) = P^*(\beta_{T,Y})\psi_T(m_{(T,d_T)})$$

which proves (10.11), hence that $\varepsilon \times \psi = \Theta(\phi)$.

So the image of Θ is exactly $\mathcal{H}\Big(\mathcal{L}_U(M), P\Big) \circ U_{|A}$, and the proposition follows. ∎

Proposition 10.8.3: Let G and H be groups, and U be a G-set-H. If A is a Green functor for H, if M is a module-A and N an A-module, then

$$\mathcal{L}_U(M\hat{\otimes}_A N) \simeq \mathcal{L}_U(M)\hat{\otimes}_{\mathcal{L}_U(A)}\mathcal{L}_U(N)$$

Proof: This result follows from the previous one by adjunction: if N is an A-module, then by proposition 6.6.2, I have

$$\mathrm{Hom}_A\Big(N, \mathcal{H}(M, P \circ U)\Big) = \mathrm{Hom}_{Mack(H)}(M\hat{\otimes}_A N, P \circ U) = \mathrm{Hom}_{Mack(G)}\Big(\mathcal{L}_U(M\hat{\otimes}_A N), P\Big)$$

On the other hand

$$\mathrm{Hom}_A\Big(N, \mathcal{H}(\mathcal{L}_U(M), P)\circ U_{|A}\Big) = \mathrm{Hom}_{\mathcal{L}_U(A)}\Big(\mathcal{L}_U(N), \mathcal{H}(\mathcal{L}_U(M), P)\Big) = \dots$$

$$\dots = \mathrm{Hom}_{Mack(G)}\Big(\mathcal{L}_U(M)\hat{\otimes}_{\mathcal{L}_U(A)}\mathcal{L}_U(N), P\Big)$$

The comparison of these two equalities now gives the claimed isomorphism. ∎

Remark: Similar tedious computations show that if N is an A-module, then

$$\mathcal{H}\Big(P, \mathcal{R}_U(N)\Big)\circ U_{|A} \simeq \mathcal{H}(P\circ U, N)$$

$$\mathcal{L}_U(N)\hat{\otimes}P)\circ U_{|A} \simeq M\hat{\otimes}(P\circ U)$$

as A-modules.

Proposition 10.8.4: Let G and H be groups, and U be a G-set-H. If A and B are Green functors for H, if (P, Q, Φ, Ψ) is a surjective Morita context for A and B, if Θ (resp. Θ') is the isomorphism from $\mathcal{L}_U(P)\hat{\otimes}_{\mathcal{L}_U(A)}\mathcal{L}_U(Q)$ to $\mathcal{L}_U(P\hat{\otimes}_A Q)$ (resp. from $\mathcal{L}_U(Q)\hat{\otimes}_{\mathcal{L}_U(B)}\mathcal{L}_U(P)$ to $\mathcal{L}_U(Q\hat{\otimes}_B P)$), then the 4-tuple $\Big(\mathcal{L}_U(P), \mathcal{L}_U(Q), \mathcal{L}_U(\Phi)\Theta, \mathcal{L}_U(\Psi)\Theta'\Big)$ is a surjective Morita context for $\mathcal{L}_U(A)$ and $\mathcal{L}_U(B)$.

Proof: This is clear, since if $P\hat{\otimes}_A Q \simeq B$, then

$$\mathcal{L}_U(B) \simeq \mathcal{L}_U(P\hat{\otimes}_A Q) \simeq \mathcal{L}_U(P)\hat{\otimes}_{\mathcal{L}_U(A)}\mathcal{L}_U(Q)$$

Moreover, the isomorphisms of proposition 10.8.3 are natural enough to give here isomorphisms of bimodules. ∎

Taking for U the set H, viewed as (1)-set-H, then the functor $\mathcal{L}_U$ is the restriction functor to the trivial subgroup, that is the evaluation functor at the trivial subgroup.

Notation: *If M is a Mackey functor for the group G, and H is a subgroup of G, I set*

$$\overline{M}(H) = (\mathrm{Res}^G_{N_G(H)}M)^H(H/H) = M(H)/\sum_{L\subset H} t^H_L M(L)$$

$$\underline{M}(H) = (\mathrm{Res}^G_{N_G(H)}M)_H(H/H) = \bigcap_{L\subset H} \mathrm{Ker}\, r^H_L$$

If A is a Green functor, then $\overline{A}(H)$ is a unitary ring for the product "." of $(\mathrm{Res}^G_{N_G(H)}A)^H$, and $\underline{A}(H)$ is a two-sided ideal of $(A(H), .)$.

Now the rings $\overline{A}(H)$ are evaluations at the trivial group of functors $(\mathrm{Res}^G_{N_G(H)}A)^H$, which are the functors $\mathcal{L}_U(A)$ for suitable sets U: indeed, let $U = H\backslash G$, viewed as a $N_G(H)/H$-set-G by

$$\bar{n}.Hx.g = Hnxg$$

where $n \mapsto \bar{n}$ is the projection from $N_G(H)$ to $N_G(H)/H$. Then for any G-set X, I have

$$U\circ_G X = X^H$$

Indeed, in $(H\backslash G)\circ_G X$, I have $(Hg,x)=(H,gx)$, and $(H,x)\in(H\backslash G)\circ_G X$ if and only if $x\in X^H$. Moreover $(H,x)=(H,x')$ if and only if $x=x'$.

Thus if M is a Mackey functor for G, I have

$$\mathcal{L}_U(M)=(\mathrm{Res}^G_{N_G(H)}M)^H$$

This gives in particular the following lemma:

Lemma 10.8.5: Let H be a subgroup of G, and A, B and C be Green functors for G. If M is a B-module-A, and N is an A-module-C, then

$$(\mathrm{Res}^G_{N_G(H)}M\hat{\otimes}N)^H\simeq(\mathrm{Res}^G_{N_G(H)}M)^H\hat{\otimes}_{(\mathrm{Res}^G_{N_G(H)}A)^H}(\mathrm{Res}^G_{N_G(H)}N)^H$$

as $(\mathrm{Res}^G_{N_G(H)}B)^H$-modules-$(\mathrm{Res}^G_{N_G(H)}C)^H$.

Now evaluating this isomorphism at the trivial $N_G(H)/H$-set, I have the following consequence:

Proposition 10.8.6: Let G and H be groups. If A and B are Green functors for G, if (P,Q,Φ,Ψ) is a surjective Morita context for A and B, then for any subgroup H of G, the 4-tuple $(\overline{P}(H),\overline{Q}(H),\overline{\Phi},\overline{\Psi})$ is a surjective Morita context for $\overline{A}(H)$ and $\overline{B}(H)$.

Chapter 11

The simple modules

11.1 Generalities

Let A be a Green functor for the group G. Theorem 3.3.5 states the equivalence between the category of A-modules and the category of representations of $\mathcal{C}_A$. I can in particular apply to A-**Mod** the generalities on representations of categories proved in [3]. In particular, the simple A-modules can be described as follows:

Proposition 11.1.1: Let A be a Green functor for the group G.

1. **If X is a G-set, and V is a simple module for the algebra $\mathrm{End}_{\mathcal{C}_A}(X) = A(X^2)$, then the module $L_{X,V}$ defined by**

$$L_{X,V}(Y) = A(YX) \otimes_{A(X^2)} V$$

has a unique maximal submodule $J_{X,V}$, defined by

$$J_{X,V}(Y) = \left\{ \sum_i a_i \otimes v_i \mid \forall \phi \in A(XY), \sum_i (\phi \circ_Y a_i).v_i = 0 \right\}$$

The quotient $S_{X,V} = L_{X,V}/J_{X,V}$ is then a simple module.

2. **Conversely, if S is a simple A-module, and if X is a G-set such that $S(X) \neq 0$, then the module $V = S(X)$ is a simple $A(X^2)$-module, and S is isomorphic to $S_{X,V}$.**

Proof: The first assertion follows from [3]. The second is not stated there explicitly, but it is a consequence of the following argument: if W is a non-zero $A(X^2)$-submodule of $S(X)$, then by adjunction there is a non-zero morphism from $L_{X,W}$ to S, which is onto since S is simple. Then as $L_{X,W}(X) = W$, it follows that $W = S(X)$, thus $S(X)$ is a simple $A(X^2)$-module. Then S is a simple quotient of $L_{X,S(X)}$, which has a unique simple quotient $S_{X,S(X)}$. Thus $S \simeq S_{X,S(X)}$. ■

11.2 Classification of the simple modules

In the special case of a functor on $\mathcal{C}_A$, I have moreover the notion of minimal subgroup: a subgroup H of G is called *minimal* for the functor F on $\mathcal{C}_A$ if $F(G/H) \neq 0$, but if $F(G/K) = 0$ for any subgroup K of G of strictly smaller order. Such a subgroup always exists if F is non-zero, since F is additive.

Let S be a simple A-module, and H be a minimal subgroup for S. Let K a subgroup of G such that $|K| < |H|$, and ϕ be an element of $A\big((G/H)^2\big)$ which factors through G/K in $\mathcal{C}_A$, i.e. such that

$$\phi = \alpha \circ_{G/K} \beta \qquad \text{for} \quad \alpha \in A\big((G/H) \times (G/K)\big),\ \beta \in A\big((G/K) \times (G/H)\big)$$

Then setting $\Gamma = G/H$, and $V = S(\Gamma)$, I have the following equality in $L_{\Gamma,V}(\Gamma) = V$ for $v \in V$

$$1_{A(\Gamma^2)} \otimes \phi.v = \phi \otimes v = (\alpha \circ_{G/K} \beta) \otimes v = S_{\Gamma,V}(\alpha)(\beta \otimes v)$$

But as H is minimal for S, the element $\beta \otimes v$ of $S(K)$ is zero. So the element $\phi.v$ is zero.

Thus the module V is annihilated by any endomorphism of Γ which factors by a set G/K such that $|K| < |H|$.

Notation: *Let A be a Green functor, and H be a subgroup of G. I denote by $\hat{A}(H)$ the quotient of $A\big((G/H)^2\big)$ by the two-sided ideal generated by the elements of the form $\alpha \circ_{G/K} \beta$, for $|K| < |H|$.*

With this definition, I have the following classification

Proposition 11.2.1: Let A be a Green functor for the group G.

1. **If S is a simple A-module, and H is a minimal subgroup for S, then $V = S(H)$ is a simple $\hat{A}(H)$-module, and S is isomorphic to $S_{G/H,V}$.**

2. **Conversely, if H is a subgroup of G, and V is a simple $\hat{A}(H)$-module, then $S_{G/H,V}$ is a simple A-module, the group H is minimal for $S_{G/H,V}$, and moreover $S_{G/H,V}(H) \simeq V$.**

3. **Let H be a subgroup of G, and V be a simple $\hat{A}(H)$-module. If X is a G-set such that $S_{G/H,V}(X) \neq 0$, then $X^H \neq \emptyset$. In particular, the minimal subgroups of $S_{G/H,V}$ are the conjugates of H in G.**

4. **Let H and K be subgroups of G. If V is a simple $\hat{A}(H)$-module, and if W is a simple $\hat{A}(K)$-module, then the modules $S_{G/H,V}$ and $S_{G/K,W}$ are isomorphic if and only if there exists $x \in G$ such that $K = {}^xH$ and $W \simeq {}^xV$.**

Remarks: a) I have identified a $\hat{A}(H)$-module V with the associated $A\big((G/H)^2\big)$-module.

b) The notation $W \simeq {}^xV$ means that V maps to W by the isomorphism $c_{x,*}$ from G/H to G/K in $\mathcal{C}_A$, deduced from the conjugation $c_x : G/H \to G/K$ defined by $c_x(g.H) = gx^{-1}.K$.

Proof: I have already proved assertion 1). For assertion 2), I already know that if V is a simple $\hat{A}(H)$-module, viewed as an $A\big((G/H)^2\big)$-module, then $S_{G/H,V}$ is a simple A-module such that $S_{H,V}(H) \simeq V$. It remains to show that H is minimal for $S_{G/H,V}$. So let K be a subgroup of G such that $|K| < |H|$. Let $a \otimes v$ be an element of $L_{G/H,V}(K)$. Then $a \otimes v \in J_{G/H,V}(K)$: indeed, if $\phi \in A\big((G/H) \times (G/K)\big)$, then $\phi \circ_{G/K} a$ factors through G/K, so annihilates V. Thus $a \otimes v \in J_{G/H,V}(K)$, and $S_{G/H,V}(K) = 0$.

Under the hypothesis of assertion 3), there exists $a \in A\big(X \times (G/H)\big)$ and $v \in V$ such that $a \otimes v$ is non-zero in $S_{G/H,V}(X)$. In particular, there exists an element $\phi \in A\big((G/H) \times X\big)$ such that $\phi \circ_X a$ has non-zero action on V, so does not factor through sets G/K such that $|K| < |H|$. But

Lemma 11.2.2: Let X and Y be G-sets, and $a \in A(XY)$. Let

$$b = A_*\begin{pmatrix} xy \\ yxy \end{pmatrix}(a) \in A(YXY) = \mathrm{Hom}_{\mathcal{C}_A}(Y, YX) \qquad c = \begin{pmatrix} yx \\ x \end{pmatrix}_* \in \mathrm{Hom}_{\mathcal{C}_A}(YX, X)$$

Then $a = c \circ_{YX} b$ in $\mathcal{C}_A$.

Admitting this lemma for a while, I see in particular that a factors in $\mathcal{C}_A$ through $(G/H) \times X$, and so does $\phi \circ_X a$. As

$$(G/H) \times X \simeq \coprod_{\substack{x \in G\backslash X \\ g \in H\backslash G/G_x}} G/(H \cap G_{gx})$$

I see that there exists x and $g \in G$ such that $|H \cap G_{gx}| = |H|$, and then $gx \in X^H$, which proves that $X^H \neq \emptyset$. Then if K is another minimal subgroup of $S_{G/H,V}$, and if $W = S_{G/H,V}(K)$, I know that $S_{G/H,V} \simeq S_{G/K,W}$. The group K has fixed points on G/H, and H has fixed points on G/K, so K is conjugate to H. This proves assertion 3).

Under the hypothesis of assertion 4), the group K is minimal for $S_{G/H,V}$, so it is conjugate to H. I can then suppose $H = K$, and the proposition follows then from the fact that $V \simeq S_{G/H,V}(H) \simeq S_{G/H,W}(H) \simeq W$. ■

Proof of lemma 11.2.2: With the notations of the lemma, I have

$$c \circ_{YX} b = A_*\begin{pmatrix} x_1y_1x_2y_2 \\ x_1y_2 \end{pmatrix} A^*\begin{pmatrix} x_1y_1x_2y_2 \\ x_1y_1x_2y_1x_2y_2 \end{pmatrix}(c \times b)$$

But

$$c \times b = A_*\begin{pmatrix} yx \\ xyx \end{pmatrix} A^*\begin{pmatrix} yx \\ \bullet \end{pmatrix}(\varepsilon) \times A_*\begin{pmatrix} xy \\ yxy \end{pmatrix}(a) = \ldots$$

$$\ldots = A_*\begin{pmatrix} y_1x_1x_2y_2 \\ x_1y_1x_1y_2x_2y_2 \end{pmatrix} A^*\begin{pmatrix} y_1x_1x_2y_2 \\ x_2y_2 \end{pmatrix}(a)$$

As the square

$$\begin{array}{ccc} XY & \xrightarrow{\begin{pmatrix} xy \\ yxy \end{pmatrix}} & YX^2Y \\ {\scriptstyle\begin{pmatrix} xy \\ xyxy \end{pmatrix}}\Big\downarrow & & \Big\downarrow{\scriptstyle\begin{pmatrix} y_1x_1x_2y_2 \\ x_1y_1x_1y_2x_2y_2 \end{pmatrix}} \\ (XY)^2 & \xrightarrow[\begin{pmatrix} x_1y_1x_2y_2 \\ x_1y_1x_2y_1x_2y_2 \end{pmatrix}]{} & (XY)^3 \end{array}$$

is cartesian, I have

$$A^*\begin{pmatrix} x_1y_1x_2y_2 \\ x_1y_1x_2y_1x_2y_2 \end{pmatrix} A_*\begin{pmatrix} y_1x_1x_2y_2 \\ x_1y_1x_1y_2x_2y_2 \end{pmatrix} = A_*\begin{pmatrix} xy \\ xyxy \end{pmatrix} A^*\begin{pmatrix} xy \\ yxxy \end{pmatrix}$$

and then

$$co_{YX}\, b = A_*\begin{pmatrix} x_1y_1x_2y_2 \\ x_1y_2 \end{pmatrix} A_*\begin{pmatrix} xy \\ xyxy \end{pmatrix} A^*\begin{pmatrix} xy \\ yxxy \end{pmatrix} A^*\begin{pmatrix} y_1x_1x_2y_2 \\ x_2y_2 \end{pmatrix}(a) = \dots$$

$$\dots = A_*\begin{pmatrix} xy \\ xy \end{pmatrix} A^*\begin{pmatrix} xy \\ xy \end{pmatrix}(a) = a$$

which proves the lemma. ■

11.3 The structure of algebras $\hat{A}(H)$

Having classified the simple A-modules, I have still to describe the structure of the algebras $\hat{A}(H)$:

Notations: *Let K be a group acting by automorphisms on an R-algebra A. I denote by $A \otimes K$ the tensor product $A \otimes_R RK$, with the following multiplication*

$$(a \otimes k).(a' \otimes k') = ak(a') \otimes kk'$$

If H is a subgroup of G, I denote by $\overline{N}_G(H)$ the quotient $N_G(H)/H$.

Proposition 11.3.1: Let A be a Green functor for the group G. If H is a subgroup of G, then $\hat{A}(H)$ identifies with $\overline{A}(H) \otimes \overline{N}_G(H)$, by the map

$$\alpha \in A\big((G/H)^2\big) \mapsto \sum_{n \in \overline{N}_G(H)} \overline{A^*(i_{H,n,H})(\alpha)} \otimes n$$

where $i_{H,n,H} : G/H \to (G/H)^2$ maps gH to (gH, gnH).

Proof: By definition, I have

$$A\big((G/H)^2\big) \simeq \bigoplus_{x \in H\backslash G/H} A(H \cap {}^xH)$$

the isomorphism being obtained by the maps

$$\alpha \in A\big((G/H)^2\big) \mapsto \sum_{x \in H\backslash G/H} A^*(i_{H,x,H})(\alpha)$$

where $i_{H,x,H}$ is the map

$$i_{H,x,H} : G/(H \cap {}^xH) \to (G/H)^2$$

defined by

$$i_{H,x,H}\big(g(H \cap {}^xH)\big) = (gH, gxH)$$

The inverse isomorphism maps the element $\beta \in A(H \cap {}^xH)$ to $A_*(i_{H,x,H})(\beta)$. But if $x \notin \overline{N}_G(H)$, then $A_*(i_{H,x,H})(\beta)$ factors through $G/(H \cap {}^xH)$:

Lemma 11.3.2: Let X and Y be G-sets. If f and g are morphisms of G-sets from X to Y, and if a and b are elements of $A(X)$, then

$$A_*\begin{pmatrix} x \\ g(x)x \end{pmatrix}(a) \circ_X A_*\begin{pmatrix} x \\ xf(x) \end{pmatrix}(b) = A_*\begin{pmatrix} x \\ g(x)f(x) \end{pmatrix}(a.b)$$

Proof: It suffices to compute

$$A_*\begin{pmatrix} x \\ g(x)x \end{pmatrix}(a) \circ_X A_*\begin{pmatrix} x \\ xf(x) \end{pmatrix}(b) = \dots$$

$$\dots = A_*\begin{pmatrix} y_1xy_2 \\ y_1y_2 \end{pmatrix} A^*\begin{pmatrix} y_1xy_2 \\ y_1xxy_2 \end{pmatrix} \left[A_*\begin{pmatrix} x_1x_2 \\ g(x_1)x_1x_2f(x_2) \end{pmatrix}(a \times b) \right]$$

The cartesian square

$$\begin{array}{ccc} X & \xrightarrow{\begin{pmatrix} x \\ xx \end{pmatrix}} & X^2 \\ \begin{pmatrix} x \\ g(x)xf(x) \end{pmatrix}\Big\downarrow & & \Big\downarrow\begin{pmatrix} x_1x_2 \\ g(x_1)x_1x_2f(x_2) \end{pmatrix} \\ YXY & \xrightarrow[\begin{pmatrix} y_1xy_2 \\ y_1xxy_2 \end{pmatrix}]{} & YX^2Y \end{array}$$

shows that

$$A^*\begin{pmatrix} y_1xy_2 \\ y_1xxy_2 \end{pmatrix} A_*\begin{pmatrix} x_1x_2 \\ g(x_1)x_1x_2f(x_2) \end{pmatrix} = A_*\begin{pmatrix} x \\ g(x)xf(x) \end{pmatrix} A^*\begin{pmatrix} x \\ xx \end{pmatrix}$$

so that

$$A_*\begin{pmatrix} x \\ g(x)x \end{pmatrix}(a) \circ_X A_*\begin{pmatrix} x \\ xf(x) \end{pmatrix}(b) = A_*\begin{pmatrix} y_1xy_2 \\ y_1y_2 \end{pmatrix} A_*\begin{pmatrix} x \\ g(x)xf(x) \end{pmatrix} A^*\begin{pmatrix} x \\ xx \end{pmatrix}(a \times b) = \dots$$

$$\dots = A_*\begin{pmatrix} x \\ g(x)f(x) \end{pmatrix}(a.b)$$

which proves the lemma. ∎

Taking $Y = G/H$ and $X = G/(H \cap {}^xH)$, and

$$f : G/(H \cap {}^xH) \to G/H \qquad f\big(u(H \cap {}^xH)\big) = uH$$

$$g : G/(H \cap {}^xH) \to G/H \qquad g\big(u(H \cap {}^xH)\big) = uxH$$

$$a = \beta \in A(H \cap {}^xH) \qquad b = \varepsilon_X = A^*\begin{pmatrix} x \\ \bullet \end{pmatrix}(\varepsilon)$$

this lemma proves that $A_*(i_{H,x,H})(\beta)$ factors through $G/(H \cap {}^xH)$ in C_A. Its image in $\hat{A}(H)$ is then zero if $x \notin N_G(H)$. If π denotes the canonical projection from $A\big((G/H)^2\big)$ onto $\hat{A}(H)$, there is a surjective morphism θ

$$\theta : \bigoplus_{n \in \overline{N}_G(H)} A(H) \to \hat{A}(H)$$

which maps the element β of the component n of the left hand side to $\pi A_*(i_{H,n,H})(\beta)$.

If K is a proper subgroup of H, and if $\beta = t_K^H(\gamma)$, for $\gamma \in A(K)$, then denoting by π_K^H the projection from G/K to G/H, I have

$$A_*(i_{H,n,H})(\beta) = A_*(i_{H,n,H} \circ \pi_K^H)(\gamma)$$

Lemma 11.2.2 shows that this element factors through G/K, and its image under π is zero. Finally I have a surjective morphism $\bar{\theta}$

$$\bar{\theta} : \bigoplus_{n \in \overline{N}_G(H)} \overline{A}(H) \to \hat{A}(H)$$

Denoting by $\beta \otimes n$ the element β of the n component, I have

$$\bar{\theta}(\beta \otimes n) = \pi A_*(i_{H,n,H})(\beta)$$

If n and n' are elements of $\overline{N}_G(H)$, and if β and β' are elements of $A(H)$, then

$$A_*(i_{H,n,H})(\beta) \circ_{G/H} A_*(i_{H,n',H})(\beta') = A_*(i_{H,nn',H})(\beta.^n\beta')$$

Indeed, setting $X = G/H$, and denoting by $x \mapsto xn$ the map from X to X which sends gH to gnH, I have

$$A_*(i_{H,n,H})(\beta) \circ_{G/H} A_*(i_{H,n',H})(\beta') = \ldots$$

$$\ldots = A_*\begin{pmatrix} x_1x_2x_3 \\ x_1x_2 \end{pmatrix} A^*\begin{pmatrix} x_1x_2x_3 \\ x_1x_2x_2x_3 \end{pmatrix} A_*\begin{pmatrix} x_1x_2 \\ x_1\ x_1n\ x_2\ x_2n' \end{pmatrix} (\beta \times \beta')$$

As the square

$$\begin{array}{ccc} X & \xrightarrow{\begin{pmatrix} x \\ x\ xn \end{pmatrix}} & X^2 \\ {\scriptstyle \begin{pmatrix} x \\ x\ xn\ xnn' \end{pmatrix}} \downarrow & & \downarrow {\scriptstyle \begin{pmatrix} x_1x_2 \\ x_1\ x_1n\ x_2\ x_2n' \end{pmatrix}} \\ X^3 & \xrightarrow[\begin{pmatrix} x_1x_2x_3 \\ x_1x_2x_2x_3 \end{pmatrix}]{} & X^4 \end{array}$$

is cartesian, I have

$$A^*\begin{pmatrix} x_1x_2x_3 \\ x_1x_2x_2x_3 \end{pmatrix} A_*\begin{pmatrix} x_1x_2 \\ x_1\ x_1n\ x_2\ x_2n' \end{pmatrix} = A_*\begin{pmatrix} x \\ x\ xn\ xnn' \end{pmatrix} A^*\begin{pmatrix} x \\ x\ xn \end{pmatrix}$$

whence

$$A_*(i_{H,n,H})(\beta) \circ_{G/H} A_*(i_{H,n',H})(\beta') = A_*\begin{pmatrix} x \\ x\ xnn' \end{pmatrix} A^*\begin{pmatrix} x \\ x\ xn \end{pmatrix} (\beta \times \beta')$$

The right hand side can also be written as

$$A_*\begin{pmatrix} x \\ x\ xnn' \end{pmatrix} A^*\begin{pmatrix} x \\ xx \end{pmatrix} A^*\begin{pmatrix} x_1x_2 \\ x_1\ x_2n \end{pmatrix} (\beta \times \beta') = \ldots$$

$$\ldots = A_*(i_{H,nn',H})A^*\begin{pmatrix} x \\ xx \end{pmatrix}\left(\beta \times A^*\begin{pmatrix} x \\ xn \end{pmatrix}(\beta')\right) = A_*(i_{H,nn',H})(\beta.^n\beta')$$

It follows that $\bar{\theta}$ is a morphism of algebras from $\overline{A}(H) \otimes \overline{N}_G(H)$ to $\hat{A}(H)$. This morphism is moreover unitary, since the unit of $\overline{A}(H) \otimes \overline{N}_G(H)$ is $\overline{\varepsilon_{G/H}} \otimes 1$, which is mapped to

$$\pi A_* \begin{pmatrix} x \\ xx \end{pmatrix} A^* \begin{pmatrix} x \\ \bullet \end{pmatrix} (\varepsilon) = \pi(1_{A((G/H)^2)})$$

Conversely, if $a \in A\big((G/H)^2\big)$, and if $n \in \overline{N}_G(H)$ I can consider the element $\overline{A^*(i_{H,n,H})(a)}$ of $\overline{A}(H)$. If a factors through $Y = G/K$, then setting $X = G/H$, there exists $b \in A(XY)$ and $c \in A(YX)$ such that $a = b \circ_Y c$. Then for $n \in N_G(H)$, I have

$$A^*(i_{H,n,H})(a) = A^* \begin{pmatrix} x \\ xn \end{pmatrix} A_* \begin{pmatrix} x_1yx_2 \\ x_1x_2 \end{pmatrix} A^* \begin{pmatrix} x_1yx_2 \\ x_1yyx_2 \end{pmatrix} (b \times c)$$

As the square

$$\begin{array}{ccc} XY & \xrightarrow{\begin{pmatrix} xy \\ x\ y\ xn \end{pmatrix}} & XYX \\ {\scriptstyle \begin{pmatrix} xy \\ x \end{pmatrix}}\Big\downarrow & & \Big\downarrow{\scriptstyle \begin{pmatrix} x_1yx_2 \\ x_1x_2 \end{pmatrix}} \\ X & \xrightarrow[\begin{pmatrix} x \\ x\ xn \end{pmatrix}]{} & X^2 \end{array}$$

is cartesian, I have also

$$A^*(i_{H,n,H})(a) = A_* \begin{pmatrix} xy \\ x \end{pmatrix} A^* \begin{pmatrix} xy \\ x\ y\ xn \end{pmatrix} A^* \begin{pmatrix} x_1yx_2 \\ x_1yyx_2 \end{pmatrix} (b \times c) = \ldots$$

$$\ldots = A_* \begin{pmatrix} xy \\ x \end{pmatrix} A^* \begin{pmatrix} xy \\ x\ y\ y\ xn \end{pmatrix} (b \times c)$$

As

$$XY = (G/H) \times (G/K) \simeq \coprod_{g \in H\backslash G/K} G/(H \cap {}^gK)$$

then for any $\alpha \in A(XY)$, the element $A_* \binom{xy}{x} (\alpha)$ is a linear combination of elements of the form $t^H_{H \cap {}^gK} a_g$. If the order of K is strictly smaller than the order of H, the groups $H \cap {}^gK$ are proper subgroups of H, and it follows that the image of $A_* \binom{xy}{x} (\alpha)$ in $\overline{A}(H)$ is zero.

So I have a morphism ϕ from $\hat{A}(H)$ to $\overline{A}(H) \otimes \overline{N}_G(H)$, defined by

$$\phi\big(\pi(a)\big) = \sum_{n \in \overline{N}_G(H)} \overline{A^*(i_{H,n,H})(a)} \otimes n$$

Moreover

$$\phi\bar{\theta}(\beta \otimes n) = \sum_{n' \in \overline{N}_G(H)} \overline{A^*(i_{H,n',H})A_*(i_{H,n,H})(\beta)} \otimes n'$$

The only non-zero product $A^*(i_{H,n',H})A_*(i_{H,n,H})$ corresponds to $n' = n$, and it is equal to the identity of $A(H)$. It follows that $\phi\bar{\theta}$ is the identity, so that $\bar{\theta}$ is injective, hence an isomorphism. Then ϕ is the inverse isomorphism, which proves the proposition. ■

11.4 The structure of simple modules

I know that the isomorphism classes of simple modules can be indexed by the conjugacy classes of couples (H, V), where H is a subgroup of G, and V is a simple $\overline{A}(H) \otimes \overline{N}_G(H)$ module.

Notation: *I will denote by $S^G_{H,V}$ or $S_{H,V}$ (instead of $S_{G/H,V}$) the module associated to the couple (H, V).*

The module V can be viewed as $\overline{N}_G(H)$-module, by the homomorphism $n \mapsto \overline{\varepsilon_{G/H}} \otimes n$ from $\overline{N}_G(H)$ to $\overline{A}(H) \otimes \overline{N}_G(H)$. Then denoting by $[Y]$ the permutation module associated to a set Y, with the set Y as a basis, I have the

Proposition 11.4.1: Let H be a subgroup of G, and V be a simple $\overline{A}(H) \otimes \overline{N}_G(H)$-module. If X is a G-set, then $[X^H]$ is a $\overline{N}_G(H)$-module, and

$$S_{H,V}(X) \simeq \mathrm{Hom}([X^H], V)_1^{\overline{N}_G(H)}$$

It is the set of morphisms of $\overline{N}_G(H)$-modules from $[X^H]$ to V which factor through a projective module.

If $f : X \to Y$ is a morphism of G-sets, then for $\alpha \in S_{H,V}(X)$ and $y \in Y^H$, I have

$$S_{H,V*}(f)(\alpha)(y) = \sum_{\substack{x \in X^H \\ f(x)=y}} \alpha(x)$$

If $g : Y \to X$ is a morphism of G-sets, then for $\alpha \in S_{H,V}(X)$ and $y \in Y^H$, I have

$$S^*_{H,V}(g)(\alpha)(y) = \alpha g(y)$$

Finally if $a \in A(X)$ and $f \in S_{H,V}(Y)$, then $a \times f$ is the morphism from $[(X \times Y)^H] = [X^H \times Y^H]$ to V defined by

$$(a \times f)(x, y) = \left(\overline{A^*(m_x)(a)} \otimes 1\right).f(y)$$

where m_x is the morphism of G-sets from G/H to X defined by $m_x(uH) = ux$.

Proof: I will first prove the isomorphism

$$S_{H,V}(X) \simeq \mathrm{Hom}([X^H], V)_1^{\overline{N}_G(H)}$$

and the other formulae will follow.

11.4.1 The isomorphism $S_{H,V}(X) \simeq \mathrm{Hom}([X^H], V)_1^{\overline{N}_G(H)}$

First I observe that X^H identifies with the set of morphisms of G-sets from G/H to X, by the map sending x to m_x. Let $a \otimes v$ be an element of $L_{G/H,V}(X)$. Then $a \in A\big(X \times (G/H)\big)$. If $x \in X^H$, then $m^*_x \in A\big((G/H) \times X\big)$, and I can consider the product $m^*_x \circ_X a$, which is an element of $A\big((G/H)^2\big)$. Its image $\pi(m^*_x \circ_X a)$ is then in

$\hat{A}(H)$. This is identified with $\overline{A}(H) \otimes \overline{N}_G(H)$, which acts on V. Now I can consider the map

$$\lambda_{a,v} : x \in X^H \mapsto \pi(m_x^* \circ_X a).v$$

Proposition 11.3.1 shows that this is also

$$\pi(m_x^* \circ_X a).v = \sum_{n \in \overline{N}_G(H)} \left[\overline{A^*(i_{H,n,H})(m_x^* \circ_X a)} \otimes n\right].v$$

But I can view a as an element of $A_{G/H}(X)$, and then

$$m_x^* \circ_X a = A_{G/H}^*(m_x)(a) = A^*(m_x \times Id)(a)$$

Thus

$$\lambda_{a,v}(x) = \sum_{n \in \overline{N}_G(H)} \left[\overline{A^*(i_{H,n,H})A^*(m_x \times Id)(a)} \otimes n\right].v = \dots$$

$$\dots = \sum_{n \in \overline{N}_G(H)} \left[\overline{A^*\begin{pmatrix} uH \\ ux\ unH \end{pmatrix}(a)} \otimes n\right].v$$

Now let $g_{a,v}$ be the linear map from $[X^H]$ to V defined by

$$x \in X^H \mapsto \left[\overline{A^*\begin{pmatrix} uH \\ ux\ uH \end{pmatrix}(a)} \otimes 1\right].v \in V$$

Then

$$Tr_1^{\overline{N}_G(H)}(g_{a,v})(x) = \sum_{n \in \overline{N}_G(H)} n \left[\overline{A^*\begin{pmatrix} uH \\ un^{-1}x\ uH \end{pmatrix}(a)} \otimes 1\right].v$$

This can also be written as

$$Tr_1^{\overline{N}_G(H)}(g_{a,v})(x) = \sum_{n \in \overline{N}_G(H)} \left[\overline{{}^nA^*\begin{pmatrix} uH \\ un^{-1}x\ uH \end{pmatrix}(a)} \otimes n\right].v$$

or

$$Tr_1^{\overline{N}_G(H)}(g_{a,v})(x) = \sum_{n \in \overline{N}_G(H)} \left[\overline{A^*\begin{pmatrix} uH \\ unH \end{pmatrix} A^*\begin{pmatrix} uH \\ un^{-1}x\ uH \end{pmatrix}(a)} \otimes n\right].v$$

whence finally

$$Tr_1^{\overline{N}_G(H)}(g_{a,v})(x) = \sum_{n \in \overline{N}_G(H)} \left[\overline{A^*\begin{pmatrix} uH \\ ux\ unH \end{pmatrix}(a)} \otimes n\right].v$$

This formula proves that $\lambda_{a,v} = Tr_1^{\overline{N}_G(H)}(g_{a,v})$, hence that $\lambda_{a,v}$ is a morphism of $\overline{N}_G(H)$-modules from $[X^H]$ to V, which factors by a projective module. I obtain that way a morphism

$$\lambda : A\big(X \times (G/H)\big) \otimes_R V \to \mathrm{Hom}([X^H], V)_1^{\overline{N}_G(H)} \qquad a \otimes v \mapsto \lambda_{a,v}$$

Moreover, if $b \in A\big((G/H)^2\big)$, then

$$\lambda_{a \circ_{G/H} b, v} = \lambda_{a, \pi(b).v}$$

Indeed, for $x \in X^H$

$$\pi\Big(m_x^* \circ_X a \circ_{G/H} b\Big).v = \pi(m_x^* \circ_X a)\pi(b).v$$

The morphism λ is then a morphism from $L_{G/H,V}(X)$ to $\mathrm{Hom}([X^H], V)_1^{\overline{N}_G(H)}$, which passes down to the quotient $S_{H,V}(X)$: indeed, if $\sum_i a_i \otimes v_i$ is an element of $J_{G/H,V}(X)$, then for any $\psi \in A\Big((G/H) \times X\Big)$, I have

$$\sum_i (\psi \circ_X a_i).v_i = \sum_i \pi(\psi \circ_X a_i).v_i = 0$$

It follows that for any $x \in X^H$

$$\sum_i \pi(m_x^* \circ_X a_i).v_i = 0$$

so that $\sum_i \lambda_{a_i,v_i} = 0$. Thus I have a morphism, that I still denote by λ, from $S_{H,V}(X)$ to $\mathrm{Hom}([X^H], V)_1^{\overline{N}_G(H)}$.

Conversely, let $f \in \mathrm{Hom}([X^H], V)_1^{\overline{N}_G(H)}$, expressed as

$$f = Tr_1^{\overline{N}_G(H)}(g) \tag{11.1}$$

where $g \in \mathrm{Hom}([X^H], V)$. If $x \in X^H$, then $m_{x,*} \in A\Big(X \times (G/H)\Big)$, and I set

$$\mu(f) = \sum_{x \in X^H} m_{x,*} \otimes g(x) \in S_{H,V}(X) \tag{11.2}$$

Then $\mu(f)$ does not depend on the choice of the element g in (11.1): this is equivalent to say that if $Tr_1^{\overline{N}_G(H)}(g) = 0$, then the expression (11.2) is zero. In $S_{H,V}(X)$, this means that for any $\psi \in A\Big((G/H) \times X\Big)$, I have

$$\sum_{x \in X^H} (\psi \circ_X m_{x,*}).g(x) = 0$$

Let g be any element of $\mathrm{Hom}_R([X^H], V)$. Then

$$\sum_{x \in X^H} \pi(\psi \circ_X m_{x,*}).g(x) = \sum_{\substack{x \in X^H \\ n \in \overline{N}_G(H)}} \Big[\overline{A^*(i_{H,n,H})\Big(\psi \circ_X m_{x,*}\Big) \otimes n}\Big].g(x)$$

But setting $Y = G/H$, I have

$$\psi \circ_X m_{x,*} = A_*\begin{pmatrix} y_1x_1y_2 \\ y_1y_2 \end{pmatrix} A^*\begin{pmatrix} y_1x_1y_2 \\ y_1x_1x_1y_2 \end{pmatrix}(\psi \times m_{x,*})$$

Moreover

$$\psi \times m_{x,*} = \psi \times A_*\begin{pmatrix} y \\ m_x(y)y \end{pmatrix} A^*\begin{pmatrix} y \\ \bullet \end{pmatrix}(\varepsilon) = A_*\begin{pmatrix} y_1x_1y_2 \\ y_1x_1m_x(y_2)y_2 \end{pmatrix} A^*\begin{pmatrix} y_1x_1y_2 \\ y_1x_1 \end{pmatrix}(\psi)$$

As the square

$$\begin{array}{ccc} Y^2 & \xrightarrow{\begin{pmatrix} y_1y_2 \\ y_1m_x(y_2)y_2 \end{pmatrix}} & YXY \\ {\scriptstyle\begin{pmatrix} y_1y_2 \\ y_1m_x(y_2)y_2 \end{pmatrix}}\Big\downarrow & & \Big\downarrow{\scriptstyle\begin{pmatrix} y_1x_1y_2 \\ y_1x_1m_x(y_2)y_2 \end{pmatrix}} \\ YXY & \xrightarrow[\begin{pmatrix} y_1x_1y_2 \\ y_1x_1x_1y_2 \end{pmatrix}]{} & YX^2Y \end{array}$$

is cartesian, I have

$$A^*\begin{pmatrix} y_1x_1y_2 \\ y_1x_1x_1y_2 \end{pmatrix} A_*\begin{pmatrix} y_1x_1y_2 \\ y_1x_1m_x(y_2)y_2 \end{pmatrix} = A_*\begin{pmatrix} y_1y_2 \\ y_1m_x(y_2)y_2 \end{pmatrix} A^*\begin{pmatrix} y_1y_2 \\ y_1m_x(y_2)y_2 \end{pmatrix}$$

and this gives

$$\psi\circ_X m_{x,*} = A_*\begin{pmatrix} y_1x_1y_2 \\ y_1y_2 \end{pmatrix} A_*\begin{pmatrix} y_1y_2 \\ y_1m_x(y_2)y_2 \end{pmatrix} A^*\begin{pmatrix} y_1y_2 \\ y_1m_x(y_2)y_2 \end{pmatrix} A^*\begin{pmatrix} y_1x_1y_2 \\ y_1x_1 \end{pmatrix}(\psi) = \ldots$$

$$\ldots = A_*\begin{pmatrix} y_1y_2 \\ y_1y_2 \end{pmatrix} A^*\begin{pmatrix} y_1y_2 \\ y_1m_x(y_2) \end{pmatrix}(\psi) = A^*(Id\times m_x)(\psi)$$

Finally

$$\sum_{x\in X^H}(\psi\circ_X m_{x,*}).g(x) = \sum_{\substack{x\in X^H \\ n\in\overline{N}_G(H)}} \left[\overline{A^*(i_{H,n,H})A^*(Id\times m_x)(\psi)\otimes n}\right].g(x) = \ldots$$

$$\ldots = \sum_{\substack{x\in X^H \\ n\in\overline{N}_G(H)}} \left[\overline{A^*\begin{pmatrix} uH \\ uH\ unx \end{pmatrix}(\psi)\otimes n}\right].g(x) = \ldots$$

$$\ldots = \sum_{\substack{x\in X^H \\ n\in\overline{N}_G(H)}} \left[\overline{A^*\begin{pmatrix} uH \\ uH\ ux \end{pmatrix}(\psi)\otimes n}\right].g(n^{-1}x) = \ldots$$

$$\ldots = \sum_{x\in X^H} \left[\overline{A^*\begin{pmatrix} uH \\ uH\ ux \end{pmatrix}(\psi)\otimes 1}\right].Tr_1^{\overline{N}_G(H)}(g)(x)$$

Thus if $Tr_1^{\overline{N}_G(H)}(g) = 0$, then the right hand side of (11.2) is zero. It follows that $\mu(f)$ is well defined by equations (11.1) and (11.2). So I have a morphism from $\mathrm{Hom}([X^H], V)_1^{\overline{N}_G(H)}$ to $S_{H,V}(X)$.

Moreover, if $a\otimes v\in S_{H,V}(X)$, then as $\lambda = Tr_1^{\overline{N}_G(H)}(g_{a,v})$, I have

$$\mu(\lambda_{a,v}) = \sum_{x\in X^H} m_{x,*}\otimes g_{a,v}(x)$$

If $\psi\in A\big((G/H)\times X\big)$, the previous computation shows that I have

$$\sum_{x\in X^H}(\psi\circ_X m_{x,*}).g_{a,v}(x) = \sum_{x\in X^H}\left[\overline{A^*\begin{pmatrix} uH \\ uH\ ux \end{pmatrix}(\psi)\otimes 1}\right].Tr_1^{\overline{N}_G(H)}(g_{a,v})(x) = \ldots$$

$$\ldots = \sum_{x\in X^H} \left[\overline{A^*\begin{pmatrix} uH \\ uH\ ux\end{pmatrix}(\psi)\otimes 1}\right].\lambda_{a,v}(x) = \ldots$$

$$\ldots = \sum_{\substack{x\in X^H\\ n\in\overline{N}_G(H)}} \left[\overline{A^*\begin{pmatrix} uH \\ uH\ ux\end{pmatrix}(\psi)\otimes 1}\right].\left[\overline{A^*\begin{pmatrix} uH \\ ux\ unH\end{pmatrix}(a)\otimes n}\right].v = \ldots$$

$$\ldots = \sum_{\substack{x\in X^H\\ n\in\overline{N}_G(H)}} \left[\overline{A^*\begin{pmatrix} uH \\ uH\ ux\end{pmatrix}(\psi).A^*\begin{pmatrix} uH \\ ux\ unH\end{pmatrix}(a)\otimes n}\right].v$$

But

$$A^*\begin{pmatrix} uH \\ uH\ ux\end{pmatrix}(\psi).A^*\begin{pmatrix} uH \\ ux\ unH\end{pmatrix}(a) = \ldots$$

$$\ldots = A^*\begin{pmatrix} uH \\ uH\ uH\end{pmatrix}\left(A^*\begin{pmatrix} uH \\ uH\ ux\end{pmatrix}(\psi)\times A^*\begin{pmatrix} uH \\ ux\ unH\end{pmatrix}(a)\right) = \ldots$$

$$\ldots = A^*\begin{pmatrix} uH \\ uH\ uH\end{pmatrix} A^*\begin{pmatrix} u_1H\ u_2H \\ u_1H\ u_1x\ u_2x\ u_2nH\end{pmatrix}(\psi\times a) = \ldots$$

$$\ldots = A^*\begin{pmatrix} uH \\ uH\ ux\ ux\ unH\end{pmatrix}(\psi\times a)$$

This gives

$$\sum_{x\in X^H}(\psi\circ_X m_{x,*}).g_{a,v}(x) = \sum_{\substack{x\in X^H\\ n\in\overline{N}_G(H)}} \left[\overline{A^*\begin{pmatrix} uH \\ uH\ ux\ ux\ unH\end{pmatrix}(\psi\times a)\otimes n}\right].v$$

On the other hand

$$(\psi\circ_X a).v = \sum_{n\in\overline{N}_G(H)} \left[\overline{A^*(i_{H,n,H})(\psi\circ_X a)\otimes n}\right].v$$

Moreover

$$A^*(i_{H,n,H})(\psi\circ_X a) = A^*(i_{H,n,H})A_*\begin{pmatrix} u_1H\ x\ u_2H \\ u_1H\ u_2H\end{pmatrix} A^*\begin{pmatrix} u_1H\ x\ u_2H \\ u_1H\ x\ x\ u_2H\end{pmatrix}(\psi\times a)$$

As the square

$$\begin{array}{ccc}
G/H\times X & \xrightarrow{\begin{pmatrix} uH\ x \\ uH\ x\ unH\end{pmatrix}} & (G/H)\times X\times(G/H) \\
\begin{pmatrix} uH\ x \\ uH\end{pmatrix}\Big\downarrow & & \Big\downarrow\begin{pmatrix} u_1H\ x\ u_2H \\ u_1H\ u_2H\end{pmatrix} \\
G/H & \xrightarrow[\begin{pmatrix} uH \\ uH\ unH\end{pmatrix}]{} & (G/H)^2
\end{array}$$

is cartesian, I have

$$A^*(i_{H,n,H})A_*\begin{pmatrix} u_1H\ x\ u_2H \\ u_1H\ u_2H\end{pmatrix} = A_*\begin{pmatrix} uH\ x \\ uH\end{pmatrix} A^*\begin{pmatrix} uH\ x \\ uH\ x\ unH\end{pmatrix}$$

and then

$$A^*(i_{H,n,H})(\psi \circ_X a) = A_*\begin{pmatrix} uH\,x \\ uH \end{pmatrix} A^*\begin{pmatrix} uH\,x \\ uH\,x\,unH \end{pmatrix} A^*\begin{pmatrix} u_1H\,x\,u_2H \\ u_1H\,x\,x\,u_2H \end{pmatrix} (\psi \times a) = \ldots$$

$$\ldots = A_*\begin{pmatrix} uH\,x \\ uH \end{pmatrix} A^*\begin{pmatrix} uH\,x \\ uH\,x\,x\,unH \end{pmatrix} (\psi \times a)$$

This gives

$$(\psi \circ_X a).v = \sum_{n \in \overline{N}_G(H)} \left[\overline{A_*\begin{pmatrix} uH\,x \\ uH \end{pmatrix} A^*\begin{pmatrix} uH\,x \\ uH\,x\,x\,unH \end{pmatrix} (\psi \times a) \otimes n}\right].v$$

But the map sending $(uH, x) \in (G/H) \times X$ to $g^{-1}x \in H\backslash X$ induces a bijection

$$G\backslash\big((G/H) \times X\big) \simeq H\backslash X$$

Then the map σ defined by

$$\sigma : u(H \cap G_x) \in \coprod_{x \in H\backslash X} G/(H \cap G_x) \mapsto \sigma_x\big(u(H \cap G_x)\big) = (gH, gx) \in (G/H) \times X$$

is an isomorphism of G-sets. In particular,

$$Id_{A((G/H)\times X)} = \sum_{x \in H\backslash X} A_*(\sigma_x) A^*(\sigma_x)$$

Then

$$(\psi \circ_X a).v = \sum_{\substack{n \in \overline{N}_G(H) \\ x \in H\backslash X}} \left[\overline{A_*\begin{pmatrix} uH\,x \\ uH \end{pmatrix} A_*(\sigma_x) A^*(\sigma_x) A^*\begin{pmatrix} uH\,x \\ uH\,x\,x\,unH \end{pmatrix} (\psi \times a) \otimes n}\right].v = \ldots$$

$$\ldots = \sum_{\substack{n \in \overline{N}_G(H) \\ x \in H\backslash X}} \left[\overline{A_*\begin{pmatrix} u(H \cap G_x) \\ uH \end{pmatrix} A^*\begin{pmatrix} u(H \cap G_x) \\ uH\,ux\,ux\,unH \end{pmatrix} (\psi \times a) \otimes n}\right].v$$

But $A_*\begin{pmatrix} u(H\cap G_x) \\ uH \end{pmatrix} = t^H_{H\cap G_x}$. Thus if $H \not\subseteq G_x$, then $\overline{A_*\begin{pmatrix} u(H\cap G_x) \\ uH \end{pmatrix}(\alpha)} = 0$ for any α. Now I can restrict the summation to $x \in H\backslash X^H = X^H$, and then

$$(\psi \circ_X a).v = \sum_{\substack{n \in \overline{N}_G(H) \\ x \in X^H}} \left[\overline{A_*\begin{pmatrix} uH \\ uH \end{pmatrix} A^*\begin{pmatrix} uH \\ uH\,ux\,ux\,unH \end{pmatrix} (\psi \times a) \otimes n}\right].v = \ldots$$

$$\ldots = \sum_{\substack{n \in \overline{N}_G(H) \\ x \in X^H}} \left[\overline{A^*\begin{pmatrix} uH \\ uH\,ux\,ux\,unH \end{pmatrix} (\psi \times a) \otimes n}\right].v$$

Finally, for any $\psi \in A\big((G/H) \times X\big)$, I have

$$(\psi \circ_X a).v = \sum_{x \in X^H} (\psi \circ_X m_{x,*}).g_{a,v}(x)$$

which proves that in $S_{H,V}(X)$, I have

$$a \otimes v = \sum_{x \in X^H} m_{x,*} \otimes g_{a,v}(x)$$

or that $\mu \circ \lambda$ is the identity.

Conversely, if $f \in \mathrm{Hom}([X^H], V)_1^{\overline{N}_G(H)}$, i.e. if f is of the form

$$f = Tr_1^{\overline{N}_G(H)}(g)$$

then

$$\mu(f) = \sum_{x \in X^H} m_{x,*} \otimes g(x)$$

so that for $y \in X^H$, I have

$$\lambda \circ \mu(f)(y) = \sum_{\substack{n \in \overline{N}_G(H) \\ x \in X^H}} \left[\overline{A^*\begin{pmatrix} uH \\ uy\ unH \end{pmatrix}(m_{x,*}) \otimes n}\right].g(x)$$

Moreover

$$A^*\begin{pmatrix} uH \\ uy\ unH \end{pmatrix}(m_{x,*}) = A^*\begin{pmatrix} uH \\ uy\ unH \end{pmatrix} A_*\begin{pmatrix} uH \\ ux\ uH \end{pmatrix} A^*\begin{pmatrix} uH \\ \bullet \end{pmatrix}(\varepsilon)$$

For given n, x and y, let $Q_{n,x,y}$, $a_{n,x,y}$ and $b_{n,x,y}$ be such that the square

$$\begin{array}{ccc} Q_{n,x,y} & \xrightarrow{\;a_{n,x,y}\;} & G/H \\ {\scriptstyle b_{n,x,y}}\big\downarrow & & \big\downarrow \begin{pmatrix} uH \\ ux\ uH \end{pmatrix} \\ G/H & \xrightarrow[\begin{pmatrix} uH \\ uy\ unH \end{pmatrix}]{} & X \times (G/H) \end{array}$$

is cartesian. Then if $Q_{x,y}$ is non-empty, there exists u and u' in G such that $ux = u'y$ and $uH = u'nH$. These equalities imply that

$$uH.x = \{ux\} = \{u'y\} = u'nH.x = \{u'nx\}$$

so $y = nx$. Thus if $y \neq nx$, then $Q_{n,x,y} = \emptyset$, and the product $A^*\begin{pmatrix} uH \\ uy\ unH \end{pmatrix} A_*\begin{pmatrix} uH \\ ux\ uH \end{pmatrix}$ is zero. And if $y = nx$, then $Q_{n,x,y}$ identifies with G/H, the map $b_{n,x,y}$ being the identity, and the map $a_{n,x,y}$ being right multiplication by n. So I can write

$$\lambda \circ \mu(f)(y) = \sum_{n \in \overline{N}_G(H)} \left[\overline{A^*(a_{n,n^{-1}y,y}) A^*\begin{pmatrix} uH \\ \bullet \end{pmatrix}(\varepsilon) \otimes n}\right].g(n^{-1}y) = \dots$$

$$\dots = \sum_{n \in \overline{N}_G(H)} \left[\overline{A^*\begin{pmatrix} uH \\ \bullet \end{pmatrix}(\varepsilon) \otimes n}\right].g(n^{-1}y) = \sum_{n \in \overline{N}_G(H)} (\overline{\varepsilon_{G/H} \otimes n}).g(n^{-1}y) = \dots$$

$$\dots = Tr_1^{\overline{N}_G(H)}(g)(y) = f(y)$$

which proves that $\lambda \circ \mu$ is also the identity, hence that λ and μ are mutual inverse isomorphisms.

11.4.2 The A-module structure of $S_{H,V}$

Let $f : X \to Y$ be a morphism of G-sets, and $h \in \mathrm{Hom}([X^H], V)_1^{\overline{N}_G(H)}$, expressed as $h = Tr_1^{\overline{N}_G(H)}(k)$. Then

$$\mu(f) = \sum_{x \in X^H} m_{x,*} \otimes k(x)$$

so that

$$S_{H,V,*}(f)\big(\mu(f)\big) = \sum_{x \in X^H} A_*(f \times Id)(m_{x,*}) \otimes k(x)$$

Under the isomorphism λ, this element is sent to the map which sends $y \in Y^H$ to the element

$$e = \sum_{\substack{x \in X^H \\ n \in \overline{N}_G(H)}} \left[\overline{A^* \begin{pmatrix} uH \\ uy\, uH \end{pmatrix} A_*(f \times Id)(m_{x,*}) \otimes n}\right] .k(x)$$

Moreover

$$A_*(f \times Id)(m_{x,*}) = A_*(f \times Id) A_* \begin{pmatrix} uH \\ ux\, uH \end{pmatrix} (\varepsilon_{G/H}) = A_* \begin{pmatrix} uH \\ uf(x)\, uH \end{pmatrix} (\varepsilon_{G/H})$$

Let $Q_{n,x,y}$, $a_{n,x,y}$ and $b_{n,x,y}$ such that the square

$$\begin{array}{ccc} Q_{n,x,y} & \xrightarrow{a_{n,x,y}} & G/H \\ {\scriptstyle b_{n,x,y}}\downarrow & & \downarrow \begin{pmatrix} uH \\ uf(x)\, uH \end{pmatrix} \\ G/H & \xrightarrow[\begin{pmatrix} uH \\ uy\, unH \end{pmatrix}]{} & Y \times (G/H) \end{array}$$

is cartesian. If $Q_{n,x,y} \neq \emptyset$, then there exists u and u' in G such that $uy = u'f(x)$ and $unH = u'H$. These equalities imply that

$$unH.f(x) = \{unf(x)\} = u'H.f(x) = \{u'f(x)\} = \{uy\}$$

Thus if $y \neq nf(x)$, then $Q_{n,x,y} = \emptyset$, and the product $A^* \begin{pmatrix} uH \\ uy\, uH \end{pmatrix} A_* \begin{pmatrix} uH \\ uf(x)\, uH \end{pmatrix}$ is zero. If $y = nf(x)$, then $Q_{x,y} \simeq G/H$, the map $b_{n,x,y}$ is the identity and the map $a_{n,x,y}$ is right multiplication n. This gives

$$e = \sum_{\substack{n \in \overline{N}_G(H) \\ x \in X^H,\, f(nx)=y}} \left[\overline{A^* \begin{pmatrix} uH \\ unH \end{pmatrix} (\varepsilon_{G/H} \otimes n}\right] .k(x) = \ldots$$

$$\ldots = \sum_{\substack{x \in X^H \\ f(x)=y}} \sum_{n \in \overline{N}_G(H)} (\overline{\varepsilon_{G/H}} \otimes n).k(n^{-1}x) = \sum_{\substack{x \in X^H \\ f(x)=y}} h(x)$$

So I have proved that

$$S_{H,V,*}(f)(h)(y) = \sum_{\substack{x \in X^H \\ f(x)=y}} h(x)$$

Now if g is a morphism of G-sets from Y to X, then

$$S^*_{H,V}\mu(h) = \sum_{x\in X^H} A^*(g\times Id)(m_{x,*})\otimes k(x)$$

The image under λ of this element is the map sending $y \in Y^H$ to

$$\sum_{\substack{x\in X^H\\ n\in \overline{N}_G(H)}} \left[\overline{A^*\begin{pmatrix} uH \\ uy\ uH\end{pmatrix} A^*(g\times Id)(m_{x,*})\otimes n}\right].k(x) = \ldots$$

$$\ldots = \sum_{\substack{x\in X^H\\ n\in \overline{N}_G(H)}} \left[\overline{A^*\begin{pmatrix} uH \\ ug(y)\ uH\end{pmatrix} (m_{x,*})\otimes n}\right].k(x) = \lambda\circ\mu(h)\big(g(y)\big) = hg(y)$$

So I have

$$S^*_{H,V}(g)(h)(y) = hg(y)$$

Finally, let X and Y be G-sets, and

$$h = Tr_1^{\overline{N}_G(H)}(k) \in \mathrm{Hom}([Y^H], V)$$

Then

$$\mu(f) = \sum_{y\in Y^H} m_{y,*}\otimes k(y)$$

Thus, if $a \in A(X)$, I have

$$a\times\mu(f) = \sum_{y\in Y^H} (a\times m_{y,*})\otimes k(y)$$

The image under λ of this element is the map sending $z \in Z^H = X^H \times Y^H$ to the element

$$e = \sum_{\substack{y\in Y^H\\ n\in \overline{N}_G(H)}} \left[\overline{A^*\begin{pmatrix} uH \\ uz\ unH\end{pmatrix} (a\times m_{y,*})\otimes n}\right].k(y)$$

Moreover

$$A^*\begin{pmatrix} uH \\ uz\ unH\end{pmatrix}(a\times m_{y,*}) = A^*\begin{pmatrix} uH \\ uz\ unH\end{pmatrix}\left(a\times A_*\begin{pmatrix} uH \\ uy\ uH\end{pmatrix} A^*\begin{pmatrix} uH \\ \bullet\end{pmatrix}(\varepsilon)\right) = \ldots$$

$$\ldots = A^*\begin{pmatrix} uH \\ uz\ unH\end{pmatrix} A_*\begin{pmatrix} x\ uH \\ x\ uy\ uH\end{pmatrix} A^*\begin{pmatrix} x\ uH \\ x\end{pmatrix}(a)$$

Let $Q_{n,z,y}$, $a_{n,z,y}$ and $b_{n,z,y}$ such that the square

$$\begin{array}{ccc} Q_{n,z,y} & \xrightarrow{a_{n,z,y}} & X\times(G/H) \\ {\scriptstyle b_{n,z,y}}\downarrow & & \downarrow\begin{pmatrix} x\ uH \\ x\ uy\ uH\end{pmatrix} \\ G/H & \xrightarrow[\begin{pmatrix} uH \\ uz\ unH\end{pmatrix}]{} & Z\times(G/H)\end{array}$$

is cartesian. If $z = (x_0, y_0)$, and if $Q_{n,z,y} \neq \emptyset$, then there exists u and u' in G such that $uz = (ux_0, uy_0) = (x, u'y)$ and $unH = u'H$. In those conditions

$$u'H.y = \{u'y\} = \{uy_0\} = unH.y = \{uny\}$$

Thus if $y_0 \neq ny$, the set $Q_{n,z,y}$ is empty, and the product $A^*\begin{pmatrix} uH \\ uz\ unH \end{pmatrix} A_*\begin{pmatrix} x\ uH \\ x\ uy\ uH \end{pmatrix}$ is zero. and if $y_0 = ny$, then $Q_{n,z,y} \simeq G/H$, the map $b_{n,z,y}$ is the identity, and the map $a_{n,z,y}$ is defined by

$$a_{n,z,y}(uH) = (ux_0, uH)$$

It follows that

$$e = \sum_{n \in \overline{N}_G(H)} \left[\overline{A^*\begin{pmatrix} uH \\ ux_0\ uH \end{pmatrix} A^*\begin{pmatrix} x\ uH \\ x \end{pmatrix}(a) \otimes n}\right] .k(n^{-1}y_0) = \ldots$$

$$\ldots = \sum_{n \in \overline{N}_G(H)} \left[\overline{A^*\begin{pmatrix} uH \\ ux_0 \end{pmatrix}(a) \otimes n}\right] .k(n^{-1}y_0) = \ldots$$

$$\ldots = \left(\overline{A^*(m_{x_0})(a)} \otimes 1\right) \sum_{n \in \overline{N}_G(H)} (\varepsilon_{G/H} \otimes n).k(n^{-1}y_0) = \ldots$$

$$\ldots = \ldots$$

$$\ldots = \left(\overline{A^*(m_{x_0})(a)} \otimes 1\right) .h(y_0)$$

which proves that

$$(a \times h)(x_0, y_0) = \left(\overline{A^*(m_{x_0})(a)} \otimes 1\right) .h(y_0)$$

and completes the proof of the proposition. ■

11.5 The simple Green functors

The previous results give a new proof of a theorem of Thévenaz (see[13]) on simple Green functors. The notion of simple Green functor relies on the notion of functorial ideal, that Thévenaz defines as follows (see [13] 1.8): a functorial ideal I of the Green functor A for the group G is a sub-Mackey functor such that for any subgroup H, the module $I(H)$ is moreover a two-sided ideal of $A(H)$. It is equivalent to say that the Green functor structure of A passes down to the quotient A/I.

Translating this definition in terms of the product $\times$, I see that for any G-sets X and Y, I must have

$$I(X) \times A(Y) \subseteq I(X \times Y) \qquad A(X) \times I(Y) \subseteq I(X \times Y)$$

Conversely, if I is a sub-Mackey functor of A such that this condition holds for any X and Y, then the product $\times$ of A passes down to the quotient A/I.

In other words, a functorial ideal is nothing but an A-submodule-A of A. Thus a Green functor A is a simple Green functor if and only if A is simple as an A-module-A.

But the A-modules-A are also the $A\hat{\otimes}A^{op}$-modules. Thus if A is a simple Green functor, then there exists a subgroup H of G and an $\overline{A\hat{\otimes}A^{op}(H)} \otimes \overline{N}_G(H)$-module V such that

$$A \simeq S_{H,V}^{A\hat{\otimes}A^{op}}$$

as $A\hat{\otimes}A^{op}$-modules (the exponent $A\hat{\otimes}A^{op}$ recalls that the simple module $S_{H,V}$ is a simple module for this Green functor). Then H is the unique minimal subgroup of A up to conjugation, and $V = A(H) = \overline{A}(H)$ is a simple $\overline{A\hat{\otimes}A^{op}}(H) \otimes \overline{N}_G(H)$-module. Now

$$(A\hat{\otimes}A^{op})^H \simeq A^H\hat{\otimes}(A^{op})^H \simeq A^H\hat{\otimes}(A^H)^{op}$$

This follows from lemma 10.8.5, since moreover for $U = H\backslash G$ as a $\overline{N}_G(H)$-set-G, I have $U/G = \bullet$, and so

$$(\mathrm{Res}^G_{N_G(H)} b)^H = \mathcal{L}_U(b) = b_{U/G} = b$$

This isomorphism is now clearly an isomorphism of Green functors, and it follows from evaluation at H the isomorphism of algebras

$$\overline{A\hat{\otimes}A^{op}}(H) \simeq \overline{A}(H) \otimes \overline{A}(H)^{op} = A(H) \otimes A(H)^{op}$$

Now say that $A(H)$ is a simple $A(H) \otimes A(H)^{op} \otimes \overline{N}_G(H)$-module, is equivalent to say that $A(H)$ is a $\overline{N}_G(H)$-algebra, or an algebra with an action of $\overline{N}_G(H)$, having no proper two-sided ideal invariant by $\overline{N}_G(H)$. Moreover, if A is non-zero, then $A(\bullet) \neq 0$. As

$$S_{H,V}^{A\hat{\otimes}A^{op}}(\bullet) \simeq \mathrm{Hom}([\bullet], V)_1^{\overline{N}_G(H)}$$

I see that $Tr_1^{\overline{N}_G(H)}\big(A(H)\big)$ has to be non-zero. As it is a two-sided invariant ideal of $A(H)$, I must then have

$$A(H) = Tr_1^{\overline{N}_G(H)}\big(A(H)\big)$$

which proves that $A(H)$ is moreover a projective $\overline{N}_G(H)$-algebra.

Conversely, if B is a $\overline{N}_G(H)$-algebra, then I have a Green functor FP_B, such that for any subgroup K of $\overline{N}_G(H)$, the ring $FP_B(H)$ is equal to B^H, the restrictions being inclusions, and the transfers being relative traces. In those conditions, for any $\overline{N}_G(H)$-set Z, I have $FP_B(Z) \simeq \mathrm{Hom}_{\overline{N}_G(H)}([Z], B)$.

If B is a projective $\overline{N}_G(H)$-algebra, having no proper two-sided invariant ideal, then let

$$A = \mathrm{Ind}_{N_G(H)}^G \mathrm{Inf}_{\overline{N}_G(H)}^{N_G(H)} FP_B$$

Then for any G-set X, I have

$$A(X) = \mathrm{Hom}_{\overline{N}_G(H)}([X^H], B)$$

and as B is projective, it is also

$$A(X) = \mathrm{Hom}([X^H], B)_1^{\overline{N}_G(H)}$$

In those conditions, the group H is a minimal subgroup for A, and $A(H) = \overline{A}(H) = B$ is a simple $A(H) \otimes A(H)^{op} \otimes \overline{N}_G(H)$-module, so that

$$A \simeq S_{H,B}^{A\hat{\otimes}A^{op}}$$

as A-modules-A. Thus A is a simple Green functor.

So I must find, for a group K, which are the algebras B with a K-action which are projective and have no proper two-sided invariant ideal. Let I be any maximal (proper) two-sided ideal of B as an algebra. Such an ideal exists by Zorn's lemma. Let L be the stabilizer of I in K. Then for any $k \in K$, the image $k(I)$ of I by k is also a maximal two sided ideal of B. In particular

$$\bigcap_{k\in K} k(I) = 0$$

since it is a two-sided invariant ideal. Let P be a subset of K of maximal cardinality such that

$$J = \bigcap_{k\in P} k(I) \neq 0$$

Then $P \neq K$, and there exists $k_0 \in K - P$. Replacing P by $k_0^{-1}P$, I can suppose $k_0 = 1 \notin P$, and then

$$I \cap J = 0$$

Moreover, the maximality of I implies $I + J = B$, thus

$$B = I \oplus J$$

Then J identifies with B/I, which is a simple B-module-B. Thus J is a minimal two-sided ideal of B. Then JI is contained in $J \cap I = 0$. So I is contained in the right annihilator of J, which is a two-sided ideal. As $J \neq 0$, the annihilator of J is equal to I.

On the other hand the sum

$$\sum_{k\in K/L} k(J)$$

is a non-zero two-sided invariant ideal of B, so it is equal to B. As $J.k(J)$ is contained in $J \cap k(J)$, which is a sub-two-sided ideal of J, I have $J.k(J) = 0$ if $k \notin L$. Thus if $(j_k)_{k\in K/L}$ is a family of elements of J such that

$$\sum_{k\in K/L} k(j_k) = 0$$

then for any $k_1 \in K$, I have

$$0 = k_1(J) \sum_{k\in K/L} k(j_k) = k_1\left(\sum_{k\in K/L} J.k_1^{-1}k(j_k)\right) = k_1(J.j_{k_1})$$

Then j_{k_1} is in the annihilator of J, hence in I, hence in $I \cap J$. Thus $j_{k_1} = 0$, and

$$B = \bigoplus_{k\in K/L} k(J)$$

In particular, the unit of B decomposes as

$$1_B = \sum_{k\in K/L} e_k$$

for some $e_k \in k(J)$. If $k_1 \in K/L$, I have

$$k_1(1_B) = 1_B = \sum_{k\in K/L} k_1(e_k)$$

and unicity of the decomposition implies $e_k = k(e_1)$. Moreover

$$e_1 = e_1 \sum_{k \in K/L} k(e_1) = e_1^2$$

and for any $b \in B$, I have

$$e_1 b = e_1 b \sum_{k \in K/L} k(e_1) = \sum_{k \in K/L} e_1 bk(e_1) = e_1 b e_1$$

because $e_1 bk(e_1) \in J.k(J) = 0$ if $k \notin L$. The same argument shows then that $be_1 = e_1 b e_1$, so that e_1 is central in B. Then $J = e_1 B$ is a simple algebra with an action of L, and B is isomorphic to the induced K-algebra

$$B \simeq \mathrm{Ind}_L^K J = RK \otimes_{RL} J$$

with product defined by

$$(k \otimes j).(k' \otimes j') = \begin{cases} 0 & \text{if } k^{-1}k' \notin L \\ k \otimes j.(k^{-1}k')(j') & \text{otherwise} \end{cases}$$

Now say that B is projective means that there exists elements j_k, for $k \in K/L$, such that

$$\sum_{k_1 \in K} k_1 \left(\sum_{k \in K/L} k(j_k) \right) = 1_B = \sum_{k \in K/L} k(e_1)$$

The product by e_1 gives

$$e_1 = \sum_{\substack{k_1 \in K \\ k \in K/L \\ k_1 k \in L}} k_1 k(j_k) = \sum_{\substack{k \in K/L \\ l \in L}} l(j_k) = Tr_1^L \left(\sum_{k \in K/L} j_k \right)$$

It follows that J is projective as L-algebra. Conversely, if

$$e_1 = Tr_1^L(j) = \sum_{l \in L} l(j)$$

then

$$Tr_1^K(j) = Tr_L^K Tr_1^L(j) = Tr_L^K(e_1) = 1_B$$

and B is projective.

Similarly, if J is a simple algebra on which L acts, then $\mathrm{Ind}_L^K J$ is a K-algebra without proper two-sided invariant ideals: if U is such an ideal, and if e_1 is the unit of J, then U is the direct sum of $k(e_1)U$, for $k \in K/L$, and $k(e_1)U$ is isomorphic to $e_1 U$, which is a two-sided ideal of J. If $e_1 U = 0$, then $U = 0$, and if $e_1 U = J$, then $U = B$.

Thus the algebra B is isomorphic to $\mathrm{Ind}_L^K J$, where L is a subgroup of K, and J is a simple algebra with an action of L, the algebra J being moreover projective as L-algebra.

Finally, I observe that the group L and the L-algebra J are entirely determined if B is known as K-algebra: indeed, if I_1 is a maximal two-sided ideal of B, then $k(e_1)I_1$

is isomorphic to $e_1 k^{-1}(I_1)$, which is a two-sided ideal of J, hence equal to zero or J, and

$$I_1 = \bigoplus_{\substack{k \in K/L \\ k(e_1)I_1 \neq 0}} k(e_1)J$$

As I_1 is maximal, there exists a unique $k_0 \in K/L$ such that

$$I_1 = \bigoplus_{k \in K/L - \{k_0\}} k(e_1)J$$

The maximal two-sided ideals of B are then conjugate by K. The group L is the stabilizer of a maximal two-sided ideal of B, and the algebra J is the unique simple quotient of B: the couple (L, J) is then unique up to conjugation by K.

Going back to the case of a simple Green functor A, and observing that induction and inflation of algebras commute with the functor FP_-, it follows that there exists a subgroup M of $N_G(H)$, containing H (such that M/H is the above group L), and a simple algebra S (equal to J) on which M/H acts projectively, such that

$$A \simeq \mathrm{Ind}_M^G \mathrm{Inf}_{M/H}^M FP_S$$

So the previous discussion gives a new proof, under slightly weaker hypothesis, of the following theorem:

Proposition 11.5.1: (Thévenaz [13] Theorem 12.11)

1. **Let A be a simple Green functor for G. Then there exists a subgroup M of G, a normal subgroup H of M, and a simple algebra S on which M/H acts projectively, such that**

$$A \simeq \mathrm{Ind}_M^G \mathrm{Inf}_{M/H}^M FP_S$$

   The triple (M, H, S) is unique up to conjugation by G (and up to isomorphism of M/H-algebras for S).

2. **Conversely, if $H \trianglelefteq M$ are subgroups of G, if S is a simple algebra on which M/H acts projectively, then $\mathrm{Ind}_M^G \mathrm{Inf}_{M/H}^M FP_S$ is a simple Green functor.**

11.6 Simple functors and endomorphisms

Let G be a group and A be a Green functor for G. In this section I will study the relations between the simple A-modules $S_{H,V}^G$ and the functors $-\hat{\otimes}-$ and $\mathcal{H}(-,-)$.

First I observe that proposition 11.4.1 gives a definition of $S_{H,V}^G$ for any $\overline{A}(H) \otimes \overline{N}_G(H)$-module V (non-necessarily simple). Moreover, the structure of Mackey functor of $S_{H,V}^G$ depends only on the restriction of V to $\overline{N}_G(H) \simeq b(H) \otimes \overline{N}_G(H)$: in other words

$$\mathrm{Res}_b S_{H,V}^G = S_{H,\mathrm{Res}_{\overline{N}_G(H)} V}^G$$

Similarly, I can define:

Definition: *Let H be a subgroup of G, and V be an $\overline{A}(H)\otimes \overline{N}_G(H)$-module. I denote by $FP_{H,V}^G = FP_{H,V}$ the A-module defined like $S_{H,V}$, but with any homomorphisms, and not only those which factor through a projective module: if X is a G-set, then*

$$FP_{H,V}^G(X) = \mathrm{Hom}_{\overline{N}_G(H)}([X^H], V)$$

If $f : X \to Y$ is a morphism of G-sets, then for $\alpha \in FP_{H,V}^G(X)$ and $y \in Y^H$, I have

$$FP_{H,V*}^G(f)(\alpha)(y) = \sum_{\substack{x\in X^H\\ f(x)=y}} \alpha(x)$$

If $g : Y \to X$ is a morphism of G-sets, then for $\alpha \in FP_{H,V}^G(X)$ and $y \in Y^H$, I have

$$FP_{H,V}^G{}^*(g)(\beta)(y) = \beta g(y)$$

Finally, if $a \in A(X)$ and $f \in FP_{H,V}^G(Y)$, then $a\times f$ is the morphism from $[(X\times Y)^H] = [X^H \times Y^H]$ to V defined by

$$(a \times m)(x,y) = \left(\overline{A^*(m_x)(a)} \otimes 1\right).f(y)$$

where m_x is the morphism of G-sets from G/H to X defined by $m_x(uH) = ux$.

I define dually $FQ_{H,V}^G$ for a G-set X by

$$FQ_{H,V}^G(X) = [X^H] \otimes_{\overline{N}_G(H)} V$$

If $f : X \to Y$ is a morphism of G-sets, then

$$FQ_{H,V*}^G(f)(x\otimes v) = f(x)\otimes v$$

If $g : Y \to X$ is a morphism of G-sets, then

$$FQ_{H,V}^G{}^*(f)(x\otimes v) = \sum_{\substack{y\in Y^H\\ g(y)=x}} y\otimes v$$

Finally, if $a \in A(X)$, and $y \otimes v \in FQ_{H,V}^G(Y)$, then

$$a \times (y\otimes v) = \sum_{x\in X^H} (x,y)\otimes (\overline{A^*(m_x)(a)} \otimes 1).v$$

It is easy to see that these definitions turn $FP_{H,V}^G$ and $FQ_{H,V}^G$ into A-modules. The notation comes from the fact that if A is the Burnside functor, then $FP_{H,V}^G$ and $FQ_{H,V}^G$ identify respectively with the functors denoted by

$$\mathrm{Ind}_{N_G(H)}^G \mathrm{Inf}_{\overline{N}_G(H)}^{N_G(H)} FP_V \qquad \mathrm{Ind}_{N_G(H)}^G \mathrm{Inf}_{\overline{N}_G(H)}^{N_G(H)} FQ_V$$

by Thévenaz and Webb. The following lemma is a way to recover this isomorphism:

Lemma 11.6.1: Let M be an A-module. Then, for any subgroup H of G, the modules $\overline{M}(H)$ and $\underline{M}(H)$ are $\overline{A}(H)\otimes \overline{N}_G(H)$-modules. Moreover

$$\mathrm{Hom}_A(FQ_{H,V}^G, M) = \mathrm{Hom}_{\overline{A}(H)\otimes\overline{N}_G(H)}\left(V, \underline{M}(H)\right)$$

$$\mathrm{Hom}_A(M, FP_{H,V}^G) = \mathrm{Hom}_{\overline{A}(H)\otimes\overline{N}_G(H)}\left(\overline{M}(H), V\right)$$

Proof: It is clear that the action of $A(H) \otimes N_G(H)$ on $M(H)$ induces an action of $\overline{A}(H) \otimes \overline{N}_G(H)$ on $\overline{M}(H)$, since

$$(t_K^H(a) \otimes n).m = t_K^H(a).n.m = t_K^H\Big(a.r_K^H(n.m)\Big)$$

Similarly, if $m \in \underline{M}(H)$, and if $K \subset H$, then

$$(t_K^H(a) \otimes n).m = t_K^H\Big(a.r_K^H(n.m)\Big) = 0$$

because $r_K^H(n.m) = n.r_{K^n}^H(m) = 0$. So $\underline{M}(H)$ is also an $\overline{A}(H) \otimes \overline{N}_G(H)$-module.

Moreover if ϕ is a morphism from $FQ_{H,V}^G$ to M, then $\phi_{G/H}$ is a morphism of $\overline{A}(H) \otimes \overline{N}_G(H)$-modules from $FQ_{H,V}^G(H)$ to $M(H)$. But it is clear that $FQ_{H,V}(H)$ is isomorphic to V, and that $FQ_{H,V}^G(K) = 0$ if K is a proper subgroup of H. Then the image of $\phi_{G/H}$ is contained in $\underline{M}(H)$, and $\phi_{G/H}$ is a morphism of $\overline{A}(H) \otimes \overline{N}_G(H)$-modules from V to $\underline{M}(H)$.

Conversely, if ψ is a morphism of $\overline{A}(H) \otimes \overline{N}_G(H)$-modules from V to $\underline{M}(H)$, and if X is a G-set, I define a morphism ψ_X from $FQ_{H,V}^G(X)$ to $M(X)$ by setting

$$\psi_X(x \otimes v) = M_*(m_x)\psi(v)$$

It is easy to see that this defines a morphism of A-modules from $FQ_{H,V}^G$ to M. The first adjunction of the lemma follows.

The second one follows from a dual proof. ■

Lemma 11.6.2: Let H be a subgroup of G, and V be a $\overline{N}_G(H)$-module. If K is a subgroup of G, then

$$\overline{S_{H,V}^G}(K) = 0 = \underline{S_{H,V}^G}(K) \quad \text{if} \quad K \neq_G H$$

$$\overline{S_{H,V}^G}(H) = V = \underline{S_{H,V}^G}(H)$$

Proof: By definition, as $(G/H)^H = N_G(H)/H$, I have

$$S_{H,V}^G(G/H) = \mathrm{Hom}([(G/H)^H], V)_1^{\overline{N}_G(H)} = \mathrm{Hom}(R\overline{N}_G(H), V)_1^{\overline{N}_G(H)} \simeq V$$

Then $v \in V$ is associated to the morphism

$$\delta_v : n \mapsto n.v = (\overline{\varepsilon_{G/H}} \otimes n).v$$

from $R\overline{N}_G(H)$ to V. Moreover, if X is a G-set, if $x_0 \in X$, and if $\delta_{x_0,v}$ is the morphism from $[X^H]$ to V defined by

$$\delta_{x_0,v}(x) = \delta_{x,x_0}.v = \begin{cases} 0 & \text{if} \quad x \neq x_0 \\ v & \text{if} \quad x = x_0 \end{cases}$$

then

$$Tr_1^{\overline{N}_G(H)}(\delta_{x_0,v})(x) = \sum_{\substack{n \in \overline{N}_G(H) \\ nx_0 = x}} n.v$$

Let m_{x_0} the morphism of G-sets from G/H to X defined by $m_{x_0}(GH) = gx_0$. It follows from proposition 11.4.1 that

$$S^G_{H,V*}(m_{x_0})(x) = \sum_{\substack{n\in\overline{N}_G(H)\\ nx_0=x}} nv$$

thus $Tr_1^{\overline{N}_G(H)}(\delta_{x_0,v}) = S^G_{H,V*}(m_{x_0})$. If K is a subgroup of G, then

$$S^G_{H,V}(K) = \mathrm{Hom}([(G/K)^H], V)_1^{\overline{N}_G(H)}$$

is zero if H is not contained in K modulo G, because then $(G/K)^H = \emptyset$. The previous argument proves moreover that

$$S^G_{H,V}(K) = \sum_{\substack{x\in G\\ H^x\subseteq K}} t^K_{H^x} S^G_{H,V}(H^x)$$

It follows that

$$\overline{S^G_{H,V}}(K) = 0$$

if $K \neq_G H$. It is clear moreover that $S^G_{H,V}(H) = \overline{S^G_{H,V}}(H) = V$. A dual argument proves the assertions on $\underline{S^G_{H,V}}(K)$, and the lemma follows. ■

Proposition 11.6.3: Let G be a group, and M and N be Mackey functors for G.

- **If for any subgroup H of G, one of the modules $\overline{M}(H)$ or $\overline{N}(H)$ is zero, then $M\hat{\otimes}N = 0$.**
- **If for any subgroup H of G, one of the modules $\overline{M}(H)$ or $\underline{N}(H)$ is zero, then $\mathcal{H}(M,N) = 0$**

Proof: Indeed, I have already seen that if M and N are Mackey functors for G, then

$$\overline{M\hat{\otimes}N}(H) \simeq \overline{M}(H) \otimes \overline{N}(H)$$

If one of the modules $\overline{M}(H)$ or $\overline{N}(H)$ is zero for any H, then $\overline{M\hat{\otimes}N}(H) = 0$ for any H. If $M\hat{\otimes}N \neq 0$, there is a minimal subgroup H for $M\hat{\otimes}N$. Then

$$0 \neq (M\hat{\otimes}N)(H) \simeq \overline{(M\hat{\otimes}N)}(H) = 0$$

This contradiction proves the first assertion of the proposition. The second one follows similarly from the equality

$$\mathcal{R}_U\mathcal{H}(M,N) = \mathcal{H}\big(\mathcal{L}_U(M), \mathcal{R}_U(N)\big)$$

for $U = H\backslash G$, viewed as a $\overline{N}_G(H)$-set-G (see proposition 10.1.2), which proves that

$$\underline{\mathcal{H}(M,N)}(H) = \mathrm{Hom}_R\big(\overline{M}(H), \underline{N}(H)\big)$$

Then if $\mathcal{H}(M,N) \neq 0$, for any minimal subgroup H of $\mathcal{H}(M,N)$, I have

$$0 \neq \underline{\mathcal{H}(M,N)}(H) = \mathcal{H}(M,N)(H) = \mathrm{Hom}_R\big(\overline{M}(H), \underline{N}(H)\big)$$

so the two modules $\overline{M}(H)$ and $\underline{N}(H)$ are non-zero. ■

Proposition 11.6.4: Let H and K be subgroups of G. If V is a $\overline{N}_G(H)$-module, and W a $\overline{N}_G(K)$-module, and if H and K are not conjugate in G, then

$$S_{H,V}^G \hat{\otimes} S_{K,W}^G = 0 \qquad \mathcal{H}(S_{H,V}^G, S_{K,W}^G) = 0$$

Moreover, if $K = H$, then

$$S_{H,V}^G \hat{\otimes} S_{H,W}^G \simeq FQ_{H,V\otimes_R W}^G \qquad \mathcal{H}(S_{H,V}^G, S_{K,W}^G) \simeq FP_{H,\mathrm{Hom}_R(V,W)}^G$$

as Mackey functors.

Proof: The first assertion follows from proposition 11.6.3 and from lemma 11.6.2. As moreover the module $S_{H,V}^G(K)$ is zero for any proper subgroup of H, it is clear that the same is true for $(S_{H,V}^G \hat{\otimes} S_{H,W}^G)(K)$, and that moreover

$$(S_{H,V}^G \hat{\otimes} S_{H,W}^G)(H) \simeq S_{H,V}^G(H) \hat{\otimes} S_{H,W}^G(H) \simeq V \hat{\otimes} W$$

Then $(S_{H,V}^G \hat{\otimes} S_{H,W}^G)(H) = \underline{(S_{H,V}^G \hat{\otimes} S_{H,W}^G)}(H)$, and lemma 11.6.1 shows that there exists a morphism Φ from $FQ_{H,V\hat{\otimes}W}^G$ to $S_{H,V}^G \hat{\otimes} S_{H,W}^G$, defined for a G-set X by

$$\Phi_X : x \otimes (v \otimes w) \in FQ_{H,V\hat{\otimes}W}^G(X) \mapsto [\delta_v \otimes \delta_w]_{(G/H,m_x)}$$

Conversely, if X and Y are G-sets, it is easy to check that the correspondence which maps

$$\phi = Tr_1^{\overline{N}_G(H)}(\phi_0) \in S_{H,V}^G(X) = \mathrm{Hom}([X^H], V)_1^{\overline{N}_G(H)}$$

$$\psi \in S_{H,W}^G(Y) = \mathrm{Hom}([Y^H], W)_1^{\overline{N}_G(H)}$$

to

$$\theta_{X,Y}(\phi,\psi) = \sum_{\substack{x\in X^H\\ y\in Y^H}} (x,y) \otimes \big(\phi_0(x) \otimes \psi(y)\big) \in FQ_{H,V}^G(X \times Y)$$

is well defined, and bifunctorial: indeed, if $\psi = Tr_1^{\overline{N}_G(H)}(\psi_0)$, I have also

$$\theta_{X,Y}(\phi,\psi) = \sum_{\substack{x\in X^H\\ y\in Y^H\\ n\in \overline{N}_G(H)}} (x,y) \otimes \big(\phi_0(x) \otimes \psi_0(n^{-1}x)\big) = \ldots$$

$$\ldots = \sum_{\substack{x\in X^H\\ y\in Y^H\\ n\in \overline{N}_G(H)}} (x,ny) \otimes \big(\phi_0(x) \otimes n\psi_0(y)\big) = \sum_{\substack{x\in X^H\\ y\in Y^H\\ n\in \overline{N}_G(H)}} (nx,ny) \otimes \big(\phi_0(nx) \otimes n\psi_0(y)\big) = \ldots$$

$$\ldots = \sum_{\substack{x\in X^H\\ y\in Y^H\\ n\in \overline{N}_G(H)}} (x,y) \otimes \big(n^{-1}\phi_0(nx) \otimes \psi_0(y)\big) = \sum_{\substack{x\in X^H\\ y\in Y^H}} (x,y) \otimes \big(\phi(x) \otimes \psi_0(y)\big)$$

It follows that $\theta_{X,Y}(\phi,\psi)$ is independent of the choice of ϕ_0 such that $\phi = Tr_1^{\overline{N}_G(H)}$.

If moreover $f : X \to X'$ and $g : Y \to Y'$ are morphisms of G-sets, then for $x' \in X'$, I have

$$S_{H,V*}^G(f)(\phi)(x') = \sum_{\substack{x\in X^H\\ f(x)=x'}} \phi(x)$$

Setting

$$\phi_0'(x') = \sum_{\substack{x \in X^H \\ f(x)=x'}} \phi_0(x)$$

I have $S_{H,V*}^G(f)(\phi) = Tr_1^{\overline{N}_G(H)}(\phi_0')$, and then

$$\theta_{X',Y'}\Big(S_{H,V*}^G(f)(\phi), S_{H,W*}^G(g)(\psi)\Big) = \sum_{\substack{x' \in X'^H \\ y' \in Y'^H}} (x',y')\otimes\Big(\phi_0'(x')\otimes S_{H,W*}^G(g)(\psi)(y')\Big) = \dots$$

$$\dots = \sum_{\substack{(x,y)\in(X\times Y)^H \\ (f\times g)(x,y)=(x',y')}} (x',y')\otimes\Big(\phi_0(x)\otimes\psi(y)\Big) = FQ_{H,V*}^G(f\times g)\theta_{X,Y}(\phi,\psi)$$

Now if f is a morphism from X' to X and g is a morphism from Y' to Y, then for $x' \in X'$

$$S_{H,V}^G{}^*(f)(\phi)(x') = \phi f(x')$$

Setting $\phi_0' = \phi_0 \circ f$, I have $S_{H,V}^G{}^*(f)(\phi) = Tr_1^{\overline{N}_G(H)}(\phi_0')$, so

$$\theta_{X',Y'}\Big(S_{H,V}^G{}^*(f)(\phi), S_{H,W}^G{}^*(g)(\psi)\Big) = \sum_{(x',y')\in X'^H\times Y'^H} (x',y')\otimes\Big(\phi_0 f(x')\otimes\psi g(y')\Big) = \dots$$

$$\dots = \sum_{(x,y)\in(X\times Y)^H} \sum_{\substack{(x',y')\in(X'\times Y')^H \\ (f\times g)(x',y')=(x,y)}} (x',y')\otimes\Big(\phi_0(x)\otimes\psi(y)\Big) = FQ_{H,V}^G{}^*(f\times g)\theta_{X,Y}(m,n)$$

Thus $\theta_{X,Y}$ is bifunctorial, and induces a bilinear morphism from $S_{H,V}^G$, $S_{H,W}^G$ to $FQ_{H,V\otimes W}^G$. This gives a morphism Θ from $S_{H,V}^G\hat{\otimes}S_{H,W}^G$ to $FQ_{H,V\hat{\otimes}W}$, defined for a G-set X by

$$\Theta_X : [\phi\otimes\psi]_{(Y,f)} \in (S_{H,V}^G\hat{\otimes}S_{H,W}^G)(X) \mapsto \sum_{y\in Y^H} f(y)\otimes\Big(\phi_0(y)\otimes\psi(y)\Big) \in FQ_{H,V\otimes W}^G(X)$$

where ϕ_0 is a morphism from $[Y^H]$ to V such that

$$Tr_1^{\overline{N}_G(H)}(\phi_0) = \phi$$

Then Φ and Θ are mutual inverse isomorphisms: if

$$x\otimes(v\otimes w) \in FQ_{H,V\otimes W}^G(X)$$

then

$$\Phi(x) = [\delta_v\otimes\delta_w]_{(G/H,m_x)}$$

Let $\delta_{0,v}$ be the map from $\overline{N}_G(H)$ to V defined by

$$\delta_{0,v}(n) = \begin{cases} v & \text{if } n=1 \\ 0 & \text{otherwise} \end{cases}$$

Then $\delta_v = Tr_1^{\overline{N}_G(H)}(\delta_{0,v})$, and

$$\Theta\Phi\Big(x\otimes(\delta_v\otimes\delta_w)\Big) = \sum_{n\in\overline{N}_G(H)} m_x(n)\otimes\Big(\delta_{0,v}(n)\otimes\delta_w(n)\Big) = m_x(1)\otimes(v\otimes w) = x\otimes(v\otimes w)$$

So $\Theta\Phi$ is the identity. Conversely, if (Y, f) is a G-set over X, then

$$\Theta_X([\phi\otimes\psi]_{(Y,f)}) = \sum_{y\in Y^H} f(y)\otimes\Big(\phi_0(y)\otimes\psi(y)\Big)$$

where ϕ_0 is such that $Tr_1^{\overline{N}_G(H)}(\phi_0) = \phi$. Thus

$$\Phi\Theta([\phi\otimes\psi]_{(Y,f)}) = \sum_{y\in Y^H} [\delta_{\phi_0(y)}\otimes\delta_{\psi(y)}]_{(G/H, m_{f(y)})} \tag{11.3}$$

Let $\phi' = \sum_{y\in Y^H} S^G_{H,V*}(m_y)(\delta_{\phi_0(y)})$. Then if $y_0 \in Y^H$, I have

$$\phi'(y_0) = \sum_{y\in Y}^{H} S^G_{H,V*}(m_y)(\delta_{\phi_0(y)})(y_0) = \sum_{\substack{y\in Y^H\\ n\in\overline{N}_G(H)\\ m_y(n)=y_0}} \delta_{\phi_0(y)}(n) = \ldots$$

$$\ldots = \sum_{\substack{y\in Y^H\\ n\in\overline{N}_G(H)\\ ny=y_0}} n\phi_0(y) = \sum_{n\in\overline{N}_G(H)} n\phi_0(n^{-1}y_0) = \phi(y_0)$$

Thus $\phi = \sum_{y\in Y^H} S^G_{H,V*}(m_y)(\delta_{\phi_0(y)})$. Then

$$[\phi\otimes\psi]_{(Y,f)} = \sum_{y\in Y^H} [\delta_{\phi_0(y)}\otimes S^G_{H,V}{}^*(m_y)(\psi)]_{(G/H, fm_y)}$$

Moreover if $n\in\overline{N}_G(H)$

$$S^G_{H,V}{}^*(m_y)(\psi)(n) = \psi\big(m_y(n)\big) = \psi(ny) = n\psi(y) = \delta_{\psi(y)}(n)$$

So

$$[\phi\otimes\psi]_{(Y,f)} = \sum_{y\in Y^H} [\delta_{\phi_0(y)}\otimes\delta_{\psi(y)}]_{(G/H, fm_y)}$$

As moreover $fm_y = m_{f(y)}$, this expression is equal to the right hand side of (11.3), thus $\Phi\Theta$ is also the identity, proving that

$$S^G_{H,V}\hat{\otimes}S^G_{H,W} \simeq FQ^G_{H,V\otimes W}$$

To prove that $\mathcal{H}(S^G_{H,V}, S^G_{H,W})) \simeq FP^G_{H,\mathrm{Hom}_R(V,W)}$, I will use the following lemma:

Lemma 11.6.5: Let H be a subgroup of G, and V be an $\overline{A}(H)\otimes\overline{N}_G(H)$-module. If X is a G-set, then $\mathrm{Hom}_R([X^H], V)$ is an $\overline{A}(H)\otimes\overline{N}_G(H)$-module by

$$\big((a\otimes n).\phi\big)(x) = (a\otimes n)\phi(n^{-1}x)$$

and there is an isomorphism of A-modules

$$(S^G_{H,V})_X \simeq S^G_{H,\mathrm{Hom}_R([X^H],V)}$$

Proof: Indeed, if $\phi \in S^G_{H,V}(X)$, and if $a\otimes n$ and $a'\otimes n'$ are elements of $\overline{A}(H)\otimes\overline{N}_G(H)$, then

$$(a'\otimes n')\Big((a\otimes n)\phi\Big)(x) = (a'\otimes n')\Big((a\otimes n)\phi(n'^{-1}x)\Big) = (a'\otimes n')(a\otimes n)\phi(n^{-1}n'^{-1}x) = \dots$$

$$\dots = \Big(a'n'(a)\otimes n'n\Big)\phi\Big((n'n)^{-1}a\Big)$$

thus $\mathrm{Hom}_R([X^H],V)$ is an $\overline{A}(H)\otimes\overline{N}_G(H)$-module. Moreover, if Y is an H-set

$$(S^G_{H,V})_X(Y) = S^G_{H,V}(YX) = \mathrm{Hom}([Y^H\times X^H],V)_1^{\overline{N}_G(H)}$$

The structure of $\overline{N}_G(H)$-module of $\mathrm{Hom}_R([X^H],V)$ is such that

$$\mathrm{Hom}([Y^H\times X^H],V)_1^{\overline{N}_G(H)} \simeq \mathrm{Hom}\Big([Y^H],\mathrm{Hom}_R([X^H],V)\Big)_1^{\overline{N}_G(H)}$$

So

$$(S^G_{H,V})_X(Y) \simeq S^G_{H,\mathrm{Hom}_R([X^H],V)}(Y)$$

It is easy to see that these isomorphisms are compatible with the structures of A-modules of both sides, and the lemma follows. ■

It follows in particular that

$$\mathcal{H}(S^G_{H,V},S^G_{H,W})(X) = \mathrm{Hom}_{Mack(G)}(S^G_{H,V},S^G_{H,\mathrm{Hom}_R([X^H],W)})$$

Then I use the following lemma:

Lemma 11.6.6: Let H be a subgroup of G, and V and W be $\overline{N}_G(H)$-modules. Then

$$\mathrm{Hom}_{Mack(G)}(S^G_{H,V},S^G_{H,W}) \simeq \mathrm{Hom}_{\overline{N}_G(H)}(V,W)$$

Proof: Indeed, if θ is a morphism of Mackey functors from $S^G_{H,V}$ to $S^G_{H,W}$, then θ_H is a morphism of $\overline{N}_G(H)$-modules from $S^G_{H,V}(H)=V$ to $S^G_{H,W}(H)=W$. Conversely, if ψ is a morphism of $\overline{N}_G(H)$-modules from V to W, and if X is a G-set, I define a morphism ψ_X from $S^G_{H,V}(X)$ to $S^G_{H,W}(X)$ by

$$\psi_X : \phi\in\mathrm{Hom}([X^H],V)_1^{\overline{N}_G(H)} \mapsto \psi\circ\phi$$

It is clear indeed that if ϕ factors through a projective module, so does $\psi\phi$.

It is easy to see that these correspondences are inverse to each other, and this proves the lemma. ■

In those conditions

$$\mathcal{H}(S^G_{H,V},S^G_{H,W})(X) \simeq \mathrm{Hom}_{\overline{N}_G(H)}\Big(V,\mathrm{Hom}_R([X^H],W)\Big) = \mathrm{Hom}_{\overline{N}_G(H)}(V\otimes[X^H],W) = \dots$$

$$\dots = \mathrm{Hom}_{\overline{N}_G(H)}\Big([X^H],\mathrm{Hom}_R(V,W)\Big) = FP_{H,\mathrm{Hom}_R(V,W)}(X)$$

Those isomorphisms are moreover natural in X, and this gives an isomorphism of Mackey functors

$$\mathcal{H}(S_{H,V},S_{H,W}) \simeq FP_{H,\mathrm{Hom}_R(V,W)}$$

and completes the proof of the proposition. ■

Remarks: 1) In the case when V and W are $\overline{A}(H)\otimes\overline{N}_G(H)$-modules, then the module $\mathrm{Hom}_{\overline{A}(H)}(V,W)$ becomes a $\overline{N}_G(H)$-module, and then I have the isomorphism of Mackey functors

$$\mathcal{H}_A(S_{H,V},S_{H,W})\simeq FP_{H,\mathrm{Hom}_{\overline{A}(H)}(V,W)}$$

2) Let H be a subgroup of G, and V be a simple $\overline{A}(H)\otimes\overline{N}_G(H)$-module. Then let

$$B=\mathcal{H}(S^G_{H,V},S^G_{H,V})\simeq FP^G_{H,\mathrm{End}_R V}$$

The Green functor B admits H as a minimal subgroup, and $B(H)\simeq\mathrm{End}_R V$. This algebra is simple if and only if R is a field k, and V is finite dimensional over k. Now the functor B is a simple Green functor if and only if it is isomorphic to $S^{B\otimes B^{op}}_{H,\mathrm{End}_k V}$. This is equivalent to say that for any G-set X, I have

$$B(X)=\mathrm{Hom}_{\overline{N}_G(H)}\Big([X^H],\mathrm{End}_k V\Big)\simeq\mathrm{Hom}\Big([X^H],\mathrm{End}_k V\Big)_1^{\overline{N}_G(H)}=S^{B\otimes B^{op}}_{H,\mathrm{End}_k V}(X)$$

The case $X=\bullet$ shows that this condition holds if and only if

$$(\mathrm{End}_k V)^{\overline{N}_G(H)}=(\mathrm{End}_k V)_1^{\overline{N}_G(H)}$$

i.e. if the $k\overline{N}_G(H)$-module V is projective. In those conditions, denoting by $V^\circ=\mathrm{Hom}_k(V,k)$ the dual of V, I have $V^\circ\otimes_k V\simeq\mathrm{End}_k V$ if and only if V is finite dimensional over k, and $S^G_{H,V^\circ}\simeq(S^G_{H,V})^\circ$. Then

$$(S^G_{H,V})^\circ\hat{\otimes}S^G_{H,V}\simeq S^G_{H,V^\circ}\hat{\otimes}S^G_{H,V}\simeq FQ^G_{H,V^\circ\otimes_k V}\simeq FQ^G_{H,\mathrm{End}_k V}$$

Moreover, it is clear that $\underline{FQ^G_{H,\mathrm{End}_k V}(H)}=\mathrm{End}_k V=\overline{FP^G_{H,\mathrm{End}_k V}(H)}$.

The morphism $FQ^G_{H,\mathrm{End}_k V}\to FP^G_{H,\mathrm{End}_k V}$ which follows from lemma 11.6.1 factors through $S^G_{H,\mathrm{End}_k V}$, which is a quotient of $FQ_{H,\mathrm{End}_k V}$ and a submodule of $FP_{H,\mathrm{End}_k V}$. It is an isomorphism if and only if

$$FQ^G_{H,\mathrm{End}_k V}\simeq S^G_{H,\mathrm{End}_k V}\simeq FP^G_{H,\mathrm{End}_k V}$$

which is equivalent to say that $\mathrm{End}_k V$ is projective as $k\overline{N}_G(H)$-module. Finally, those considerations prove the following proposition:

Proposition 11.6.7: Let $R=k$ be a field. Let G be a group, and A be a Green functor for G. Let moreover H be a subgroup of G, and V be a simple $\overline{A}(H)\otimes\overline{N}_G(H)$-module. The following conditions are equivalent:

1. **The Green functor $\mathcal{H}(S^G_{H,V},S^G_{H,V})$ is simple.**
2. **As Mackey functors**

$$\mathcal{H}(S^G_{H,V},S^G_{H,V})\simeq(S^G_{H,V})^\circ\hat{\otimes}S^G_{H,V}$$

3. **The module V is finite dimensional over k and projective as $k\overline{N}_G(H)$-module.**

I will say then that $S_{H,V}^G$ is *endosimple*. The number of endosimple modules is related to Alperin's conjecture in the following way: let $A = b_p$ be the p-part of the Burnside functor, for a prime number p: if X is a G-set, then $b_p(X)$ is the Grothendieck group of the full subcategory of G-sets over X, formed of G-sets Y such that for any $y \in Y$, the stabilizer of y in G is a p-group. Then $\overline{b_p}(H)$ is zero if H is not a p-group, and isomorphic to k otherwise.

Let k be an algebraically closed field of characteristic p. Then if Q is a p-subgroup of G, and V is a simple $\overline{N}_G(Q)$-module (hence finite dimensional over k), the b_p-module $S_{H,V}^G$ is endosimple if and only if the module V is simple and projective. Thus Alperin's conjecture (in its global form) can be expressed by saying that the number of endosimple b_p-modules is equal to the number of simple kG-modules.

Chapter 12

Centres

12.1 The centre of a Green functor

Definition: *Let G be a group, and A be a Green functor for G. I call centre of A, and I denote by $Z(A)$, the commutant of A in A: if X is a G-set, then*

$$Z(A)(X) = \{\alpha \in A(X) \mid \forall Y,\ \forall \beta \in A(Y),\ \alpha \times \beta = \alpha \times^{op} \beta\}$$

With this definition, the functor $Z(A)$ is a sub-Green functor of A, which is clearly commutative.

Definitions: *If A and B are Green functors for the group G, then the direct sum $A \oplus B$ of A and B is the direct sum of A and B as Mackey functors, with the product defined for G-sets X and Y, and elements $a \in A(X)$, $b \in B(X)$, $c \in A(Y)$ and $d \in B(Y)$ by*

$$(a \oplus b) \times (c \oplus d) = (a \times c) \oplus (b \times d) \in (A \oplus B)(X \times Y)$$

The unit of $A \oplus B$ is the element $\varepsilon_A \oplus \varepsilon_B$ of $(A \oplus B)(\bullet)$.

If M is an A-module, and if $z \in Z(A)(\bullet)$, I denote by $z \times M$ the A-module defined for a G-set X by

$$(z \times M)(X) = z \times M(X)$$

If e is an idempotent of $Z(A)(\bullet)$, I denote by $e \times A$ the subfunctor of A defined for a G-set X by

$$(e \times A)(X) = e \times A(X) \subseteq A(X)$$

Then $e \times A$ is a sub-Green functor of A (the inclusion being not unitary in general), with $e = e \times \varepsilon_A \in (e \times A)(\bullet)$ as unit. Moreover, if M is an A-module, then $e \times M$ is a $e \times A$-module

The module $z \times M$ is an A-module, because if X and Y are G-sets, if $\alpha \in A(X)$ and $m \in M(Y)$, then

$$\alpha \times z \times m = (\alpha \times^{op} z) \times m = (z \times \alpha) \times m$$

If e is an idempotent of $Z(A)(\bullet)$, the functor $e \times M$ is an $e \times A$-module, since

$$(e \times a) \times (e \times m) = e \times (a \times^{op} e) \times m = e \times e \times a \times b = e \times a \times m$$

With these definitions:

Lemma 12.1.1: Let A, B, and C be Green functors for the group G. Then A is isomorphic to $B \oplus C$ if and only if there exists orthogonal idempotents e and f of $Z(A)(\bullet)$, such that $e + f = \varepsilon_A$, and $B \simeq e \times A$ and $C \simeq f \times A$.

Proof: Indeed, the unit of B is in $Z(B \oplus C)(\bullet)$, since for a G-set X and elements $b \in B(X)$ and $c \in C(X)$, I have

$$\varepsilon_B \times (b \oplus c) = (\varepsilon_B \oplus 0) \times (b \oplus c) = (\varepsilon_B \times b) \oplus 0 = b$$

and similarly

$$(b \oplus c) \times \varepsilon_B = (b \times \varepsilon_B) \oplus 0 = b$$

Then if $A \simeq B \oplus C$, the image e of ε_B and the image f of ε_C in $A(\bullet)$ are orthogonal idempotents, and $e + f = \varepsilon_A$. Conversely, if e and f are orthogonal idempotents of $Z(A)(\bullet)$ such that $e + f = \varepsilon_A$, then it is clear that the maps

$$a \in A(X) \mapsto (e \times a) \oplus (f \times a) \in (e \times A)(X) \oplus (f \times A)(X)$$

define an isomorphism from A to $(e \times A) \oplus (f \times A)$.

The main interest of this lemma is that since $Z(A)$ is a Green functor, there is always a (unique) unitary morphism of Green functors from b to $Z(A)$. Thus any decomposition of ε_b as a sum of orthogonal idempotents of $b(\bullet)$ induces a decomposition of A as a direct sum of Green functors.

Lemma 12.1.2: Let G and H be groups, and A be a Green functor for G. Let i be the injection from $Z(A)$ into A.

- **If U is a G-set-H, then**

$$(i \circ U)\big(Z(A) \circ U\big) \subseteq Z\big(A \circ U\big)$$

- **If V is an H-set-G, then**

$$\mathcal{L}_V(i)\big(Z(A)\big) \subseteq Z\big(\mathcal{L}_V(A)\big)$$

Proof: Indeed, if X and Y are H-sets, if

$$z \in \big(Z(A) \circ U\big)(X) = Z(A)(U \circ_H X)$$

and if $a \in (A \circ U)(Y)$, then their product for $A \circ U$ is

$$z \times^U a = A^*(\delta^U_{X,Y})(z \times a)$$

Similarly, in $(A \circ U)^{op}$

$$z \times^{U\ op} a = (A \circ U)_* \begin{pmatrix} yx \\ xy \end{pmatrix} A^*(\delta^U_{Y,X})(a \times z)$$

But $a \times z = A^* \begin{pmatrix} (u_1,y)(u_2,x) \\ (u_2,x)(u_1,y) \end{pmatrix} (z \times a)$, and moreover

$$\begin{pmatrix} (u_1,y)(u_2,x) \\ (u_2,x)(u_1,y) \end{pmatrix} \circ \delta^U_{Y,X} = \delta^U_{X,Y} \circ \left(U \circ_H \begin{pmatrix} yx \\ xy \end{pmatrix}\right)$$

Thus

$$z \times^{U\,op} a = A_*\left(U \circ_H \begin{pmatrix} yx \\ xy \end{pmatrix}\right) A^*\left(U \circ_H \begin{pmatrix} yx \\ xy \end{pmatrix}\right) A^*(\delta^U_{X,Y})(z \times a) = z \times^U a$$

and the first assertion follows.

Similarly, for the second assertion, if X and Y are H-sets, and (Z,f) and (T,g) are H-sets respectively over X and Y, and if $z \in Z(A)(H\backslash V.Z)$ and $a \in A(H\backslash V.T)$, then in $\mathcal{L}_V(A)(X \times Y)$, I have

$$z_{(Z,f)} \times a_{(T,g)} = A^*(\kappa^V_{Z,T})(z \times a)_{(Z.T,f.g)}$$

On the other hand

$$z_{(Z,f)} \times^{op} a_{(T,g)} = \mathcal{L}_V(A)_* \begin{pmatrix} yx \\ xy \end{pmatrix} \left[A^*(\kappa^V_{T,Z})(a \times z)_{(T.Z,g.f)}\right]$$

But $a \times z = A^*\begin{pmatrix} G.v_1.t\ G.v_2.z \\ G.v_2.z\ G.v_1.t \end{pmatrix}(z \times a)$, and

$$\begin{pmatrix} G.v_1.t\ G.v_2.z \\ G.v_2.z\ G.v_1.t \end{pmatrix} \circ \kappa^V_{T,Z} = \kappa^V_{Z,T} \circ \left(G\backslash V. \begin{pmatrix} tz \\ zt \end{pmatrix}\right)$$

Thus

$$z_{(Z,f)} \times^{op} a_{(T,g)} = \mathcal{L}_V(A)_* \begin{pmatrix} yx \\ xy \end{pmatrix} \left[A^*\left(G\backslash V. \begin{pmatrix} tz \\ zt \end{pmatrix}\right) A^*(\kappa^V_{Z,T})(z \times a)_{(T.Z,g.f)}\right]$$

As $\begin{pmatrix} tz \\ zt \end{pmatrix}$ is bijective, in $\mathcal{L}_V(A)(X \times Y)$, I have

$$A^*\left(G\backslash V. \begin{pmatrix} tz \\ zt \end{pmatrix}\right) A^*(\kappa^V_{Z,T})(z \times a)_{(T.Z,g.f)} = A^*(\kappa^V_{Z,T})(z \times a)_{(Z.T,(g.f)\circ\binom{zt}{tz})}$$

Moreover

$$\mathcal{L}_V(A)_* \begin{pmatrix} yx \\ xy \end{pmatrix} \left[A^*(\kappa^V_{Z,T})(z \times a)_{(Z.T,(g.f)\circ\binom{zt}{tz})}\right] = A^*(\kappa^V_{Z,T})(z \times a)_{(Z.T,\widetilde{\binom{yx}{xy}}\circ(g.f)\circ\binom{zt}{tz})}$$

Finally as $\widetilde{\begin{pmatrix} yx \\ xy \end{pmatrix}} \circ (g.f) \circ \begin{pmatrix} zt \\ tz \end{pmatrix} = f.g$, I have

$$z_{(Z,f)} \times^{op} a_{(T,g)} = A^*(\kappa^V_{Z,T})(z \times a)_{(Z.T,f.g)} = z_{(Z,f)} \times a_{(T,g)}$$

which proves the second assertion of the lemma. ■

I have already used an application of these lemmas, when I have built for an $\mathcal{L}_U(A)$-module N the module $N \circ U_{|A} = \lambda_{A,\bullet}(\varepsilon_A) \times (N \circ U)$: indeed, the element $\lambda_{A,\bullet}(\varepsilon_A)$ is in $Z\big(\mathcal{L}_U(A) \circ U\big)(\bullet)$.

Notation: *I will denote by $\pi = \pi_R(G)$ the set of prime factors of the order of G which are not invertible in R.*

I will suppose for simplicity that R is contained in $A(\bullet)$ (otherwise I can replace R by its image in $A(\bullet)$). Then the Burnside ring $b(\bullet) = b_R(G)$ with coefficients in

R has a family of idempotents f_H^G indexed by π-perfect subgroups of G (see theorem (9.3) of Thévenaz and Webb [15], or chapter 5.4 of Benson's book [1]): the primitive idempotents of $b_{\mathbf{Q}}(G)$ are indexed by the subgroups L of G, up to conjugation, and given by the following formulae of Gluck (see [7])

$$e_L^G = \frac{1}{|N_G(L)|}\sum_{K\subseteq L} |K|\tilde{\chi}]K, L[.G/K$$

The idempotent e_L^G is characterized by the fact that for any G-set X, I have

$$e_H^G.X = |X^H|e_H^G$$

and in particular $|(e_H^G)^K|$ is zero if H and K are not conjugate in G, and equal to 1 otherwise.

Then the idempotent f_H^G corresponding to a π-perfect subgroup H of G is given by

$$f_H^G = \sum_{\substack{O^\pi(L)=H \\ L\, mod. N_G(H)}} e_L^G$$

where $O^\pi(L)$ is the smallest normal subgroup N of L such that the quotient L/N is a solvable π-group.

Now a theorem of Dress (see [1] 5.4.7 and 5.4.8) shows that the idempotents f_H^G, as H runs through a system of representatives of the conjugacy classes of π-perfect subgroups of G (i.e. subgroups H such that $H = O^\pi(H)$, or equivalently subgroups having no non-trivial p-quotient, for $p \in \pi$), are mutual orthogonal idempotents of sum G/G in $b_R(G)$.

It follows that for any Green functor A for G, I have

$$A \simeq \bigoplus_H f_H^G \times A$$

where the sum runs on π-perfect subgroups of G up to conjugation by G. In this expression I also denote by f_H^G the image of the element f_H^G of $b(G)$ in $Z(A)(\bullet)$.

Now observe that $f_H^G = \mathrm{Ind}_{N_G(H)}^G f_H^{N_G(H)}$, and it seems natural to compare the Green functors $f_H^G \times A$ and $f_H^{N_G(H)} \times \mathrm{Res}_{N_G(H)}^G A$. It is actually easier to compare their associated categories:

Lemma 12.1.3: Let $B = f_H^G \times A$ and $C = f_H^{N_G(H)} \times \mathrm{Res}_{N_G(H)}^G A$. Let moreover $\Gamma = G/N_G(H)$. Then

1. **For any G-set X**

$$(\mathrm{Res}_{N_G(H)}^G A)(\mathrm{Res}_{N_G(H)}^G X) \simeq A(\Gamma \times X)$$

Moreover, if Y is a G-set, if $\alpha \in A(\Gamma \times X)$ and $\beta \in A(\Gamma \times Y)$, then their product $\alpha \times^r \beta$ for the functor $\mathrm{Res}_{N_G(H)}^G A$ is given by

$$\alpha \times^r \beta = A^*\begin{pmatrix} \gamma xy \\ \gamma x\gamma y \end{pmatrix}(\alpha \times \beta)$$

2. For any G-set X

$$C(\mathrm{Res}^G_{N_G(H)}X) = A^*\begin{pmatrix}\gamma x\\ \gamma\gamma x\end{pmatrix}\left(f_H^{N_G(H)} \times A(\Gamma\times X)\right) \subseteq A(\Gamma\times X)$$

3. The correspondence $\mathcal{R}$ mapping the G-set X to $\mathrm{Res}^G_{N_G(H)}X$, and the element $u \in B(YX)$ to $f_H^{N_G(H)} \times u$ is a fully faithful functor from $\mathcal{C}_B$ to $\mathcal{C}_C$.

4. Moreover, any object of $\mathcal{C}_C$ is isomorphic to a direct summand of an object in the image of $\mathcal{R}$.

Proof: Set $N = N_G(H)$. The first assertion is clear, since for any G-set X

$$(\mathrm{Res}^G_N A)(\mathrm{Res}^G_N X) = A(\mathrm{Ind}^G_N \mathrm{Res}^G_N X)$$

Moreover $\mathrm{Ind}^G_N\mathrm{Res}^G_N X \simeq \Gamma\times X$, and assertions 1) and 2) follow from the definition of the product for the functor $\mathrm{Res}^G_N A$.

Now if $u \in A(YX)$, then $A^*\binom{\gamma yx}{yx}(u) \in A(\Gamma YX)$, and

$$A^*\begin{pmatrix}\gamma yx\\ \gamma\gamma yx\end{pmatrix}\left(f_H^N \times A^*\begin{pmatrix}\gamma yx\\ yx\end{pmatrix}(u)\right) = A^*\begin{pmatrix}\gamma yx\\ \gamma\gamma yx\end{pmatrix}A^*\begin{pmatrix}\gamma\gamma' yx\\ \gamma yx\end{pmatrix}(f_H^N \times u) = \ldots$$

$$\ldots = A^*\begin{pmatrix}\gamma yx\\ \gamma yx\end{pmatrix}(f_H^N \times u) = f_H^N \times u$$

This shows that $\mathcal{R}(u) \in C(\mathrm{Res}^G_N Y \times \mathrm{Res}^G_N X)$.

Now if X, Y and Z are G-sets, if $u \in A(\Gamma YX)$ and $v \in A(\Gamma ZY)$, then the composition of $\mathcal{R}(v)$ and $\mathcal{R}(u)$ is given by

$$\mathcal{R}(v)\circ^r \mathcal{R}(u) = A_*\begin{pmatrix}\gamma zyx\\ \gamma zx\end{pmatrix}A^*\begin{pmatrix}\gamma zyx\\ \gamma zyyx\end{pmatrix}\left(\mathcal{R}(v)\times^r \mathcal{R}(u)\right) = \ldots$$

$$\ldots = A_*\begin{pmatrix}\gamma zyx\\ \gamma zx\end{pmatrix}A^*\begin{pmatrix}\gamma zyx\\ \gamma zy\gamma yx\end{pmatrix}(f_H^N \times v \times f_H^N \times u)$$

As f_H^N is in the center, this is also

$$\mathcal{R}(v)\circ^r \mathcal{R}(u) = A_*\begin{pmatrix}\gamma zyx\\ \gamma zx\end{pmatrix}A^*\begin{pmatrix}\gamma zyx\\ \gamma\gamma zyyx\end{pmatrix}(f_H^N \times f_H^N \times v \times u) = \ldots$$

$$\ldots = A_*\begin{pmatrix}\gamma zyx\\ \gamma zx\end{pmatrix}\left(A^*\begin{pmatrix}\gamma\\ \gamma\gamma\end{pmatrix}(f_H^N \times f_H^N)\times A^*\begin{pmatrix}zyx\\ zyyx\end{pmatrix}(v\times u)\right)$$

But $A^*\binom{\gamma}{\gamma\gamma}(f_H^N \times f_H^N) = f_H^N . f_H^N = f_H^N$, so

$$\mathcal{R}(v)\circ^r \mathcal{R}(u) = A_*\begin{pmatrix}\gamma zyx\\ \gamma zx\end{pmatrix}\left(f_H^N \times A^*\begin{pmatrix}zyx\\ zyyx\end{pmatrix}(v\times u)\right) = \ldots$$

$$\ldots = f_H^N \times A_*\begin{pmatrix}zyx\\ zx\end{pmatrix}A^*\begin{pmatrix}zyx\\ zyyx\end{pmatrix}(v\times u) = \mathcal{R}(v\circ u)$$

Moreover

$$\mathcal{R}(f_H^G \times 1_{A(X^2)}) = f_H^N \times f_H^G \times 1_{A(X^2)} = f_H^N \times f_H^G \times A_*\begin{pmatrix} x \\ xx \end{pmatrix} A^*\begin{pmatrix} x \\ \bullet \end{pmatrix}(\varepsilon) = \ldots$$

$$\ldots = A_*\begin{pmatrix} \gamma x \\ \gamma xx \end{pmatrix} A^*\begin{pmatrix} \gamma x \\ \gamma \end{pmatrix}(f_H^N \times f_H^G)$$

Furthermore

$$f_H^N \times f_H^G = f_H^N . A^*\begin{pmatrix} \gamma \\ \bullet \end{pmatrix}(f_H^G) = f_H^N . \mathrm{Res}_N^G f_H^G$$

But $\mathrm{Res}_N^G f_H^G$ is the sum of the f_K^N, where K runs through the different conjugates of H in G which are contained in N. So $f_H^N \times f_H^G = f_H^N$, and

$$\mathcal{R}(f_H^G \times 1_{A(X^2)}) = A_*\begin{pmatrix} \gamma x \\ \gamma xx \end{pmatrix} A^*\begin{pmatrix} \gamma x \\ \gamma \end{pmatrix}(f_H^N)$$

which it the unit of $C(\mathrm{Res}_N^G X^2)$. So $\mathcal{R}$ is a functor from $\mathcal{C}_B$ to $\mathcal{C}_C$.

Note that the previous argument shows that

$$\mathcal{R}(f_H^G \times u) = f_H^N \times f_H^G \times u = f_H^N \times u$$

Now if X and Y are G-sets, and if $v \in A(\Gamma \times Y \times X)$, then define

$$\mathcal{S}(v) = A_*\begin{pmatrix} \gamma yx \\ yx \end{pmatrix}(v) \in A(YX)$$

The equalities

$$\mathcal{S}(f_H^N \times u) = A_*\begin{pmatrix} \gamma \\ \bullet \end{pmatrix}(f_H^N) \times u = f_H^G \times u$$

show that $\mathcal{R}$ and $\mathcal{S}$ are mutual inverse bijections between $B(YX)$ and $C\big(\mathcal{R}(Y)\mathcal{R}(X)\big)$, so the functor $\mathcal{R}$ is fully faithful.

Finally, as any N-set Y is a sub-N-set of $\mathrm{Res}_N^G \mathrm{Ind}_N^G Y$, it is a direct summand in $\mathcal{C}_C$ of an object in the image of $\mathcal{R}$. ∎

Corollary 12.1.4: The functors

$$M \mapsto f_H^{N_G(H)} \times \mathrm{Res}_{N_G(H)}^G M \qquad \text{and} \qquad L \mapsto \mathrm{Ind}_{N_G(H)}^G L$$

are mutual inverse equivalences of categories between B-Mod and C-Mod.

Proof: If $\mathcal{C}$ is an R-additive category, define the category $\hat{\mathcal{C}}$ in the following way: the objects of $\hat{\mathcal{C}}$ are the couples (X, i), where X is an object of $\mathcal{C}$, and i and idempotent endomorphism of X in $\mathcal{C}$. A morphism in $\hat{\mathcal{C}}$ from (X, i) to (Y, j) is a morphism f from X to Y in $\mathcal{C}$, such that $f \circ i = f = j \circ f$. In other words

$$\mathrm{Hom}_{\hat{\mathcal{C}}}\big((X,i),(Y,j)\big) = j \circ \mathrm{Hom}_{\mathcal{C}}(X,Y) \circ i$$

The composition of morphisms is the composition in $\mathcal{C}$. The category $\hat{\mathcal{C}}$ is obtained by "adding direct summands" of objects of $\mathcal{C}$. It is also an R-additive category (the direct sum of (X, i) and (Y, j) is of course $(X \oplus Y, i \oplus j)$).

There is a canonical functor $X \mapsto (X, Id_X)$ from $\mathcal{C}$ to $\hat{\mathcal{C}}$. So if $\hat{F}$ is an additive functor from $\hat{\mathcal{C}}$ to R-**Mod**, it gives by composition an additive functor from $\mathcal{C}$ to R-**Mod**. Conversely, if F is an additive functor from $\mathcal{C}$ to R-**Mod**, then F admits a unique extension $\hat{F}$ to $\hat{\mathcal{C}}$, defined by

$$\hat{F}(X,i) = F(i)\big(F(X)\big)$$

This is because if (X,i) is an object of $\hat{\mathcal{C}}$, then $i : X \to X$ defines two morphisms in $\hat{\mathcal{C}}$

$$i^+ : (X,i) \to (X, Id_X) \qquad\qquad i^- : (X, Id_X) \to (X,i)$$

Moreover $i^- \circ i^+$ is equal to i, which is the identity morphism of (X,i) in $\hat{\mathcal{C}}$. So (X,i) is a direct summand of (X, Id_X) in $\hat{\mathcal{C}}$.

This shows that the categories of representations of $\mathcal{C}$ and $\hat{\mathcal{C}}$ are equivalent. Now if $\Phi : \mathcal{C} \to \mathcal{D}$ is a functor between additive categories, it induces a functor $\hat{\Phi} : \hat{\mathcal{C}} \to \hat{\mathcal{D}}$. It is clear moreover that if Φ is fully faithful, then $\hat{\Phi}$ is fully faithful: indeed, the bijection

$$F : \mathrm{Hom}_{\mathcal{C}}(X,Y) \to \mathrm{Hom}_{\mathcal{D}}\big(F(X), F(Y)\big)$$

restricts to a bijection $\hat{F}$ from $\mathrm{Hom}_{\hat{\mathcal{C}}}\big((X,i),(Y,j)\big) = j \circ \mathrm{Hom}_{\mathcal{C}}(X,Y) \circ i$ to

$$\mathrm{Hom}_{\hat{\mathcal{D}}}\big(F(X,i), F(Y,j)\big) = F(j) \circ \mathrm{Hom}_{\mathcal{D}}\big(F(X), F(Y)\big) \circ F(i)$$

If moreover any object Y of $\mathcal{D}$ is a direct summand of an object $F(X)$ in the image of F, then there are morphisms

$$\alpha : F(X) \to Y \qquad\qquad \beta : Y \to F(X)$$

such that $\alpha \circ \beta = Id_Y$. If i is any idempotent in $\mathrm{End}_{\mathcal{D}}(Y)$, then $\beta \circ i \circ \alpha$ is an idempotent endomorphism of $F(X)$. So it is equal to $F(\gamma)$, for a unique $\gamma \in \mathrm{End}_{\mathcal{C}}(X)$. This implies that γ is an idempotent. Now the morphisms

$$i \circ \alpha : \big(F(X), F(\gamma)\big) \to (Y,i) \qquad\qquad \beta \circ i : (Y,i) \to \big(F(X), F(\gamma)\big)$$

are mutual inverse isomorphisms in $\hat{\mathcal{D}}$. So $\hat{F}$ is essentially surjective, hence it is an equivalence of categories.

Thus in the situation of the lemma, the functor $\hat{\mathcal{R}}$ is an equivalence of categories: this implies in particular that $\mathcal{R}$ induces an equivalence of categories between the categories of representations of $\mathcal{C}_B$ and $\mathcal{C}_D$. It remains to check that this equivalence is as stated in the corollary, which is clear since if L is a C-module, then for any G-set X

$$f_H^G \times (\mathrm{Ind}_{N_G(H)}^G L)(X) = f_H^G \times L(\mathrm{Res}_{N_G(H)}^G X) = \ldots$$

$$\ldots = f_H^G \times f_H^{N_G(H)} \times L(\mathrm{Res}_{N_G(H)}^G X) = f_H^{N_G(H)} \times L(\mathrm{Res}_{N_G(H)}^G X) = L(\mathrm{Res}_{N_G(H)}^G X)$$

which proves that $f_H^G \times \mathrm{Ind}_{N_G(H)}^G L = \mathrm{Ind}_{N_G(H)}^G L$. ■

Lemma 12.1.5: Let H be a normal subgroup of G. If A is a Green functor for G such that $f_H^G \times A = A$, then $f_1^{G/H} \times A^H = A^H$, and the correspondence $X \mapsto X^H$ induces an equivalence of categories from $\mathcal{C}_A$ to $\mathcal{C}_{A^H}$.

Proof: Let Y be a (G/H)-set. Then

$$A^H(Y) = A(\mathrm{Inf}_{G/H}^G Y)/\sum_{(Z,f)} A_*(f)\big(A(Z)\big)$$

where (Z,f) runs through the G-sets over $\mathrm{Inf}_{G/H}^G Y$ such that $Z^H = \emptyset$. Now since f_H^G is the unit of A, and as $\mathrm{Res}_K^G f_H^G = 0$ if $H \not\subseteq K$, it follows that $A(K) = 0$ if $H \not\subseteq K$, so $A(Z) = 0$ if $Z^H = \emptyset$. Thus $A^H(Y) = A(\mathrm{Inf}_{G/H}^G Y)$.

Let i_X denote the injection from $\mathrm{Inf}_{G/H}^G(X^H)$ into X. Then

$$A^H(X^H) = A(\mathrm{Inf}_{G/H}^G X^H)$$

The unit $A(X) \to (\mathrm{Inf}_{G/H}^G A^H)(X) = A^H(X^H)$ is the map $A^*(i_X)$.

Now as

$$X = (\mathrm{Inf}_{G/H}^G X^H) \coprod Z$$

with $Z^H = \emptyset$, it follows that $A(Z) = 0$, and the map $A^*(i_X)$ induces an isomorphism $A(X) \to A^H(X^H)$.

Now it is clear that the correspondence

$$X \mapsto X^H \qquad\qquad f \in A(YX) \mapsto A^*(i_{YX})(f) \in A^H(Y^H X^H)$$

is a fully faithful functor from $\mathcal{C}_A$ to $\mathcal{C}_{A^H}$. As any (G/H)-set Y is isomorphic to $(\mathrm{Inf}_{G/H}^G Y)^H$, this functor is essentially surjective, hence it is an equivalence of categories. Finally, the unit of A^H is $f_1^{G/H}$, because $f_H^G - \mathrm{Inf}_{G/H}^G f_1^{G/H}$ is a linear combination of G/K, for $H \not\subseteq K$, so it acts by zero on A. ■

Corollary 12.1.6: Let H be a normal subgroup of G. If A is a Green functor for G such that $f_H^G \times A = A$, then the functors

$$M \mapsto M^H \qquad \text{and} \qquad L \mapsto \mathrm{Inf}_{G/N}^G L$$

are mutual inverse equivalences of categories between A-Mod and A^H-Mod.

Proof: This follows from lemma 12.1.5, and from the definition

$$(\mathrm{Inf}_{G/H}^G L)(X) = L(X^H)$$

So $\mathrm{Inf}_{G/H}^G$ is an equivalence of categories, and its inverse is the left adjoint functor $M \mapsto M^H$. ■

Lemma 12.1.7: Let H be a subgroup of G, and M be a Mackey functor for $\overline{N}_G(H)$. If Z is an element of $b(\bullet)$, then

$$Z \times \mathrm{Ind}_{N_G(H)}^G \mathrm{Inf}_{\overline{N}_G(H)}^{N_G(H)} M \simeq \mathrm{Ind}_{N_G(H)}^G \mathrm{Inf}_{\overline{N}_G(H)}^{N_G(H)} (Z^H \times M)$$

Proof: If Z is a G-set, I know that $Z = b_*\binom{z}{\bullet} b^*\binom{z}{\bullet}(\varepsilon_b)$. Thus if X is a G-set, and L is a Mackey functor for G, then for $l \in L(X)$ I have

$$Z \times l = L_*\binom{zx}{x} L^*\binom{zx}{x}(l)$$

Set $N = N_G(H)$ and $\overline{N} = \overline{N}_G(H)$. If $L = \mathrm{Ind}_N^G \mathrm{Inf}_{\overline{N}}^N M$, I have $L(X) = M(X^H)$, and setting $Z' = Z^H$ and $X' = X^H$, the maps $L^*\binom{zx}{x}$ and $L_*\binom{zx}{x}$ are respectively equal to $M^*\binom{z'x'}{x'}$ and $M_*\binom{z'x'}{x'}$. It follows that the image of $L_*\binom{zx}{x} L^*\binom{zx}{x}$ is the image of $M_*\binom{z'x'}{x'} M^*\binom{z'x'}{x'}$, that is $Z^H \times M(X^H)$. In other words, the element Z acts on $M(X^H)$ via Z^H, for any $Z \in b(G)$. It follows that

$$(Z \times \mathrm{Ind}_N^G \mathrm{Inf}_{\overline{N}}^N M)(X) = Z^H \times M(X^H) = \mathrm{Ind}_N^G \mathrm{Inf}_{\overline{N}}^N (Z^H \times M)(X)$$

This proves the lemma. ∎

Lemma 12.1.8: Let H be a π-perfect normal subgroup of G. Then in $b_R(G/H)$, I have

$$(f_H^G)^H = f_1^{G/H}$$

Proof: Let $\overline{K} = K/H$ be a π-perfect subgroup of G/H. If $(f_H^G)^H . f_{K/H}^{G/H}$ is non-zero, then there exists a subgroup L of G with $O^\pi(L) = H$ and a subgroup $\overline{L'} = L'/H$ of G/H with $O^\pi(L') = K$ such that

$$0 \neq (e_L^G)^H . e_{\overline{L'}}^{G/H} = \left(e_L^G . \mathrm{Inf}_{G/H}^G e_{\overline{L'}}^{G/H}\right)^H = |(\mathrm{Inf}_{G/H}^G e_{\overline{L'}}^{G/H})^L|(e_L^G)^H = |(e_{\overline{L'}}^{G/H})^{L/H}|(e_L^G)^H$$

Thus L/H is conjugate of $\overline{L'}$ in G/H. Then L is conjugate to L' in G, and then $O^\pi(L) = H$ is conjugate in G to $O^\pi(L') = K$. Thus $H = K$, and the lemma follows. ∎

Corollary 12.1.9: If H is a π-perfect normal subgroup of G, then for any A-module M and any A^H-module L

$$(f_H^G \times M)^H \simeq f_1^{G/H} \times M^H \qquad f_H^G \times \mathrm{Inf}_{G/H}^G L \simeq \mathrm{Inf}_{G/H}^G (f_1^{\overline{N}_G(H)} \times L)$$

Finally, corollary 12.1.4, 12.1.6 and 12.1.9 show that if A is a Green functor for G, and H is a π-perfect subgroup of G, then the functor

$$L \mapsto \mathrm{Ind}_{N_G(H)}^G \mathrm{Inf}_{\overline{N}_G(H)}^{N_G(H)} L$$

is an equivalence of categories from $f_1^{\overline{N}_G(H)} \times (\mathrm{Res}_{N_G(H)}^G A)^H$-**Mod** to $f_H^G \times A$-**Mod**. The inverse equivalence is the adjoint functor

$$M \mapsto f_1^{\overline{N}_G(H)} \times (\mathrm{Res}_{N_G(H)}^G M)^H$$

In particular, the unit morphism

$$f_H^G \times A \to \mathrm{Ind}_{N_G(H)}^G \mathrm{Inf}_{\overline{N}_G(H)}^G \left(f_1^{\overline{N}_G(H)} \times (\mathrm{Res}_{N_G(H)}^G A)^H\right)$$

is an isomorphism of A-modules. As it is a morphism of Green functors, it is an isomorphism of Green functors.

The following lemma gives a characterization of the Green functors $f_1^G \times A$ as the functors which are projective relative to the set $\Sigma_\pi(G)$ of solvable π-subgroups of G (i.e. relative to the set $\coprod_H G/H$, where H runs through $\Sigma_\pi(G)$):

Proposition 12.1.10: Let G be a group, and A be a Green functor for G over R. Let π the set of prime factors of $|G|$ which are not invertible in R. The following conditions are equivalent:

1. **The functor A is projective relative to the set of solvable π-subgroups.**

2. **The idempotent f_1^G acts as the identity on A, i.e. $A = f_1^G \times A$.**

Proof: It is clear that 2) implies 1), because if $A = f_1^G \times A$, then ε_A is the image in $A(\bullet)$ of f_1^G. This element is a sum of e_L^G, for $L \in \Sigma_\pi(G)$, hence a linear combination of G/K, for $K \subseteq L$, thus $K \in \Sigma_\pi(G)$. Then the identity map of $A(\bullet) = A(G)$ is a linear combination of $t_K^G r_K^G$, for $K \in \Sigma_\pi(G)$. Thus A is projective relative to the set of solvable π-subgroups.

Conversely, if A is projective relative to the set of solvable π-subgroups of G, then ε_A can be written as

$$\varepsilon_A = \sum_{K \in \Sigma_\pi(G)} t_K^G(\alpha_K)$$

Then

$$f_1^G.\varepsilon_A = \sum_{K \in \Sigma_\pi(G)} t_K^G(r_K^G f_1^G.\alpha_K)$$

and it is easy to see that for any subgroup K of G

$$\mathrm{Res}_K^G f_1^G = f_1^K$$

and that $f_1^K = K/K$ if K is π-solvable. Thus $f_1^G = \varepsilon_A$, and 2) holds. ∎

Finally, I have proved the following proposition:

Proposition 12.1.11: Let G be a group, and A be a Green functor for G over R. Let π the set of prime factors of $|G|$ which are not invertible in R. Then there is an isomorphism of Green functors

$$A \simeq \bigoplus_H f_H^G \times A$$

where the summation runs through a set of representatives of the conjugacy classes of π-perfect subgroups of G. If H is such a subgroup, then

$$f_H^G \times A \simeq \mathrm{Ind}_{N_G(H)}^G \mathrm{Inf}_{\overline{N}_G(H)}^{N_G(H)}\left(f_1^{\overline{N}_G(H)} \times (\mathrm{Res}_{N_G(H)}^G A)^H\right)$$

The functor $f_1^{\overline{N}_G(H)} \times (\mathrm{Res}_{N_G(H)}^G A)^H$ is projective relative to the set of solvable π-subgroups, and the functors

$$M \mapsto \mathrm{Ind}_{N_G(H)}^G \mathrm{Inf}_{\overline{N}_G(H)}^{N_G(H)} M \qquad L \mapsto f_1^{\overline{N}_G(H)} \times (\mathrm{Res}_{N_G(H)}^G L)^H$$

are mutual inverse equivalences of categories between $(f_H^G \times A)$-Mod and $f_1^{\overline{N}_G(H)} \times (\mathrm{Res}_{N_G(H)}^G A)^H$-Mod.

Remarks: 1) In the case of Mackey functors (i.e. the case $A = b$), this is essentially theorem 10.1 of Thévenaz and Webb ([15]).
2) In the case $\pi = \emptyset$, i.e. when the order of G is invertible in R, the solvable π-subgroups are trivial. In those conditions, I have $f_1^G = \frac{1}{|G|}G/1$. Thus if $f_1^G \times A = A$, then the maps

$$a \in A(H) \mapsto r_1^H a \in A(1)^H \qquad b \in A(1)^H \mapsto \frac{1}{|H|} t_1^H b$$

are mutual inverse isomorphisms between A and $FP_{A(1)}$. Thus if the order of G is invertible in R, then the isomorphism of the proposition gives

$$A \simeq \bigoplus_H \mathrm{Ind}_{N_G(H)}^G \mathrm{Inf}_{\overline{N}_G(H)}^{N_G(H)} FP_{\overline{A}(H)}$$

This is theorem 13.1 of Thévenaz (see [13]).

12.2 The functors ζ_A

12.2.1 Another analogue of the centre

The definition of the centre of a Green functor given in the previous section is certainly natural, and mimics the definition of the centre of a ring. However, one can also define the centre of a ring from its category of modules, as the endomorphism ring of the identity functor ([1] prop. 2.2.7).

If A is a Green functor for the group G, then the endomorphisms ring of the identity functor of A-**Mod** is just a ring, isomorphic to the centre of $A(\Omega^2)$. I would like to have a similar construction, using the category A-**Mod**, and producing a Green functor ζ_A. The idea is then to mimic the construction of the Green functors $\mathcal{H}(M, M)$, setting for a subgroup H of G

$$\zeta_A(H) = \mathrm{End}_{Funct}(\mathrm{Res}_H^G)$$

where $\mathrm{End}_{Funct}(\mathrm{Res}_H^G)$ denotes the endomorphisms of the restriction functor from A-**Mod** to $\mathrm{Res}_H^G A$-**Mod**: such an endomorphism is determined by giving, for any A-module M, a morphism θ_M of $\mathrm{Res}_H^G A$-modules from $\mathrm{Res}_H^G M$ to itself, such that if $\phi : M \to N$ is a morphism of A-modules, the square

$$\begin{array}{ccc} \mathrm{Res}_H^G M & \xrightarrow{\theta_M} & \mathrm{Res}_H^G M \\ {\scriptstyle \mathrm{Res}_H^G \phi} \downarrow & & \downarrow {\scriptstyle \mathrm{Res}_H^G \phi} \\ \mathrm{Res}_H^G N & \xrightarrow[\theta_N]{} & \mathrm{Res}_H^G N \end{array}$$

is commutative. With this definition, it is clear that if A is a Green functor on R, then $\zeta_A(H)$ has a natural R-algebra structure. It is natural to try to turn ζ_A into a Green functor.

First I observe that if $\mathcal{I}$ denotes the identity functor of A-**Mod**, and $\mathcal{I}_{G/H}$ the functor $M \mapsto M_{G/H}$ from A-**Mod** to A-**Mod**, then

$$\zeta_A(H) \simeq \mathrm{Hom}_{Funct}(\mathcal{I}, \mathcal{I}_{G/H})$$

Indeed, I know that $\mathcal{I}_{G/H}(M) \simeq \mathrm{Ind}_H^G \mathrm{Res}_H^G M$, functorially in M, and that Ind_H^G is right adjoint to Res_H^G. The above isomorphism follows then from the next lemma:

Lemma 12.2.1: Let $\mathcal{C}$ and $\mathcal{D}$ be categories, and F be a functor from $\mathcal{C}$ to $\mathcal{D}$ having a right adjoint F'. Then if $\mathcal{E}$ is a category, if A is a functor from $\mathcal{E}$ to $\mathcal{C}$, and B is a functor from $\mathcal{E}$ to $\mathcal{D}$, I have

$$\mathrm{Hom}_{Funct}(F \circ A, B) \simeq \mathrm{Hom}_{Funct}(A, F' \circ B)$$

Proof: Indeed, a morphism of functors from $F \circ A$ to B is determined by giving, for any object X of $\mathcal{E}$, a morphism θ_X from $F \circ A(X)$ to $B(X)$, such that if ϕ is a morphism in $\mathcal{E}$ from X to Y, the square

$$\begin{array}{ccc} F \circ A(X) & \xrightarrow{\theta_X} & B(X) \\ {\scriptstyle F \circ A(f)} \downarrow & & \downarrow {\scriptstyle B(f)} \\ F \circ A(Y) & \xrightarrow[\theta_Y]{} & B(Y) \end{array}$$

is commutative. Denoting by $u \mapsto u^o$ the bijection

$$\mathrm{Hom}_{\mathcal{D}}\big(F(Z), T\big) \to \mathrm{Hom}_{\mathcal{C}}\big(Z, F'(T)\big)$$

deduced of the adjunction, I obtain for any X a morphism θ_X^o from $A(X)$ to $F' \circ B(X)$. Moreover the square

$$\begin{array}{ccc} A(X) & \xrightarrow{\theta_X^o} & F' \circ B(X) \\ {\scriptstyle A(f)} \downarrow & & \downarrow {\scriptstyle F \circ B(f)} \\ A(Y) & \xrightarrow[\theta_Y^o]{} & F' \circ B(Y) \end{array}$$

is commutative, because its diagonal $h : A(X) \to F' \circ B(Y)$ is both equal to $\big(B(f) \circ \theta_X\big)^o$ and to $\big(\theta_Y \circ (F \circ A)(f)\big)^o$.

Thus the maps θ_X^o define a morphism of functors from A to $F' \circ B$. There is an obvious inverse construction, which gives the isomorphism of the lemma. ∎

Now for any G-set X, I can define the endofunctor $\mathcal{I}_X$ of A-**Mod** by $\mathcal{I}_X(M) = M_X$. Then, for any G-set X, I define

$$\zeta_A(X) = \mathrm{Hom}_{Funct}(\mathcal{I}, \mathcal{I}_X)$$

If $f : X \to Y$ is a morphism of G-sets, then f determines a natural transformation from $\mathcal{I}_X$ to $\mathcal{I}_Y$: if M is an A-module, and Z is a G-set, then the maps

$$M_*(Id_Z \times f) : M_X(Z) \to M_Y(Z)$$

define a natural transformation from M_X to M_Y, which is itself natural in M.

Similarly, the maps

$$M^*(Id_Z \times f) : M_Y(Z) \to M_X(Z)$$

define a natural transformation from $\mathcal{I}_Y$ to $\mathcal{I}_X$.

This gives by composition morphisms

$$\zeta_{A,*}(f) : \zeta_A(X) \to \zeta_A(Y) \qquad \zeta_A^*(f) : \zeta_A(Y) \to \zeta_A(X)$$

Finally, if $\alpha \in \zeta_A(X)$ and $\beta \in \zeta_A(Y)$, then for any A-module M, I have a morphism $\alpha_M : M \to M_X$, and a morphism $\beta_M : M \to M_Y$. I set then

$$(\alpha \times \beta)_M = \alpha_M \times \beta_M$$

the product in the right hand side being computed in $\mathcal{H}(M, M)$. In other words, if Z is a G-set, I have for $m \in M(Z)$

$$(\alpha \times \beta)_{M,Z}(m) = M_* \begin{pmatrix} zyx \\ zxy \end{pmatrix} \circ \alpha_{M,ZY} \circ \beta_{M,Z}(m)$$

Then if $\theta : M \to N$ is a morphism of Mackey functors, I have

$$(\alpha \times \beta)_{N,Z}\theta_Z = N_* \begin{pmatrix} zyx \\ zxy \end{pmatrix} \circ \alpha_{N,ZY} \circ \beta_{N,Z} \circ \theta_Z(m) = \ldots$$

$$\ldots = N_* \begin{pmatrix} zyx \\ zxy \end{pmatrix} \circ \alpha_{N,ZY} \circ \theta_{ZY} \circ \beta_{M,Z} = N_* \begin{pmatrix} zyx \\ zxy \end{pmatrix} \circ \theta_{ZYX} \circ \alpha_{M,ZY} \circ \beta_{M,Z}(m) = \ldots$$

$$\ldots = \theta_{ZXY} \circ M_* \begin{pmatrix} zyx \\ zxy \end{pmatrix} \circ \alpha_{M,ZY} \circ \beta_{M,Z}(m) = \theta_{ZXY} \circ (\alpha \times \beta)_{M,Z}$$

which proves that $\alpha \times \beta \in \zeta_A(X \times Y)$.

Lemma 12.2.2: The above definitions turn ζ_A into a Green functor, such that

$$\zeta_A(\bullet) = \mathrm{End}_{Funct}(\mathcal{I}) \simeq Z\big(A(\Omega^2)\big)$$

Proof: The verifications to make are not difficult, and similar to those made for the functors $\mathcal{H}(M, M)$. The equality $\zeta_A(\bullet) \simeq Z\big(A(\Omega^2)\big)$ follows from the fact that the category A-**Mod** is equivalent to $A(\Omega^2)$-**Mod**. Thus $\zeta_A(\bullet)$ identifies to the endomorphisms of the identity functor in this category, hence to the centre of the algebra $A(\Omega^2)$. ∎

Let X be a G-set, and $\alpha \in \zeta_A(X)$. Then for any A-module M, I have a morphism $\alpha_M : M \to M_X$, such that if ϕ is a morphism of A-modules from M to N, the square

$$\begin{array}{ccc} M & \xrightarrow{\alpha_M} & M_X \\ \phi \downarrow & & \downarrow \phi_X \\ N & \xrightarrow[\alpha_N]{} & N_X \end{array} \qquad (C)$$

is commutative.

In particular, if Y is a G-set, I can take $M = A_Y$. Then $\mathrm{Hom}_A(A_Y, M) \simeq M(Y)$ (see proposition 3.1.3): the element $m \in M(Y)$ is associated to the morphism $\phi_m : A_Y \to M$, defined for any G-set Z by

$$a \in A_Y(Z) = A(Z \times Y) \mapsto a \circ_Y m = M_* \begin{pmatrix} zy \\ z \end{pmatrix} M^* \begin{pmatrix} zy \\ zyy \end{pmatrix} (a \times m)$$

It follows in particular that the morphism $\alpha_{A_Y} : A_Y \to (A_Y)_X = A(XY)$ is determined by an element

$$a_Y \in A_{XY}(Y) = A(YXY)$$

Let $n \in N(Y)$. I must have a commutative square

$$\begin{array}{ccc} A_Y & \xrightarrow{\alpha_{A_Y}} & A_{XY} \\ {\scriptstyle\phi_n}\downarrow & & \downarrow{\scriptstyle(\phi_n)_X} \\ N & \xrightarrow[\alpha_N]{} & N_X \end{array}$$

Then for any G-set Z and any $a \in A_Y(Z) = A(ZY)$, I must have

$$\alpha_{N,Z}(a \circ_Y n) = \big((\phi_n)_X\big)_Z (a \circ_Y a_Y) = (a \circ_Y a_Y) \circ_Y n$$

The case $Z = Y$ and $a = 1_{A(Y^2)}$ gives then

$$\alpha_{N,Y}(n) = a_Y \circ_Y n$$

Conversely, if this equality is used to define $\alpha_{N,Y}$ for any N and any Y, then the commutativity of the square (C) is equivalent to

$$a_Y \circ_Y \phi_Y(m) = \phi_{XY}(a_Y \circ_Y m)$$

fo any Y and any $m \in M(Y)$. But

$$\phi_{XY}(a_Y \circ_Y m) = \phi_{XY} M_* \begin{pmatrix} y_1xy_2 \\ y_1x \end{pmatrix} M^* \begin{pmatrix} y_1xy_2 \\ y_1xy_2y_2 \end{pmatrix} (a_Y \times m)$$

As ϕ is a morphism of Mackey functors, the right hand side is

$$N_* \begin{pmatrix} y_1xy_2 \\ y_1x \end{pmatrix} N^* \begin{pmatrix} y_1xy_2 \\ y_1xy_2y_2 \end{pmatrix} \phi_{YXY^2}(a_Y \times m)$$

As ϕ is a morphism of A-modules, it is also

$$N_* \begin{pmatrix} y_1xy_2 \\ y_1x \end{pmatrix} N^* \begin{pmatrix} y_1xy_2 \\ y_1xy_2y_2 \end{pmatrix} \big(a_Y \times \phi_Y(m)\big) = a_Y \circ_Y \phi_Y(m)$$

Thus the square (C) is commutative.

The only condition on the elements a_Y comes then from the fact that the morphism α_Y is determined twice by the previous argument: I must have for any Z and any $a \in A_Y(Z) = A(ZY)$

$$\alpha_Y(a) = a \circ_Y a_Y = a_Z \circ_Z a$$

Finally, if those conditions are satisfied, then the morphisms $\alpha_{N,Y}$ are morphisms of A-modules: indeed, if $a \in A(Z)$ and $n \in N(Y)$, then

$$a \times \alpha_{N,Y}(n) = a \times (a_Y \circ_Y n)$$

whereas

$$\alpha_{N,XY}(a \times n) = a_{ZY} \circ_{ZY} (a \times n)$$

But

$$a \times n = A_* \begin{pmatrix} zy \\ zyy \end{pmatrix} A^* \begin{pmatrix} zy \\ z \end{pmatrix} (a) \circ_Y n$$

As $A_* \binom{zy}{zyy} A^* \binom{zy}{z} (a) \in A(ZYY)$, I have

$$a_{ZY} \circ_{ZY} (a \times n) = \left(a_{ZY} \circ_{ZY} A_* \begin{pmatrix} zy \\ zyy \end{pmatrix} A^* \begin{pmatrix} zy \\ z \end{pmatrix} (a) \right) \circ_Y n = \ldots$$

$$\ldots = (A_* \begin{pmatrix} zy \\ zyy \end{pmatrix} A^* \begin{pmatrix} zy \\ z \end{pmatrix} (a) \circ_Y a_Y) \circ_Y n = A_* \begin{pmatrix} zy \\ zyy \end{pmatrix} A^* \begin{pmatrix} zy \\ z \end{pmatrix} (a) \circ_Y (a_Y \circ_Y n) = \ldots$$

$$\ldots = a \times (a_Y \circ_Y n)$$

So I have

$$a \times \alpha_{N,Y}(n) = \alpha_{N,XY}(a \times n)$$

If $f : X \to X'$ is a morphism of G-sets, then the sequence (a_Y) determines a morphism from $\mathcal{I}$ to $\mathcal{I}_{X'}$, defined by composition for an A-module M and a G-set Z by

$$M(Z) \longrightarrow M_X(Z) = M(ZX) \xrightarrow{M_*(Id_Z \times f)} M(ZX')$$

In particular, if $M = A_Y$ and $Z = Y$, the image of $1_{A(Y^2)}$ in $A_Y(YX') = A(YX'Y)$ is

$$A_{Y,*}(Id_Y \times f)(a_Y) = A_*(Id_Y \times f \times Id_Y)(a_Y)$$

In other words, if s is the sequence (a_Y), I have

$$\zeta_{A,*}(f)(s)_Y = A_*(Id_Y \times f \times Id_Y)(a_Y)$$

It is clear similarly that if $s' = (a'_Y)$ is a sequence defining an element of $\zeta_A(X')$, then

$$\zeta_A^*(f)(s')_Y = A^*(Id_Y \times f \times Id_Y)(a'_Y)$$

If $s = (a_Y)$ defines the element $\alpha \in \zeta_A(X)$, and $t = (b_Y)$ defines the element $\beta \in \zeta_A(Y)$, then for any A-module M, any G-set Z and any $m \in M(Z)$, I have

$$(\alpha \times \beta)_{M,Z}(m) = M_* \begin{pmatrix} zyx \\ zxy \end{pmatrix} \circ \alpha_{M,ZY} \circ \beta_{M,Z}(m) = M_* \begin{pmatrix} zyx \\ zxy \end{pmatrix} \circ \alpha_{M,ZY}(b_Z \circ_Z m) = \ldots$$

$$\ldots = M_* \begin{pmatrix} zyx \\ zxy \end{pmatrix} (a_{ZY} \circ_{ZY} b_Z \circ_Z m)$$

Moreover as $b_Z \in A(ZYZ)$, I have

$$a_{ZY} \circ_{ZY} b_Z = b_Z \circ_Z a_Z$$

Thus

$$(\alpha \times \beta)_{M,Z}(m) = M_* \begin{pmatrix} zyx \\ zxy \end{pmatrix} (b_Z \circ_Z a_Z \circ_Z m)$$

In particular, if $M = A_Z$ and $m = 1_{A(Z^2)}$, this equality shows that $\alpha \times \beta$ is defined by the sequence $s \times t$ such that

$$(s \times t)_Y = A_* \begin{pmatrix} z_1 y x z_2 \\ z_1 x y z_2 \end{pmatrix} (b_Z \circ_Z a_Z)$$

Finally, I have proved the following proposition:

Proposition 12.2.3: Let A be a Green functor for the group G, and X be a G-set. Then $\zeta_A(X)$ identifies with the set of sequences $s = (s_Y)$, indexed by the G-sets, such that $s_Y \in A(YXY)$ and

$$a \circ_Y s_Y = s_Z \circ_Z a$$

for any G-sets Y and Z and any $a \in A(ZY)$.

If $f : X \to X'$ is a morphism of G-sets, if $s \in \zeta_A(X)$ and $s' \in \zeta_A(X')$, then

$$\zeta_{A,*}(f)(s)_Y = A_*(Id_Y \times f \times Id_Y)(s_Y) \qquad \zeta_A^*(f)(s')_Y = A^*(Id_Y \times f \times Id_Y)(s'_Y)$$

If $s \in \zeta_A(X)$ and $t \in \zeta_A(Y)$, then for any Z

$$(s \times t)_Z = A_* \begin{pmatrix} z_1 y x z_2 \\ z_1 x y z_2 \end{pmatrix} (t_Z \circ_Z s_Z)$$

Let X and Y be G-sets, such that X divides a multiple of Y in $\mathcal{C}_A$. This is equivalent to say that there exists elements $\alpha_i \in A(XY)$ and $\beta_i \in A(YX)$, for $1 \leq i \leq n$, such that

$$1_{A(X^2)} = \sum_i \alpha_i \circ_Y \beta_i$$

Then let Z be a G-set, and $s \in \zeta_A(Z)$. I can write

$$s_X = s_X \circ_X 1_{A(X^2)} = \sum_i s_X \circ_X \alpha_i \circ_Y \beta_i = \sum_i \alpha_i \circ_Y s_Y \circ_Y \beta_i$$

This formula shows that s_X is determined by s_Y. The element s_Y is such that $s_Y \circ_Y u = u \circ_Y s_Y$ for any $u \in A(Y^2)$.

Conversely, if T is a G-set such that any G-set divides in $\mathcal{C}_A$ a multiple of T, if I choose an element $s \in A(TZT)$, such that $s \circ_T u = u \circ_T s$ for any $u \in A(T^2)$, and if I choose for any X elements $\alpha_{i,X} \in A(XT)$ and $\beta_{i,X} \in A(TX)$, for $1 \leq i \leq n_X$, such that

$$1_{A(X^2)} = \sum_{i=1}^{n_X} \alpha_{i,X} \circ_T \beta_{i,X}$$

I can set

$$s_X = \sum_{i=1}^{n_X} \alpha_{i,X} \circ_T s \circ_T \beta_{i,X} \in A(XZX)$$

This element does not depend on the choices of $\alpha_{i,X}$ and $\beta_{i,X}$: indeed, if

$$1_{A(X^2)} = \sum_{j=1}^{n'_X} \alpha'_{j,X} \circ_T \beta'_{j,X}$$

then

$$s_X \circ_X \alpha'_{j,X} = \sum_{i=1}^{n_X} \alpha_{i,X} \circ_T s \circ_T \beta_{i,X} \circ_X \alpha'_{j,X}$$

As $\beta_{i,X} \circ_X \alpha'_{j,X} \in A(T^2)$, it is also

$$s_X \circ_X \alpha'_{j,X} = \sum_{i=1}^{n_X} \alpha_{i,X} \circ_T \beta_{i,X} \circ_X \alpha'_{j,X} \circ_T s = 1_{A(X^2)} \circ_X \alpha'_{j,X} \circ_T s = \alpha'_{j,X} \circ_T s$$

It follows that

$$s_X \circ_X \alpha'_{j,X} \circ_T \beta'_{j,X} = \alpha'_{j,X} \circ_T s \circ_T \beta'_{j,X}$$

and summing over j

$$s_X = \sum_{j=1}^{n'_X} \alpha'_{j,X} \circ_T s \circ_T \beta'_{j,X}$$

Moreover, the sequence s_X is an element of $\zeta_A(Z)$: indeed, if X and Y are G-sets, and if $a \in A(XY)$, then

$$a = \sum_{j=1}^{n_Y} a \circ_Y \alpha_{j,Y} \circ_T \beta_{j,Y}$$

Thus

$$s_X \circ_X a = \sum_{\substack{1 \leq i \leq n_X \\ 1 \leq j \leq n_Y}} \alpha_{i,X} \circ_T s \circ_T \beta_{i,X} \circ_X a \circ_Y \alpha_{j,Y} \circ_T \beta_{j,Y}$$

As $\beta_{i,X} \circ_X a \circ_Y \alpha_{j,Y} \in A(T^2)$, it is also

$$s_X \circ_X a = \sum_{\substack{1 \leq i \leq n_X \\ 1 \leq j \leq n_Y}} \alpha_{i,X} \circ_T \beta_{i,X} \circ_X a \circ_Y \alpha_{j,Y} \circ_T s \circ_Y \beta_{j,Y} = \sum_{j=1}^{n_Y} a \circ_Y \alpha_{j,Y} \circ_T s \circ_Y \beta_{j,Y} = a \circ_Y s_Y$$

Then I see that $\zeta_A(Z)$ identifies with the set of elements s in $A(TZT)$ such that $s \circ_T u = u \circ_T s$ for any $u \in A(T^2)$. But

$$A(TZT) \simeq \mathrm{Hom}_A(A_T, A_{ZT}) = \mathrm{Hom}_A\Big(A_T, (A_T)_Z\Big) = \mathcal{H}_A(A_T, A_T)(Z)$$

and I have also $A(T^2) \simeq \mathcal{H}_A(A_T, A_T)(\bullet)$. Moreover, the product of $u \in \mathcal{H}_A(A_T, A_T)(\bullet)$ and $s \in \mathcal{H}_A(A_T, A_T)(Z)$ for the functor $\mathcal{H}_A(A_T, A_T)$ is precisely $u \circ_T s$. It follows that ζ_A is the commutant of $\mathcal{H}(A_T, A_T)(\bullet)$ in $\mathcal{H}(A_T, A_T)$.

Now say that any X divides a multiple of T in $\mathcal{C}_A$ is equivalent to say that A_T is a progenerator of A-**Mod**. A natural question is then to know if, whenever P is a progenerator of A-**Mod**, I have

$$\zeta_A \simeq C_{\mathcal{H}_A(P,P)}\Big(\mathcal{H}_A(P,P)(\bullet)\Big)$$

First it is clear that if $\alpha \in \zeta_A(X)$, then α_P is a element of $\mathrm{Hom}_A(P, P_X) = \mathcal{H}_A(P,P)(X)$. I obtain that way a natural morphism $\zeta_A(X) \to \mathcal{H}_A(P,P)(X)$, which induces a morphism

$$\Theta : \zeta_A \to \mathcal{H}_A(P,P)$$

If $f \in \mathrm{End}_A(P) = \mathcal{H}_A(P,P)(\bullet)$, then the square

$$\begin{array}{ccc} P & \xrightarrow{\alpha_P} & P_X \\ {\scriptstyle f}\downarrow & & \downarrow{\scriptstyle f_X} \\ P & \xrightarrow[\alpha_P]{} & P_X \end{array}$$

has to be commutative. This expresses exactly the fact that the image of Θ is contained in the commutant of $\mathcal{H}_A(P,P)(\bullet)$.

Let $P^{(I)}$ be a direct sum of copies of P. It is easy to see that the commutativity of the squares

$$\begin{array}{ccc} P^{(I)} & \xrightarrow{\alpha_{P^{(I)}}} & P_X^{(I)} \\ {\scriptstyle f_i}\downarrow & & \downarrow{\scriptstyle (f_i)_X} \\ P & \xrightarrow[\alpha_P]{} & P_X \end{array}$$

where f_i is the projection on the component of index $i \in I$, forces $\alpha_{P^{(I)}} = (\alpha_P)^{(I)}$. Then if $\alpha_P = 0$, I have $\alpha_{P^{(I)}} = 0$ for any i. As any A-module M is a quotient of some $P^{(I)}$, for a suitable (I), the commutativity of the square

$$\begin{array}{ccc} P^{(I)} & \xrightarrow{\alpha_{P^{(I)}}} & P_X^{(I)} \\ {\scriptstyle \sigma}\downarrow & & \downarrow{\scriptstyle \sigma_X} \\ M & \xrightarrow[\alpha_M]{} & M_X \end{array}$$

forces then $\alpha_M \circ \sigma = 0$, thus $\alpha_M = 0$ if σ is surjective. Then $\alpha = 0$, which proves that Θ is injective.

Conversely, if $\alpha \in \mathrm{Hom}_A(P, P_X)$ is such that all the squares

$$\begin{array}{ccc} P & \xrightarrow{\alpha} & P_X \\ {\scriptstyle f}\downarrow & & \downarrow{\scriptstyle f_X} \\ P & \xrightarrow[\alpha]{} & P_X \end{array}$$

are commutative, then setting $\alpha_{P^{(I)}} = (\alpha)^{(I)}$, it is easy to see that all the squares

$$\begin{array}{ccc} P^{(I)} & \xrightarrow{\alpha_{P^{(I)}}} & P_X^{(I)} \\ {\scriptstyle f}\downarrow & & \downarrow{\scriptstyle f_X} \\ P^{(J)} & \xrightarrow[\alpha_{P^{(J)}}]{} & P_X^{(J)} \end{array} \qquad (C)$$

obtained for index sets I and J and a morphism $f : P^{(I)} \to P^{(J)}$, are commutative. In particular, if L is the image of f, I see that $\alpha_{P^{(J)}}(L) \subseteq L_X$. So if M is an arbitrary A-module, and σ a surjective morphism from $P^{(J)}$ to M, then there exists a unique morphism α_M such that the square

$$\begin{array}{ccc} P^{(J)} & \xrightarrow{\alpha_{P^{(J)}}} & P_X^{(J)} \\ \sigma \downarrow & & \downarrow \sigma_X \\ M & \xrightarrow[\alpha_M]{} & M_X \end{array}$$

is commutative. The commutativity of the squares (C) and the projectivity of $P^{(I)}$ shows that α_M does not depend on (J) or on σ. It is then clear that the α_M define an element of $\zeta_A(X)$. Thus Θ is surjective. Finally, I have proved the following proposition:

Proposition 12.2.4: Let A be a Green functor for the group G. If P is a progenerator of A-Mod, then

$$\zeta_A \simeq C_{\mathcal{H}_A(P,P)}\Big(\mathrm{End}_A(P)\Big)$$

In particular, if T is a G-set such that any G-set divides a multiple of T in C_A, then for any G-set Z

$$\zeta_A(Z) = \{s \in A(TZT) \mid \forall u \in A(T^2),\ s \circ_T u = u \circ_T s\}$$

12.2.2 Endomorphisms of the restriction functor

Let H be a subgroup of G. The first definition of $\zeta_A(H)$ is

$$\zeta_A(H) = \mathrm{End}_{Funct}(\mathrm{Res}_H^G)$$

where Res_H^G is the restriction functor from A-**Mod** to $\mathrm{Res}_H^G A$-**Mod**. Setting

$$\Omega = \coprod_{K \subseteq G} G/K \qquad \Omega' = \coprod_{K \subseteq H} H/K$$

the evaluation $M \mapsto M(\Omega)$ is an equivalence of categories from A-**Mod** to $A(\Omega^2)$-**Mod**, and the evaluation $M' \mapsto M'(\Omega')$ is an equivalence of categories from $\mathrm{Res}_H^G A$-**Mod** to $(\mathrm{Res}_H^G A)(\Omega'^2)$-**Mod**.

Moreover

$$(\mathrm{Res}_H^G A)(\Omega'^2) = A(\mathrm{Ind}_H^G \Omega'^2)$$

and the product of the elements α and β of $(\mathrm{Res}_H^G A)(\Omega'^2)$ is defined by

$$\alpha \circ'_{\Omega'} \beta = (\mathrm{Res}_H^G A)_* \begin{pmatrix} \omega_1'\omega_2'\omega_3' \\ \omega_1'\omega_2' \end{pmatrix} (\mathrm{Res}_H^G A)^* \begin{pmatrix} \omega_1'\omega_2'\omega_3' \\ \omega_1'\omega_2'\omega_2'\omega_3' \end{pmatrix} (\alpha \times' \beta)$$

where the products $\circ'$ and $\times'$ are those of the Green functor $\mathrm{Res}_H^G A$. Thus

$$\alpha \circ'_{\Omega'} \beta = A_* \left(\mathrm{Ind}_H^G \begin{pmatrix} \omega_1'\omega_2'\omega_3' \\ \omega_1'\omega_2' \end{pmatrix} \right) A^* \left(\mathrm{Ind}_H^G \begin{pmatrix} \omega_1'\omega_2'\omega_3' \\ \omega_1'\omega_2'\omega_2'\omega_3' \end{pmatrix} \right) A^*(\delta)(a \times \beta)$$

where δ is the map from $\mathrm{Ind}_H^G\Omega'^4$ to $(\mathrm{Ind}_H^G\Omega'^2)^2$ defined by

$$\delta(g,\omega_1',\omega_2',\omega_3',\omega_4') = \big((g,\omega_1',\omega_2'),(g,\omega_3',\omega_4')\big)$$

It follows that

$$\alpha \circ'_{\Omega'} \beta = A_*\begin{pmatrix}(g,\omega_1',\omega_2',\omega_3')\\ (g,\omega_1',\omega_2')\end{pmatrix} A^*\begin{pmatrix}(g,\omega_1',\omega_2',\omega_3')\\ (g,\omega_1',\omega_2')(g,\omega_2',\omega_3')\end{pmatrix}(\alpha\times\beta)$$

Let i be the map from $\mathrm{Ind}_H^G\Omega'^2$ to Ω^2, defined by

$$i(g,\omega_1',\omega_2') = (g\omega_1', g\omega_2')$$

where the expression $g\omega'$ is equal to ghK if $\omega' = hK \in \Omega'$. The map i is injective: indeed, if

$$i(g,\omega_1',\omega_2') = i(g',\omega_1'',\omega_2'')$$

then $g\omega_1' = g'\omega_1''$ and as ω_1' and ω_1'' are contained in H, I have $g'H = gH$. Then there exists $h \in H$ such that

$$g' = gh \qquad \omega_1' = h\omega_1'' \qquad \omega_2' = h\omega_2''$$

Then $(g',\omega_1'',\omega_2'') = (g'h,\omega_1,\omega_2) = (g,\omega_1',\omega_2')$, and i is injective.

It follows that $A_*(i)$ is an injective morphism of $(\mathrm{Res}_H^G A)(\Omega'^2)$ into $A(\Omega^2)$. Moreover

$$A_*(i)(\alpha \circ_{\Omega'} \beta) = A_*\begin{pmatrix}(g,\omega_1',\omega_2',\omega_3')\\ g\omega_1', g\omega_2'\end{pmatrix} A^*\begin{pmatrix}(g,\omega_1',\omega_2',\omega_3')\\ (g,\omega_1',\omega_2')(g,\omega_2',\omega_3')\end{pmatrix}(\alpha\times\beta)$$

On the other hand

$$A_*(i)(\alpha)\circ_\Omega A_*(i)(\beta) = A_*\begin{pmatrix}\omega_1\omega_2\omega_3\\ \omega_1\omega_3\end{pmatrix} A^*\begin{pmatrix}\omega_1\omega_2\omega_3\\ \omega_1\omega_2\omega_2\omega_3\end{pmatrix}\big(A_*(i)(\alpha)\times A_*(i)(\beta)\big) = \ldots$$

$$\ldots = A_*\begin{pmatrix}\omega_1\omega_2\omega_3\\ \omega_1\omega_3\end{pmatrix} A^*\begin{pmatrix}\omega_1\omega_2\omega_3\\ \omega_1\omega_2\omega_2\omega_3\end{pmatrix} A_*\begin{pmatrix}(g_1,\omega_1',\omega_2')(g_2,\omega_3',\omega_4')\\ g_1\omega_1'\, g_1\omega_2'\, g_2\omega_3'\, g_2\omega_4'\end{pmatrix}(\alpha\times\beta)$$

Let (C) be the square

$$\begin{array}{ccc}
\mathrm{Ind}_H^G\Omega'^3 & \xrightarrow{\begin{pmatrix}(g,\omega_1',\omega_2',\omega_3')\\ (g,\omega_1',\omega_2')(g,\omega_2',\omega_3')\end{pmatrix}} & (\mathrm{Ind}_H^G\Omega'^2)^2\\
{\scriptstyle\begin{pmatrix}(g,\omega_1',\omega_2',\omega_3')\\ g\omega_1'\, g\omega_2'\, g\omega_3'\end{pmatrix}}\Big\downarrow & & \Big\downarrow{\scriptstyle\begin{pmatrix}(g_1,\omega_1',\omega_2')(g_2,\omega_3',\omega_4')\\ g_1\omega_1'\, g_1\omega_2'\, g_2\omega_3'\, g_2\omega_4'\end{pmatrix}}\\
\Omega^3 & \xrightarrow[\begin{pmatrix}\omega_1\omega_2\omega_3\\ \omega_1\omega_2\omega_2\omega_3\end{pmatrix}]{} & \Omega^4
\end{array}$$

This square is cartesian: indeed, if $(\omega_1,\omega_2,\omega_3)$ and $\big((g_1,\omega_1',\omega_2')(g_2,\omega_3',\omega_4')\big)$ are such that

$$(\omega_1,\omega_2,\omega_2,\omega_3) = (g_1\omega_1', g_1\omega_2', g_2\omega_3', g_2\omega_4')$$

then $g_1\omega'_2 = g_2\omega'_3$. As ω'_2 and ω'_3 are contained in H, this forces $g_1H = g_2H$, and there exists $h \in H$ such that $g_1 = g_2h$. Then $\omega'_3 = h\omega'_2$, and

$$(g_2, \omega'_3, \omega'_4) = (g_1h^{-1}, h\omega'_2, \omega'_4) = (g_1, \omega'_2, h^{-1}\omega'_4)$$

Let then $u = (g, \omega'_1, \omega'_2, h^{-1}\omega'_4) \in \mathrm{Ind}_H^G\Omega'^3$. I have

$$\begin{pmatrix} (g,\omega'_1,\omega'_2,\omega'_3) \\ (g,\omega'_1,\omega'_2)(g,\omega'_2,\omega'_3) \end{pmatrix}(u) = \big((g_1,\omega'_1,\omega'_2),(g_2,\omega'_3,\omega'_4)\big)$$

and

$$\begin{pmatrix} (g,\omega'_1,\omega'_2,\omega'_3) \\ g\omega'_1\ g\omega'_2\ g\omega'_3\ g\omega'_4 \end{pmatrix}(u) = (g_1\omega'_1, g_1\omega'_2, g_1\omega'_3, g_1h^{-1}\omega'_4) = (\omega_1,\omega_2,\omega_3,\omega_4)$$

On the other hand, the map $i = \begin{pmatrix} (g,\omega'_1,\omega'_2,\omega'_3) \\ (g,\omega'_1,\omega'_2)(g,\omega'_2,\omega'_3) \end{pmatrix}$ is injective: indeed, if

$$i(g,\omega'_1,\omega'_2,\omega'_3) = i(g',\omega''_1,\omega''_2,\omega''_3)$$

then there exists $h \in H$ such that

$$g' = gh^{-1} \qquad \omega''_1 = h\omega'_1 \qquad \omega''_2 = h\omega'_2 \qquad \omega''_3 = h\omega'_3$$

and then

$$(g',\omega''_1,\omega''_2,\omega''_3) = (gh^{-1}, h\omega'_1, h\omega'_2, h\omega'_3) = (g,\omega'_1,\omega'_2,\omega'_3)$$

So the square (C) is cartesian, and then

$$A_*(i)(\alpha) \circ_\Omega A_*(i)(\beta) = \dots$$

$$\dots = A_*\begin{pmatrix} \omega_1\omega_2\omega_3 \\ \omega_1\omega_3 \end{pmatrix} A_*\begin{pmatrix} (g,\omega'_1,\omega'_2,\omega'_3) \\ g\omega'_1\ g\omega'_2\ g\omega'_3 \end{pmatrix} A^*\begin{pmatrix} (g,\omega'_1,\omega'_2,\omega'_3) \\ (g,\omega'_1,\omega'_2)(g,\omega'_2,\omega'_3) \end{pmatrix}(\alpha\times\beta) = \dots$$

$$\dots = A_*\begin{pmatrix} (g,\omega'_1,\omega'_2,\omega'_3) \\ g\omega'_1\ g\omega'_3 \end{pmatrix} A^*\begin{pmatrix} (g,\omega'_1,\omega'_2,\omega'_3) \\ (g,\omega'_1,\omega'_2)(g,\omega'_2,\omega'_3) \end{pmatrix}(\alpha\times\beta) = \dots$$

$$\dots = i(\alpha \circ_{\Omega'} \beta)$$

It follows that $A_*(i)$ is a morphism of algebras (non-unitary in general) from $\mathrm{Res}_H^G A(\Omega'^2)$ to $A(\Omega^2)$. In terms of generators, if K and L are subgroups of H, if $x \in H$, and if $i_{K,x,L}$ is the map from $H/(K \cap {}^xL)$ to Ω'^2 defined by

$$i_{K,x,L}\big(h(K\cap {}^xL)\big) = (hK, hxL)$$

then $i \circ \mathrm{Ind}_H^G i_{K,x,L}$ is the map

$$g(K\cap {}^xL) \in G/(K\cap {}^xL) \mapsto (gK, gxL) \in \Omega^2$$

It follows that

$$A_*(i)(t'^K_{K\cap {}^xL}\lambda'_{a,K\cap {}^xL}xr'^L_{K^x\cap L}) = t^K_{K\cap {}^xL}\lambda_{a,K\cap {}^xL}xr^L_{K^x\cap L}$$

In other words, the morphism $A_*(i)$ is just the inclusion

$$(\mathrm{Res}_H^G A)(\Omega'^2) = \bigoplus_{\substack{K,L\subseteq H \\ x\in K\backslash H/L}} A(K\cap {}^xL) \hookrightarrow \bigoplus_{\substack{K,L\subseteq G \\ x\in K\backslash G/L}} A(K\cap {}^xL) = A(\Omega^2)$$

I have finally proved the

Lemma 12.2.5: Let H be a subgroup of G. Then the inclusion i from $\mathrm{Ind}_H^G\Omega'^2$ into Ω^2 induces an injective morphism of algebras

$$A_*(i) = (\mathrm{Res}_H^G)(A)(\Omega'^2) \hookrightarrow A(\Omega^2)$$

Remark: This is the only point on which I disagree with Thévenaz and Webb (see [15] 5.3 to 5.4): this lemma shows in particular that the morphism $b_*(i)$ from the Mackey algebra of H to the Mackey algebra of G *is* injective.

If M is an A-module, then

$$\mathrm{Res}_H^G M(\Omega') = \bigoplus_{K \subseteq H} M(K) = \sum_{K \subseteq H} t_K^K \circ_\Omega M(\omega)$$

In other words, the restriction functor can be translated in terms of the algebras $B = A(\Omega^2)$ and $C = (\mathrm{Res}_H^G A)(\Omega'^2)$ as

$$(\mathrm{Res}_H^G M)(\Omega') = A_*(i)(1_C) \circ_\Omega M(\Omega)$$

Let e be the idempotent $A_*(i)(1_C)$ of B. Then $e \circ_\Omega B$ is a C-module-B, and

$$\mathrm{Res}_H^G M(\Omega') \simeq (e \circ_\Omega B) \otimes_B M(\Omega)$$

The restriction functor is then given by tensoring with a bimodule, and then I can apply the following lemma:

Lemma 12.2.6: Let A and B be R-algebras, and M be an A-module-B. Then

$$\mathrm{End}_{Funct}(M \otimes_B -) \simeq \mathrm{End}_{A \otimes B^{op}}(M)$$

In particular, if $f : A \to B$ is a morphism of algebras, if $e = f(1_A)$, and $M = eB$, then $eM \otimes_B -$ is the functor Res_f of restriction along f, and

$$\mathrm{End}_{Funct}(\mathrm{Res}_f) \simeq \{b \in eBe \mid \forall a \in A,\ bf(a) = f(a)b\}$$

Proof: An endomorphism ϕ of the functor $M \otimes_B -$ is determined by giving, for any B-module L, a morphism of A-modules ϕ_L from $M \otimes_B L$ to itself, such that for any morphism of B-modules $\psi : L \to L'$, the square

$$\begin{array}{ccc} M \otimes_B L & \xrightarrow{\phi_L} & M \otimes L \\ {\scriptstyle M \otimes_B \psi}\downarrow & & \downarrow{\scriptstyle M \otimes_B \psi} \\ M \otimes_B L' & \xrightarrow[\phi_{L'}]{} & M \otimes L' \end{array}$$

is commutative. In particular ϕ_B is an endomorphism of $M \otimes_B B$, which is isomorphic to M by the map θ defined by $\theta(m \otimes b) = mb$. Then $\Phi = \theta\phi_B\theta^{-1}$ is an endomorphism of M. Moreover $\mathrm{Hom}_B(B, L) \simeq L$ for any L, the element $l \in L$ defining the morphism $\mu_l : b \mapsto bl$ from B to L. The commutativity of the square

$$\begin{array}{ccc} M \otimes_B B \simeq M & \xrightarrow{\phi_B} & M \otimes_B B \simeq M \\ {\scriptstyle M \otimes_B \mu_l}\downarrow & & \downarrow{\scriptstyle M \otimes_B \mu_l} \\ M \otimes L & \xrightarrow[\phi_L]{} & M \otimes L \end{array} \qquad (C)$$

forces then

$$\phi_L(m\otimes l) = (M\otimes\mu_l)\phi_B(m\otimes 1) = (M\otimes\mu_l)\theta^{-1}\Phi(m) = (M\otimes\mu_l)(\Phi(m)\otimes 1) = \Phi(m)\otimes l$$

Thus Φ determines ϕ. Moreover, if $a\in A$ and $b\in B$, I must have

$$\Phi(amb)\otimes l = \phi_L(amb\otimes l) = a\phi_L(m\otimes bl) = a\Phi(m)\otimes bl = a\Phi(m)b\otimes l$$

The case $L = B$ now implies that Φ is an endomorphism of M as A-module-B.

Conversely, being given Φ, I can define Φ_L by $\phi_L(m\otimes l) = \Phi(m)\otimes l$. This definition makes sense, because for $b\in B$

$$\phi_L(mb\otimes l) = \Phi(mb)\otimes l = \Phi(m)b\otimes l = \Phi(m)\otimes bl = \phi_L(m\otimes bl)$$

If moreover Φ is a morphism of A-modules, then for any $a\in A$, I have

$$\phi_L(am\otimes l) = \Phi(am)\otimes l = a\Phi(m)\otimes l = a\phi_L(m\otimes l)$$

Thus ϕ_L is a morphism of A-modules from M to M_X. With this definition of ϕ_L, it is clear that all squares (C) are commutative. This proves the first assertion of the lemma.

For the second assertion, I must find the endomorphisms of eB as an A-module-B. Such an endomorphism ϕ is entirely determined by the image of $e\in eB$, since

$$\phi(eb) = \phi(e)b$$

Then I must have $\phi(e^2) = \phi(e) = \phi(e)e$ and also

$$\phi(e) = \phi(1_A e) = 1_A\phi(e) = e\phi(e)$$

So $\phi(e)\in eBe$. Moreover, if $a\in A$, then

$$f(a)e = f(a)f(1_A) = f(a) = f(1_A)f(a) = ef(a)$$

Conversely, if $p\in eBe$ commutes with any element in the image of f, then setting

$$\phi(eb) = pb$$

I obtain an endomorphism of eB as A-module-B, since

$$\phi(a.eb.b') = \phi\big(f(a)ebb'\big) = \phi\big(ef(a)bb'\big) = pf(a)bb' = f(a)pbb' = a.(pb).b'$$

This proves the lemma. ■

This lemma applies to the morphism $A_*(i)$ from $(\mathrm{Res}_H^G A)(\Omega'^2)$ to $A(\Omega^2)$, and this gives the

Proposition 12.2.7: Let A be a Green functor for G, and H be a subgroup of G. Then $\zeta_A(H)$ identifies with the set of elements in

$$\sum_{K\subseteq H} t_K^K A(\Omega^2) t_K^K$$

which commute to all the elements

$$t_{K\cap {}^xL}^K \lambda_{K\cap {}^xL,a} x r_{K^x\cap L}^L$$

for $K, L\subseteq H$, for $a\in A(K)$ and $x\in H$. In particular, the ring $\big(\zeta_A(1),.\big)$ is isomorphic to the centralizer of $A(1)$ in $A(1)\otimes G$.

Proof: The first assertion is only a reformulation of the previous results: if

$$z \in (\sum_{K\subseteq H} t_K^K)A(\Omega^2)(\sum_{K\subseteq H} t_K^K)$$

and if z commutes with all the t_K^K, for $K \subseteq H$, then

$$z = \sum_{K,L\subseteq H} t_K^K z t_L^L = \sum_{K,L\subseteq H} t_K^K t_L^L z = \sum_{K\subseteq H} t_K^K z = \sum_{K\subseteq H} t_K^K z t_K^K$$

thus $z \in \sum_{K\subseteq H} t_K^K A(\Omega^2) t_K^K$, and the first assertion holds.

The second one follows from the case $H = \{1\}$: the algebra $t_1^1 A(\Omega^2) t_1^1$ is formed of the elements

$$t_1^1 \lambda_{1,a} x r_1^1$$

for $a \in A(1)$ and $x \in G$. Hence it is isomorphic to $A(1) \otimes G$. The image of $A(\Omega'^2)$ corresponds to the elements such that $x = 1$. Thus it is $A(1)$, and the proposition follows. ■

Let $z \in \zeta_A(H)$. Then $z = \sum_{K\subseteq H} z_K$, where $z_K = t_K^K z t_K^K$. Say that z commutes with all the $t_{K\cap {}^xL}^K \lambda_{K\cap {}^xL,a} x r_{K^x\cap L}^L$ is equivalent to say that

$$z_K t_{K\cap {}^xL}^K \lambda_{K\cap {}^xL,a} x r_{K^x\cap L}^L = t_{K\cap {}^xL}^K \lambda_{K\cap {}^xL,a} x r_{K^x\cap L}^L z_L \tag{12.1}$$

which gives in particular, for $L \subseteq K \subseteq H$, for $a \in A(K)$ and $x \in H$

$$z_K t_L^K = t_L^K z_L \qquad r_L^K z_K = z_L r_L^K \qquad z_K \lambda_{K,a} = \lambda_{K,a} z_K \qquad x z_K x^{-1} = z_{{}^xK}$$

Conversely, these relations imply equalities (12.1).

A computation similar to the one made to identify the functors $\mathcal{H}(M,N)$ (see proposition 1.4.1) shows then the following proposition:

Proposition 12.2.8: Let A be a Green functor for G. If H is a subgroup of G, then $\zeta_A(H)$ identifies with the set of sequences (z_K) indexed by the subgroups of H, such that $z_K \in t_K^K A(\Omega^2) t_K^K$, and

$$z_K t_L^K = t_L^K z_L \qquad r_L^K z_K = z_L r_L^K \qquad z_K \lambda_{K,a} = \lambda_{K,a} z_K \qquad x z_K x^{-1} = z_{{}^xK}$$

for any subgroups $K \supseteq L$ of H, and any $a \in A(K)$ and $x \in H$.

If $H' \subseteq H$ and $z = (z_K) \in \zeta_A(H)$, then

$$r_{H'}^H(z)_K = z_K \quad \forall K \subseteq H'$$

If $H \subseteq H'$ and $z = (z_K) \in \zeta_A(H)$, then

$$(t_H^{H'} z)_K = \sum_{x\in K\backslash H'/H} t_{K\cap {}^xH}^K x z_{K^x\cap H} x^{-1} r_{K\cap {}^xH}^K \quad \forall K \subseteq H'$$

Finally if $g \in G$ and $z = (z_K) \in \zeta_A(H)$, then

$$({}^gz)_K = {}^g(z_{K^g}) \quad \forall K \subseteq {}^gH$$

The product $z.z'$ of two elements of $\zeta_A(H)$ is given by

$$(z.z')_K = z_K z'_K \quad \forall K \subseteq H$$

12.2.3 Induction and inflation

Let G and H be groups, and U be a G-set-H. If A is a Green functor for G, and X is an H-set, then an element $a \in \zeta_A(U \circ_H X)$ is a sequence (a_Y), indexed by the G-sets, such that $a_Y \in A(Y \times (U \circ_H X) \times Y)$. Then if T is an H-set, I set

$$b_T = A^*(\delta^U_{T,X,T})(a_{U\circ_H T}) \in (A \circ U)(TXT)$$

Lemma 12.2.9: If $U/H = \bullet$, the above construction is a unitary morphism of Green functors from $\zeta_A \circ U$ to $\zeta_{A\circ U}$.

Proof: First, if $U/H = \bullet$, and if X and Y are G-sets, then

$$U \circ_H (X \times Y) \simeq (U \circ_H X)(U \circ_H Y) = (U \circ_H X) \times (U \circ_H Y)$$

so that the maps $\delta^U_{X,Y}$ are bijective.

Then the maps

$$\delta^U_{T,X,T} = (\delta^U_{T,X} \times Id_{U\circ_H T}) \circ \delta^U_{TX,T}$$

are also bijective. Let then S be an H-set, and $\beta \in (A \circ U)(TS)$. Then denoting by $\circ'$ and $\times'$ the products $\circ$ and $\times$ for the functor $A \circ U$, I have

$$b_T \circ'_T \beta = (A\circ U)_* \begin{pmatrix} t_1xt_2s \\ t_1xs \end{pmatrix} (A\circ U)^* \begin{pmatrix} t_1xt_2s \\ t_1xt_2t_2s \end{pmatrix} \big(A^*(\delta^U_{T,X,T})(a_{U\circ_H T}) \times' \beta\big) = \dots$$

$$\dots = (A\circ U)_* \begin{pmatrix} t_1xt_2s \\ t_1xs \end{pmatrix} (A\circ U)^* \begin{pmatrix} t_1xt_2s \\ t_1xt_2t_2s \end{pmatrix} A^*(\delta^U_{TXT,TS})\big(A^*(\delta^U_{T,X,T})(a_{U\circ_H T}) \times \beta\big)$$

Moreover, setting $\beta' = A_*(\delta^U_{T,S})(\beta)$, I have

$$A^*(\delta^U_{T,X,T})(a_{U\circ_H T}) \times \beta = A^*(\delta^U_{T,X,T})(a_{U\circ_H T}) \times A^*(\delta^U_{T,S})(\beta') = A^*(\delta^U_{T,X,T,T,S})(a_{U\circ_H T} \times \beta')$$

Setting

$$T' = U \circ_H T \qquad X' = U \circ_H X \qquad S' = U \circ_H S$$

it is then clear that

$$(A \circ U)^* \begin{pmatrix} t_1xt_2s \\ t_1xt_2t_2s \end{pmatrix} A^*(\delta^U_{T,X,T,T,S}) = A^*(\delta^U_{T,X,T,S}) A^* \begin{pmatrix} t'_1x't'_2s' \\ t'_1x't'_2t'_2s' \end{pmatrix}$$

Similarly

$$(A \circ U)_* \begin{pmatrix} t_1xt_2s \\ t_1xt_2 \end{pmatrix} A^*(\delta^U_{T,X,T,S}) = (A \circ U)_* \begin{pmatrix} t_1xt_2s \\ t_1xs \end{pmatrix} A_*(\delta^U_{T,X,T,S}{}^{-1}) = \dots$$

$$\dots = A_*(\delta^U_{T,X,S}{}^{-1}) A_* \begin{pmatrix} t'_1x't'_2s' \\ t'_1x's' \end{pmatrix}$$

so that finally

$$b_T \circ'_T \beta = A^*(\delta^U_{T,X,S})(a_{T'} \circ_{T'} \beta')$$

As $a \in \zeta_A(X')$, and as $\beta' \in A(T'S')$, I have also

$$b_T \circ'_T \beta = A^*(\delta^U_{T,X,S})(\beta' \circ_{S'} a'_S)$$

The same argument applied to $\beta \circ_S' b_S$ shows that

$$b_T \circ_T' \beta = \beta \circ_S b_S$$

hence that the sequence b_T defines an element of $\zeta_{A\circ U}(X)$.

Now the lemma follows from proposition 8.4.1, which says that the correspondence

$$X \mapsto U \circ_H X \qquad a \in A(YX) \mapsto A_*(\delta^U_{X,Y})(a) \in A\big((U \circ_H X) \times (U \circ_H Y)\big)$$

is a functor from $\mathcal{C}_{A\circ U}$ to $\mathcal{C}_A$. ■

In some cases, the morphism from $\zeta_A \circ U$ to $\zeta_{A\circ U}$ is an isomorphism:

Proposition 12.2.10: Let $L \trianglelefteq K$ be subgroups of G, and A be a Green functor for the group K/L. Then

$$\zeta_{\mathrm{Ind}_K^G \mathrm{Inf}_{K/L}^K A} \simeq \mathrm{Ind}_K^G \mathrm{Inf}_{K/L}^K \zeta_A$$

Proof: I will show that if G and H are groups, if U is a G-set-H such that $U/H = \bullet$ and such that G acts freely on U, then

$$\zeta_A \circ U \simeq \zeta_{A\circ U}$$

This assertion is equivalent to the proposition, because with the hypothesis on U, there exists a subgroup P of $G \times H$ such that

$$p_1(P) = G \qquad k_1(P) = \{1\} \qquad U \simeq (G \times H)/P$$

In those conditions, if $K = p_2(P)$ and $L = k_2(P)$, I have an isomorphism θ from K/L on G, and if Z is an H-set, I have

$$U \circ_H X \simeq \theta\big((\mathrm{Res}_K^H Z)^L\big)$$

Then if $G = K/L$ and $\theta = Id$, and if A is a Green functor for G, I have

$$A \circ U = \mathrm{Ind}_K^H \mathrm{Inf}_{K/L}^K A$$

With the notations of lemma 12.2.9, if $b_T = 0$ for any T, then $a_Y = 0$ for any Y: the hypothesis implies indeed that $a_Y = 0$ if Y is of the form $U \circ_H T$, since $A_*(\delta^U_{T,X,T})$ is an isomorphism. But if Y is a G-set, I have the morphism

$$\nu_Y : Y \to U \circ_H (G\backslash U.Y)$$

defined in this case by

$$\nu_Y(y) = \big(u, G(u,y)\big)$$

for an arbitrary element $u \in U$ (indeed $U.Y = U \times Y$ if $U/H = \bullet$). If moreover G acts freely on U, then ν_Y is injective: indeed, if

$$\nu_Y(y) = \big(u, G(u,y)\big) = \nu_Y(y') = \big(u, G(u,y')\big)$$

then there exists $h \in H$ and $g \in G$ such that

$$u = uh \qquad g(u,y) = (gu, gy) = (uh, y')$$

Whence $gy = y'$ and $gu = uh = u$. Then $g = 1$ and $y' = y$.

Thus any G-set is a subset of a set of the form $U \circ_H T$. Then in $\mathcal{C}_A$, any object divides an object of this type. But I know that if Y divides Y' in $\mathcal{C}_A$, and if $a \in \zeta_A(X)$, then $a_{Y'}$ determines a_Y. In those conditions, I have then $a_Y = 0$ for any Y, and the morphism from $\zeta_A \circ U$ to $\zeta_{A \circ U}$ is injective.

It is also surjective: if $b \in \zeta_{A\circ U}(X)$, and if T is an H-set, I have an element

$$b_T \in (A \circ U)(TXT)$$

I have thus an element

$$b'_T = A_*(\delta_{T,X,T})(b_T) \in A\Big((U \circ_H T) \times (U \circ_H X) \times (U \circ_H T)\Big)$$

and it is easy to see by proposition 8.4.1 that if $\alpha \in A\Big((U \circ_H T) \times (U \circ_H T')\Big)$, then setting

$$\alpha' = A^*(\delta^U_{T,T'})(\alpha) \in (A \circ U)(TT')$$

I have

$$b'_T \circ_{U \circ_H T} \alpha = A_*(\delta^U_{T,X,T})(b_T) \circ_{U \circ_H T} A_*(\delta^U_{T,T'})(\alpha') = A_*(\delta^U_{T,X,T'})(b_T \circ'_T \alpha') = \ldots$$

$$\ldots = A_*(\delta^U_{T,X,T'})(\alpha' \circ'_{T'} b_{T'}) = A_*(\delta^U_{T,T'})(\alpha') \circ_{U \circ_H T'} A_*(\delta^U_{T',X,T'})(b_{T'}) = \alpha \circ_{U \circ_H T'} b'_{T'}$$

If Y is a G-set, then Y maps into $U \circ_H (G \backslash U.Y)$ via ν_Y, and I set for $T = G\backslash U.Y$

$$a_Y = \nu^*_Y \circ_{U \circ_H T} b'_T \circ_{U \circ_H T} \nu_{Y,*}$$

If Y' is another G-set, if $T' = G\backslash U.Y'$, and if $\alpha \in A(YY')$, then

$$a_Y \circ_Y \alpha = \nu^*_Y \circ_{U \circ_H T} b_T \circ_{U \circ_H T} \nu_{Y,*} \circ_Y \alpha \circ_{Y'} \nu^*_{Y'} \circ_{U \circ_H T'} \nu_{Y',*}$$

But $\nu_{Y,*} \circ_Y \alpha \circ_{Y'} \nu^*_{Y'} \in A\Big((U \circ_H T) \times (U \circ_H T')\Big)$. So this gives

$$a_Y \circ_Y \alpha = \nu^*_Y \circ_{U \circ_H T} \nu_{Y,*} \circ_Y \alpha \circ_{Y'} \nu^*_{Y'} \circ_{U \circ_H T'} b'_{T'} \circ_{U \circ_H T'} \nu_{Y',*} = \alpha \circ_{Y'} a_{Y'}$$

So the sequence a_Y defines an element of $\zeta_A(U \circ_H X)$. If moreover $Y = U \circ_H T$, then for $T' = G\backslash U.Y$, I have

$$a_Y = \nu^*_Y \circ_{U \circ_H T'} b'_{T'} \circ_{U \circ_H T'} \nu_{Y,*}$$

But $\nu_{Y,*}$ is an element of

$$A\Big((U \circ_H T') \times Y\Big) = A\Big((U \circ_H T') \times (U \circ_H T)\Big)$$

Then

$$a_Y = \nu^*_Y \circ_{U \circ_H T'} \nu_{Y,*} \circ_{U \circ_H T} b_T = b'_T$$

and then

$$A^*(\delta^U_{T,X,T})(a_{U \circ_H T}) = A^*(\delta^U_{T,X,T})(b'_T) = b_T$$

which proves the proposition. ∎

12.3 Examples

Let A be a Green functor for G. If M is an A-module, if Y is a G-set, and if $z \in \zeta_A(Y)$, then z induces by definition a morphism z_M of A-modules from M to M_Y. It is clear that I obtain that way a morphism of Mackey functors from ζ_A to $\mathcal{H}(M, M_Y)$. The construction of the product on ζ_A, by the formulae

$$(\alpha \times \beta)_M = \alpha_M \times \beta_M$$

where the product of the right hand side is the product of $\mathcal{H}(M, M)$, shows that this morphism is compatible with the product. It is moreover clearly unitary. Thus M is a ζ_A-module, and the image of ζ_A in $\mathcal{H}(M, M)$ is contained in $\mathcal{H}_A(M, M)$ (since z_M is a morphism of A-modules). Thus M is an $A\hat{\otimes}\zeta_A$-module.

Proposition 12.3.1: Let A be a Green functor for the group G. If M is an A-module, and if X, Y and Z are G-sets, then the product

$$a \in A(X),\ b \in \zeta_A(Y),\ m \in M(Z) \mapsto a \times b \times m = a \times M_* \begin{pmatrix} zy \\ yz \end{pmatrix} (b_Z \circ_Z m)$$

turns M into an $A\hat{\otimes}\zeta_A$-module.

In particular, the primitive idempotents of $\zeta_A(\bullet) = Z\big(A(\Omega^2)\big)$ lead to a decomposition of A-modules in blocks: the block j is formed of the A-modules M such that $j \times M = M$.

12.3.1 The functors FP_B

Let B be a G-algebra, and $A = FP_B$ be the fixed points functor on B, defined for a G-set X by

$$FP_B(X) = \mathrm{Hom}_G([X], B)$$

Then if X and Y are G-sets, the module $FP_B(XY)$ identifies with the set of matrices $m(x, y)$ indexed by $X \times Y$, with coefficients in B, which are invariant by G, i.e. such that for any $g \in G$

$$m(gx, gy) = {}^g m(x, y)$$

A similar computation as in proposition 4.5.2 shows that if Z is a G-set, if $p(y, z)$ is a matrix in $FP_B(YZ)$, then the product $m \circ_Y p$ is obtained as a product of matrices, i.e.

$$(m \circ_Y p)(x, z) = \sum_{y \in Y} m(x, y) p(y, z)$$

where the expression in the right hand side is computed in the algebra B.

Let X be a G-set, and $m \in \zeta_A(X)$. Then for any G-set Y, I have an element $m_Y \in A(YXY)$, that is a "matrix" $m_Y(y, x, y')$ indexed by YXY, and invariant by G, i.e. such that for any $g \in G$, I have

$$m_Y(gy, gx, gy') = {}^g m_Y(y, x, y')$$

Say that $m \in \zeta_A(X)$ is equivalent to say that for any G-sets Y and Z, for any $a \in A(YZ)$, and for any $(x,y,z) \in XYZ$, I have

$$\sum_{y' \in Y} m_Y(y,x,y')a(y',z) = \sum_{z' \in Z} a(y,z')m_Z(z',x,z) \tag{12.2}$$

First I will take $Z = G/1$. I set

$$m(x,z) = m_{G/1}(1,x,z) \quad \text{for} \quad x \in X,\ z \in G = G/1$$

so that $m_{G/1}(z',x,z) = {}^{z'}m(z'^{-1}x, z'^{-1}z)$. Similarly, if ϕ is a map from Y to B, and if I set

$$a(y,z) = {}^{z}\phi(z^{-1}y)$$

I obtain an element of $A\big(Y \times (G/1)\big)$, and any element of $A\big(Y \times (G/1)\big)$ is of this form.

In particular, if $\phi(y) = 1$ for $y = y_0$ and $\phi(y) = 0$ otherwise, equality (12.2) becomes

$$m_Y(y,x,zy_0) = \sum_{\substack{z' \in G \\ z'^{-1}y = y_0}} {}^{z'}m(z'^{-1}x, z'^{-1}z)$$

or for $y' = zy_0$

$$m_Y(y,x,y') = \sum_{\substack{z' \in G \\ z'^{-1}y = z^{-1}y'}} {}^{z'}m(z'^{-1}x, z'^{-1}z) \tag{12.3}$$

Then taking $Y = G/1$, this equality forces

$$m_{G/1}(y,x,y') = {}^{y}m(y^{-1}x, y^{-1}y') = {}^{yy'^{-1}z}m(z^{-1}y'y^{-1}x, z^{-1}y'y^{-1}z)$$

which can also be written as

$$m(y^{-1}x, y^{-1}y') = {}^{y'^{-1}z}m(z^{-1}y'y^{-1}x, z^{-1}y'y^{-1}z)$$

This relation must hold for any $x \in X$, and any y, y', z and z' in G. Changing x to yx and z to $y'z$, this gives

$$m(x, y^{-1}y') = {}^{z}m(z^{-1}x, z^{-1}y^{-1}y'z)$$

or changing furthermore y' to yy'

$$m(x,y') = {}^{z}m(z^{-1}x, z^{-1}y'z)$$

which gives finally, changing z to z^{-1} and y' to y

$${}^{z}m(x,y) = m(zx, {}^{z}y)$$

In those conditions, equation (12.3) becomes

$$m_Y(y,x,y') = \sum_{\substack{z' \in G \\ z'^{-1}y = z^{-1}y'}} m(x, zz'^{-1})$$

which gives

$$m_Y(y,x,y') = \sum_{\substack{g\in G\\ gy=y'}} m(x,g)$$

Using this relation in equation (12.2) gives

$$\sum_{\substack{y'\in Y\\ g\in G\\ gy=y'}} m(x,g)a(y',z) = \sum_{\substack{z'\in Z\\ g\in G\\ gz'=z}} a(y,z')m(x,g)$$

or

$$\sum_{g\in G} m(x,g)a(gy,z) = \sum_{g\in G} a(y,g^{-1}z)m(x,g)$$

that is

$$\sum_{g\in G} m(x,g).{}^g a(y,g^{-1}z) = \sum_{g\in G} a(y,g^{-1}z).m(x,g) \tag{12.4}$$

Let again $Y = Z = G/1$. Let $b \in B$, and ϕ be the function from Y to B which is equal to b in y_0 and to 0 elsewhere. If $a(y,z) = {}^z\phi(z^{-1}y)$, equation (12.4) gives

$$m(x, zy_0y^{-1}).{}^z b = {}^{yy_0^{-1}} b.m(x, zy_0y^{-1})$$

This equation gives for $y = y_0 = 1$

$$m(x,z).{}^z b = b.m(x,z)$$

Conversely, it is clear that this relation implies equation (12.4). Thus $\zeta_A(X)$ is isomorphic to the set of matrices $m(x,z)$, indexed by $X \times G$, and satisfying the following conditions

$${}^z m(x,y) = m(zx, {}^z y) \quad \forall x \in X, \forall z, y \in G$$

$$m(x,z).{}^z b = b.m(x,z) \quad \forall x \in X, \forall z \in G, \forall b \in B$$

Then let $\tilde{m}(x) = \sum_{z\in G} m(x,z) \otimes z \in B \otimes G$. The second condition is equivalent to the fact that $\tilde{m}(x)$ commutes with all the elements of B. I have then a map

$$\tilde{m} : X \to C_{B\otimes G}(B)$$

The first condition shows that for $z \in G$

$$(1\otimes z)\tilde{m}(x)(1\otimes z^{-1}) = \sum_{y\in G} {}^z m(x,y) \otimes {}^z y = \sum_{y\in G} m({}^z x, {}^z y) \otimes {}^z y = \tilde{m}({}^z x)$$

In other words, if $C = C_{B\otimes G}(B)$ with its natural structure of G-algebra (induced by the morphism $g \mapsto 1 \otimes g$ from G to $B \otimes G$), I see that $\tilde{m}$ is just a morphism of RG-modules from $[X]$ to C.

On the other hand, the product in ζ_A of the element $m \in \zeta_A(X)$ by the element $m' \in \zeta_A(X')$ is defined by

$$(m \times m')_Y = A_* \begin{pmatrix} z_1 x' x z_2 \\ z_1 x x' z_2 \end{pmatrix} (m'_Y \circ_Y m_Y)$$

Then if m corresponds to the matrix $m(x,z)$, and m' to the matrix $m'(x',z)$, I have

$$(m\times m')_Y(y,x,x',y') = \sum_{y"\in Y} m'_Y(y,x',y")m_Y(y",x,y') = \sum_{\substack{gy=y"\\ g'y"=y'}} m'(x',g)m(x,g') = \dots$$

$$\dots = \sum_{\substack{g"\in G\\ g"y=y'}} \sum_{\substack{g,g'\in G\\ g'g=g"}} m'(x',g)m(x,g')$$

It follows that

$$\widetilde{m\times m'}(x,x') = \sum_{g,g'\in G} m'(x',g)m(x,g')\otimes g'g$$

But

$$\tilde{m}(x)\tilde{m}'(x') = \sum_{g,g'\in G} m(x,g).{}^g m'(x',g')\otimes gg' = \sum_{g,g'\in G} m'(x',g')m(x,g)\otimes gg'$$

Then $\zeta_A(X)\simeq \mathrm{Hom}_G([X],C)$, and it is clear that this isomorphism induces an isomorphism of Green functors from ζ_A to FP_C.

Now let $b\in Z(B)(X)$. Then b is a morphism of RG-modules from $[X]$ to B, such that for any Y, and any morphism c of RG-modules from Y to B

$$b(x)c(y) = c(y)b(x)$$

The case $Y = G/1$ shows then that $b(x)\in Z(B)$, and it follows an isomorphism of Green functors $Z(FP_B)\simeq FP_{Z(B)}$. Finally, I have the following proposition:

Proposition 12.3.2: Let B be a G-algebra, and $C = C_{B\otimes G}(B)$. Then

$$Z(FP_B)\simeq FP_{Z(B)} \qquad \zeta_{FP_B}\simeq FP_C$$

as Green functors. In particular $\zeta_{FP_B}(\bullet)\simeq Z(B\otimes G)$ and

$$\zeta_{FP_R}\simeq FP_{RG}$$

Thus the centre of Yoshida algebra (see proposition 4.5.2) is isomorphic to the centre of RG.

Remark: If the algebra B is an interior algebra (see [11]), then the algebra C is isomorphic to $Z(B)G$: indeed, if $g\mapsto \rho(g)$ is a morphism from G to the group of invertible elements of B, and if g acts on B by conjugation by $\rho(g)$, then say that $\sum_g a_g\otimes g$ is in C is equivalent to say that for any g, the element $a_g\rho(g)^{-1}$ is in $Z(B)$, hence that $\sum_g (a_g\rho(g)^{-1})g\in Z(B)G$. This correspondence is moreover an isomorphism of algebras $C\simeq Z(B)G$.

12.3.2 The blocks of Mackey algebra

Let $R = k$ be a field of characteristic $p>0$ and $A = b_p$ the p-part of the Burnside functor, with coefficients in k. Then $A(1)\simeq k$, and $\zeta_A(1)\simeq kG$. On the other hand, $\zeta_A(G)$ is the centre of the p-part $\mu_1(G)$ of the Mackey algebra (i.e. the piece of the Mackey algebra corresponding to the central idempotent f_1^G, or the subalgebra of the Mackey algebra formed of elements $t_{P}^H x r_{P^x}^K$, for any subgroups H and K of G, any element x of G, and any p-subgroup P of $H \cap {}^xK$). The $\mu_1(G)$-modules are exactly the Mackey functors which are projective relative to p-subgroups.

Then I have the map $r_1^G : Z\big(\mu_1(G)\big) \to (kG)^G = ZkG$. It is a morphism of algebras, it is unitary, and surjective: indeed, ZkG is generated as k-module by the elements of the form

$$Tr_{C_G(x)}^G(x)$$

for $x \in G$. Let for $K \subseteq C_G(x)$

$$\tilde{x}_K = t_K^K x r_K^K$$

It is easy to see that this defines an element of $\zeta_A\big(C_G(x)\big)$. Moreover

$$r_1^G t_{C_G(x)}^G(\tilde{x}) = t_1^1 Tr_{C_G(x)}^G(x) r_1^1$$

Thus r_1^G induces a surjection from $Z\big(\mu_1(G)\big)$ to ZkG. The kernel of this surjection is moreover nilpotent, because if e is an idempotent of the kernel, then $r_1^G(e) = 0$. The module $e\mu_1(G)$ is then a projective Mackey functor, which is projective relative to p-subgroups, and equal to zero at $\{1\}$. Hence it is zero (see [15] corollary 12.2) and $e = 0$.

It follows that r_1^G induces a bijection between the blocks of kG and those of $\mu_1(G)$ (see [15] Theorem 17.1).

Bibliography

[1] D. Benson. *Representations and cohomology I*, volume 30 of *Cambridge studies in advanced mathematics*. Cambridge University Press, 1991.

[2] S. Bouc. Construction de foncteurs entre catégories de G-ensembles. *J. of Algebra*, 183(0239):737–825, 1996.

[3] S. Bouc. Foncteurs d'ensembles munis d'une double action. *J. of Algebra*, 183(0238):664–736, 1996.

[4] C. Curtis and I. Reiner. *Methods of representation theory with applications to finite groups and orders*, volume 1 of *Wiley classics library*. Wiley, 1990.

[5] A. Dress. *Contributions to the theory of induced representations*, volume 342 of *Lecture Notes in Mathematics*, pages 183–240. Springer-Verlag, 1973.

[6] L. G. Lewis, Jr. The theory of Green functors. Unpublished notes, 1981.

[7] D. Gluck. Idempotent formula for the Burnside ring with applications to the p-subgroup simplicial complex. *Illinois J. Math.*, 25:63–67, 1981.

[8] J. Green. Axiomatic representation theory for finite groups. *J. Pure Appl. Algebra*, 1:41–77, 1971.

[9] H. Lindner. A remark on Mackey functors. *Manuscripta Math.*, 18:273–278, 1976.

[10] S. Mac Lane. *Categories for the working mathematician*, volume 5 of *Graduate texts in Mathematics*. Springer, 1971.

[11] L. Puig. Pointed groups and construction of characters. *Math. Z.*, 176:265–292, 1981.

[12] H. Sasaki. Green correspondence and transfer theorems of Wielandt type for G-functors. *J. Algebra*, 79:98–120, March 1982.

[13] J. Thévenaz. Defect theory for maximal ideals and simple functors. *J. of Algebra*, 140(2):426–483, July 1991.

[14] J. Thévenaz and P. Webb. Simple Mackey functors. In *Proceedings of the 2nd International group theory conference Bressanone 1989*, volume 23 of *Rend. Circ. Mat. Palermo*, pages 299–319, 1990. Serie II.

[15] J. Thévenaz and P. Webb. The structure of Mackey functors. *Trans. Amer. Math. Soc.*, 347(6):1865–1961, June 1995.

[16] P. Webb. A split exact sequence for Mackey functors. *Comment. Math. Helv.*, 66:34–69, 1991.

[17] T. Yoshida. On G-functors (II): Hecke operators and G-functors. *J.Math.Soc.Japan*, 35:179–190, 1983.

Index